Routledge International Handbook of Adventure Tourism

This handbook presents the latest research, industry trends, developments and initiatives in adventure tourism. It explores conceptualisations of adventure tourism, historical and intellectual developments, insights into adventure tourists and the supply side of adventure tourism, and sustainable and inclusive adventure tourism.

With contributions from leading international researchers, 28 chapters are organised into four thematic parts to provide a comprehensive overview of adventure tourism. The book presents core topics within the field as well as contemporary themes, debates and challenges within the industry. It adopts a multidisciplinary approach which draws on and applies current research from disciplines including tourism, recreation, sport and psychology to adventure tourism. As such, it presents different ways of examining this form of tourism, expands knowledge on recent developments and discusses the major claims in this field. It explores topics such as instantly accessible adventures, the increasing use of technology by adventure tourists and industry, and the well-being of tourists, destinations and communities. The handbook informs the reader of how literature translates into practice across different adventure tourism operations. It also investigates some of the key challenges affecting the adventure tourism industry and presents practical solutions and initiatives to overcome these. Case studies and vignettes are embedded throughout the handbook to illustrate practitioner perspectives, and each chapter includes learning outcomes and review questions to encourage readers to further consolidate their understanding.

The handbook is intended for undergraduates, postgraduates, doctoral candidates and early-career and more established researchers interested in the fields of adventure tourism and related disciplines, such as adventure recreation, outdoor leadership and outdoor education. It is useful for industry organisations, policymakers, professionals and those working towards outdoor activity qualifications. It is also a valuable resource for supporting related modules on sustainable tourism, consumer behaviour and marketing in tourism.

Gill Pomfret, PhD, is an associate professor in tourism at Sheffield Hallam University (UK). Her research focuses on consumer behaviour in outdoor tourism and recreation, primarily the psychological aspects of adventure tourists, and different types including mountaineer tourists and family adventure tourists.

Adele Doran, PhD, is a researcher and consultant of adventure tourism and outdoor recreation. Her research focuses on the experiences of outdoor adventure participants, including their motivations, barriers, and well-being; equality, diversity, and inclusivity in outdoor adventure; responsible behaviour in and custodianship of the outdoors; adventure media and marketing; and decent work within the outdoor sector.

Carl Cater, PhD, is an associate professor in tourism at Swansea University in Wales, a visiting professor at the University of Highlands and Islands, Scotland and President's International Fellow at the Chinese Academy of Sciences. A geographer at heart, his research centres on adventure tourism and ecotourism and their contribution to development.

Routledge International Handbook of Adventure Tourism

Edited by Gill Pomfret, Adele Doran and Carl Cater

LONDON AND NEW YORK

Designed cover image: gettyimages.ca/guvendemir

First published 2025
by Routledge
4 Park Square, Milton Park, Abingdon, Oxon OX14 4RN

and by Routledge
605 Third Avenue, New York, NY 10158

Routledge is an imprint of the Taylor & Francis Group, an informa business

© 2025 selection and editorial matter, Gill Pomfret, Adele Doran and Carl Cater; individual chapters, the contributors

The right of Gill Pomfret, Adele Doran and Carl Cater to be identified as the authors of the editorial material, and of the authors for their individual chapters, has been asserted in accordance with sections 77 and 78 of the Copyright, Designs and Patents Act 1988.

British Library Cataloguing-in-Publication Data
A catalogue record for this book is available from the British Library

Library of Congress Cataloging-in-Publication Data
Names: Pomfret, Gill, editor. | Doran, Adele, editor. | Cater, Carl, editor.
Title: Routledge international handbook of adventure tourism / edited by Gill Pomfret, Adele Doran and Carl Cater.
Other titles: International handbook of adventure tourism
Description: First edition. | Abingdon, Oxon ; New York : Routledge, 2025.
| Includes bibliographical references and index. |
Identifiers: LCCN 2024030606 (print) | LCCN 2024030607 (ebook) |
ISBN 9781032316925 (hbk) | ISBN 9781032888668 (pbk) |
ISBN 9781003393153 (ebk)
Subjects: LCSH: Adventure travel--Handbooks, manuals, etc.
Classification: LCC G516 .R68 2025 (print) | LCC G516 (ebook) |
DDC 338.4/791--dc23/eng/20241009
LC record available at https://lccn.loc.gov/2024030606
LC ebook record available at https://lccn.loc.gov/2024030607

ISBN: 978-1-032-31692-5 (hbk)
ISBN: 978-1-032-88866-8 (pbk)
ISBN: 978-1-003-39315-3 (ebk)

DOI: 10.4324/9781003393153

Typeset in Sabon
by SPi Technologies India Pvt Ltd (Straive)

For Pete, Matthew, Adam and my mum (Gill Pomfret)
For Rob and my doglets (Adele Doran)
For Tiffany, Lochlan and Finnley (Carl Cater)

Contents

Figures

Tables

Boxes

Foreword

Ralf Buckley

Adventure is a state of mind, an activity, an addiction, an industry, and all these change over time, at spatial scales from individuals to regions. Adventure gives us thrills, which make us aware of being alive and grateful for it. Many adventure thrills are only possible after learning adventure skills, and the thrill–skill combination creates outdoor sports and recreation. Adventure tourism is the commercialised component.

The evolution of adventure as a sector of commercial tourism is comparatively recent. The first commercial raft trip on the Colorado Grand Canyon was in 1955, and the first commercial trek in Nepal was in 1965. At that time, outdoor skills were widespread. For later generations, however, outdoor skills were less easily accessible, and this created a market for the fully equipped and guided trips that form the mainstay of the modern adventure tourism industry.

The adventure industry still includes a wide range of risk levels, from low-risk, high-volume highly social mass adventure tourism products to high-risk, low-volume custom products and independent adventure expeditions. Adventure remains a large component of the global tourism sector, with at least a trillion-dollar annual turnover, still increasing. Urbanisation of wealthy nations is continuing so that fewer individuals grow up with outdoor experience, and more rely on commercial adventure tourism enterprises to provide specialised equipment, guide and training skills; expert knowledge of activities and terrain; risk management and safety; and permits to access particular areas.

Cultural and economic changes are continually bringing new countries into the international adventure tourism market and taking others out or introducing new geopolitical risks. There are large specialised domestic adventure tourism sectors in many countries, notably China and India, and some of these differ considerably in design from corresponding international products, so international comparisons are especially interesting. Ageing populations in many wealthy countries are creating not only a new driver, the adventure bucket-list phenomenon, but also new constraints, such as managing adventure tourism products for older clientele with different safety considerations and comfort expectations. Adventure tourism for older clients is a growth opportunity for those enterprises that can adapt to these.

Outdoor activities in natural environments, including adventure tourism, are increasingly recognised for their contributions to mental health, which have a substantial economic value, several times greater than direct tourism expenditure. This will be important for the political positioning of adventure tourism in future. It has also generated two major new research themes: first, on practical policies to bring adventure activities into mainstream health care and, second, how adventure activities, characterised by a thrill component, contribute to mental resilience against future life obstacles.

Many people now face decreasing risks in their day-to-day domestic and workplace activities but increasing risks from uncontrollable world-scale existential risks: climate change, disease pandemics, economic disruption and global geopolitical upheavals. In developed nations, the incidence of poor mental health has increased greatly over the past few years, especially for women and younger age brackets. In such a world, the opportunity to undertake adventure activities with controllable risk challenges and the associated thrill, triumph, and self- and peer-esteem is an increasingly important contribution to public health and individual well-being. This new edited volume on adventure tourism, bringing in a new generation of adventure researchers, is thus a timely update.

Ralf Buckley was born in the UK in 1954. In the 1960s, like many country kids, he learned to run, swim, ride and use ropes, axes, boats, skis and 4WDs. UK schools had rifle training and opportunities to pilot aerobatic aircraft. In 1971, he travelled for 3 months through Malaysia, Thailand and Laos, later popularised by Lonely Planet. In 1974, he travelled by foot and 4WD through East and Southern Africa. In 1975 and 1976, he drove solo across Australia's Simpson and Gibson Deserts as part of his PhD fieldwork.

During the next 25 years, adventure was pushed aside by jobs, but in the early 1980s, he visited the deserts of Inner Mongolia and Xinjiang, and in the mid-1990s, paddled as a safety kayaker on raft descents of rivers in western China. In 1993, he set up the International Centre for Ecotourism Research at Griffith University in Australia. Together with Carl Cater, he led ecotourism student field trips in Australia, New Zealand, Nepal, Ecuador, Papua New Guinea and Finland. In the mid-1990s, he started the bid for the Cooperative Research Centre for Sustainable Tourism, where he was Director of Nature and Adventure Tourism. That led to two books on adventure tourism, one of case studies in 2006 and one on management in 2010. In many ways, those were predecessors of this volume, which is published just as Ralf turns 70.

Figure 0.1 Professor Buckley in his element (!). Photograph by Carl Cater.

Contributors

Tahir Albayrak, PhD, works as a professor at the Department of Tourism Management, at Akdeniz University, Antalya, Türkiye. He has published many international conference papers and refereed articles about destination marketing, adventure tourist behaviour and service quality in marketing and tourism journals. He served as the associate editor of *Tourism Management Perspectives* and currently works as the editorial advisory board member of the *International Journal of Contemporary Hospitality Management* and the *Journal of Destination Marketing & Management*. He is also a review member of some other tourism, travel and hospitality management journals.

Sigmund Andersen is an IFMGA mountain guide who has settled in the high Arctic islands of Svalbard. He took part in developing and establishing the Arctic Nature Guide study program in Svalbard and is an associate professor at UiT The Arctic University in Norway. His research is related to nature-based tourism, the guide role, learning and experiences in nature and risk and safety in a commercial context.

Claire Backhouse is a lecturer in sport and adventure management at Solent University, UK, and a former product manager in the adventure tourism industry. In her free time, she works as a freelance outdoor instructor and enjoys hiking, skiing and kayaking. Her research interests are in sustainable tourism and climate change and the role of interpretation in promoting pro-environmental behavioural change.

Simon Beames, PhD, is a professor of friluftsliv at the Norwegian School of Sport Sciences. He is co-author of the books *Learning Outside the Classroom, Adventurous Learning, Adventure & Society* and *Outdoor Learning across the Curriculum*.

Bart Bloem Herraiz, PhD. Bart's research focuses on the experiences of trans people in outdoor recreation and adventure activities in the natural environment. They also hold a master's in management of adventure tourism activities and the UIMLA Mountain Guide Certification. Further research interests are queer, feminist, and outdoor methodologies and the intersections between queer and outdoor pedagogies. They are part of the project Petricor Aventuras, which aims to create more diverse and inclusive outdoor spaces. They are passionate about travelling by bike, their cat Tjena (who always comes on biking and hiking trips), hiking, climbing and skiing.

Eric Brymer, PhD, is a behavioural scientist specialising in lived experience with a background in adventure, tourism, sport and business. He is interested in human potential and designing effective interventions for facilitating human potential and psychological health and well-being. He is especially interested in the human–nature relationship and performance in extreme environments.

Meltem Caber, PhD, a professor at Akdeniz University, Antalya, Türkiye. Her research interests are adventure tourist behaviour, service quality and destination management. In some of the well-known scientific journals, she has been the co-editor and guest editor. She has published several international conference papers and refereed articles about tourism marketing and management. She currently serves as the editorial review board member of the *Journal of Hospitality and Tourism Management* and editorial board member of *Current Issues in Tourism*. She is a review member of some other tourism and travel-related journals.

Carl Cater, PhD, is an associate professor in tourism at Swansea University in Wales and a visiting professor at the University of Highlands and Islands, Scotland. A geographer at heart, he centres his research on adventure tourism and ecotourism and their contribution to development. He has written more than 50 papers and book chapters, is co-author of *Marine Ecotourism: Between the Devil and the Deep Blue Sea* (2007) and is co-editor of the *Encyclopaedia of Sustainable Tourism* (2015). Carl has travelled to more than 80 countries and undertaken field research, supervision, curriculum oversight and teaching worldwide. He is a fellow of the Royal Geographical Society and a qualified pilot, diver, sailor, lifesaver, mountain and tropical forest leader and maintains an interest in both the practice and pursuit of sustainable outdoor tourism activity.

Jenny Cave, PhD, is an experienced cultural heritage manager, evaluator and tourism enterprise developer. As a craftsperson, she weaves practical insights and creative thinking. With more than 25 years of experience in design, cultural heritage management and operations, program evaluation, market research and governance, Jenny is a systems thinker and activator. Jenny is skilled in community engagement and women's entrepreneurship. Her work includes projects for international aid agencies, federal and local government, not-for-profits and the private sector in Canada, the United Kingdom, New Zealand and Australia. Jenny resides in the Waikato region, New Zealand, where she is engaged in various regenerative projects with her partner.

Kristen Chmielewski, PhD, is an assistant professor of recreation management and leadership at Western Washington University, USA. She has a PhD in educational policy and leadership studies and is a certified therapeutic recreation specialist. Her research agenda broadly concerns the formation of disability stigmatization and how that stigmatization affects the lived experiences of disabled individuals, and her research has been published in *History of Education Quarterly*, *Paedagogica Historica* and *Social Education*.

Tove I. Dahl, PhD, is a professor of educational psychology in the Department of Psychology at UiT The Arctic University of Norway. Her research focus has always been on some aspect of learning, more recently in relation to what we can learn from adventure. Her most recent work has therefore focused on the nature and value of interest and courage and their place in facilitating learning and change.

Tim Dassler is currently working on a PhD on human decision- and sense-making in uncertain and wicked environments at the Psychology Department at UiT The Arctic University of Norway. With a background in philosophy, pedagogy, explorative learning and extended risk assessment and education, he now focuses his research on how we can facilitate the learning of how to make better decisions on adventures in snow-covered mountains and other environments. In his free time, Tim works as an avalanche instructor in the breathtakingly beautiful mountains of North Norway.

Amy DiRenzo, PhD, is an adjunct faculty at the State University of New York College at Cortland. Her research is primarily focused on social-psychological aspects of adventure and outdoor learning, particularly regarding the development of empowerment, resilience and well-being.

Adele Doran, PhD, is a researcher and consultant of adventure tourism and outdoor recreation. Her research focuses on the experiences of outdoor adventure participants, including their motivations, barriers, and well-being; equality, diversity, and inclusivity in outdoor adventure; responsible behaviour in and custodianship of the outdoors; adventure media and marketing; and decent work within the outdoor sector. She has worked with both private and public organisations to collaborate on research and consult on projects. Adele employs a range of methodologies, including survey research, interviews, focus groups, ethnography, social media research and systematic literature reviews.

Alan Ewert, PhD, is a professor emeritus at Indiana University, Bloomington, Indiana, USA. His research involves motivations for outdoor adventure recreation and tourism, the relationship between natural environments and human health and the use of outdoor recreation and adventure tourism activities for personal growth and development.

Jelena Farkić, PhD, is a lecturer at Breda University of Applied Sciences, Netherlands. She is an active member of Forest Therapy South Eastern Europe and is its certified forest wellness mediator. Jelena has acted as a steering group member of the Adventure Tourism Research Association, has undertaken organising the annual International Adventure Conference since 2015 and is an associate editor for the *Journal of Adventure Education and Outdoor Learning*. Jelena currently explores the ways in which human (tourist) subjective well-being can be enhanced through concepts and practices such as slow adventure, natural selfness, forest bathing (*shinrin-yoku*) and idleness. She focuses mainly on the interrogation of embodied experiences, which entails utilising qualitative, mostly ethnographic, methodologies and phenomenological philosophy to critically interrogate bodily presence, senses, memories of and reflections on the encounters within hospitality and tourism contexts.

Jon Frankel, MS, is the regional experiences manager for REI (Recreational Equipment Inc.) in the Rocky Mountain Region of the United States. He has worked as a guide and administrator in the experiential adventure and travel industry for 25 years.

Takafumi Fukuyama is a doctoral researcher at Hokkaido University Center of Advanced Tourism Studies, Japan. His main research field is tourism resource development with a focus on the utilisation of snow and ice in the North. He has conducted comparative research including cases (such as drift ice/ice-breaking tourism and ice hotels) from Hokkaido in northern Japan and Finnish and Swedish Lapland and initiated studies concentrating on (Indigenous) culture-based adventure tourism.

Sarah Gardiner, PhD, is a professor in tourism and the director of the Griffith Institute for Tourism at Griffith University, Australia. Her research interests include travel consumer behaviour, travel trends, experience design, tourism technology and innovation, with a particular interest in youth tourism.

Steven Gillespie, PhD, is a senior lecturer and head of the subject of environmental science and sustainability at the University of Glasgow, UK. His research interests span sustainable tourism, sustainable rural development and restoration ecology.

Jasmine Goodnow, PhD, is an associate professor of recreation management and leadership at Western Washington University, USA. She has a PhD in parks, recreation and tourism. She studies the liminal qualities of adventure travel and microadventure. Furthermore, her research agenda focuses on how to create more sustainable, inclusive and sacred adventure travel experiences. Her research has been published in *Leisure Sciences*, *Tourism Review International*, the *International Journal of Religious Tourism and Pilgrimage*, the *Journal of Travel and Tourism Research* and *Leisure/Loisir: The Journal of the Canadian Association of Leisure Studies*.

Tamara Griffiths, PhD, is a lecturer at the University of Highlands and Islands within the School of Adventure Studies, Scotland. Tamara's research interests focus on human relationships with nature (feminist anthropology), multispecies ethics, and post productive experience economies in adventure. She is the author of an open-access monograph in press: *Foraging for a Future*.

Sven Gross, PhD, is the deputy director of the Institute for Tourism Research at Harz University, Wernigerode, Germany. He was also a member of the New Zealand Tourism Research Institute. His main research areas include tourism and transport and business travel management, as well as tourist market research, destination management and adventure tourism. He has authored more than 130 publications on these subjects, including 18 books.

Matt Groves is a programme leader at the University of the Highlands and Islands, based in Scotland, for undergraduate and postgraduate programmes in the School of Adventure Studies – BA (Hons) Adventure Tourism Management, BA (Hons) Adventure Education and PGCert Outdoor and Adventure Therapeutic Practice. His research and teaching specialism is in professional adventure competencies and factors in risk perception and decision-making.

Ashley Hardwell, PhD, is a senior lecturer in outdoor and adventurous activities at Leeds Beckett University, UK. He is passionate about the outdoors both personally and professionally. His ways of working are underpinned by 30 years of leisure, tourism and outdoor education teaching and research. His PhD considered UK rock-climbing culture which helped make personal sense of the changing nature of a life-defining activity. Through his wealth of outdoor experience over more than 40 years, he continues to examine and capture what makes the outdoors such a unique learning environment.

Jon Heshka, BSc, BA, BEd, MEd, LL.M, is a professor specializing in the legal liability and risk management of adventure and extreme sport. He teaches at Thompson Rivers University (TRU) in Kamloops, British Columbia, in both the law school and the adventure studies departments. Jon was Associate Dean of Law at TRU from 2013 to 2017. He has written numerous book chapters and more than 70 articles, and his work has been published in several law journals. Jon has climbed and led expeditions worldwide and trained and coordinated search and rescue full-time for 7 years in British Columbia. He has also worked as an expert witness in Canada, the US and Europe in matters relating to adventure sports risk management.

Susan Houge Mackenzie, PhD, is an associate professor at the University of Otago School of Business Tourism Department, New Zealand. Her research, which focuses on links between nature-based adventure and psychological well-being, has been featured in more than 50 academic publications and translated into a range of applications for public and private organisations and communities. Previously an adventure guide in New Zealand, the US and Chile and an avid footballer, Susan is passionate about sport, adventure and translational research that enhances the well-being of individuals and organisations. Her applied work includes sport psychology for business workshops, risk management for adventure tourism and consulting with the US Forest Service, NZ Riverboarding, the Ministry of Tourism and the History Channel. She serves as an associate editor for two leading research journals and on advisory boards for Tourism Central Otago and the Adventure Tourism Research Association. Her current translational project focuses on how regional tourism organisations can promote community well-being and flourishing via tourism.

Ming Feng Huang, PhD, was born in Taiwan and completed his doctorate in tourism at Aberystwyth University on mountain trekking in Taiwan. Since completing his PhD, he has been based at Surrey International Institute, Dongbei University of Finance and Economics, China. His research interests are adventure tourism, mountain tourism, sustainable tourism and ecotourism.

David Huddart, PhD, taught at Trinity College, University of Dublin and has extensive experience teaching/researching at Liverpool John Moores University, UK. He was the course leader for the undergraduate outdoor education programme and postgraduate advanced diploma. He was the associate dean for research and is currently an emeritus professor.

Ingo Janowski, PhD, is a member of the Griffith Institute for Tourism at Griffith University, Australia. His research interests include adventure tourism, consumer behaviour and cross-cultural communication.

Anna Kwek, PhD, is an associate professor in the Department of Tourism, Sport, and Hotel Management at Griffith University, Australia, with interests in geopolitically sustainable tourism, Chinese tourism and tourism destination image.

Jason King is a lecturer and PhD researcher in the School of Sport at Leeds Beckett University, UK. His research integrates an ecological framework with leadership and links this to adventure, leisure, tourism and education contexts. His primary research focuses on how leaders create and maintain relationships with new groups in challenging environments. Jason's interest in leadership stems from 15 years of leading groups in a range of mountainous and water environments across the world.

Young Sook Lee, PhD, is a professor at UIT The Arctic University of Norway. She is Korean Australian and completed her PhD in sociology at the University of Queensland, Australia. She worked as a tourism academic at Griffith University for over 15 years. In 2014, she relocated to UiT The Arctic University of Norway, where she acted as an expert adviser at the Nordic Expert Group for the Prime Minister's Office of Finland and the Idea Lab for the Future of Tourism by the Research Council of Norway. She also serves as an expert panel member for the United Nations World Tourism Organization. Her areas of research include East Asian cultural philosophies in nature and tourism, tourist behaviour, Arctic tourism and sustainability. Continuing her research and teaching, she contributes towards better understanding and connecting between the changing Arctic and the global world.

Monkgogi Lenao, PhD, is a senior lecturer in the Faculty of Business at the University of Botswana where he teaches tourism studies and handles both undergraduate- and graduate-level research supervision. Lenao holds a PhD in geography (tourism) from the University of Oulu in Finland. His research interests centre around sustainable tourism, cultural and heritage tourism, community-based tourism and rural tourism. Lenao has made presentations on tourism-related topics at several local and international conferences covering Southern and East Africa as well as Europe, Asia and the USA. He has authored and co-authored more than 20 pieces of academic work, including journal articles in regional and international peer-reviewed journals as well as chapters and case studies in edited book projects and a co-edited book. One of Lenao's latest research escapades involves serving as a co-guest editor for a special edition of CABI Tourism Cases.

Melissa Jeanette Lötter, PhD, holds a Doctor Technologiae in Adventure Tourism Management. She is a client experience analyst with teaching and research interests in adventure tourism management and marketing.

Kevin Lyons, PhD, is an emeritus professor in tourism at Newcastle Business School, University of Newcastle, Australia. His research focuses on volunteer tourism, tourism in national parks and adventure tourism.

Joseph E. Mbaiwa, PhD, is a professor of tourism studies at the Okavango Research Institute, University of Botswana. He holds a PhD in recreation, park and tourism sciences from Texas A&M University (USA). Joe Mbaiwa has more than 20 years' experience in tourism and natural resource management research and consultancy experience in Botswana. He is widely published on issues of tourism development, rural livelihoods and conservation.

Maggie C. Miller, PhD, is a professor in the School of Hospitality and Tourism Management at George Brown College, Canada. She holds a PhD in recreation and leisure studies, and much of her work focuses on socio-cultural dimensions of sustainability and enhancing the overall quality of life for communities engaged in tourism and recreation. She is informed by relational ways of knowing and her research draws on diverse and innovative qualitive methodologies.

Soumya Mitra is currently working as a program coordinator at Outdoor Education Group in Australia. He has a background in outdoor adventure tourism, recreation and education working across India, Tanzania, the US, Vietnam and Australia. His primary interest is to help grow the Indian outdoor education community by mentoring new practitioners. He is an emerging academic researcher, currently working on documenting the history of Indian outdoor education.

Chiedza N. Mutanga is a Senior Research Fellow in Sustainable Tourism at the Okavango Research Institute, University of Botswana. She holds a DPhil in tourism and wildlife management from Chinhoyi University of Technology, Zimbabwe. Her research interests are in sustainable tourism development, with a focus on wildlife tourism, nature-based tourism, community-based tourism and community livelihoods, ecotourism, protected area tourism, community-based natural resource management and wildlife resource conservation.

Cam O'Beirne is the owner and operator of The Margaret River Adventure Co, a bespoke adventure operator in remote Australia. Before becoming involved in the tourism industry, Cam was a rescue swimmer with the local emergency rescue helicopter for many years and was employed as a Hollywood stuntman and marine coordinator in various movies. He has spent numerous years as a reluctant academic providing him with extensive experience in teaching, course design, and leading e-commerce, web marketing and information literacy-related courses at Australian universities. Cam has authored chapters in two books on sport management, as well as the key author in the future of surf lifesaving in Australia. Cam was also employed as a policy adviser to successive state governments, managing a wide range of grants and multi-million-dollar projects, and providing strategic advice to various government ministers on the construction of the world's first artificial recreational surf reef. He also managed operations for the Emirate of Dubai water safety and tourism project aiming to prevent drownings and increase tourism visitation and coordinating risk management and operations for lifesaving services in Western Australia.

Gill Pomfret, PhD, is an associate professor in tourism at Sheffield Hallam University, UK. Her research focuses on consumer behaviour in tourism, primarily psychological aspects of adventure tourists, and different types, including mountaineer tourists and family adventure tourists. The key themes of her research include motives encouraging adventure tourism participation, experiences of adventure activities, benefits gained from participation, constraint negotiation journeys and gender in adventure. Gill's most recent research focuses on subjective well-being and family well-being through adventure tourism and recreation. Related to this, she is currently exploring the role of adventure in coping, resilience, and mental health. Gill works with researchers across the globe on projects concerned with adventure tourism. Her latest work is in collaboration with researchers in Germany, the Netherlands and Australia, as well as with researchers based in the UK. Gill is an active committee member of the Adventure Tourism Research Association and Adventure Mind.

Outi Rantala, PhD, is a professor of responsible Arctic tourism at the University of Lapland, Finland, and an adjunct professor of environmental humanities at the University of Turku, Finland. Outi has focused in her research activities on creating critical, reflective and alternative narratives on northern tourism. Currently she leads a research profiling project on multispecies hospitality (funded by the Academy of Finland). She is an active member of several research networks, such as the Adventure Tourism Research Association, UArctic Thematic Network on Northern Tourism and Sustainable Change Research Network.

Arild Røkenes, is a professor (Docent) of Tourism at UiT The Arctic University of Norway Alta campus. He is both a guide and a specialist in the guide role, focusing on how it affects the customer experience. He also works on risk management with a particular emphasis on understanding risk behaviour and the peculiarities of managing risk in complex natural environments where customers expect spectacular experiences.

Manuel Sand, PhD, is the academic director of the Adventure Campus at the University for Applied Management and course administrator for the degree in outdoor sports and adventure management. His research interests include adventure tourism, mountain bike tourism, well-being and adventure activities, nature protection and outdoor sports and the effects of experiential learning in the outdoors

Juha Saunavaara, PhD, is an assistant professor at Hokkaido University Arctic Research Center (Social Science and Humanities Research Group), Japan, researching regional development, tourism, infrastructure development and international and interregional relations in the Arctic and northern contexts. He has earned his PhD from the University of Oulu and holds the title of Docent at the University of Turku.

Vinathe Sharma-Brymer, PhD, is a lecturer in social work at the University of Sunshine Coast, Australia. She is interested in interdisciplinary approaches to the facilitation of human agency, drawing from psychology, environmental studies, education and social work disciplines. Her research focuses on exploring the nature–human relationship for improving human health and well-being through valuing nature for what it is and seeing humans as part of the natural world. She is also a qualified Forest Schools (UK) practitioner.

Steve Taylor, PhD, is an associate professor in tourism and the director of the Centre for Recreation and Tourism Research at the University of the Highlands & Islands in Scotland. In his 10 years working at the centre he has led or been a partner in 20 transnational tourism or recreation projects. A flagship project was Slow Adventure in Northern Territories, which saw the development of nature-based recreation experiences across northern Europe and that is currently being commercialised by a spin-off company, Slow Adventure Ltd. Steve's scholarly output has most recently been on the slow-adventure marketing concept, cultural heritage tourism and the development of mountain biking, while the regenerative tourism concept is an area of developing interest. He continues to have research interests in nature-based tourism and the psychological dimensions of being and recreating in wild spaces.

Peter Varley, PhD, is a professor of tourism in the Department of Entrepreneurship, Innovation and Strategy in Newcastle Business School, UK. He has written on diverse topics centring on outdoor experience, including slow adventure, non-places and the commodification of adventure. Recently, he also has written about gastronomy, the bohemian imaginary and 'thickening' places via mundane interactions.

Elía Velo Camacho has a bachelor's in physical activity and sport sciences, a master's in management of adventure tourism activities and a master's in high sport performance from the University Pablo de Olavide (Spain). They also hold the UIMLA Mountain Guide Certification. Elía started researching Outdoor Pedagogies in university contexts as a member of the research group "Motiva2: Sport, Organization and Security" at the University Pablo de Olavide. Later, because of personal experiences with gender non-normativities, Elía felt the necessity to research the intersections of gender studies and outdoor studies, which is now their field of work. They are also a member of the Spanish Network of Outdoor Physical Education – REEFNAT. Elía loves outdoor sports, cooking and taking care of their vegetable garden.

May Kristin Vespestad, PhD, is an associate professor of tourism marketing at the School of Business and Economics, at UIT The Arctic University of Norway. Her main teaching and research areas are within tourism marketing, tourism experiences, adventure tourism, consumer behavior and experience co-creation. She studies amongst others marketing aspects and consumer behaviour in areas such as nature-based tourism and adventure tourism. She has published in several international journals and serves on editorial boards.

Po-Yu Wang, PhD, is an associate professor in the Department of Recreational Sport, National Taiwan University of Sport. His research focuses on adventure tourism, outdoor recreation management, adventure education and tourism in national parks.

Christopher Webber is a research fellow exploring behaviour change in relation to physical activity and a PhD student investigating learning transfer from adventure education residentials at Leeds Beckett University, UK. Previous studies have considered implementing outdoor adventure within primary education, and teaching and learning adventure education within post-graduate teacher training. Chris has worked as an outdoor educator around the UK for more than a decade and led several expeditions to developing world countries.

Christopher Woodfield is passionate about facilitating transformational change and is the co-initiator of Venture Shift, a new systems transformation initiative within the outdoor industry. He is Better Business and Futures Lead at TYF Adventure and delivers sustainability and impact consultancy, along with Climate Fresk and Carbon Literacy training. Chris believes in the power of play, nature connection and taking a compassionate, collaborative and curious approach to making change happen.

Peter White is a lecturer in the School of Adventure Studies at the University of the Highlands and Islands, Scotland. He has experience in residential outdoor education, instructor training programmes and commercial guiding. He has a research interest in professionalism and training within the outdoor adventure sector and has a particular speciality in paddle sports, delivering a range of British Canoeing qualifications.

Tamara Young, PhD, is an associate professor in tourism, Newcastle Business School, University of Newcastle, Australia. Her research focuses on the role of tourism media as an interface between traveller cultures and travelled cultures, Indigenous tourism and culturally sustainable tourism in Australia.

Acknowledgements

Much like any other reference volumes, this volume must acknowledge a huge number of people who have laid the foundations for this work, both past and present. We truly feel that we are able to stand on the shoulders of giants and were particularly pleased that Ralf Buckley, eternal adventurer, agreed to write the foreword to this handbook. Similarly, other key researchers in adventure and tourism would recognise their work as underpinning the insights within these pages. These include Paul Beedie, Bill Bramwell, Anna Carr, Kay Dimmock, Alan Ewert, David Fennell, Barbara Humberstone, Ghaz Musa, Reidar Mykletun and John Swarbrooke.

At Routledge, we would like to thank Faye Leerink for commissioning this book, and Prachi Priyanka for their continued support and guidance.

We would also like to thank the hosts of all the wonderful International Adventure Conferences who have provided such a fertile environment for adventure tourism research and shared their local opportunities for adventure with us. In particular, Jelena Farkić, Steve Taylor and Pete Varley have provided the much-needed continuity in supporting these events and ensuring their scientific rigour through the Adventure Tourism Research Association (ATRA).

We are also inspired by the many adventure operators and organisations who seek to promote responsible and sustainable adventure tourism and recognise its importance for both people and the planet. Whilst it would be impossible to acknowledge them all, many have also assisted us in various ways over the years. These include Ed Chard (Jagged Globe), the British Mountaineering Council, Martin Chester (guide and consultant), David Croy (Raasay Outdoor Centre), Chris Doyle (Adventure Travel Trade Association), Pete Raines (Coral Cay Conservation), Sean Taylor (Zipworld), Simon Turner (Aber Adventures), Shane and Nigel Winser (Royal Geographical Society) and Belinda Kirk (Explorers Connect and Adventure Mind). We would also like to acknowledge everyone working within the adventure industry. We have had some truly fantastic instructors and guides over the years who often go way beyond expectations because they love what they do and care deeply about our world.

We have also been supported by many colleagues, some of whom have had the fortune (or misfortune!) of sharing our adventures: Robert Bowen, David Dowell, Daniel Funk, Dave Gallagher, Brian Garrod, Glen Hornby, Ian Keirle, Tiffany Low, Scott McCabe, Carola May, Pete Merriman, Maggie Miller, Manuel Sand and Kieron Stanley.

We would also like to thank our student researchers – Athena Barbas-Marcroft and Caitlin O'Neill – who carried out some really useful research for the original book proposal and gave us their perspectives on the book's appeal to the student market, and Louisa Hardwick and Jessica Schwittay, who compiled some literature and the index.

Gill would like to thank her parents, who encouraged her to enjoy family hikes with her three sisters as soon as she could walk, with no map, no fixed plan and no time limit, no matter what the weather or the amount of mud. This instilled in her a sense of adventure and exploration from an early age. Thanks also go to her husband, Pete, and her sons, Matthew and Adam, for being a huge part of her adventures over the last 30-plus years! Also, thanks to Gill's wonderful friends with whom she has shared so many great adventure experiences. Gill would particularly like to acknowledge Bill Bramwell for providing unlimited support and mentoring in the early stages of her writing career and guiding her through the challenges of publishing peer-reviewed journal articles. She would also like to thank John Swarbrooke for getting her into academic writing as one of the editors of the very first textbook on adventure tourism, *Adventure Tourism: The New Frontier*.

Adele would like to thank her husband Rob for being the partner she needed (in many ways!) to get her outdoors when none of her friends were interested. The multi-day hiking trips and snowboarding seasons that they did together triggered an adventure bug within her, and he has since had to persevere 20-plus years of being dragged up mountains! Challenging for someone who hates heights and exposure. Adele would also like to thank her adventurous friends who have shared many incredible and challenging experiences with her – you know who you are. Finally, a special thank you to her dogs – the best research buddies that anyone could ask for – who make her smile and laugh every day and force her outside to stretch her legs, all of which are good for her mind, body and soul.

Carl would particularly like to acknowledge the late professor Paul Cloke, who together with professor Harvey Perkins pioneered academic interest in the commercialisation of adventure tourism in the 1990s and set Carl on his own journey of adventure tourism research. He hopes that he has continued to support his own adventure tourism doctoral students in a similar way, as he feels that they have also contributed to this work in many ways. These include Claudia Ollenburg, Scott Richardson, Jinchin Wang, Ming (Tim) Feng Huang, Mandy Talbot, Greg Dash, Olga Garcia, Sam Dabamona, Caglar Bideci, Rhiannon Rees and Jessica Schwittay. Personally, he must thank his parents and sister for fostering a spirit of adventure and his lifelong adventure companions, his wife, Tiffany, and sons, Lochlan and Finnley.

Introduction

Happiness is in the journey not the destination. This is one thing that climbing has taught me and that I know is very real. For me it's all about fitting in as much experience as possible. Some of my yuppie friends used to say whoever dies with the most toys wins. For me it's whoever dies with the most experiences wins.

The late Rob Hall, who perished on Mount Everest in 1996 looking after his clients wrote these lines to Helen Wilton, his long-time basecamp manager in 1989. Whilst working at the literal pinnacle of the adventure tourism industry, Rob was aware that not all would achieve their goals and so strove to make their time enjoyable and a real experience, whatever the outcome. In many ways, this encapsulates what adventure tourism is all about, providing happiness and enjoyment through incredible experiences, irrespective of ability, and this is the very essence of what we aim to achieve within this handbook. We wanted to capture the ethos of adventure tourism and present a comprehensive book which showcases how this form of tourism has evolved, its eclectic conceptualisations and intellectual development, who its consumers are, how the adventure industry is structured and organised and the fundamental priorities of sustainability and inclusivity.

The adventure tourism industry has enjoyed exponential expansion over the past few decades, benefitting from heightened commercialisation, and a continuous influx of new adventure products, destinations, and activity experiences. Correspondingly, during and since the global pandemic, more individuals are participating in outdoor nature-based adventure activities locally as well as on holiday to maintain or strengthen their well-being. These adventure enthusiasts have heterogeneous demographic and psychographic profiles and are motivated by these diverse industry offerings, or by designing their own unique adventure experiences on holiday. These are all key ingredients to support a continued positive growth trajectory for adventure tourism.

Undoubtedly, adventure tourism has the power to be a significant contributor to the sustainability agenda as it can and does support several Sustainable Development Goals (SDGs) including good health and well-being (SDG3), gender equality (SDG5), decent work and economic growth (SDG8), reduced inequality (SDG10), responsible consumption and production (SDG12) and climate action (SDG13). In a nutshell, the adventure tourism industry is making its mark as a global player in the wider leisure industry.

But what is adventure tourism and who participates in it? The plethora of meanings attached to the term *adventure tourism*, the rich insights on this topic provided by researchers to date, and the fundamental shift in how and where we participate in tourism post-COVID-19, have prompted the writing of this handbook. In other words, 'it is time to reimagine adventure' (Houge Mackenzie & Goodnow, 2020, p.1).

DOI: 10.4324/9781003393153-1

This comprehensive handbook offers a major reference work within the field of adventure tourism. It adopts a multidisciplinary approach which draws on and applies theories, concepts and current research from different disciplines such as tourism, recreation, sport, psychology and sociology to adventure tourism. As such, it presents different ways of examining this form of tourism, establishes new concepts and develops existing ones, expands knowledge on recent developments, and discusses the key claims in this field. It identifies current and future adventure industry trends, development, solutions and initiatives. It introduces contemporary themes and pressing issues and debates within the current milieu, for instance accessible adventures, climate change and the increasing use of technology by adventure tourists and the industry. It provides rigorous academic underpinning while also applying significant theories and concepts to industry practice. Accordingly, it informs the reader of how literature translates into practice across different adventure tourism operations. It proposes new approaches, outlines practical applications and shares best practice in adventure tourism. It documents contemporary research and issues in adventure tourism and offers guidance to scholars on pertinent themes, or a "road map" for future investigations.

We, as the editorial team, have a strong track record of research expertise and high-impact publications. We are internationally known for our research in adventure tourism and have well-established industry and research networks. We are committee members of the Adventure Tourism Research Association (ATRA), a thriving community of academics and practitioners which organise the International Adventure Conference.

In preparing this handbook, we have drawn on this experience and our networks to invite an amazing set of authors to contribute to this cutting-edge handbook. It offers contributions from a broad range of scholars from early-career researchers and emerging new talent to renowned academic and industry experts in the field, including professionals working in different sectors of the adventure tourism industry. We are really pleased and lucky to have contributors from across the globe and from different geographical regions present their latest adventure tourism research and insights. We have authors based in 15 countries: Turkey, Norway, Spain, Australia, the UK, New Zealand, the US, the Netherlands, Japan, Germany, Canada, China, Botswana, South Africa and Finland. The differing perspectives of adventure tourism that they offer highlight the diversity of this topic and its industry. These varied contributions will help drive the adventure tourism research agenda forward by proposing future investigations.

Primarily, the handbook is intended as a study resource for students on undergraduate and postgraduate courses in adventure tourism and related disciplines, such as adventure recreation, outdoor leadership and outdoor education, and for tourism students who study adventure tourism modules as part of their course. Furthermore, the handbook may be used to support related modules on, for example, sustainable tourism, consumer behaviour and marketing in tourism. The handbook comprises international case studies and examples and, therefore, will appeal to students globally and reflect their interests.

As the handbook provides insights into recent adventure tourism literature and proposes future research themes, it will also have considerable scholarly appeal. It is intended to inspire doctoral candidates and early-career and more established researchers to investigate the diverse themes pertaining to adventure tourism. In this way, the handbook will stimulate further knowledge creation and consolidate knowledge and theoretical development within this field. From a practitioner's perspective, this handbook comprises many practical examples, case studies, key industry challenges, potential solutions and future trends. Accordingly, it is useful for industry organisations, policymakers, professionals and those working towards outdoor activity qualifications.

The handbook is divided into four parts. Each chapter includes learning outcomes and review questions to encourage readers to further consolidate their understanding. The handbook provides a theoretical introduction to the key concepts within each chapter before continuing to discuss how these translate into practice. Case studies and vignettes are embedded throughout the handbook to illustrate practitioner perspectives. Furthermore, each chapter presents future implications related to its key themes.

Part I: Adventure tourism: An introduction. This part sets the scene by introducing theoretical perspectives on adventure tourism and different conceptualisations, including relational and reciprocal forms. It examines the historical development of adventure tourism from its beginnings in colonial exploration to the highly commercialised industry it is today. It explores the philosophy of adventure and its intellectual development before turning to a discussion of emerging and transformational trends in the industry. The last two chapters focus on the application of research methods in adventure tourism.

Part II: Consumers of adventure tourism. This part focuses on the motives and benefits encouraging adventure tourism participation, particularly those related to well-being. It explores the constraints adventure tourists potentially face and how they negotiate these to engage in adventure activities. It also examines the segmentation of adventure tourists and the close association between adventure destinations and activities. It focuses on culturally diverse and inclusive adventure experiences, introduces the South Korean "leiports" adventure concept and discusses the ways in which adventure tourists increasingly use technology during their activity experiences.

Part III: The supply of adventure tourism. This part provides an overview of the structure and organisation of the adventure tourism industry and how it is integrated into the broader tourism sector. It examines the diverse array of adventure tourism products and destinations, including slow adventure tourism. It discusses safety and risk management considerations, which continue to be key to success in the adventure tourism industry. The growing professionalisation of the industry, including leadership qualities, training and skill development and careers are discussed. The last two chapters examine the experiential marketing of adventure tourism and how adventure organisations use technology.

Part IV: Sustainability and inclusivity. This final part comprises chapters which explore the complexity of social and environmental sustainability, inclusive adventure tourism and some of the pertinent challenges facing the adventure tourism industry. It presents chapters which focus on climate change, environmental impacts and their management, and host community involvement in adventure tourism. In particular, the focus is on the potential for adventure tourism to restore, protect and promote indigenous cultures. This part also examines different genders and the LGBTQ community and the power adventure tourism providers have in designing accessible adventure products and experiences.

Reference

Houge Mackenzie, S., & Goodnow, J. (2020). Adventure in the age of COVID-19: Embracing microadventures and locavism in a post-pandemic world. *Leisure Sciences*, 43(1–2), 62–69.

Part I
Adventure tourism
An introduction

This first section of this handbook presents a theoretical perspective on both the origins and the evolution of adventure tourism research. As Outi Rantala, Gill Pomfret and Adele Doran note in the first chapter, adventure tourism has undergone somewhat of a reconceptualisation, both theoretically and in its application. Whilst adventure tourism was historically framed around certain dichotomies which are still relevant, contemporary interest takes a greater interest in more relational and reciprocal forms. Interest in the importance of non-human actors, well-being (linking also to a later discussion in Chapter 12), and adventure that might actually "do good" are shown to be prevalent in the contemporary discussion, and we should recognise that a definition of adventure tourism is dynamic and as elusive as ever.

DOI: 10.4324/9781003393153-2

Whilst being more descriptive, the historical overview provided by Carl Cater in Chapter 2 illustrates how adventure tourism evolved from its roots in colonial exploration and individual activity to the highly commercialised industry that we see today. Technological innovation has been important, for example bungee cords and ziplines, but societal trends have been equally powerful in the shaping of modern adventure tourism. This discussion is continued in Chapter 3, where Simon Beames and Pete Varley show how the philosophy of adventure has historically been quite narrow. The commodification of the sector has simultaneously led to the refinement of theoretical models of adventure, but also problematised prior understandings, particularly related to the place of risk and escape. The chapter concludes by returning to a discussion that acknowledges the non-human, offering a path out of these contradictions. They argue that appreciative adventure must re-examine our relationships with nature, acknowledging our insignificance in the universe, prioritising being-with as the very essence of adventure. The following case study by Christopher Woodfield illustrates the more appreciative approach taken by an adventure operator in Wales and their efforts to gain "b-corp" certification.

Representing Alan Ewert, Amy DiRenzo and Jon Frankel's decades of experience in the adventure recreation sector, Chapter 4 presents an overview of changes in the industry, continuing some of the points made in the foreword. The significant growth of the sector and participant patterns from the US and the UK are positioned alongside the disruptive forces of technology, climate change, and economic impacts, including those of the COVID-19 pandemic. They highlight the acute staffing issues currently being faced in the industry, including education, recruitment, communication, training, job satisfaction and retention and turnover, drawing on a stress-continuum projection to identify issues. In addition, there are particular challenges for tertiary and higher education for the adventure sector, including risk aversion and outsourcing of delivery.

The final two chapters in this section focus on the application of research methods in adventure tourism, which we feel will be particularly useful to students and researchers in the area. Whilst adventure, and tourism more broadly, has come in for some critique regarding its theoretical sophistication, the current breadth of empirical research studies on adventure tourism refutes this claim. In Chapter 5, Tahir Albayrak and Meltem Caber, in addition to having contributed several important quantitative papers on climbing, whitewater rafting and scuba diving themselves, review the extant literature on adventure tourism that has used a quantitative approach. Whilst the methodological divide is a fallacy, the discussion of research methods is continued in Chapter 6, which examines qualitative research in adventure tourism to date. Indeed Tove Dahl, Tim Dassler and Arild Røkenes show how all research starts with asking questions, and we should be aware of the foundations, or thinking tools, these are based on. The authors demonstrate a wide range of methods but note that there has been quite limited use of what is possible in the study of adventure tourism, calling for open access and greater methodological imagination.

1 Reconceptualising adventure tourism

Outi Rantala, Gill Pomfret and Adele Doran

Chapter learning outcomes

1 Assess current conceptualisations of adventure tourism.
2 Examine emerging research directions to demonstrate the importance of continually reconceptualising adventure tourism.
3 Become sensitive towards diversity in adventure tourism and attentive towards non-human voices in adventure.
4 Appreciate the role of subjective well-being in adventure tourism and its reconceptualisation
5 Define adventure philanthropy and appreciate its increase in popularity.

Introduction

What is adventure tourism really about? Shared images of adventure tourism often capture someone conquering a mountaintop, rafting down a river or visiting a remote village and engaging in cultural activities with the locals. Lately, this imaginary of 'active doing' has been complemented and extended with products and services related to slowing down, for example forest bathing (discussed in Chapter 14). The COVID-19 pandemic has also broadened our ideas of adventure tourism. For instance, sleeping in a tent in your garden or geocaching in your neighbourhood are now regarded as adventure activities (see Gaetano, 2023). Reflecting this wide-ranging view of adventure, the Adventure Travel Trade Association ATTA, 2020 suggests that this form of tourism comprises three elements: being in nature, doing physical activity, and engaging with culture. Individuals can be classified as adventure tourists if they participate in one or all of these three elements. Therefore, adventure tourism organisations should focus on delivering 'specific types of experiences for travellers who are motivated by goals such as transformation, challenge and wellness' (ATTA, 2020). Unlike conventional delineations of tourism (e.g., UNWTO, 2023), the definition adopted by this association does not consider travel distance from tourists' homes and the time taken to reach the adventure destination. It therefore recognises that adventure tourism can take place "on your own doorstep" a view which aligns with the concept of microadventures (Houge Mackenzie & Goodnow, 2021).

Even though the ATTA's definition has been widely used in the research and practice of adventure tourism (e.g., Adventurati Outdoor, 2023; Business Finland, 2023; Pomfret

DOI: 10.4324/9781003393153-3

et al., 2023), scholars acknowledge a need to understand more extensively the scope and dimensions of adventure tourism. For example, a recent review by Janowski et al. (2021) suggests that aside from the dimensions of natural environment and physical activity, risk, danger, thrill, excitement, and challenge are at the core of adventure tourism. Sand and Gross (2019) emphasise the growing importance of health and well-being in adventure tourism. Furthermore, Janowski et al. (2021) claim that even though the ATTA endorses cultural experience as a key adventure tourism dimension, this is not fully reflected in academic research. Therefore, the aims of this chapter are to, first, outline some of the most commonly used definitions and conceptualisations of adventure tourism. Second, we introduce some new research directions in adventure tourism and, third, conclude why we need to continually conceptualise adventure tourism.

Current conceptualisations and definitions of adventure tourism

Definitions and conceptualisations of adventure tourism have primarily evolved from the fields of adventure recreation and adventure education (Varley & Taylor, 2013). Aside from tourism, adventure is also strongly connected with leisure studies, outdoor education, sociology, and consumer behaviour (Varley, 2006). There has been plentiful discussion on the scope and dimensions of adventure tourism since the 1990s (e.g., see Hall, 1992; Sung, 1997; Walle, 1997), and this topic continues to provoke considerable debate (e.g. see Janowski et al., 2021; Sand & Gross, 2019). One way to capture the differing perspectives of adventure tourism conceptualisations which feature in academic research is to categorise these into three dichotomies. These are (1) hard and soft, (2) commercial and "original", and (3) home and away dimensions of adventure tourism (Rantala et al., 2018).

First, the dichotomy between hard and soft dimensions of adventure tourism has been addressed in many studies (see e.g., Buckley, 2012; Cater, 2013; Pomfret & Bramwell, 2016). Within this dichotomy, soft adventure 'refers to activities with a perceived risk but low levels of real risk, requiring minimal commitment and beginning skills' (Hill, 1995, p. 63). Motives include escapism, experiencing new environments, self-discovery, novelty, excitement, and socialising (Lipscombe, 1995; Pomfret, 2006). In contrast, hard adventure activities involve 'high levels of risk, requiring intense commitment and advanced skills' (Hill, 1995, p. 63). Hard adventure motives include risk and challenge (Lipscombe, 1995; Pomfret, 2006). Hard and soft dimensions of adventure tourism have also been distinguished by referring to 'real' risks in hard adventures and 'perceived' risks in soft adventures (Williams & Soutar, 2005, p. 250). See Chapters 15 and 16 for more detailed accounts of the role of risk in adventure tourism and the risk management process.

Second, the dichotomy between commodified and "original" adventure tourism is prominent in academic studies. For instance, Buckley (2010) proposed that adventure tourism has become commodified from originally being a set of activities that only a minority of "hard core" individuals engaged in, to a form of mass tourism which almost anyone can participate in. Further reflecting this perspective, Varley (2006) constructed an adventure commodification continuum, which stretches from the realm of the original, "deep" adventures to the realm of highly saleable, "shallow" adventures. This continuum is discussed in more depth in Chapter 3. Here the dichotomy lies between "ideal-original" adventures and the "bulk" adventures made to be consumed.

Third, the dichotomy between home and away is something that is pertinent to any tourism study. For instance, Weber (2001) suggested that the study of adventure tourism should deal more clearly with the tourism aspects of adventure. Cater (2013) referred to adventure tourism as something different to everyday life, and Varley and Taylor (2013) discussed the inner journey related to adventure tourism. This third dichotomy is highly relevant to the current era of environmental crisis and climate change. Furthermore, recent conceptualisations such as microadventures or proximity tourism (Höckert et al., 2022) aim to challenge the notions of nearness and farness by emphasising the diverse scales of movement and different possibilities for the spatial ordering of tourism practices.

It should be noted that categorisations tend to build borders instead of highlighting connections. Hence, rather than these three dichotomies being clearly defined categories, we should view them as interconnected with overlapping dynamics. These dynamics have tended to focus on the production and consumption of adventure tourism, whereas tourism employees, host communities, non-human participants and, for example, philanthropy have often been given limited attention in previous conceptualisations. Therefore, in the following section, we introduce some new research directions that complement the previous discussions on the conceptualisation of adventure tourism and illustrate why the three earlier described dichotomies are still relevant today.

New research directions in reconceptualising adventure tourism

Relational approaches to adventure tourism

It could be argued that even though the natural environment or our interaction with nature has been, and is, at the core of adventure tourism, nature itself has been regarded more as a resource rather than as something which would have agency within adventure. When we see nature as a resource, we conceive of it as a manageable entity, which can be managed in a sustainable way (Jóhannesson, 2019). However, in this current environmental crisis epoch, during which we are experiencing a mass extinction of species, climate change, increased pollution, and overuse of resources, we should no longer perceive nature as a mere resource for adventure activities. Instead, we need to understand how inextricably intertwined humans and non-humans are and the implications of this strong connection (see Case study 3.1). Relational approaches, which suggest that the world exists in a continuous state of "becoming", adopt this viewpoint. Such approaches propose that humans and non-humans make their way in the world in unfolding relations with each other (Ingold, 2011; West et al., 2021). Relational approaches often enhance post-anthropocentric thinking – that is, thinking that is not human-centred – since they decentre the human as the "master of nature" and observe instead the agency of multiple other beings. They acknowledge the way that all earthly creatures are entangled with each other and how they live together (Valtonen et al., 2020).

Within tourism research, relational approaches were originally used in conceptualising the ordering of 'non-human objects, systems, machines, bureaucratic processes, times, timetables, sites, photographs, tents, flows, desires, visitors, businesses, locals' (Franklin, 2004, p. 284) and the constantly shifting networks of relationships in making tourism, for example by describing how cultural traditions, touristic commodification, modern mass production, and small-scale entrepreneurship come together in food in tourist destinations (Ren, 2011; van Der Duim, 2007; van der Duim et al., 2017). Since then, tourism scholars have applied relational approaches to a range of adventure tourism contexts, for example by discussing the role of weather in wilderness guiding work (Rantala et al., 2011), the intertwinedness of nature and culture in nature experiences (Lund, 2013), the healing agency of trekking (Olafsdottir, 2013), the materiality of camping (Rantala & Varley, 2019), the ethical relations enacted in mosquito-tourist encounters (Valtonen et al., 2020), and the agency of sled dogs in making tourism landscapes (Äijälä, 2021).

Case study 1.1 provides an illustrative example of a relational approach using the Sleeping Outdoors campaign.

Acknowledging and addressing the entangled relations of humans and non-humans means that we do not see nature as separate. We do not view it as something humans go to visit and consume while having an adventure holiday or as something they can fully control during their adventures (see Lund, 2013). Instead, the core idea of relationality – the intertwinedness of lives – opens up new possibilities for conceptualising adventure tourism. For example, relational approaches enable us to address issues related to safety, skills, ethics, responsibility, and animal rights in new ways, which is very much needed now that the environmental crisis is deepening.

For example, related to the previously mentioned dichotomy between soft and hard adventures, Hanna et al. (2019) have applied the relational approach within outdoor adventure tourism in the context of pro-environmentalism and eco-psychology. They argue that 'much of the research concerning outdoor adventure activities in the leisure context has generally theorised "nature" as something to "conquer" or "overcome" by the participants' (p. 1356). Their findings demonstrate that the dominant notions of risk, excitement, and thrill are clearly central to outdoor adventure tourists' understanding of their experience and are the driving motivations for future engagement with adventure activities. However, Hanna and colleagues illustrate how these notions also enable reciprocal understanding and active resistance to the dominant discourse of conquering nature.

Both hard and soft adventure can lead to feelings of becoming part of nature and include elements of reciprocity. Thus, the well-being of the non-human world may become a central issue when our practical concerns with, and care for, the non-human world are made tangible in outdoor contexts (Rantala, 2019). This can happen, for example, through becoming ready to interrupt one's own learned ways of doing things and being open towards different ways of being, doing, and knowing with nature (Höckert, 2018) or by acknowledging how one cultivates care towards the equipment and other non-human beings involved in adventure. Furthermore, adventure and tourism should not be seen as escaping from everyday responsibilities but rather as being anchored in the materialities and social responsibilities through concentrating on the 'micro-geographies', and on the very intense moments of being-with (Olafsdottir, 2013, p. 228; Rantala & Varley, 2019). This means that rather than engaging in romantic ideas of the wild, the surrounding world with its materiality and beings is experienced and regarded as an active co-habitant (Olafsdottir, 2013).

In line with this, in her doctoral thesis, Farkić (2018) combines adventure tourism research with hospitality studies and explores the commodification of nature on guided multi-day journeys in the outdoors. She examines, for example, the sensitivities – such as belonging, cosiness, sharing, intimacy, and togetherness – that are constructed by engaging with the diverse accommodations, when preparing food together, or when handling technologies, such as GoPro cameras, kayaks, wetsuits, and waterproof clothes. The research applies a relational approach by focusing on the embodied experiences of the outdoors and on the entanglements of human bodies with human and non-human worlds. Farkić shows how the range of human and non-human beings and objects contributes to humans becoming comfortable through the engagement with diverse materialities (e.g., equipment mentioned earlier) as well as through their interactions with one another.

The relational approaches enable us to examine critically the capitalist tourism systems and practices inherent in the development of adventure tourism, both near and far. For example, Jukes and Reeves (2020) have challenged the human-centred gaze in the context of outdoor environmental education by exploring possibilities for prompting students to understand how human and non-human relationships currently influence and might shape the future of different places. They do this by engaging students in more-than-human storytelling, with an aim to illustrate how places become co-storytellers – and with an intention to expose the students to the messiness of relationality. They wish to show how a new understanding of cultural, political, and environmental relations impacting place development can be created by writing and drawing stories from, for example a ski tourism journey. In a similar vein, Rantala and Höckert (2024) explore possibilities to include non-human voices in place-making through storytelling and show how slowing down and attuning to alternative rhythms and tempos of familiar places can mobilise activism and hope, and provide alternative perspectives for capitalist tourism development – such as highlighting justice instead of growth. They also warn that storytelling does not offer an easy solution. Certainly, finding creative possibilities to include non-human voices or relational perspectives does not provide straightforward alternatives for existing conceptualisations of adventure tourism since these tend to aim to understand and describe rather than categorise. However, they can help scholars become sensitive towards diversity when conducting research on adventure tourism and planning the future of adventure tourism.

Case study 1.1   Sleeping outdoors

A master's thesis by Henna Nevala (2021) on the Sleeping Outdoors campaign illustrates well the possibilities of relational approaches in addressing the contemporary issues related to environmental crises and consumerism in the context of adventure tourism. The campaign is an annual, recreation-oriented initiative organised by Suomen Latu – The Outdoor Association of Finland. It aims to encourage and challenge people to sleep outdoors for one night in their desired location, in whichever way they prefer. This can entail sleeping in a national park, a campsite, or in one's own backyard in a tent, a hammock, or under the night sky (Figure 1.1). The organisers especially want to support first-time campers since the aim of the campaign is to help people discover their preferred way to enjoy nature and make outdoor camping as accessible as possible for everyone.

According to Nevala (2021), the campaign contributes to the development of alternative tourism practices in several ways. It brings forth incommensurable 'commons' such as well-being, communality, and biodiversity by forming an economic space where these commons can be shared, maintained, and created. The term *commons* refers to a wealth of valuable assets that belong to everyone (Walljasper, 2010; see also Nevala, 2021). Commons can be both material and immaterial assets that arise whenever a community decides to share and manage a resource in a collective way with an emphasis on equitable access and use (Bollier, 2014, p. 11). These can include things like soil and minerals of public lands, clean air, sunlight, the atmosphere, water, fisheries, and nature preserves

Figure 1.1 Original microadventure? Camping in one's own backyard in a tent. Photograph by Outi Rantala.

as well as knowledge and cultural resources (Bollier, 2014; Gibson-Graham et al., 2013; Walljasper, 2010). Overall, the core idea of commons is that anyone can use them, as long as they are managed in a way that they will maintain themselves and offer equal benefits for people in the future (Walljasper, 2010).

Thus, the campaign illustrates how non-governmental organisations are altering the economic structures of tourism by establishing reciprocal connections between people and nature in local communities, sharing alternative ideologies, and teaching people about sustainable use of commons, as well as fostering alternative, low-threshold outdoor tourism practices. Their work is based on premises of care, cooperation, and nature-appreciation and their pursuits show how treating humans and non-humans with respect and care supports their well-being. In addition, the campaign contributes to the rise of proximity tourism and microadventures as an example of alternative adventure practices, which support rethinking what adventure could entail in local contexts while also shifting the focus from personal benefits to the well-being of others by strengthening, for example reciprocal ways of being and thinking. All in all, Nevala's study shows how seemingly marginalised actions can have profound effects.

Subjective well-being and adventure tourism conceptualisations

Given the growing importance of well-being-related motives and benefits of adventure tourism referred to in the introduction (Sand & Gross, 2019) and in Case study 1.1., it is useful to outline the well-being concept here. The word *well-being* is commonly used in everyday society, particularly by health services, to refer to a person's state of being healthy and happy, both mentally and physically (https://wellbeingpeople.com/). Well-being is also concerned with contributing to the community and maintaining a flourishing society in which citizens are healthy, happy, engaged, and capable (Marks & Sha, 2004). Scholars (e.g., Diener, 2006; Dodge et al., 2012; Ryff, 2014) propose various definitions of *well-being*, yet they disagree on how it should be defined. This is most likely due to the complex nature of well-being and the plethora of other terms, such as *happiness, quality of life*, and *life satisfaction*, which are used interchangeably to describe this concept.

Subjective well-being is an umbrella term which describes different types of individual well-being, for instance psychological, emotional, and social well-being (Sirgy, 2010). Diener and Suh (1997, p. 200) suggest that subjective well-being comprises three elements: 'life satisfaction, pleasant affect, and unpleasant affect. Affect refers to pleasant and unpleasant moods and emotions, whereas life satisfaction refers to a cognitive sense of satisfaction with life'. Attention has also been given to the distinctive dimensions of subjective well-being (Dodge et al., 2012). These include positive emotions, engagement, relationships, meaning, and accomplishment for a flourishing life (Seligman, 2011) and self-acceptance, positive functioning, autonomy, and equilibrium (Linley & Joseph, 2004). Subjective well-being involves interactions between multifaceted elements, including psychological resources, coping styles to manage challenges, personality, personal circumstances, and cognitive evaluations of life events (Ryff, 2014; Sirgy, 2010).

Adventure tourism researchers tend to examine subjective well-being as a set of activity participation motives and beneficial outcomes (e.g., Buckley, 2020; Pomfret, 2021), and, therefore, focus on the consumer. The increased scholarly interest in the role of adventure tourism in subjective well-being represents a move away from traditional research approaches, which only investigate aspects such as risk, conquering nature, and deviant personalities (Houge Mackenzie & Brymer, 2018), to exploring the 'holistic social nature of the [adventure tourism] experience' (Varley & Semple, 2015, p.77). A growing positive tourism movement has also prompted more interest in subjective well-being, alongside a departure from established recreation theories (Cheng et al., 2016), which argue that risk and danger are integral to adventure experiences (Priest & Gass, 2018).

While this chapter focuses on subjective well-being from a consumer perspective, Houge Mackenzie (Chapter 12) extends the discussion by providing a comprehensive overview of extant research (e.g., Brymer & Houge Mackenzie, 2015; Houge Mackenzie et al., 2021) on the role that adventure tourism plays in the psychological well-being of consumers, guides, adventure destinations, and their host communities. Chapter 12 proposes key mechanisms underpinning the psychological well-being outcomes of adventure activity participation and makes recommendations for future research and practice in this field. Both chapters serve to illustrate how subjective well-being is starting to gain traction as an integral element of adventure tourism conceptualisations (Houge Mackenzie & Hodge, 2020).

Notably, Janowski et al.'s (2021) conceptualisation of adventure tourism, referred to in the introduction, found that well-being was a recent consumer-based dimension

Figure 1.2 A scuba diver experiencing ill-being. Photograph by Carl Cater.

which featured in existing research. Yet, although there were many positive effects of adventure tourism cited in the literature, participation could also negatively impact individual well-being, for example over exposure to real risk, leading to injury or abject terror. This could lead to ill-being, which receives much less attention in adventure literature, but it is an important consideration in conceptualising this form of tourism. Ill-being can be triggered by stressful events or circumstances (see Figure 1.2) and dissatisfaction with one's health. It involves degrees of negative affect, such as feeling nervous, lonely, or helpless; depressive symptoms; feelings of anger and anxiety; and somatic complaints (Diener, 2006). Janowski et al. (2021), therefore, concluded that the 'diverse range of adventure-related catalysts and deterrents for well-being thus reflect the multidimensional nature of adventure tourism' (p. 7). This discussion reflects the three dichotomies in that adventure tourists potentially encounter feelings of ill-being and well-being before, during, and after their activity experiences, whether they are participating in hard or soft adventure pursuits, commodified or original experiences, or at home or away.

The value of adventure tourism-related literature reviews in conceptualising adventure tourism

Extant systematic literature reviews which examine subjective well-being in the context of adventure (e.g., Boudreau et al., 2020; Pomfret, Sand & May, 2023) can be helpful in conceptualising adventure tourism from a well-being perspective. Boudreau et al. (2020, p. 2) define adventure recreation as self-initiated nature-based physical activities that generate heightened bodily sensa-tions (e.g., vestibular sensations arising from quick acceleration in varying dimensions of space) and require skill development to manage unique perceived and objective risks.

(p. 2)

This definition includes, or at least hints at, elements of subjective well-being as it palpa-bly reflects hedonic well-being, which involves short bursts of positive affect and momen-tary emotional states (Diener, 1984). Notably, this review examines adventure recreation flow states (Csikszentmihalyi, 1975), such as happiness and natural highs. *Flow* in an adventure context refers to an optimal state of happiness that participants enjoy when engaged in activities. It represents 'the state in which people are so involved in an activity that nothing else seems to matter' (Csikszentmihalyi, 1992, p.4). Therefore, it is unsur-prising that 'heightened bodily sensations' (Boudreau et al., 2020, p.2) feature in the adventure recreation definition. The authors found that the flow outcomes of adventure recreation participation facilitate numerous positive consequences, such as improvements in well-being, motivation, skills, and performance. They conclude that there is a 'positive relationship between happiness (a form of hedonic well-being) and flow in adventure recreation' (p. 10). This supports the inclusion of subjective well-being in reconceptualis-ing adventure tourism.

Another systematic literature review (Pomfret et al., 2023) explored the interplay between adventure activity participation and subjective well-being. The authors defined adventure as 'participating in nature-based outdoor adventure tourism or recreation activities which may be challenging and may include perceived or real risk' (p. 2). Although this definition reflects a more traditional perspective of adventure, it suggests that challenge and risk are optional elements. Additionally, the literature search key-words adopt a broader view of adventure and include well-being, wellness, optimal experience, and *friluftsliv* (the Nordic concept of free-air-life; Gurholt, 2015). The study established that much of the extant research does not specifically refer to subjec-tive well-being in an adventure context. Instead, it alludes to this concept when discuss-ing the beneficial outcomes of adventure tourism or recreation participation. Figure 1.3. illustrates the five core subjective well-being themes and 16 subthemes of subjective well-being that appear in extant adventure-related research. The vertical lines under-neath each meta-theme and adjacent to each subtheme illustrate the most frequently cited themes within the reviewed literature, and the status of research in this field. Figure 1.3. shows that personal development (25%) is the most frequently cited core theme, followed by immersion and transformation (24%), physical and mental balance (22%), extraordinary experiences (18%), and community (12%). However, to date, scholars have directed more attention towards core themes such as personal develop-ment, and specific subthemes such as individual identity, Figure 1.3. highlights the need to address under-researched areas in the future. For example, the immersion and trans-formation core theme, and its subtheme, human/nature, are particularly pertinent to our discussion around the reconceptualisation of adventure tourism and reciprocity between humans and non-humans: 'Natural environments facilitate our convergence with nature during activity engagement and we interact with it in real time using all our senses' (Pomfret et al., 2023, p.7). It is the third-least cited subtheme out of four within the immersion and transformation meta-theme, palpably signalling the need for more research on this topic.

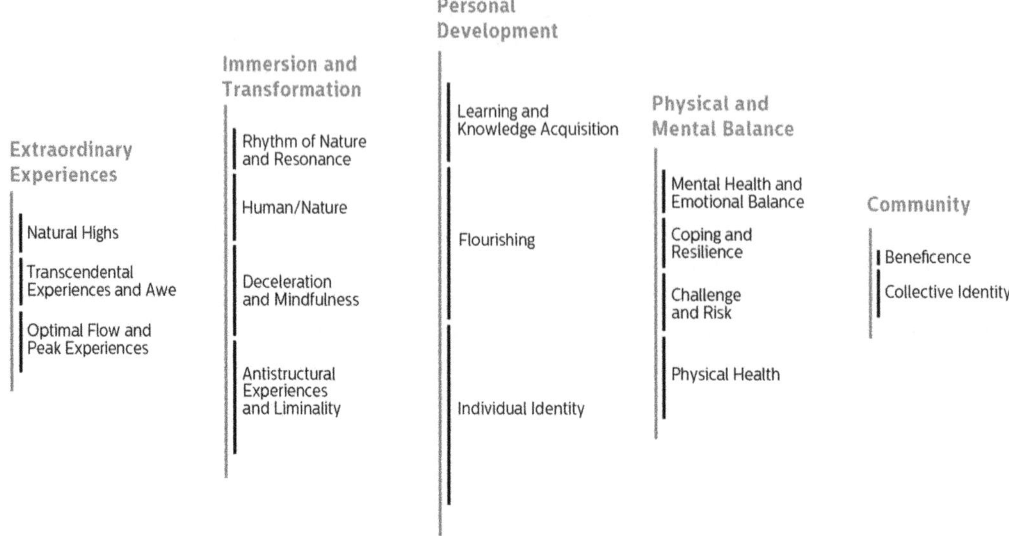

Figure 1.3 The subjective wellbeing metathemes and subthemes of outdoor adventure activity participation.

Source: Pomfret et al. (2023).

While we are not suggesting here that adventure tourism is only concerned with subjective well-being, as it can lead to feelings of ill-being also, we propose that well-being should form an integral element of adventure tourism conceptualisations. Integrating subjective well-being into adventure tourism conceptualisations can help to develop 'more robust, integrated models of adventure participation that can inform policy and practice' (Houge Mackenzie & Hodge, 2020, p.4). Relatedly, this reflects the third dichotomy and the notion that consumers can enjoy adventure tourism in their home regions, in the form of microadventures and proximity tourism (Höckert et al., 2022), as well as away, in tourist destinations with adventure resources. Ideally, therefore, tourists would participate in adventure activities both when at home and when away to continuously gain, maintain and enhance their subjective well-being. Researchers suggest this to be the case. For instance, Pomfret and Varley (2021) found that 'families embraced outdoor active lifestyles, regularly partaking in different activities at home' (p. 501) and holidays 'provided opportunities to extend their hobbies through participating in their usual activities in different destinations' (p. 501).

As noted earlier, some researchers do not specifically investigate subjective well-being benefits in relation to adventure tourism but rather allude to these in their study findings. For instance, Gstaettner et al. (2018) conducted a systematic literature review on the interplay between risk and nature-based tourism and recreation. Although this research focused on how scholars conceptualised risk, it also explored the potential benefits of participating in risky nature-based pursuits. Findings revealed a broad range of intrinsic benefits, for example skill development, insight, self-reflection, and self-actualisation, as well as extrinsic benefits, for example identity formation, being part of a group, and social status. Such benefits are labelled as beneficial well-being outcomes of adventure activity participation in other work (e.g., Pomfret et al., 2023). Therefore, exploring research which does not directly examine adventure tourism and subjective well-being also offers valuable insights into the inextricable links between the two concepts.

Another example is in the area of health-related research. Scholars (Britton et al., 2020) examined the literature on the health benefits of blue space therapeutic interventions for facilitating or restoring psychological well-being and physical health. Although not specifically focused on adventure tourism, the reviewed studies include adventure activities such as canoeing, surfing, and sailing. The authors note that the most frequently assessed indicators of well-being in extant research 'included self-esteem, self-efficacy, social confidence, resilience and other psychological indicators (e.g., stress, mood) using self report measures' (p. 60). Furthermore, improved psychosocial well-being and mental health were the most commonly cited positive outcomes. This research therefore confirms the value of blue space water-based adventure activities for subjective well-being and provides further evidence for the inclusion of well-being in emergent conceptualisations of adventure tourism.

Philanthropy through adventure tourism

Classic understandings of philanthropy refer to goodwill towards fellow humans (Novelli et al., 2016). In contrast, contemporary meanings of the term equate to helping humankind through donating money and time, and charitable gift-giving. The notion that one can "do good" through "giving back" whilst travelling has increased in popularity, especially through volunteering (Novelli et al., 2016) and has given rise to the wider term *travel philanthropy* (Honey, 2011). *Travel philanthropy* is defined as donating money, in-kind resources (office equipment, flights, accommodation) or time (mentoring or volunteering), occasioned or facilitated by travel (Goodwin et al., 2009). Both individual tourists and tourism businesses can be travel philanthropists, and three distinct practices are commonplace (Goodwin et al., 2009):

1 *Monetary donations* by travellers (either directly or via an intermediary such as a tour operator) and travel companies whereby the charitable donation is secondary to the primary purpose (travel/holidaymaking and the commercial operation of travel businesses). For example, after travelling to locations affected by poverty or major health and environmental problems, a tourist might be inspired to sponsor a school place or assist a community or species in danger of extinction (Novelli, 2015; Novelli et al., 2016).
2 *Corporate and/or personal charity fundraising.* For example challenges where travel is primarily motivated by the intention to raise funds for a charitable cause in their home country or in a destination.
3 *Volunteering* abroad or away from home, such as spending time on a conservation project whilst on holiday.

In each case, agents of travel philanthropy devote their skills, professional expertise and/or financial resources to individual community projects and/or organisations to foster environmental stewardship, development solutions, and sustainable social change (Goodwin et al., 2009; Honey, 2011; Novelli et al., 2016).

Adventure tourism as a vehicle for philanthropy is, for the most part, absent in scholarly literature. Adventure philanthropy is defined as combining altruism and humanitarianism with adventure travel (Lyons & Wearing, 2008). Therefore, whilst it could incorporate all three travel philanthropy practices, the limited academic literature has focused on the traveller's personal charity fundraising. For example, when arguing that adventure travel is part of broader trends in philanthropy, scholars (Coghlan & Filo, 2013; Goodwin et al., 2009; Rickly & Clouser, 2019) refer to the employment of alternative means of fundraising, specifically charity challenges conducted by individual participants (see Case study 1.2)

There are similarities between the adventure philanthropy practice of charity challenges and *fitness philanthropy*, which is defined as using sports events to raise money for, and awareness of health and social problems (Palmer, 2016; Palmer & Dwyer, 2020). Indeed, for some time now, charitable organisations have turned to sport-based events to engage communities and attract monetary donations (Ruperto & Kerr, 2009), whereas charity challenge is a relatively new phenomenon by comparison. The inclusion of more adventurous activities in the examples of sports-based events provided by Palmer and Dwyer (2020), such as epic hikes and long bike rides, implies there is a convergence between these two forms of philanthropy. However, the exclusion of adventure travel and the inclusion of and focus on traditional sports such as swimming, fun runs, and multi-sport challenges highlight divergence. Like adventure philanthropy, fitness philanthropy is also experiencing a paucity of scholarly interest, despite its emergence as a social or sociological phenomenon (Palmer, 2016; Palmer & Dwyer, 2020).

This lack of academic interest is perhaps neglectful as demand for adventure philanthropy is buoyant, therefore presenting good business opportunities for tour operators, and it offers charities an additional fundraising strategy (Goodwin et al., 2009). Correspondingly, we know little about the negative impacts of adventure philanthropy, and its sustainability has been questioned. For example, travel philanthropy tends to be

Case study 1.2   Charity challenges

Charity challenges involve individuals partnering with charitable organisations to set travel goals and raise funds for specific causes in the originating market or destination. For example, trekking and paddling in the Amazon to raise funds for clean water in destinations, trekking to Everest Base Camp to support children's education, or cycling around Sri Lanka to raise funds for homelessness in the originating market (Charity Challenge, 2023). The challenge duration is typically 1–14 days; it usually takes place in a tourist destination, although many participate in challenges close to home such as the Three Peaks Challenge in the UK (dichotomy 3), and they are usually commercially organised by an adventure tour operator (dichotomy 2). Whilst the challenges are of varying difficulty, for the most part, they are based on soft activities, such as trekking and cycling, they require basic skills and are low in risk. However, when done over consecutive days, they become more demanding and require commitment (dichotomy 1). Therefore, this form of adventure philanthropy exemplifies all three dichotomies explained earlier in the chapter.

Charity challenges may be specific to a charity, designed bespoke for a group of people to raise money for a chosen charity, or an open challenge where each of the participants is raising money for different causes (Goodwin et al., 2009). This can take the form of an event, whereby 'the charity promotes a pre scheduled event, chooses the amount of adrenaline, or creates a challenge which fits its charitable purposes' (p. 66).

reactive, short-term, and often generates incidental charitable contributions and relies on individuals who are constantly seeking new experiences (Novelli et al., 2016). Furthermore, it can be inequitable, with specific geographic locations benefitting more than others from donations and increased tourism, which may, in turn, lead to an overreliance on the financial and human resources of foreign visitors (*ibid.*). Major charity challenge events can also attract large numbers of participants and can result in negative environmental and social impacts (Goodwin et al., 2009). Therefore, there is significant scope for research in this emerging strand of adventure tourism. Furthermore, its increasing popularity implies that it has an important role to play in reconceptualising adventure tourism.

Adventure philanthropic motivations

Knowledge about adventure philanthropic motivations is largely limited to those participating in charity challenges, as this is where the scant literature lies. What is noticeable is that the charitable cause and the challenge are not independent of one another but, rather, work together in forming a mutually reinforced motivation (Palmer & Dwyer, 2020). For example, the challenge is used as a motivation to get fit, yet the charity cause itself is used as a motivation to complete the physical task. Bunds et al. (2016) coined this embodied philanthropy. The body's activities become a means of raising money (*ibid.*) and the challenge an act of solidarity with those who are in need (Rickly & Clouser, 2019). Therefore, like other humanitarians, adventure philanthropists are motivated by a sense of responsibility, that they can personally make a difference and implore others to do the same (Rickly & Clouser, 2019).

Philanthropic motivations related to doing good and giving back are missing in conceptualisations of adventure tourism, probably due to its scholarly neglect, yet adventure philanthropy can lead to enhanced psychological well-being (Coghlan, 2015). Improving mental health whilst on an adventure holiday is now almost as important as improving physical fitness (Worthington, 2022); therefore, the inclusion of well-being in contemporary adventure tourism conceptualisations (see Janowski et al., 2021) is paramount. We would further argue for the inclusion of adventure philanthropy in emergent conceptualisations, as it provides an avenue to meet the growing demand to do good and give back whilst on holiday and at home, as discussed later in this chapter. Finally, and undeniably, adventure philanthropy often presents a travel opportunity of a lifetime (Coghlan, 2015; Goodwin et al., 2009).

The commodification of adventure philanthropy

Drawing on Debord's (2009) theory of the society of the spectacle, which argues that commodification has transformed social life from *being*, to *having*, to *appearing*, Rickly and Clouser (2019) argue that the production of spectacle, and therefore commodification, is necessary in adventure philanthropy. To garner public attention, spread awareness, and raise funds, adventure philanthropists must use social media to create a spectacle and reproduce the spectacular images that inspired their participation in the trip. This then turns attention away from aid recipients towards the personal journey of the adventurer, thereby commodifying the event. Utilising alternative means of travel in the challenge is interpreted as liberating and transgressive, thus setting the individual apart from the crowd of other humanitarians and initiating an adventure story that enhances self-image and expression of values. Arrival at the destination is the culmination of the spectacle event, and the charity and adventurer implore the local resident and aid recipient to partake in the spectacle and to be a "guest" in their humanitarian performance. Thus, the recipients' poverty is commodified. This normalises inequality and reproduces the status quo, raising questions as to the effectiveness and the results of adventure philanthropy as a means of humanitarian tourism. They argue that such questions are relevant to stakeholders such as tourists/participants, audiences (e.g., those following the participant's challenge on social media or who has sponsored them), recipient communities, corporate sponsors, and humanitarian organisations (i.e., the charities that are benefitting).

Through the application of the society of the spectacle theory, Rickly and Clouser (2019) exemplify the second dichotomy which we found in adventure tourism conceptualisations. By participating in a saleable and shallow (Varley, 2006) commercially organised adventure challenge, the participants are consuming the experience. Yet, to achieve their philanthropic intentions, they must further commodify their experience, as well as the aid recipients' plight. A more genuinely altruistic example of doing good through adventure might be found in the example of TYF in Case study 3.1.

Finding demand in a crowded philanthropy marketplace

Nevertheless, market research shows that doing good and giving back is becoming increasingly important in everyday life and in people's holiday decisions. For example, in 2020, 62% of people in the UK gave to charity via donation or sponsorship (Charities Aid Foundation Report, 2021). Relatedly, preliminary desk research has found at least 20 companies in the UK that either specialise in or offer bespoke charity challenges trips, which are funded through sponsorship to raise funds for charity. This has, however, resulted in a crowded marketplace with many charities competing for the public's charitable interests and limited philanthropic spending. This is more acute when economies are stagnant or shrinking and consumer spending is constrained (Goodwin et al., 2009).

To attract philanthropic spending, charities must develop close relationships with philanthropists. Philanthropists, including travel philanthropists, want to see the impact and results of their charitable support to ensure their funds are invested well (Goodwin et al., 2009). This accountability and transparency are also demanded in the philanthropy of travel companies. Consumers are increasingly seeking ethically minded travel operators who make a difference to the communities in which they serve, as consumers look to give back through the travel products they buy and activities they take part in (Bryans, 2023). Further, the impact of the rising cost of living and consumers' preference for quality over the sustainability of a holiday means that it falls on operators to implement a range of sustainable practices, which include philanthropy, on tourists' behalf (Alexander, 2023). Therefore, it would seem now, more than ever, a tour operator's travel philanthropy presents a business opportunity to gain greater market share and it strengthens its inclusion of contemporary conceptualisations of adventure tourism.

Further, despite the crowded charity marketplace, the desire for volunteering is rising amongst younger travellers (16–24-year-olds) as many view it as an opportunity to become well-rounded global citizens (Alexander, 2023). Therefore, travel brands which appeal to the philanthropic consumer by providing experiences that combine volunteering with other holiday activities could access latent demand. Younger travellers are also increasingly interested in adventure holidays that enable them to achieve a personal challenge, make iconic journeys or expeditions, and spend time in the natural environment for its therapeutic value and to conserve and protect it. Adventure philanthropy offers this and operators that can leverage these motivations and activate philanthropic interests could access further latent demand.

Conclusion – reconceptualising adventure tourism

It could be due to the multiple crises that we have been facing in recent years, including ecological calamity, pandemics, and political instability, that the focus on adventure tourism research seems to be shifting towards relational and reciprocal approaches and themes. Until recently, emphasis within adventure tourism research has been on discussing the qualitative differences between hard and soft activities, commodified and "original" experiences and the role of tourism in adventure. Lately, the impacts of engaging in adventure have gained more space in these discussions. Hence, in future, we need to more firmly acknowledge the intertwinedness of adventure tourism with the surrounding human and non-human worlds. For instance, we need to understand better the reciprocal nature of adventure tourism's well-being impacts. We should facilitate the positive tourism movement, which supports the notion of adventure tourism for well-being. We need to comprehend the valuable contribution that adventure tourism makes to individual, community, societal and environmental well-being. For instance, previous research (Kaplan, 2001; Kuo & Sullivan, 2001; Maller, 2009) has shown that spending time in nature, e.g., increases self-esteem and mood, reduces anger, and improves general psychological well-being, with positive effects on emotions and behaviour. Despite these well-being benefits, we also need to acknowledge that adventure tourism can trigger feelings of ill-being. In some cases, negative emotions are integral to the adventure experience, particularly when there is an element of challenge within the activity. In other cases, they are unwanted and result in feelings of fear, despondency, and an unenjoyable experience.

These impacts are not based on a one-dimensional relationship with nature, but rather, they require active engagement with the natural environment. The notion of reciprocity draws attention to interpersonal relations between the self and the other and helps us reflect on our responsibilities and openness towards the other (Viken et al., 2021). While the idea of reciprocity can be understood as the mutual exchange of gifts, Viken et al. (2021) approach it as a more fundamental care relationship, where both hosts and guests care for each other's well-being. Thus, we are only starting to understand the relationships of, for instance, the human body and mind with diverse non-human species, such as microbes, plants, and fungi. A step further is to also comprehend the impacts of human engagement on non-human species, and how mutual care should form the basis of all our relations (Puig de la Bellacasa, 2017).

As discussed in this chapter, the environmental crisis requires us to care for the surrounding natural environment when experiencing and producing adventure tourism. Furthermore, increasing urbanisation and the hecticness of everyday life have underlined the need to care for oneself and resulted in rising interest and awareness of the well-being impacts of adventure tourism on oneself. Recently, new themes such as philanthropy have pointed out the willingness to care for other human beings while engaging in adventure tourism. Indeed, adventure tourism can enable us to ethically engage with the culture and materiality of the destination or community, which furthermore offers a potential to transform the habituated ways of practising adventure tourism from within (Ren & Rantala, 2022). The ethical engagement is based on increased sensitivity to one's surroundings, be it the non-human world, one's own ordinary life, other human beings, or the inclusion of all genders in adventure tourism (Hall & Brown, 2022).

Review questions

1 Select one of the new research direction themes presented within this chapter: (1) relational approaches to adventure tourism, (2) subjective well-being and adventure tourism conceptualisations, and (3) philanthropy through adventure tourism. Design a table which includes examples of adventure tourism organisations that promote activity experiences and holidays on this theme. What can you conclude about how this theme is represented by the adventure tourism industry?

2 Reflect on how you take into account the environmental crisis when practising, studying, or researching adventure tourism? How do you care for the non-human environment when practising, studying, or researching adventure tourism?

3 Think about an adventure tourism experience that you have enjoyed, either home or away. What subjective well-being benefits did you experience both during and after taking part in this activity? Did you experience any ill-being effects? If so, would you consider these to be an important part of your adventure experience?

4 Attention has largely been given to adventure tourists as travel philanthropists, but how can adventure tourism businesses be travel philanthropists? Conduct desk research and provide examples.

Further reading

Goodwin, H., McCombes, L., & Eckardt, C. (2009). *Advances in travel philanthropy: Raising money through the travel and tourism industry for charitable purposes.* WTM Responsible Tourism Day Report 2. https://www.haroldgoodwin.info/uploads/OP13.pdf

Pomfret, G., Sand, M. & May, C. (2023). Conceptualising the power of outdoor adventure activities for subjective well-being: A systematic literature review. *Journal of Outdoor Recreation and Tourism, 42,* 100641. https://doi.org/10.1016/j.jort.2023.100641

Rantala, O., Kinnunen, V. & Höckert, E. Eds. (2024). *Researching with proximity: Relational Methodologies for the Anthropocene.* Palgrave Macmillan.

References

Adventurati Outdoor. (2023). Retrieved August 2, 2023, from https://www.adventurati-outdoor.com/en

Adventure Travel Trade Association (ATTA). (2020). *Adventure tourism development index.* ATTA.

Äijälä, M. (2021). Mobile video ethnography for evoking animals in tourism. *Annals of Tourism Research, 89.* https://doi.org/10.1016/j.annals.2021.103203

Alexander, G. (2023). Sustainability in travel – UK – 2022. Mintel. https://reports.mintel.com/display/1102655/?fromSearch=%3Ffreetext%3Dsustainability%2520in%2520travel%26resultPosition%3D2

Bollier, D. (2014). Think like a commoner. In *A short introduction to the life of the commons.* New Society Publishers.

Boudreau, P., Houge Mackenzie, S., & Hodge, K. (2020). Flow states in adventure recreation: A systematic review and thematic synthesis. *Psychology of Sport and Exercise, 46,* 101611. https://doi.org/10.1016/j.psychsport.2019.101611

Britton, E., Kindermann, G., Domegan, C., & Carlin, C. (2020). Blue care: A systematic review of blue space interventions for health and wellbeing. *Health Promotion International, 35*(1), 50–69. https://doi.org/10.1093/heapro/day103

Bryans, J. (2023). Sustainability in travel – UK – 2023. Mintel. https://reports.mintel.com/display/1156431/?fromSearch=%3Ffreetext%3Dsustainability%2520in%2520travel%26resultPosition3D1

Brymer, E., & Houge Mackenzie, S. (2015). The impact of extreme sports on host communities' psychological growth and development. In Y. Reisinger (Ed.) *Transformational tourism: Host benefits* (pp. 129–140). CABI.

Buckley, R. (2010). *Adventure tourism management*. Butterworth-Heinemann.

Buckley, R. (2012). Rush as a key motivation in skilled adventure tourism: Resolving the risk recreation paradox. *Tourism Management, 33*(4), 961–970. http://doi.org/10.1016/j.tourman.2011.10.002

Buckley, R. (2020). Nature tourism and mental health: Parks, happiness, and causation. *Journal of Sustainable Tourism, 28*(9), 1409–1424. https://doi.org/10.1080/09669582.2020.1742725

Bunds, K., Brandon-Lai, S., & Armstrong, C. (2016). An inductive investigation of participants attachment to charity sport events: The case of team water charity. *European Sport Management Quarterly, 16*(3), 364–383. https://doi.org/10.1080/16184742.2016.1164212

Business Finland. (2023). Accelerating Finland's sustainable future in adventure tourism. At Outdoors Finland. Retrieved May 23, 2023. Helsinki, Finland. https://www.businessfinland.fi/globalassets/finnish-customers/news/events/2018/visit-finland-2018/of230518-christopher-doyle-adventure-travel-trade-association_web.pdf

Cater, C. (2013). The meaning of adventure. In S. Taylor, P. Varley, & T. Johnston (Eds.), *Adventure tourism: Meanings, experience and learning: Contemporary geographies of leisure, tourism and mobility* (pp.7–18). Routledge.

Charities Aid Foundation. (2021). UK Giving Report 2021. https://www.cafonline.org/docs/default-source/about-usresearch/uk_giving_report_2021.pdf

Charity Challenge (2023, Oct 7). Choose your challenge. https://www.charitychallenge.com/choose-your-challenge?gclid=CjwKCAjwg4SpBhAKEiwAdyLwvOrVliCshTx9XMoCfGB4vOn dHI-P3KYwjxomXjWVkLJKTxAb30U2KRoCwNUQAvD_BwE

Cheng, M., Edwards, D., Darcy, S., & Redfern, K. (2016). A tri-method approach to a review of adventure tourism literature: Bibliometric analysis, content analysis and a quantitative systematic literature review. *Journal of Tourism and Hospitality Research, 42*(6), 997–1020. https://doi.org/10.1177/1096348016640588

Coghlan, A. (2015). Tourism and health: Using positive psychology principles to maximise participants' wellbeing outcomes–A design concept for charity challenge tourism. *Journal of Sustainable Tourism, 23*(3), 382–400. https://doi.org/10.1080/09669582.2014.986489

Coghlan, A., & Filo, K. (2013). Using constant comparison method and qualitative data to understand participants' experiences at the nexus of tourism, sport, and charity events. *Tourism Management, 35*(1), 122–131. https://doi.org/10.1016/j.tourman.2012.06.007

Csikszentmihalyi, M. (1975). Play and intrinsic rewards. *The Journal of Humanistic Psychology, 15*(3), 41–63.

Csikszentmihalyi, M. (1992). *Flow: The psychology of happiness*. Rider.

Debord, G. (2009). *Society of the spectacle*. Soul Bay Press Ltd.

Diener, E. (1984). Subjective well-being. *Psychological Bulletin, 95*(3), 542–575.

Diener, E. (2006). Guidelines for national indicators of subjective well-being and ill-being. *Applied Research in Quality of Life, 1*(2), 151–157. https://doi.org/10.1007/s11482-006-9007-x

Diener, E., & Suh (1997). Measuring quality of life: Economic, social, and subjective indicators. *Social Indicators Research, 40*, 189–216. http://doi.org/10.1023/A:1006859511756

Dodge, R., Daly, A. P., Huyton, J., & Sanders, L. D. (2012). The challenge of defining wellbeing. *International Journal of Wellbeing, 2*(3), 222–235. https://doi.org/10.5502/ijw.v2i3.4

Farkić, J. (2018). *Hospitality in adventure tourism: Being, dwelling, guiding*. Doctoral Thesis. The University of Edinburgh. https://pure.uhi.ac.uk/en/studentTheses/hospitality-inadventuretourism

Franklin, A. (2004). Tourism as an ordering: Towards a new ontology of tourism. *Tourist Studies, 4*(3), 277–301. https://doi.org/10.1177/1468797604057328

Gaetano, R. (2023). Tips for your next adventure: Backyard winter camping. NEMO Equipment. Retrieved August 2, 2023, from. https://www.nemoequipment.com/blogs/journal/tips-for-your-next-adventure-backyardwintercamping

Gibson-Graham, J. K., Cameron, J., & Healy, S. (2013). *Take back the economy: An ethical guide for transforming our communities*. University of Minnesota Press.

Goodwin, H., McCombes, L., & Eckardt, C. (2009). *Advances in travel philanthropy: Raising money through the travel and tourism industry for charitable purposes*. WTM Responsible Tourism Day Report 2. https://www.haroldgoodwin.info/uploads/OP13.pdf

Gstaettner, A. M., Lee, D., & Rodger, K. (2018). The concept of risk in nature-based tourism and recreation–A systematic literature review. *Current Issues in Tourism, 21*(15), 1784–1809. https://doi.org/10.1080/13683500.2016.1244174

Gurholt, K. P. (2015). Friluftsliv: Nature-friendly adventures for all. In B. Humberstone, H. Prince, & Karla A. Henderson (Eds.), *Routledge international handbook of outdoor studies* (pp. 288–296). Routledge.

Hall, C. M. (1992). Adventure, sport and health tourism. In B. Weiler & C. M. Hall (Eds.), *Special interest tourism* (pp. 141–158). Belhaven.

Hall, J., & Brown, K. M. (2022). Creating feelings of inclusion in adventure tourism: Lessons from the gendered sensory and affective politics of professional mountaineering. *Annals of Tourism Research*, 97, 103505. https://doi.org/10.1016/j.annals.2022.103505

Hanna, P., Wijesinghe, S., Paliatsos, I., Walker, C., Adams, M., & Kimbu, A. (2019). Active engagement with nature: Outdoor adventure tourism, sustainability and wellbeing. *Journal of Sustainable Tourism*, 27(9), 1355–1373. https://doi.org/10.1080/09669582.2019.1621883

Hill, B. J. (1995). A guide to adventure travel. *Parks & Recreation*, 30(9), 56–65.

Honey, M. (2011). Origin and overview of travelers' philanthropy. In M. Honey (Ed.), *Travelers' philanthropy handbook* (pp. 3–12). Center for Responsible Travel.

Houge Mackenzie, S., & Brymer, E. (2018). Conceptualizing adventurous nature sport: A positive psychology perspective. *Annals of Leisure Research*, 23(1), 79–91. https://doi.org/10.1080/11745398.2018.1483733

Houge Mackenzie, S., & Hodge, K. (2020). Adventure recreation and subjective wellbeing: A conceptual framework. *Leisure Studies*, 39(1), 26–40. https://doi.org/10.1080/02614367.2019.1577478

Houge Mackenzie, S., Hodge, K., & Filep, S. (2021). How does adventure sport tourism enhance well-being? A conceptual model. *Tourism Recreation Research*, 48(1), 3–16. https://doi.org/10.1080/02508281.2021.1894043

Höckert, E. (2018). *Negotiating hospitality. Ethics of tourism development in the Nicaraguan Highlands*. Routledge.

Höckert, E., Rantala, O., & Jóhannesson, G. T. (2022). Sensitive communication with proximate messmates. *Tourism Culture and Communication*, 22(2), 181–192. https://doi.org/10.3727/109830421X16296375579624

Ingold, T. (2011). *Being alive: Essays on movement, knowledge and description*. Routledge.

Janowski, I., Gardiner, S., & Kwek, A. (2021). Dimensions of adventure tourism. *Tourism Management Perspectives*, 37, 100776. https://doi.org/10.1016/j.tmp.2020.100776

Jóhannesson, G. T. (2019). Looking down, staying with and moving along: Towards collaborative ways of knowing with nature in the Anthropocene. *Matkailututkimus*, 15(2), 9–17. https://doi.org/10.33351/mt.88264

Jukes, S., & Reeves, Y. (2020). Morethanhuman stories: Experimental coproductions in outdoor environmental education pedagogy. *Environmental Education Research*, 26(9–10), 1294–1312. https://doi.org/10.1080/13504622.2019.1699027

Kaplan, R. (2001). The nature of the view from home. Psychological benefits. *Environment and Behavior*, 33 (4), 507–542. https://doi.org/10.1177/00139160121973115

Kuo, F. E., & Sullivan, W. C. (2001). Aggression and violence in the inner city. Effects of environment via mental fatigue. *Environment and Behavior* 33(4), 543–571. https://doi.org/10.1177/00139160121973124

Linley, P. A., & Joseph, S. (2004). (Eds.) *Positive psychology in practice*. John Wiley & Sons, Inc.

Lipscombe, N. (1995). Appropriate adventure for the aged. *Australian Parks and Recreation*, 31(2), 41–51.

Lund, K. A. (2013). Experiencing nature in nature-based tourism. *Tourist Studies*, 13(2), 156–171. https://doi.org/10.1177/1468797613490373

Lyons, K., & Wearing, S. (2008). All for a good cause? The blurred boundaries between volunteering and tourism. In K. Lyons & S. Wearing (Eds.), *Journeys of discovery in volunteer tourism* (pp. 147–154). CAB International.

Mackenzie, H., & Goodnow, J. (2021). Adventure in the age of COVID-19: Embracing microadventures and locavism in a postpandemic world, *Leisure Sciences*, 43(1–2), 62–69. https://doi.org/10.1080/01490400.2020.1773984

Marks, N., & Sha, H. (2004). A well-being manifesto for a flourishing society. *Journal of Public Mental Health*, 3(4), 9–15.

Maller, C. J. (2009) Promoting children's mental, emotional and social health through contact with nature: A model. *Health Education* 109(6), 522–543. https://doi.org/10.1108/09654280911001185

Nevala, H. (2021). *NGOs fostering alternative tourism economies: Sleeping outdoors campaign as a case of proximity tourism*. Master thesis, University of Lapland. https://urn.fi/URN:NBN:fi-fe2021102952942

Novelli, M. (2015). Are travel philanthropists doing more harm than good? *The Conversation*. https://theconversation.com/are-travel-philanthropists-doing-more-harm-than-good-44629

Novelli, M., Morgan, N., Mitchel, G., & Ivanov, K. (2016). Travel philanthropy and sustainable development: The case of the Plymouth-Banjul Challenge. *Journal of Sustainable Tourism*, 24(6), 824–845. https://doi.org/10.1080/09669582.2015.1088858

Olafsdottir, G. (2013). On nature-based tourism. *Tourist Studies*, 13(2), 127–138. https://doi.org/10.1177/1468797613490370

Palmer, C. (2016). Research on the run: Moving methods and the charity 'thon'. *Qualitative Research in Sport, Exercise and Health*, 8(3), 225–236. https://doi.org/10.1080/2159676X.2015.1129641

Palmer, C., & Dwyer, Z. (2020). Good running? The rise of fitness philanthropy and sports-based charity events. *Leisure Sciences*, 42(5-6), 609–623. https://doi.org/10.1080/01490400.2019.1656122

Pomfret, G. (2006). Mountaineering adventure tourists: A conceptual framework for research. *Tourism Management*, 27(1), 113–123. https://doi.org/10.1016/j.tourman.2004.08.003

Pomfret, G. (2021). Family adventure tourism: Towards hedonic and eudaimonic wellbeing. *Tourism Management Perspectives*, 39, 100852. https://doi.org/10.1016/j.tmp.2021.100852

Pomfret, G., & Bramwell, B. (2016). The characteristics and motivational decisions of adventure tourists: A review and analysis. *Current Issues in Tourism*, 19(14), 1447–1478. https://doi.org/10.1080/13683500.2014.925430

Pomfret, G., Sand, M., & May, C. (2023). Conceptualising the power of outdoor adventure activities for subjective well-being: A systematic literature review. *Journal of Outdoor Recreation and Tourism*, 42, 100641. https://doi.org/10.1016/j.jort.2023.100641

Pomfret, G., & Varley, P. (2021). Families at leisure outdoors: Well-being through adventure. *Leisure Studies*, 38(4), 494–508.

Priest, S., & Gass, M. (2018). *Effective leadership in adventure programming*. 3rd ed. Human Kinetics.

Puig De La Bellacasa, M. (2017). *Matters of care: Speculative ethics in more than human worlds*. University of Minnesota Press.

Rantala, O. (2019). With the rhythm of nature: Reordering everyday life through holiday living. In H. Halkier, L. James, & C. Ren (Eds.) *Theories of practice in tourism* (pp.58–76). Routledge.

Rantala, O., & Höckert, E. (2024). Multispecies stories from the margins. In B. Thorsteinsson, K. A. Lund, G. T. Jóhannesson, & G. R. Jóhannesdóttir (Eds.) *Mobilities on the margins. Creative processes of place-making*. Palgrave.

Rantala, O., Rokenes, A., & Valkonen, J. (2018). Is adventure tourism a coherent concept? A review of research approaches on adventure tourism. *Annals of Leisure Research*, 21(5), 539–552. https://doi.org/10.1080/11745398.2016.1250647

Rantala, O., & Varley, P. (2019). Wild camping and the weight of tourism. *Tourist Studies*, 19(3), 295–312. https://doi.org/10.1177/1468797619832308

Rantala, O., Valtonen, A., & Markuksela, V. (2011). Materializing tourist weather: Ethnography on weather-wise wilderness guiding practices. *Journal of Material Culture*, 16(3), 285–300. https://doi.org/10.1177/1359183511413646

Ren, C. (2011). Non-human agency, radical ontology and tourism realities. *Annals of Tourism Research*, 38(3), 858–881. https://doi.org/10.1016/j.annals.2010.12.007

Ren, C., & Rantala, O. (2022). Arctic adventure. In D. Buhalis (Ed.) *Encyclopedia of tourism management and marketing* (pp. 174–176). Edward Elgar Publishing.

Rickly, J. M., & Clouser, R. (2019). Spectacle and adventure philanthropy. *Annals of Tourism Research*, 77, 158–160. https://doi.org/10.1016/j.annals.2018.12.008

Ruperto, A. R., & Kerr, G. (2009). A study of community events held by not-for-profit organizations in Australia. *Journal of Nonprofit and Public Sector Marketing*, 21(3), 298–308. https://doi.org/10.1080/10495140802644547

Ryff, C. D. (2014). Psychological well-being revisited: Advances in the science and practice of eudaimonia. *Psychotherapy and Psychosomatic*, 83(1), 10–28. https://doi.org/10.1159/000353263

Sand, M., & Gross, S. (2019). Tourism research on adventure tourism–Current themes and developments. *Journal of Outdoor Recreation and Tourism*, 28, 100261. https://doi.org/10.1016/j.jort.2019.100261

Seligman, M. E. P. (2011). *Flourish*. Random House.

Sirgy, M. J. (2010). Toward a quality-of-life theory of leisure travel satisfaction. *Journal of Travel Research*, 49(2), 246–260. https://doi.org/10.1177/0047287509337416

Sung, H. (1997). Classification of adventure travelers: Behaviour, decision making, and target markets. *Journal of Travel Research*, 42(4), 343–356.

UNWTO. (2023). World Tourism Organization. Glossary of tourism terms. Retrieved August 2, 2023, from https://www.unwto.org/glossary-tourism-terms

Valtonen, A., Salmela, T., & Rantala, O. (2020). Living with mosquitoes. *Annals of Tourism Research*, 102945. https://doi.org/10.1016/j.annals.2020.102945

van der Duim, R. (2007). Tourismscapes an actor-network perspective. *Annals of Tourism Research*, 34(4), 961–976. https://doi.org/10.1016/j.annals.2007.05.008

van der Duim, R., Ren, C., & Jóhannesson, G. T. (2017). ANT: A decade of interfering with tourism. *Annals of Tourism Research*, 64, 139–149. https://doi.org/10.1016/j.annals.2017.03.006

Varley, P. (2006). Confecting adventure and playing with meaning: The adventure commodification continuum. *Journal of Sport & Tourism*, 11(2), 173–194. https://doi.org/10.1080/14775080601155217

Varley P., & Semple, T. (2015). Nordic slow adventure: Explorations in time and nature, *Scandinavian Journal of Hospitality and Tourism*, 15(1–2), 73–90. https://doi.org/10.1080/15022250.2015.1028142

Varley, P., & Taylor, S. (2013). Introduction. In S. Taylor, P. Varley, & T. Johnston (Eds.), *Adventure tourism. Meanings, experience and learning. Contemporary geographies of leisure, tourism and mobility* (pp. 1–3). Routledge.

Viken, A., Höckert, E., & Grimwood, B. S. R. (2021). Cultural sensitivity: Engaging difference in tourism. *Annals of Tourism Research*, 89, 103223. https://doi.org/10.1016/j.annals.2021.103223

Walle, A. H. (1997). Pursuing risk or insight: Marketing adventures. *Annals of Tourism Research*, 24(2), 265–282.

Walljasper, J. (2010). *All that we share: A field guide to the commons*. The New Press.

Weber, K. (2001). Outdoor adventure tourism. *Annals of Tourism Research*, 28(2), 360–377.

West, S., Haider, L. J., Stålhammar, S., & Woroniecki, S. (2021). A relational turn for sustainability science? Relational thinking, leverage points and transformations. *Ecosystems and People*, 16(1), 304–325. https://doi.org/10.1080/26395916.2020.1814417

Williams, P., & Soutar, G. N. (2005). Value, satisfaction and behavioral intentions in an adventure tourism context. *Annals of Tourism Research*, 36(3), 413–438. https://doi.org/10.1016/j.annals.2009.02.002

Worthington, J. (2022). Physical activity holidays – UK – 2022. Mintel. https://reports.mintel.com/display/1101381/

2 History of adventure tourism

Carl Cater

Chapter learning outcomes

1 Understand the biological and evolutionary basis for adventure.
2 Recognise the historical origins of adventure tourism in exploration.
3 Understand how innovation has shaped product development in adventure tourism.
4 Explain how adventure tourism has become progressively commodified and widespread.

Introduction

As Chapter 1 has shown, definitions of adventure tourism are at best, diverse. A further enigma lies in the fact that many of the activities that we might today put under the umbrella of adventure tourism, for example mountaineering, skiing, or scuba diving, significantly predate the term by decades or even centuries. This chapter looks at the origins of adventure tourism by examining the emergence of these activities and the changing nature of adventure itself.

Although the term adventure tourism was in general use by government and operators in the 1990s, it was not until the end of the decade that there was any mention of the term in an academic sense. For example, in New Zealand, a pioneer of much adventure activity, a seminar by the Hillary Commission took place in 1987, but the earliest academic work seems to be the seminal paper on adventure tourism in Queenstown by Cloke and Perkins in 1998. Of course, this does not mean that adventure tourism was not happening, and as is often the case for scholars, the activities were already firmly established on a global scale. In this chapter, we will explore how this came about.

Ancestral adventure

Our need for adventure is to a large degree rooted in our biology and begins at the dawn of our evolution. Our prehistoric ancestors faced some form of challenge to their very existence on a regular basis, whether it be the search for food, shelter, or avoiding predators. Whilst these challenges might seem dissimilar to modern adventure, there are a number of parallel elements that underpinned this quest. In *The Songlines*, Bruce Chatwin (1987) describes the relationship to one of our last known predators, a leopard-like

DOI: 10.4324/9781003393153-4

creature called *Dinofelis*, which hunted our ancestors in Africa, the Australopithecines, some 1.2 million years ago. Dealing with such a threat required the production of various hormones within the body. Sensing a threat, a phenomenon recognised as *fight-or-flight* involves the 'aminergic system of the brain stem preparing the brain-mind for action by heightening all the functions normally associated with waking' (Hobson, 1994, p.161). This includes the production of serotonin which serves to make the brain more alert, adrenaline to prepare for physical and mental exertion, and dopamine in order to prepare the body for injury.

Whilst a grisly end with a sabre-toothed tiger might not immediately seem to have much in common with adventure, it is these naturally produced drugs that drive the thrill-seeking side to adventure participation. These hormones heighten the experience and are often central to satisfaction in more active adventure. Colloquially, the feelings that result from these activities are described in non-medical terms, for example as a "buzz" or "rush" (Buckley, 2012). However, even in an adventure that is less adrenaline-based, the completion of a challenge is likely to lead to the same sense of well-being. Thus, this desire for a connection to our own biological nature runs parallel to connection to the natural world, which has been described as biophilia (Wilson, 1993). Indeed, Chatwin argues that there was a sense of sublime intimacy with our predators and that there is, in some senses, a 'nostalgia for the Beast we have lost' (Chatwin, 1987, p. 254). Adventurous ecotourism experiences like shark cage diving or polar bear tracking are evidence of a desire to reconnect with these 'beasts'. The point here, then, is that adventure is part of who we are, and whilst its form has changed in terms of participation, risk, and place, adventure is a factor throughout all human development.

Adventure and exploration

A human pursuit that might be more readily recognised as adventurous is the long history of exploration. In earlier times, exploration may have been driven by the survival needs described earlier. Humans are by their nature nomadic, and our more modern sedentary lifestyles have only become dominant over the past 10,000 years with the advent of agriculture. Paradoxically, as these societies grew and developed, state-sponsored exploration became more organised and often fulfilled the expansionist desires of states forming the basis of the colonial period. Whilst Christopher Columbus and Captain Cook may not have thought of themselves as adventure tourists, the desire to explore the unknown and new frontiers is certainly one that is well aligned to adventure, and it is a theme that is well used in contemporary adventure marketing.

Therefore, culturally and historically the dominant Western meaning of adventure has been shaped by the European exploration and colonisation of the world. Phillips shows how the production of the earliest world maps in particular 'established an intellectual framework in which contemporary geographical knowledge was logged, and ignorance made visible' (1997, p. 6). The blank hearts of continents like those in the Mercator map of 1587 acted as an incentive both to the imagination and hence exploration of them. These "hearts of darkness" continue to do so, as noted in Macfarlane's (2004) descriptions of the limited maps available to him of the Tien Shan mountains. Of course, the exploration and further mapping of these spaces was initially bound up with the scramble to 'colonise and consolidate imperial power' (Phillips, 1997, p. 7). The irreducible links that lie between are highlighted by how 'science, exploration, and indeed "adventure" occupied (and, by implication, continue to occupy) a common space'

(Gregory, 1994, p. 20). This rhetoric continues with organisations such as the *Royal Geographical Society*, founded in 1830 and behind many famous expeditions, still closely linked to the promotion of adventure and exploration, through funding for adventures and annual training workshops such as *Explore*.

The fact that these explorations of unmapped space are inherently adventurous is also linked to the generation of adventure tales, both real and literary, founded in this other space. Indeed, the binary between objective science and emotional romanticism is a terrain where adventure may be seen to be negotiated and defined. The romantic view of the natural spaces which are often the setting for adventure began to emerge from the late 17th century and was clear in Victorian accounts of growing tourism to the European Alps. Glaciers began to attract significant crowds, and early guiding, with descriptions of the Mer de Glace printed in the influential Baedecker guidebook to Switzerland. By 1785, there were three sizeable inns in Chamonix catering to the 1,500 tourists who visited the glacier each summer (Macfarlane, 2004). A 1927 Stanford guide to Norway (Willson, 1927) includes a guide to glacier climbing for visitors, with 26 points of helpful advice, noting the 'large increase in the number of tourists who cross this greatest snowfield in Europe' (p. 261). Amongst the guidance regarding the use of nails in one's boots and traversing of crevasses, point 25 emphasises 'drink no aquavit or cognac until you have finished your day's work' (!). Of particular interest is that these suggestions were aimed at the novice and were;

> not written for experienced mountaineers, but rather for the benefit of the many strong and active young men and girls [*sic*] whose love of adventure naturally tempts them at their first opportunity to leave the ease and luxury of the valleys for the joys of the mountains.
>
> (Willson, 1927, p. 262)

The popularity of adventure in the public imagination also led to the explorers themselves becoming a spectacle for tourists. For almost as long as climbers have been attempting to summit the Eiger in the Bernese Alps, tourists have been watching their progress from relative safety. Sometimes this spectacle took a turn to the darker side, as in the 1936 Eiger disaster when five climbers died ascending the north face. In the late 1950s, the body of an Italian climber, Stefano Longhi, hung for two years at the end of a rope, as a 'grotesque and macabre tourist attraction for the Eiger watchers at the telescopes in Alpiglen and Kleine Scheidegg' (Simpson, 2003, p.195).

Similarly, the exploits of famous polar explorers such as Franklin, Scott, and Shackleton were very popular with the public, despite their significant failures. John Franklin disappeared searching for the North-West Passage in 1847 but was a very popular public figure at the time. Arguably the most famous polar explorer of all, Robert Scott was beaten to the South Pole by the lesser known Norwegian Roald Amundsen in 1912 and died on the return voyage to the coast. Yet these men remained heroes in the public imagination, with significant funding for these trips coming from books, serialisations, and later movies. Although the infamous advertisement Shackleton placed in the *Times* asking for volunteers for 'a hazardous journey from which safe return was doubtful' probably never really existed, he certainly received several thousand applications for places on the expedition (Fiennes, 2021). Whilst the expedition was ultimately a failure, with the loss of *Endurance* in pack ice, it is celebrated for the successful return of all the crew under Shackleton's leadership. An 800-mile journey through the tempestuous Southern Ocean in an open lifeboat, followed by a perilous crossing of the mountains of South Georgia to a whaling station is still seen as an incredible feat.

When Edward Whymper designated Mount Everest (Chomolungma/Sagarmāthā) as the 'third pole' in 1894, it underlined a desire for expeditionary attempts to conquer the summit (Macfarlane, 2004). A series of British expeditions in the early 1920s took a circuitous route through Sikkim and Tibet to attempt ascents from the Rongbuk glacier. These ended in tragedy in 1924 when George Mallory and Andrew Irvine made an attempt on the North East ridge but disappeared short of the summit. When questioned by a New York reporter as to why he kept returning to the mountain, Mallory famously said 'because it is there'. As Macfarlane (2004) notes, their deaths were not seen as folly or a waste of life but were seen as "heroic" and "glorious" in the print media at the time. It would take nearly thirty years before the mountain was finally conquered, this time from the southern Nepali side. News of Tenzing Norgay and Edmund Hillary's successful ascent in 1953 reached London on the morning of the coronation of Queen Elizabeth II, cementing the links between mountaineering achievements and national pride.

It is also undeniable that historically, adventure has an underlying masculinist imperative, and this has existed throughout history in one form through wars. It is clear that adventure in this case was not always entirely voluntary, although an idea of duty has a powerful, albeit externally influenced, part to play. However, it is apparent that many of the concepts that can be seen to emerge around ideas of pure "adventure" are closely entwined with those of conflict. Often it enabled the only means of escape for the ordinary man from a more traditional life, and the opportunity to travel, and then return with tales of glory and heroism, was an attractive one. World War II was an important watershed because huge numbers of people, mostly men, had undergone some form of temporary or permanent displacement. Whilst this was not necessarily tourism, it opened up the possibilities and desires for travel in a way not before witnessed. Furthermore, the post-war period was an important one in terms of technology and adventure, when large amounts of surplus military equipment, such as jeeps, rafts, backpacks, scuba gear, camping equipment, skis, and guns could be redeployed in the search for adventure in inhospitable places (Adventure Travel Society, 1999). River navigation, and whitewater rafting in particular, owes an early debt to the availability of this hardware. Much of this surplus equipment deteriorated, however, and it was not until the 1970s that the availability of cheap new materials like aluminium, rubber, and plastics enabled rafts to be marketed at a reasonable price again (Maritime Safety Authority, 1995).

Selling mountains

The post-war period and the growth of both leisure time and leisure practice saw the emergence of activities that we would more readily recognise as adventure tourism, even if the term itself was not in common use. Following the conquest of Everest, there was a rush to conquer the other great peaks, but it was not until the 1980s that commercial mountaineering really took off. It took 30 years after Tenzing and Hillary for there to be a further 150 summits, but during the 1980s and 1990s, guided ascents accelerated. In 2019, a record 891 climbers successfully climbed the world's highest mountain, and pictures of huge queues during the short climbing window circulated in the media. Whilst there were no ascents of Everest in 2020 during the COVID-19 pandemic, similar images of congestion emerged at much lower elevations, for example on Snowdon in Wales, when people were able to access the outdoors after repeated lockdowns (BBC, 2021).

The commercialisation of mountain environments over the past decades was well described by Johnston and Edwards (1994, p.468).

> Corporate sponsorship has shaped mountain experiences and even the fantasy of a mountain experience in order to sell commodities to a consuming culture. [As a result,] … many more well-equipped, stylishly dressed holiday consumers are travelling to mountain regions … sent by an ever-growing legion of adventure travel companies who advertise their services in (an also growing number) "Adventure Travel" magazines and guides. They arrive carrying clothing and equipment purchased at outdoor shops staffed by adventure enthusiasts; and they are guided through their mountain adventure by mountaineers turned tour guides.

This phenomenon was also evident in another long-standing mountain adventure activity, that of skiing. Whilst existing for several centuries in Nordic regions as a means of transport, it was not until the 20th century that it became recognised as a leisure activity. Evolving from the Nordic disciplines, which are still popular as cross-country skiing, Alpine, or downhill, skiing required the development of new types of skis and significant infrastructure development in terms of chair lifts (first introduced in Idaho in 1936) and ski resorts. The latter initially developed from mountain locations which had existing summer tourism, but as the sport grew in popularity, resorts were increasingly built from scratch. These ski resorts have been instrumental in the development of adventure tourism in particular, as a wide range of adventure activities in addition to skiing have been developed in these locations in recent decades. These have developed either organically or strategically but, in many mountain locations, form a year-round adventure tourism product.

Winter sports themselves have also continued to evolve. Snowboarding emerged in the late 1960s and became particularly popular in North America, although in recent years it seems to be in decline (Kelechava, 2021). Heliskiing, where experienced skiers access pristine off-piste runs via helicopter, continues to be popular, particularly in the Canadian Rockies (Buckley 2006; Richins & Hull, 2016). Artificial snowmaking equipment, pioneered in the 1950s in New York state is now present at a majority of ski resorts to lengthen the season and ensure the quality of snow cover. New locations and new markets are continuing to develop, for example the Shahdag Mountain Resort Complex in Azerbaijan, which opened in 2012. However, many ski resorts, particularly those at lower altitudes or latitudes are currently grappling with the impacts of climate change (as described in Chapter 21).

(Under)water adventure

Mountainous environments were not the only places where adventure was becoming popular. For example, the development of the aqualung by Jaques Cousteau in the 1940s allowed exploration of the underwater realm. The world's leading certification agency has certified over 29 million people since 1967 (PADI, 2022), and the number of amateur divers in particular has increased dramatically over the past three decades. The recent annual growth rate of the dive travel market is now 16% worldwide, and the most important source markets for diving tourism are identified as the US, Italy, Germany, France, and the UK (CBI, 2019). Whilst scuba diving was once seen as a niche activity, it is now a popular adventure tourism activity offered in many coastal tourism resorts. New markets have been important in this growth, with emerging economies such as China, Turkey, and Russia showing significant interest in scuba diving activity. In addition, the gender balance of scuba divers has gradually become more equal, with a split of 37% female and 63% male certifications in 2021. However, the dive market is also getting progressively younger, with a tripling of under-20-years-old certifications between 2017 and 2021 (PADI, 2022). This may be a result of changes to the qualification structures that enable younger participation.

The dive industry itself has changed in parallel with the experiential turn in tourism more broadly. In its early development, scuba divers were often certified in their home country and did much of the activity there. Since 2000, many exotic destinations such as Egypt and Thailand have developed their diving infrastructure, altering the global strategy of the industry to promote diving trips to the most stunning marine ecosystems on Earth (Dimmock & Cummins, 2013). Today, diving is deeply bonded to tourism through visiting places underwater (Garcia & Cater, 2020). Therefore, diving activity is no longer just a sport (Lemke & Olech, 2011), and over the last decade, many Mediterranean destinations have developed a distinct scuba tourism product. For example, Malta has scuttled a number of wrecks and developed entry facilities around the islands since identifying this opportunity in the 2006 national tourism plan (Bideci & Cater, 2019; Garrod & Gossling, 2008).

Given the preference of diving tourists for warmer water destinations, much of this growth in these regions has also been at the expense of destinations in cooler places. For example, diving numbers at Skomer marine reserve, Pembrokeshire, Wales, a cooler, temperate dive location, have halved over recent decades (NRW, 2017). However, the proliferation of such experiences has largely eroded the connection to place, with many scuba diving products being globally substitutable. Indeed, Garrod and Gossling (2008, p. 5) highlight some of the difficulties in defining scuba diving tourism, noting that the 'degree to which going diving forms part of an individual's travel motivation tends to vary considerably'. Whilst some individuals plan holidays exclusively around scuba diving, much of the recent growth in dive tourism is attributed to those who 'intend to dive only when diving conditions are particularly favourable or when they have no other leisure activities planned for the day' (2008, p. 5). These tourists who dive during their holiday have been termed resort divers (Davis & Tisdell, 1995) or sideliner divers (WTO, 2001).

Above-the-water activities have also demonstrated the development of adventure tourism, for example white water rafting. Evidence from North America and New Zealand suggested that initially, passengers were usually passive, enjoying the trip, whilst the river guides did all the work. However, a desire to be more actively involved meant paddles were introduced for the rafters, and the experience became a more active one of working as a team under the leadership of the river guide. This change from passive to active tourism was recognised by industry commentators in the 1980s:

There has been a change in tourism. Not so long ago tourists used to come to New Zealand, they would come in a bus and sit and look and look and look, get back in the bus, jump in the plane and go home, and say what a lovely place it was. This has changed, and the people now not only come to look. ... Consequently, the change is also one of moving from a passive industry to an active industry. That is what we must promote. We must promote getting them here and get them to do things.

(Gebbie, 1982, p. 194)

This call was certainly taken up by the industry in New Zealand, as statistics demonstrated a significant increase in activities undertaken whilst on holiday, many of which were adventurous. Although they were likely to be at the softer end of the scale, the International Visitors Survey showed that by 1995, 'visitors were participating in more than twice as many (127% increase) activities, and/or attractions as in 1992' (New Zealand Tourism Board, 1996, p.18).

The growth of whitewater rafting also prompted the development of adventure tourism in less developed countries, such as at Victoria Falls (Zimbabwe/Zambia) and in Nepal. Beginning in the 1980s the first operator on large Himalayan rivers in Nepal, Ultimate Descents, had, by the 2000s, been joined by more than 30 other operators (Buckley, 2006; Richins & Hull, 2016). In marine areas, sea kayaking was also becoming more popular. For example, in New Zealand's Abel Tasman National Park, visitors hiking the traditional coastal path in the 2000s became outnumbered by kayakers on organised trips.

However, the growth of commercial adventure tourism did not come without risks, particularly as these new adventurers did not always have the skills to complete the activity. Accidents, such as a spate of whitewater rafting drownings in New Zealand in the mid-1990s, led to an inquest by the Maritime Safety Authority, which placed the blame squarely at the door of the rafting company management:

> Personnel in the industry are predominantly young men with a desire for excitement and adventure. Many are quite immature and live life "on the edge." This severely affects safety judgements and assessment of client capability. There is a significant difference between the guides' perceptions of an exciting trip; the guides' perceptions of client expectations; and the actual clients' concept and expectations of the trip. This leads to guides running trips for their own entertainment without due regard for the needs of their clients.
>
> (cited in Morgan, 1998, p. 4)

Unfortunately, these lessons were not learnt internationally as four British tourists drowned during a whitewater rafting trip in Austria in June 1999 (*The Times*, 8 June 1999). Barely a month later, 21 people were killed whilst canyoning in Interlaken in Switzerland. A torrential flash flood swept down the valley, wiping out almost half of those on the trip. The company running the trip, Adventure World, was severely criticised for ignoring storm warnings on the day of the trip (*The Guardian*, 28 July 1999), but an article at the time noted the real issue at the heart of this growth in adventure tourism:

> Most people don't have the time or want to reach a level of independent competence in many activities, so in effect they go shopping for the expertise, buying in experience, handing the duty of care to someone who does have the right certificate or training.
>
> (The Guardian, 1999)

Commercialisation of adventure

Perhaps the adventure activity most emblematic of the commercialisation of adventure, with little skill required by the actual participant, is that of bungy jumping, emerging in New Zealand in the late 1980s. It seems the inspiration for bungy jumping came originally from the land divers of Pentecost Island in Vanuatu (Figure 2.1). An ancient ritual for male members of the community to guarantee a good yam harvest, the practice involves the use of specially prepared vines secured around the ankles to jump off giant wooden towers (Cater et al., 2015). Filmed by David Attenborough in 1960, the practice received further worldwide attention when a demonstration was put on for a visit of Queen Elizabeth II in 1974. Prompted by this display, together with a number of magazine articles that appeared in *National Geographic*, the Oxford University Dangerous Sports Club took the idea and made a jump off the Clifton suspension bridge, Bristol, in 1979 (Kierke, 2007).

When New Zealanders AJ Hackett and friend Henry Van Asch saw the footage of the Bristol jump, and a similar one performed in San Francisco, they were taken with the adrenaline possibilities of the idea. With the help of a student at Auckland University, they developed a more scientific method of calculating bungy elasticity within weight parameters. A series of jumps off the Auckland Harbour Bridge were completed in 1986, and the AJ Hackett name began to grow in publicity. The real attention grabber, however, was a jump off the Eiffel Tower in Paris in 1987. Following a cold night encamped on the tower itself, Hackett carried out the stunt down the centre of the structure, dressed in a dinner suit and was promptly arrested, adding to bungy jumping's aura of breaking the rules.

Despite this original genesis of bungy as a stunt for publicity, it was only a matter of time before other people would want to get in on the gravity-defying fun. As a result, Van

Figure 2.1 Land diver at Naghol ceremony, Pentecost, Vanuatu. Photograph by Carl Cater.

Asch and Hackett set up a small operation in the winter of 1988 on the North Island. However, the site was too small, and the pair were soon looking for a site in New Zealand that would offer better long-term potential. Having spent a lot of time in Queenstown as competitive skiers, the pair decided that the site at Kawarau Bridge was ideal. Not only was it on a major road, but it also had the opportunity to tap into the significant Queenstown tourist market as a sustained customer base. The historical bridge, originally built in 1880 on the main route into Queenstown had been superseded by a newer bridge just upstream. Nestled in the Kawarau Gorge, with the vibrant blue waters swirling below, the site seemed perfect. On 11 November 1988, commercial operations began at the site, heralding the start of New Zealand's most famous homegrown adventure activity.

The only lasting bungy competition came with Pipeline bungy, which for several years held the title of the world's highest bungy jump. Meanwhile, the AJ Hackett brand was busily expanding overseas, starting operations in Cairns in Australia, France, the US, and Bali. Indeed, it was the success of the Queenstown Hackett operations that effectively bankrolled the overseas jumps into viability. A foray into the heli-bungy market in 1994, with jumps of up to 300 metres, was stopped by the Civil Aviation Authority in 1996. In the mid-1990s, the company split into two, with Van Asch taking over the Australasian operations and Hackett, now resident in France, taking on responsibility for the rest of the global operations. Hackett himself was still very much the public face of the company, performing a record-breaking jump off Auckland's sky tower in 1998, and starring in numerous adverts, for items as diverse as cameras (Minolta, France) and credit cards (ANZ bank, New Zealand).

By the turn of the millennium, close to 1 million people had bungy-jumped in Queenstown. A new chapter in the development of bungy jumping in Queenstown began in 1999 with the opening of the Nevis highwire site by AJ Hackett. In addition to restoring to the company the all-important "highest" accolade, at 134 metres, the jump also introduced some new concepts to the commercial bungy product. A cable car, specially built on a private station a half-hour ride by four-wheel drive from the Kawarau site, formed the base for this new jump. The use of a large amount of glass, including in the floor, guaranteed a significant amount of adrenaline production. In addition, the harness is specially adapted to allow the jumper to swivel upright after the initial fall. This innovation of the existing bungy product led to AJ Hackett being granted two decorations in the 2000 New Zealand Tourism Awards. However, the advent of this new site markedly changed the geography of bungy operations in Queenstown. The popularity of bungy jumping not only as an adventure activity but also as an adventure *spectacle* (much like the Eiger watchers of the previous century mentioned earlier) prompted the development of a purpose-built visitor centre (Figure 2.2) in 2013 on the original Kawarau site that prior to the pandemic attracted more than 400,000 visitors a year (Destination Queenstown, 2022).

One of the interesting impacts of the development of bungy jumping in Queenstown was the positive impact it had on other adventure tourism activities, both existing and consequently. The operators of the long-established Shotover Jet, canyon jetboat ride, noted how bungy actually made the jetboat seem less extreme and opened to a broader, and older, market with a 50% increase in customers in 1 year. The innovation apparent in bungy jumping also spawned a range of other temporary and permanent adventure activities such as river surfing, canyoning and a suspended flying machine called fly-by-wire. Operators began to package some of these adventure activities together so that

Figure 2.2 AJ Hackett Kawarau Bungy Centre, Queenstown, New Zealand. Photograph by Carl Cater.

they could be experienced in a day of adrenaline (Cloke & Perkins, 1998). However, a lack of adventurous promotion from the local destination management organisation at the time led to the formation of the Adventure Capital of the World Group of leading adventure tourism operators. This pushed the *Adventure Capital of the World* moniker (Figure 2.3) which put Queenstown on the global adventure tourism map and spawned a host of worldwide imitators from Victoria Falls to Fort William (Scotland) (Buckley 2006).

Over the past decade, we have seen further growth in adventure tourism activity. In some areas, this has truly become mass adventure tourism. In the Turkish Mediterranean resort of Antalya, more than 600,000 tourists do whitewater rafting annually in the nearby Koprulu Canyon (Figure 2.4; Albayrak & Caber, 2018). The recent popularity of ziplines (see Case study 2.1) has also meant that virtually all these whitewater rafting operators have built ziplines for their guests at various lunch stops. Very high numbers of tourists also try scuba diving in the region, with perhaps two thirds of them opting for a 20-minute discovery diving experience which does not require any training or ongoing commitment (Albayrak et al., 2019). Some of these established adventure activities have been adapted for different cultures, for example whitewater rafting in China which has been developed for a more passive market as river drifting or *piaoliu*. This is 'characterised by short trips, in small unguided rafts without paddles, on heavily modified watercourses with exclusive control of access, receiving up to 10,000 clients per site per day' (Buckley et al., 2014, p. 5), meaning it is estimated that there have been a staggering 80 million participants to date. A slower form of adventure

Figure 2.3 Queenstown, Adventure Capital of the World, 1990s brochure picture. Photograph by Carl Cater.

Figure 2.4 Whitewater rafting operators at Koprulu Canyon, Antalya, Türkiye. Photograph by Carl Cater.

Case study 2.1   Zip World

Highlighting the importance of product innovation, a story somewhat similar to the history of bungy jumping has played out in the development of ziplines in the UK, led by the Welsh Zip World group. In 2011 the operator of an existing high ropes activity explored the potential for building a much larger zip line in an old slate quarry. Following a £400,000 investment, Zip World Velocity opened in 2013 as the world's largest and fastest zipline, to immediate success. Much like AJ Hackett, the significant global media interest meant that marketing spend was minimal, and celebrity endorsement and social media were used to a large degree.

The commercial success of the initial zipline allowed Zip World to develop a range of other innovative attractions including a parallel zipline (Zip World Titan); giant underground trampoline (Bounce Below) and underground ziplines (Cavern); Luge (Coaster), as well as developments to the original high ropes course; and a £2.5 million visitor centre, restaurant, and call centre.

A sophisticated online booking system ensures high turnover. As noted by the owner, the adventure market is usually low volume and high price or high volume and low price. However, in this case, the product is both high volume and high price (the signature zipline costs over £100). The recent introduction of dynamic pricing for activities allows more efficient demand management. The integration of the latest photo and video technology into the product allows participants to share their experiences instantly on social media, providing more free advertising.

By 2018, Zip World had brought in over £250 million into the North Wales economy and created over 450 jobs in this economically deprived area. Paying above living wage and enhanced staff training allow for higher staff retention and customer service. In 2021, Zip World opened a new site in South Wales, Zip World Tower, built at the site of an old coal colliery. Built in partnership with the trustees of this historically and politically significant site, the attraction has a strong heritage component that Zip World are seeking to develop and align with their product offering.

website: https://www.zipworld.co.uk/

participation is found in Taiwan, where mountain hikers put more emphasis on the aesthetic qualities of peaks than simply their height (Huang, 2015).

The significance of adventure tourism, as discussed in Chapter 10, has led to governments taking the sector seriously, both for destination marketing (see Chapter 14) and well-being of populations (see Chapter 12). Adventure tourism was the backbone of the successful 100% Pure tourism marketing campaign pursued by New Zealand in the 2000s. Nonetheless, the interpretation of adventure tourism has become very broad by both governments and industry associations. In 2016, Wales began a thematic marketing approach, selecting the *Year of Adventure* to frame all its tourism products. However, this included such diverse interpretations of adventure as castles and the stories of Roald

Dahl (linking to the narratives of adventure described earlier). In a similar vein, the privately owned, membership-based Adventure Travel Trade Association has evolved from an organisation specifically focused on pure adventure operators to one that includes ecotourism and cultural tourism companies, for the simple reason that bigger is better, and having more clout with supranational organisations such as the UNWTO. Nevertheless, proof of the significance of adventure tourism at the highest levels of policymaking is shown in the current Welsh Government International Strategy in seeking to 'establish our reputation for sustainable adventure tourism', placed alongside major national industrial and investment projects (Welsh Government, 2020, p.7).

The COVID-19 pandemic had a major impact on the tourism sector, and the close contact required by many adventure activities meant these were among the hardest hit businesses. As a result of the widespread border restrictions, international adventure tourism slowed to all but a trickle, posing particular problems for adventure operators in less developed countries reliant on an international market. Even financially strong operators like AJ Hackett in New Zealand had to rely on short-term government support and product discounting for the domestic market to get them through a difficult period. However, with an increased interest in the outdoors following lengthy lockdowns, as well as adventure tourism being, by its nature, less risk-averse, recovery is predicted to be relatively swift (CBI, 2021).

Conclusion

This chapter has shown that modern adventure tourism is based on the long history of adventure itself and is even part of who we are as humans. However, it is apparent that there are far more participants in modern adventure tourism compared to the formerly narrow elite, so in that sense, it is vastly more democratised. Whilst the pyramid of participation suggested by Varley (2006; discussed in Chapter 3) is larger at all levels, the majority of this participation is at the softer end of adventure tourism, for this is where opportunities for commercialisation and innovation are greatest. Indeed, product innovation, whether it be with bungy jumping or ziplines, has often driven adventure tourism participation more broadly, despite only representing an unskilled part of the sector. Nevertheless, the social media integration described later in this handbook in Chapter 11 means that adventure tourism remains "cool", and we are likely to see its continued popularity.

Review questions

1 Which activities might predate the term adventure tourism, and why did they become popular?
2 Looking at the websites of Arctic/Antarctic tourism operators, what evidence can you see of a theme of polar exploration being used in marketing?
3 How has technological innovation influenced the development of adventure tourism activity?
4 In what ways has adventure tourism become more commercial over time?

Further Reading

Beedie, P. (2015). A history of mountaineering tourism. In G. Musa, J. Higham, & A. Thompson-Carr, (Eds.). *Mountaineering tourism*. Routledge.

Buckley, R. (2006). *Adventure tourism*. CABI.

Dimmock, K., & Cummins, T. (2013). History of scuba diving tourism. In G. Musa & K. Dimmock (Eds.), *Scuba diving tourism* (pp. 14–28). Routledge.

References

Adventure Travel Society. (1999, September 10). The importance of adventure travel and ecotourism. http://www.adventuretravel.com/seminar_home.htm

Albayrak, T., & Caber, M. (2018). A motivation-based segmentation of holiday tourists participating in white-water rafting. *Journal of Destination Marketing & Management, 9*, 64–71.

Albayrak, T., Caber, M., & Cater, C. (2019). Mass tourism underwater: A segmentation approach to motivations of scuba diving holiday tourists, *Tourism Geographies, 23*(5–6), 985–1000.

BBC. (2021, August 23). Snowdon: Call to respect mountain amid spike in visitors. BBC. https://www.bbc.co.uk/news/uk-wales-58283816

Bideci, C., & Cater, C. (2019). In search of underwater atmosphere: A new diving world on artificial reefs. In M. Volgger & D. Pfister (Eds.), *Atmospheric turn in culture and tourism: Place, design and process impacts on customer behaviour, marketing and branding* (pp. 245–257). Emerald Publishing Limited.

Buckley, R. (2006). *Adventure tourism*. CABI.

Buckley, R. (2012). Rush as a key motivation in skilled adventure tourism: Resolving the risk recreation paradox. *Tourism Management, 33*(4), 961–970.

Buckley, R., McDonald, K., Duan, L., Sun, L., & Chen, L. X. (2014). Chinese model for mass adventure tourism. *Tourism Management, 44*, 5–13.

Cater, C., Albayrak, T., Caber, M., & Taylor, S. (2020). Flow, satisfaction and storytelling: A causal relationship? Evidence from scuba diving in Turkey. *Current Issues in Tourism, 24*(12), 1749–1767.

Cater, C. Garrod, B. and Low, T. (Eds.). (2015). *The encyclopaedia of sustainable tourism*. CABI.

CBI. (2019, April 10). Dive tourism from Europe. Centre for the Promotion of Imports from Developing Countries. https://www.cbi.eu/market-information/tourism/dive-tourism/europe

CBI. (2021, October 10). The European market potential for adventure tourism. https://www.cbi.eu/market-information/tourism/adventure-tourism/adventure-tourism/market-potential

Chatwin, B (1987). *The Songlines*. Vintage.

Cloke, P., & Perkins, H. C. (1998). "Cracking the canyon with the awesome foursome": Representations of adventure tourism in New Zealand. *Environment and Planning D: Society and Space, 16*(2), 185–218.

Davis, D., & Tisdell, C. (1995). Recreational scuba-diving and carrying capacity in marine protected areas. *Ocean & Coastal Management, 26*(1), 19–40.

Destination Queenstown. (2022, October 20). AJ Hackett Bungy Queenstown Kawarau Bungy Centre. https://www.queenstownnz.co.nz/listing/aj-hackett-bungy-queenstown/3371/

Dimmock, K., & Cummins, T. (2013). History of scuba diving tourism. In G. Musa & Dimmock, K. (Eds.), *Scuba diving tourism* (pp.14–28). Routledge.

Fiennes, R. (2021). *Shackleton: A Biography*. Penguin.

Garcia, O., & Cater, C. (2020). Life below water; challenges for tourism partnerships in achieving ocean literacy. *Journal of Sustainable Tourism, 30*(10), 2427–2447.

Garrod, B., & Gossling, S. (Eds.). (2008). *New frontiers in marine tourism*. In *Diving Experiences, Sustainability & Management*. Routledge.

Gebbie, M. (1982). *The Adventure Industry Proceedings of the National Outdoor Education Conference 1982 New Zealand Council for Recreation and Sport*, Wellington.

Gregory, D. (1994). *Geographical Imaginations*. Blackwells.

Hillary Commission for Recreation and Sport. (1987). *Proceedings Adventure Tourism Seminar: Current Issues and Future Management*. North Island Seminars, Massey University, Palmerston North.

Hobson, J. A. (1994). *The chemistry of conscious states*. Little, Brown and Company.

Huang, M. F. (2015). Mountaineering tourism in Taiwan: Hiking the '100 mountains'. In G. Musa, J. Higham, & A. Thompson-Carr (Eds.). *Mountaineering tourism*. Routledge.

Johnston, B. R., & Edwards, T. (1994). The commodification of mountaineering. *Annals of Tourism Research*, 21, 459–477.

Kelechava, B. (2021). The decline of snowboarding. https://blog.ansi.org/decline-of-snowboarding-lower-participation

Kierke, D. (2007). Risk and reward: The Dangerous Sports Club. Anthropology Today 23:23 Royal Anthropological Institute.

Lemke, L., & Olech, L. (2011). Dive tourism. In A. Papathanassis (Eds). *The long tail of tourism* (pp. 105–114). Gabler.

Macfarlane, R. (2004). *Mountains of the mind: A history of a fascination*. Granta.

Maritime Safety Authority. (1995). *Review of commercial whitewater rafting safety standards*; Final Advisory Group Report MSA.

Morgan, D. J.(1998). The adventure tourism experience on water: Perceptions of risk and competence and the role of the operator Unpublished MA thesis, Lincoln University.

New Zealand Tourism Board. (1996). *International Visitors Survey 1995–1996 NZTB, Wellington*.

NRW. (2017). Skomer marine conservation zone annual report 2017. Natural Resources Wales.

PADI. (2022, September 10). *Statistics*. https://www.padi.com/corporate/company-info

Phillips, R. (1997). *Mapping men and empire: A geography of adventure*. Routledge.

Richins, H., & Hull, J. (Eds.). (2016). *Mountain tourism: Experiences, communities, environments and sustainable futures*. Cabi.

Simpson, J. (2003). *The beckoning silence*. Vintage.

The Guardian. (1999, July 28). Criticism grows of disaster gorge trip. *The Guardian*, G2, 3.

The Times. (1999, June 8). *Four Britons die in a raft accident. The Times*. 1

Varley, P. (2006). Confecting adventure and playing with meaning: The adventure commodification continuum. *Journal of Sport & Tourism*, 11(2), 173–194.

Welsh Government. (2020). *International Strategy*. Cardiff.

Willson, T. B. (1927). *The handy guide to Norway* 7th ed. Stanford.

Wilson, E. O. (1993). 'Biophilia and the conservation ethic. In S. Kellert & E. O. Wilson (Eds.), *The biophilia hypothesis*. Island Press and Shearwater Books.

WTO. (2001). *Tourism 2020 vision*. (Vol. 7) *Global forecasts and profiles of market segments*. WTO.

3 The idea of adventure

From human conquest to being-with the world

Peter Varley and Simon Beames

Chapter learning outcomes

1   The study of adventure tourism and how it has evolved over time.
2   The key contributions in adventure practice and theory to date.
3   The interdisciplinary nature of adventure tourism and its philosophical and practical futures.

Introduction

Going on an adventure to scratch some existential itch is a notion that has only emerged in *Homo sapiens'* recent, modern history. Yet human beings have been embarking on adventures in search of food, mates, resources, treasure and land for hundreds, if not thousands, of years. More latterly, humans have used adventure as a vehicle for countries to boost their geopolitical standing and for individuals to exercise, go on holiday, find inner peace and increase their social recognition.

As a point of departure, we contend, following Varley (2006), that the intellectual and practical roots of adventure in urbanised industrialised societies in modernity emanate partly from a perceived distancing or alienation from immediate experience, from nature, from each other, and from ourselves. Such feelings have been the source of great theorising and contention for classical and romantic philosophers and sociologists from the start of the 20th century and have since found specific critical application as theoretical positions shift from masculinised, anthropocentric and instrumentalist paradigms to Romanticism and then to reconceptualisations of humans at play and learning in nature. In due course, this chapter explores these positions as it illuminates ways in which scholars have applied such ideas to outdoor experiences, commercial and educational adventures and leisure practices. Beyond this, in our concluding thoughts, we suggest that the current direction of critical thinking in this fascinating realm of human practice, through acknowledging the post-materialist, post-human and chthulu-centric ideas of Barad (2003), Braidotti (2006) and Haraway (2016) in particular. This last move, inspired by feminist philosophical critique, positions the doings and 'meanings' made by humans as being relatively small and recognises the huge planetary time and material agency contexts in which *Homo sapiens* scuttles about, pausing only occasionally to celebrate its own logic, self-importance and imagined power.

DOI: 10.4324/9781003393153-5

Early meanings of adventure

As also noted in Chapter 2, adventure has a long-standing narrative history. Zweig (1974) referred to the story form of 'adventure myths' from ancient times that comprise 'perilous journeys, encounters with inhuman monsters, ordeals of loneliness and hunger, descents into the underworld' (p. 3). As with Luke Skywalker and Frodo Baggins, the heroes of these tales were often not the instigators of the adventurous circumstances which befell them.

The origins of the word *adventure* have been traced back to the 12th century, and over time, the English word emerged. The earliest recorded use of the word adventure was in French: *aventure*. *Aventure* was coined in the Middle Ages (around the late 1100s) as part of a 'knightly' ideology (Nerlich, 1987, p. 5). Knights went on *aventures* as a rite of passage and later to earn a living. This is not only an early example of the attractiveness of the aura of adventure but also one in which *not* being adventurous might be a distinct disadvantage, as nothing was gained and one's personal position in life was not improved (Nerlich, 1987).

Commerce and capitalism began to take hold of the idea of adventure in the 1400s. Merchant adventurers sailed the seas in search of wealth, through resource-rich lands and peoples to enslave and exploit. As noted in Chapter 2, at this point these quests intertwined with commerce and the intentions of the government, as well as with the intentions of the individual. Furthermore, some of these expeditions would be blessed by the Crown and thus provide an additional moral rationale for the 'exploration, subjugation, and exploitation' inherent in such adventures (Nerlich, 1987, p. 129).

Michael Nerlich (1987) linked these journeys to his notion of an 'ideology of adventure' that was gaining traction at this point in human history. In essence, Nerlich suggests that adventuring developed in the Middle Ages from the human desire to accrue capital and that this morphed into capitalism itself over time. Indeed, to this day, we might speak of 'venture capitalism' and see clearer links to earlier periods in time, when the essential aim of adventuring was to make money, with high risks and correspondingly high rewards. Boje and Luhmann (1999) show how Nerlich's ideology of adventure makes the Industrial Revolution and Enlightenment possible as a project of capitalism. Nerlich also stressed the role of stories, literature, poems and plays in promoting an 'adventure-mentality', which was learned in childhood and became applied in commerce and production as adventure practices.

These myths, histories and meanings of adventure can be associated with what Campbell (1949) called the rather masculinised concept of the 'hero's journey'. Loynes (2003) explained how Campbell identified 17 stages of the hero's journey that are familiar to this day in popular storytelling forms. Bell (2016) further explains how today's adventurous and heroic individual is created from narratives of 'histories of imperialism and enforced inequalities' (p. 8). From this perspective, we see that the normative concept of what it means to be adventurous is built on stories, images and ideas which are enmeshed within dominant, masculine cultural forces. As such, historical notions of adventure have, to a large extent, marginalised women (see also Chapter 26). Writing more than 20 years ago, Warren (1996, p. 16) warned us that;

> the heroic quest is a metaphor that has little meaning to women. Each stage of a woman's journey in the wilderness is a direct contradiction of the popular quest model. A woman rarely hears a call to adventure; in fact, she is more often dissuaded ... from leaving home to engage in adventurous pursuits.

Early adventuring not only oppressed women and anyone (irrespective of gender) who was not part of this dominant adventurous milieu. Thus, the roots of adventure are also entwined with centuries of male capitalist endeavour, exploitation and colonisation.

Defining adventure

Fast-forward to the 20th century, when, following ideas from Kurt Hahn and others, adventure had become a way to develop scarce personal qualities among young moderns. Colin Mortlock (1984), a British adventurer and educator, claimed in his early works that adventures should involve 'a degree of uncertainty' (p. 14) and 'demand the best of our capabilities – physically, mentally, emotionally' (p. 19). Indeed, most scholars seem to contend that uncertainty is a central feature of adventures.

Some cultures might consider the notion of seeking adventure for its own sake to be a rather odd concept. Those living an agrarian life with the threat of drought or crop failure or those having fled war zones may well not seek experiences that are especially daring, exciting or unpredictable; they seek stability, peace, shelter, clean water, food, jobs and schools. However, when we are fortunate enough to live in relative comfort and stability, the opposite of the everyday holds a special draw. It is not unusual for people in the dark of a northern hemisphere winter to dream of the Mediterranean sun. It is relatively common for office workers accustomed to nine-to-five schedules to seek uncertainty, spontaneity and physical activity in fresh air.

Back at the turn of the 20th century, Georg Simmel (1919), suggested that adventure has to do with breaking routines and doing something 'alien, untouchable, out of the ordinary' (p. 2). Peter Becker (2016), further explained that at the 'structural core' of adventure is the 'interplay between crisis and routine' (p. 26). The space between the polar opposites of reasonably predictable day-to-day life and immediate crisis is significant, but there is some middle territory which attracts many of us. However, our accession to adventure, and our interpretation of it, is subjective and occasionally intersubjective, and we become influenced by others around us. The individual relativity of adventure is also important here, as 'what one person deems adventurous may not be to another person' (Beames & Pike, 2013, p. 2). This simple notion is wonderfully liberating for the individual seeking their own personal adventurous 'sweet spot'. By contrast, in the context of commercial adventure tourism, it is incredibly challenging to ensure that every participant feels a sense of adventure and challenge through the same programme.

There was relatively little academic work in the 20th century that focused directly on the subject of adventure tourism, and the key scholarly contributions were therefore unsurprisingly drawn from a wide and interdisciplinary field of academic inquiry (Vester, 1987). The subject areas that connect in some way with the concept of adventure include leisure studies (Rojek, 1995, 2000; Vester, 1987), outdoor education (Loynes, 1998, 2003; Mortlock, 1984; Martin & Priest, 1986), sociology (Heywood, 1994; Lyng, 1990; Mitchell, 1983; Simmel, 1919), tourism (Elsrud, 2001; Swarbrooke, 2003; Weber, 2001) and consumer behaviour (Arnould & Price, 1993; Celsi et al., 1993). This chapter therefore attempts to critically synthesise the salient aspects of this diverse body of work as a means to develop a definitive 'ideal' concept of adventure against which subjective interpretations of the term might be considered.

We therefore proceed with further critical review of the early work in this regard, initially drawn from the field of adventure education, in which a number of conceptual models of adventure experiences are offered. Models created by Mortlock (1984) and Martin and Priest (1986) provide a helpful starting point for discussing the stages of adventure experiences. The argument then broadens out its concerns, through a discussion of the late modern fascination with death and "limit" experiences, to consider an important intellectual lineage, particularly inspired by Nietzsche, that runs through the work of those authors whose ideas touch on the subject of adventure. That lineage stems from Nietzsche's use of the *Dionysiac* (chaos, disruption and ecstasy) versus the *Apollonian* aspects of life (control and ordered beauty – management and aesthetisation).

Simmel's ideas on the subject stem from his appreciation of the Nietzschean concept of tragedy and are evidenced in his comparison of art and adventure. Roughly speaking, there is a necessary sense of potential loss, or even death, inherent in adventurous activity – the 'risk and danger' parts, if you will, which lend the pursuit its life-affirming character (explored in more detail in Part 3 of the book). Other themes from Simmel's essay, such as the adventure as a 'dropping out' of the continuity of everyday life are instead predominantly accounted for here from their original source in Nietzsche's *Birth of Tragedy* (1967).

Models of adventure

Working through Nietzsche's Dionysian themes of disorder, ecstasy, and emotion and their impact on the various ideas developed by subsequent authors about the individual in modernity, a definition of the 'ideal-original' adventure was developed from the concepts considered by Varley (2006). Varley used this grounding to offer an adventure commodification continuum, which stretches between the realm of the original adventurers, and that of the post-adventurers, who pay to consume adventure-like products and services (see Figure 3.1). This model was an attempt to show the progression from the 'ideal-type' conception of

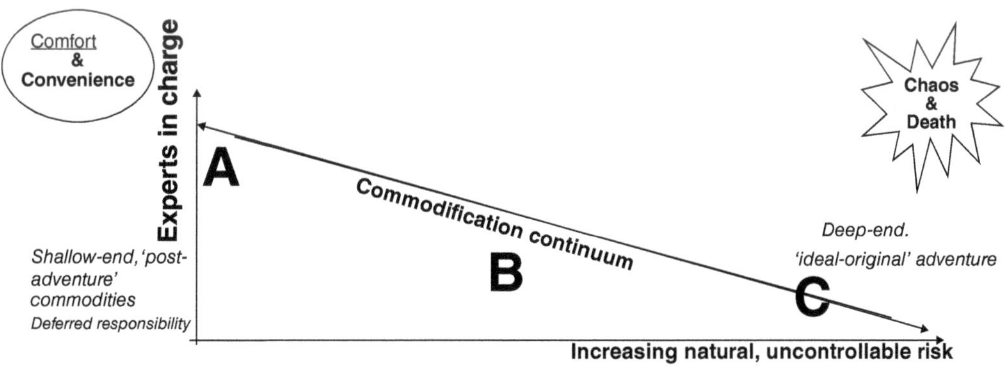

Figure 3.1 The adventure commodification continuum.
A. Highly saleable; not very adventurous at all: tourist products, staged events, balloon rides, bungy jumps, safari tours and so on. B. The terrain of the outdoor pursuits market: skills courses, guiding, expert-led expeditions. C. Generally too 'deep end' to ever become a product: solo mountaineering, independent expeditions: folk doing it for themselves, backpacking, sea kayaking, accidental survival situations, death by (mis)adventure.

Source: Varley (2006).

adventure towards its commodified form of a tourist/leisure adventure product. Varley's continuum is in effect a provocative conversational device and lays a platform upon which taken-for-granted assumptions of the meanings of the idea of adventure can be confronted. Deliberately imperfect, and always open to modification, the model might yet be useful precisely because of its challenge to common sense, partial and pre-conceived notions about what adventure *is*, particularly in consumer society, and what it might be in the future.

Building on this theoretical foundation, and returning to themes of Apollonian order and beauty, we can expand our working definition of adventures to also include a certain element of desirability. The modern identity project – to 'be' adventurous – is said to be a positive quality, and worth promoting; indeed, some are able to base a lifestyle around curated social media profiles, lecture tours and equipment sponsors, which support their explorations. Adventures, then, are actively sought and therefore planned to an extent. Norwegian scholar Gunnar Breivik (2010, p. 262) suggests that adventure sports;

- have elements of challenge, excitement and (in most sports) risk
- take place in demanding natural or artificially constructed environments
- are more loosely organized than mainstream sports
- represent a freedom from or opposition to the dominant sport culture
- are individualistic pursuits but tend to build groups and subcultures around the activity

Beyond Breivik, we can look back to Mortlock's seminal work in 1984 which proposes stages of adventure. Although written through the lens of outdoor education, its conceptual importance stretches to adventure tourism as well.

Before Mortlock's (1984) study, *The Adventure Alternative*, many of those who had approached the subject tended to provide descriptions of specific components of the adventure form, where there had been a particular focus on risk, rather than on a more general account of adventure's other characteristics. Mortlock began his account by explaining that he developed a firm belief, after some 20 years as an outdoor pursuits instructor and as a lifelong adventurer himself, in the enormous potential benefits of adventure in the great outdoors for people of all ages. He then provided a descriptive model of what he believes are the four (non-sequential) stages of adventure, as paraphrased here (Mortlock, 1984, p. 22–23):

Stage 1: *play*. The person is operating well below their capabilities, with minimal involvement of emotions, skill, mental control and concentration. There is no fear of physical harm and subjective responses to the experience may range from 'pleasant' and 'fun' to 'boring' and 'a waste of time'.

Stage 2: *adventure*. The person feels in control of the situation but is using their skills and experience to overcome a particular problem. This experience may take place in a strange environment, and as a result, fear and unease may lurk just below the surface. It is important to note that, for Mortlock, this stage is a crucial part of the preparation for a demanding adventure/journey. Skills may readily be learned at this stage, as the participant is not preoccupied with survival.

Stage 3: *frontier adventure*. This stage is what many adventurers arguably seek, though for some, it is often only 'just beyond' stage 2. The person no longer feels complete mastery of the situation and has fears of physical and/or psychological harm; however, they feel that they can overcome the situation without serious incident through careful concentration and sustained effort. Undoubtedly, in this phase, there is a strong element of uncertainty, and the individual may feel poised on a knife edge between success and failure. Feelings of intense satisfaction (or disappointment) may settle, depending to some extent on the scale and intensity of the experience.

Stage 4: *misadventure*. In this stage, the degree of challenge is beyond the control or capabilities of the adventurer. There are degrees of misadventure, which extend from a slightly out-of-control 'frontier adventure', where the journey fails and there are minor mental and/or physical grazes to serious injury or death at the extreme end. There may be learning experiences of great value to the participant, but there may also be such damage to the individual's confidence that they vow never to engage in such a journey again.

In the remaining sections of his work, Mortlock (1984) deals with the social value and applications of the adventure experience, both as an ingredient of modern life and as a possible social education tool. An example of this is Beames and Brown's (2016) work on *Adventurous Learning*, which highlights the four features of uncertainty, authenticity, agency and mastery, as tools for educators and learners to employ. Of special significance to this study is Mortlock's (1984) chapter titled 'The Instinct for Adventure' (p. 46–53), where he argues that the urge to adventure is found in all sectors and ages of the population, and acknowledges that there are adventures of an active (e.g., wilderness canoeing) and passive (e.g., watching an action movie) type, and of a social (e.g., group expeditions) and anti-social (e.g., joy riding in stolen cars) nature. What he does not identify in this early work, however, is the misplacement of the original qualities of adventure in its sanitised, risk- and quality-assured commercial forms.

Together with Breivik's five defining characteristics of contemporary adventure sports and Mortlock's four stages of adventure, we have conceptual language that will enable us to discuss adventure tourism with greater precision.

Adventure as escape metaphor

Cohen and Taylor (1992) discuss the different forms of escape attempts that people make to cope with the constraining, predictable routines of everyday living. This is more complicated than it seems, Rojek (1993) argues, because our values are so tightly intertwined with society that our escape attempts themselves are artificial: complete escape is not possible! Another way of explaining this is while we might feel like our adventures provide freedom, they are still 'tightly determined and modified by cultural, social and economic settings' (Lynch & Moore, 2004, p. 2).

What cannot be emphasised enough is that most – but by no means all – of the adventure practices that we are interrogating are those undertaken by people in positions of relative privilege. Quite often, it is those who have high enough paying jobs to have the money to buy specialist technical equipment, coupled with the accrued holiday time to use it. Seen this way, adventures are 'romanticised as escape' (p. 2) from, and a 'psychological palliative' (p. 3) against, the trials of late modern life.

Lynch and Moore (2004) refer to an adventure paradox, that features, on one hand, 'the crucial role of adventure ideology in the historical development of the modern, industrialised world and economy' and, on the other hand, 'the current promotion of adventure as the romantic escape from that world' (p. 2). While these and other narratives are about temporarily leaving behind one reality in order to visit another one (Becker, 2016) or enter the unknown (Nerlich, 1987), there is clearly more to it than this.

Playing with risk

Four key elements of adventure experiences constantly arise in Mortlock's (1984) book, and these provide the core of the critical model of commodified adventure generated in this chapter. These Mortlockian elements are risk (p. 32–37), responsibility (p. 50, 68–73), uncertainty (p. 85–92) and commitment (p. 68–73).

The significance of these key qualities is that they were precisely the ones so often omitted from the work of those authors who offer their analyses of the commodified "adventure" form (e.g., Arnould & Price, 1993; Holyfield, 1999; Walle, 1997) and yet which seem so integral to the qualities of adventure as a way of experiencing. For example, as Holyfield (1999) and Arnould and Price (1993) describe their experiences, it is clear that the white-water rafting participants in their studies enjoy a considerable communion with each other, with the guides, and with nature. There appears to be an element of risk, too, as Holyfield describes how she fell into the foaming waters of the river and experienced considerable discomfort, fear and uncertainty. However, the responsibility for her safety rested with the raft guide.

Whilst Holyfield's is a convincing account of a tourist experience and is titled *Manufacturing Adventure* to begin with, what seems lacking is a commitment to the action. Cater's (2006) important work similarly focused on this element of adventure, as he describes adventure tourism as a dalliance with risk itself. However, at this stage, there is little in the literature which considers the agency of the adventure consumer, and as such, the label "adventure" somehow seems inappropriate: the customers do as they are told and are closely guided through their experiences. Crucially, we now see the emergence of professionals for whom the pursuit of so-called marginal activities has metamorphosed into something that can be provided for clients at a price. It may be assumed that, from the professionals' perspective, creativity, chaos and freedom are undesirable elements in the supply of commodified, risk-assessed experiences. As Holyfield (1999) has illustrated, it is the stage-managed delivery of these factors – what Beedie (2003), writing about mountain guiding, labelled 'the choreography of the experience' (p.52) – that lend the glamour and excitement to the tourist's experiences.

Play, flow, edgework and peak experiences

Theorist-practitioners Martin and Priest (1986), also from the outdoor adventure education field, sought to extend Mortlock's original framework through building on the contribution of studies on play and optimal arousal levels (Ellis, 1973, cited in Martin & Priest, 1986). These studies tended to focus on the psychosocial dramas that occur during what Mortlock termed the adventure, frontier adventure and early misadventure experiences. Martin and Priest's work offers a model of adventure that uses ascending risk as the vertical axis and increasing competence along the horizontal (Figure 3.2). Their paradigm allows an accommodation of different personality types, such as those at the extremes: the 'timid-fearful' and the 'arrogant-fearless' (1986, p. 20). They explain that whilst the former type may struggle to push themselves beyond the play stage (which they term 'experimentation'), the latter may charge toward misadventure or even death due to an overconfident mismatch between actor competence and the risks posed by the

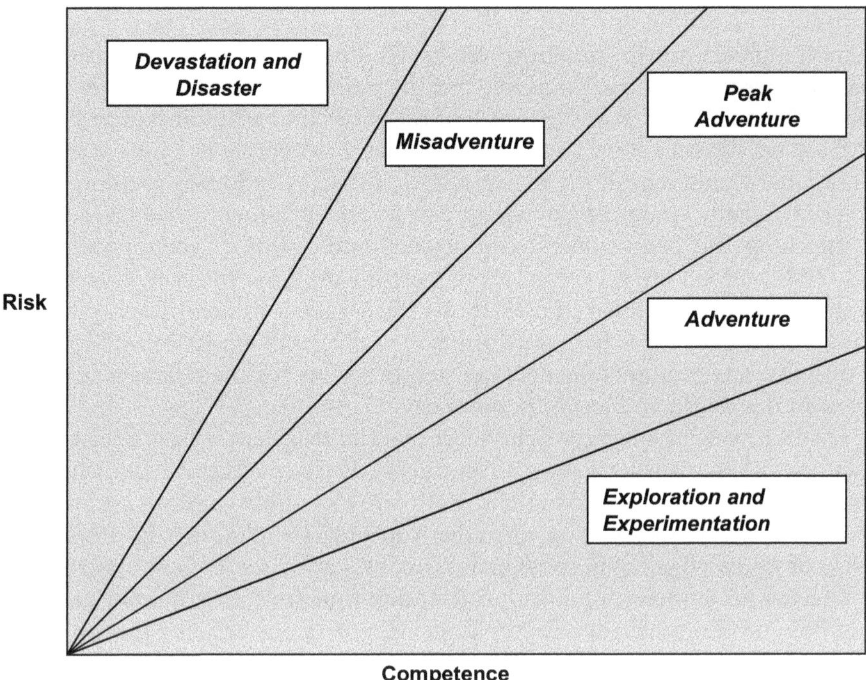

Figure 3.2 The adventure experience paradigm adapted from Priest and Bunting (1993).

situation (Martin & Priest, 1986). The proposed ideal stage for any would-be adventurer to attempt to reach is what Martin and Priest term "peak adventure", which borrows from Csikszentmihalyi (1975; 2000) work on flow and peak experience. This is the equivalent of Mortlock's "frontier" stage, where challenge, competence, confidence and tension all converge to create the flow experiences at the core of the adventure.

However, whilst it seems reasonable to argue that the exciting moments of an overall adventure might well be remembered as those involving flow experiences, an adventure is surely far more than that. The flow situation is really just the negotiation of a situation's adventurous potential, and it might never happen. Rock climbers, for example, discuss with amazement the climbs that have given them far greater challenges than they had expected, but they also acknowledge that some climbs do not live up to the grade or description. Flow, then, is a bonus experience in some adventures, but it is not an essential element. Work on flow experiences (Csikszentmihalyi, 2000), and edgework (Lyng, 1990) do, however, provide valuable contributions to an understanding of certain aspects of the components of the adventure tourist's experience.

Csikszentmihalyi (2000, p. 36) explains the personal transcendence and intrinsic rewards of flow thus:

Flow refers to the holistic sensation present when we act with total involvement. It is the subconscious state in which action proceeds according to an internal logic. We experience it as a unified flowing from one second to the next, time passes imperceptibly, and in which self and environment blur. In flow, there is no past, present and future. There is only now.

Csikszentmihalyi's (2000) and Mitchell's (1983) use of the play categories of competition, chance, mimicry and vertigo (inspired by Caillois, 1961) may all be seen as elements of the late modern adventurer's actions, but the authors stress that it is the experience's self-induced "chaos", that is at the centre of the rock climber/mountaineer's flow experiences. "Flow" is often experienced as the adventurer attempts to make sense of natural hazards or 'chaos' inherent in the tensions and physical conditions presented in the outdoor arena. Crucially, states of flow are created and sustained by the levels of challenge closing matching the practitioner's competence and ability (Csikszentmihalyi, 1975). Mitchell (1983) added another important motif to the flow phenomenon and borrows from Mortlock, freedom. Here, he refers to the 'process of creativity' (p. 154) during which flow might occur. It is here that Mitchell is effectively underlining Mortlock's contention that true adventurers take personal responsibility for their actions in adventurous leisure – even if it might end in injury or death.

Lyng's (1990) concept of edgework shares this challenge/competence relationship with flow discourses. The focus here is on adventurers (skydivers, in this case) who, due to the fact that one small error can lead to their death, are incredibly cautious and well prepared in their practices. Following Lyng, the edge can only be inhabited by those possessing high levels of knowledge, skill, judgement and, crucially, the mental control to execute the task at hand under pressure. For the adventure tourism sector, helping people achieve a state of flow or find their edge is very difficult, unless one is being guided one on one. For this reason, most operators and participants must accept what Fletcher (2010) explains as a paradox in which participants need to feel at risk, but safe, at the same time; Martínková and Parry (2017) refer to this as "safe danger". It is impression management and product placement, where in this case the "product" is the curated client experience *and* the essential attendant social media component of momentary instantaneous adventure identity.

As suggested at the start of his chapter and the previous one, war, colonialism, performative masculinity and associated narratives coloured adventure theorising and practice for many years, and as such, we have dominant accounts of "conquering" Everest and motifs of overcoming challenge and risk. We also have niche business related to these motifs: super jeeps traversing icy wildernesses, heli-skiing for wealthy clients and the ever-online identity project of social media – the spectacularisation of the world and our celebrated place in it. Indeed, it is perhaps this bundle of commodity characteristics that has ensured the enduring popularity of adventure-type experiences in our identity-obsessed, post-industrial societies to this day. But the times, they are a-changing.

Beyond risk to the more-than-human

Early discussions on the subject of adventure tourism at the end of the 20th and the start of the 21st century tended to continue a rather frustrating over-focus on risk and its effects, its management and its policy implications. This is necessary discourse indeed, but it threatened to drown out other voices adopting more critical positions which were embedded in post-structural feminist theory and which acknowledged the over-emphasised "power" of human beings in nature simply as a way of "coming home" – back to origin.

Some of this thinking stems from the much earlier work of thinkers such as Arne Næss and Nils Faarlund in the *friluftsliv* (Nordic concept of outdoor leisure) and philosophical realm, but also from the pens of Humberstone (2000) and Pederson Gurholt (2008), who were looking at things from an (initially) female emancipatory perspective. Mortlock, too produced a superb later treatise on this in 2002 – *Beyond adventure: An inner journey* – and was followed by myriad related works, including Ewert's (2013, p. 91) 'Beyond because it's there' thesis and Imboden's (2012, p. 310) 'Between risk and comfort'. So, if it's not all about heroism, conquering and self-development, what else is going on here?

The answer to this question and its effects build on Varley's (2011) notions of our return to nature, where leisure pursuits lead us to a liminal state in which our dominant human-ness is shrunk as we immerse ourselves in wild seas, on rock faces or in the snow/forest/desert and so forth. These ideas come to the fore in Varley and Semple's (2015, p. 73) paper 'Nordic Slow Adventure' and in Rantala et al.'s (2018) subsequent discussion on the softening of adventure tourism. Both follow Næss, Faarlund and others in suggest-ing that being there, being-with, being-in nature, as an immersive dwelling-in-place is a significant departure from the sensual assault that characterises late modern, post-industrial life for many in advanced developed, societies … *without* necessarily engaging with risk.

More recently, building on Deleuze and Guattari's (1988) influential ideas in *A Thousand Plateaus*, came further philosophical work encouraging a recognition of for-gotten voices and ignored presences, which sought to celebrate the role of alternative agencies in the field of outdoor leisure and adventure tourism. Recent work building on the feminist philosophical subjects, of post humanism (Barad, 2003; Braidotti, 2006), and post-materialism and non-human subjects and their agency (Barad, 2003; Haraway, 2016), are emerging, for example from Ren, Jóhannesson and Van der Duim (2023) and Rantala et al. (2020, p. 3948), who write about 'proximity tourism', as explored through the feminist new materialism literature. They accord agency to;

the ongoing common worlding of all matter – including but not limited to humans – explor[ing] the potential of proximity tourism in ways that intertwine non-living and living matter, science stories, history, local communities and tourism.

Figure 3.3 Being with the world. Photograph by Matt Groves.

In so doing, they propose an alternative narrative of tourism after the Anthropocene, and instead use Haraway's term *chthulucene,* where humans acknowledge and feel their presence on the earth as just *part-of-the-earth,* which of course has particular resonance with the contemporary concerns of adventure tourism (Figures 3.3 & 14.2). In parallel, work in outdoor environmental education has used new materialities, post-humanism and ideas partly drawn from Ingold's (2021) 'Being Alive' thesis, to explore ways in which our adventures are in fact explorations of ways of being and are a confirmation of our place in the world (e.g., Clarke & McPhie, 2020; Mikaels & Asfeldt, 2017; McPhie & Clarke, 2015; Stewart, 2020).

And so, as adventure tourism shifts and changes, it seems crucial that scholars in the field continue to recognise the relative smallness of humanity and consider how the adventurous practices of individuals and groups of tourists present an opportunity to engage deeply with the realisation that the planet does not cling to us as we do to it. After billions of years in development, the momentary emergence of the adventure tourism phenomenon, and indeed of human beings in general, registers as so little on the temporal and material progress of the earth – but we can do much better in enacting kin and care What current writers are exploring, then, is how we make sense of our place in this fleeting moment of humanity, just as the seas, mountains, deserts and forests try to make sense of us. There is a recognition of the intense aesthetic and mindful potential of letting go of the conceits of the Anthropocene and seeking the embrace (being with) of the *chthulucene,* in which Haraway (2016) suggests simply, 'human beings are not the only important actors' (p. 55). Adventure tourism, if it is to have any purpose at all, beyond its role as another fragment of advanced capitalism, must foster an understanding of the intimate immensities of life on our planet and an ethic that enables all beings to thrive on it as micro parts of it (Brady, 2021). We must learn to think *with.*

Review questions

1 How has adventure tourism come about?
2 How have the ideals of adventure tourism shifted in emphasis, and what global issues have influenced those shifts?
3 What considerations should new businesses in the field make in their marketing, product design and communications?

Further reading

Beames, S., & Brown, M. (2016). *Adventurous learning: A pedagogy for a changing world.* Routledge.
Varley, P. (2006). Confecting adventure and playing with meaning: The adventure commodification continuum. *Journal of Sport & Tourism, 11*(2), 173–194.

References

Arnould, E. J., & Price, L. L. (1993). River magic: Extraordinary experience and the extended service encounter. *Journal of Consumer Research, 20*(1), 24–45.
Barad, K. (2003). Posthumanist performativity: Toward an understanding of how matter comes to matter. *Signs: Journal of Women in Culture and Society, 28*(3), 801–831.
Beames, S., & Brown, M. (2016). *Adventurous learning.* Routledge.

Beames, S., & Pike, E. (2013). Outdoor adventure and social theory. In E. Pike & S. Beames (Eds.), *Outdoor adventure and social theory* (pp. 1–9). Routledge.

Becker, P. (2016). A German theory of adventure: A view on the Erlebnispedagogik. In B. Humberstone, H. Prince, & K. Henderson (Eds.), *Routledge international handbook of outdoor studies* (pp. 20–29). Routledge.

Beedie, P. (2003). Mountain guiding and adventure tourism: Reflections on the choreography of the experience. *Leisure Studies, 22*(2), 147–167.

Bell, M. (2016). The romance of risk: Adventure's incorporation in risk society. *Journal of Adventure Education and Outdoor Learning, 17*(4), 280–293.

Boje, D., & Luhman, J. (1999). The knight errant's ideology of adventure. Presentation to the *Academy of Management session on 'Reclaiming Indigenous Knowledge'*. https://business.nmsu.edu/~dboje/knight.html

Brady, R. (2021). Slow & light. Perspectives for place responsiveness. Unpublished paper. The Swedish School of Sport & Health Sciences (GIH). https://www.researchgate.net/profile/Ryan_Brady10/

Braidotti, R. (2006). Posthuman, all too human: Towards a new process ontology. *Theory, Culture and Society, 23*(7–8), 197–208.

Breivik, G. (2010). Trends in adventure sports in a post-modern society. *Sport in Society, 13*(2), 260–273.

Caillois, R. (1961). *Man, play and games* (M. Barash, Trans.). Free Press of Glencoe.

Campbell, J. (1949). *The hero with a thousand faces*. Princeton University Press.

Cater, C. I. (2006). Playing with risk? Participant perceptions of risk and management implications in adventure tourism. *Tourism Management, 27*(2), 317–325.

Celsi, R. L., Rose, R. L., & Leigh, T. W. (1993). An exploration of high-risk leisure consumption through skydiving. *Journal of Consumer Research, 20*(1), 1–23.

Clarke, D. A., & Mcphie, J. (2020). Tensions, knots, and lines of flight: Themes and directions of travel for new materialisms and environmental education. *Environmental Education Research, 26*(9–10), 1231–1254.

Cohen, S., & Taylor, L. (1992). *Escape attempts: The theory and practice of resistance to everyday life* (2nd ed.). Routledge.

Csikszentmihalyi, M. (2000). *Beyond boredom and anxiety: Experiencing flow in work and play.* Jossey- Bass. (Original work published 1975)

Csikszentmihalyi, M. (1975). *Beyond boredom and anxiety.* Jossey- Bass.

Deleuze, G., & Guattari, F. (1988). *A thousand plateaus: Capitalism and schizophrenia.* Bloomsbury.

Elsrud, T. (2001). Risk creation in traveling: Backpacker adventure narration. *Annals of Tourism Research, 28*(3), 597–617.

Ewert, A., Gilbertson, K., Luo, Y. C., & Voight, A. (2013). Beyond "because it's there" motivations for pursuing adventure recreational activities. *Journal of Leisure Research, 45*(1), 91–111.

Fletcher, R. (2010). The emperor's new adventure: Public secrecy and the paradox of adventure tourism. *Journal of Contemporary Ethnography, 39*(1), 6–33.

Haraway, D. J. (2016). 2. Tentacular Thinking. In Durham (Ed.) *Staying with the Trouble* (pp. 30–57). Duke University Press.

Heywood, I. (1994). Urgent dreams: Climbing, rationalization and ambivalence. *Leisure Studies, 13*(3), 179–194.

Holyfield, L. (1999). Manufacturing adventure: The buying and selling of emotions. *Journal of Contemporary Ethnography, 28*(1), 3–32.

Humberstone, B. (2000). The 'outdoor industry' as social and educational phenomena: Gender and outdoor adventure/education. *Journal of Adventure Education & Outdoor Learning, 1*(1), 21–35.

Imboden, A. (2012). Between risk and comfort: Representations of adventure tourism in Sweden and Switzerland. *Scandinavian Journal of Hospitality and Tourism, 12*(4), 310–323.

Ingold, T. (2021). *Being alive: Essays on movement, knowledge and description.* Routledge.

Loynes, C. (1998). Adventure in a bun. *The Journal of Experimental Education, 21*(1), 35–39.

Loynes, C. (2003). Narratives of agency: The hero's journey as a construct for personal development through outdoor adventure. In J. Koch, L. Rose, J. Schirp, & J. Vieth (Eds.), *Bewegungs: Und korperorientierte ansatze in der sozialen arbeit* (pp. 133–143). BSJ/Springer VS.

Lynch, P., & Moore, K. (2004). Adventures in paradox. *Australian Journal of Outdoor Education, 8*(2), 3–12.

Lyng, S. (1990). Edgework: A social psychological analysis of voluntary risk taking. *American Journal of Sociology, 95*, 851–856.

Martin, P., & Priest, S. (1986). Understanding the adventure experience. *Journal of Adventure Education, 3*(1), 18–21.

Martínková, I., & Parry, J. (2017). Safe danger: On the experience of challenge, adventure and risk in education. *Sport, Ethics and Philosophy, 11*(1), 75–91.

McPhie, J., & Clarke, D. A. G. (2015). A walk in the park: Considering practice for outdoor environmental education through an immanent take on the material turn. *The Journal of Environmental Education, 46*(4), 230–250.

Mikaels, J., & Asfeldt, M. (2017). Becoming-crocus, becoming-river, becoming-bear: A relational mate- rialist exploration of place(s). *Journal of Outdoor and Environmental Education, 20*(2), 2–13.

Mitchell Jr, R. G. (1983). *Mountain experience: The psychology and sociology of adventure.* University of Chicago Press.

Mortlock, C. (1984). *The adventure alternative.* Cicerone Press.

Mortlock, C. (2002). *Beyond adventure: An inner journey.* Cicerone.

Nerlich, M. (1987). *Ideology of adventure: Studies in modern consciousness, 1100–1750.* Vols. 1 and 2. University of Minnesota Press.

Nietzsche, F. (1967). *The birth of tragedy* (W. Kaufmann, Trans.). Vintage Books.

Pedersen Gurholt, K. (2008). Norwegian friluftsliv and ideals of becoming an 'educated man'. *Journal of Adventure Education & Outdoor Learning, 8*(1), 55–70.

Priest, S., & Bunting, C. (1993). Changes in perceived risk and competence during whitewater canoeing. *Journal of Applied Recreation Research, 18*(4), 265–280.

Rantala, O., Hallikainen, V., Ilola, H., & Tuulentie, S. (2018). The softening of adventure tourism. *Scandinavian Journal of Hospitality and Tourism, 18*(4), 343–361.

Rantala, O., Salmela, T., Valtonen, A., & Höckert, E. (2020). Envisioning tourism and proximity after the anthropocene. *Sustainability, 12*(10), 3948.

Rojek, C. (1995). *Decentring Leisure: Rethinking leisure theory.* SAGE.

Rojek C. (1993). *Ways of escape: Modern transformations in leisure and travel.* Palgrave Macmillan.

Rojek, C. (2000). *Leisure and culture.* Macmillan.

Simmel, G. (1919/n.d.). The adventurer (D. Kettler, Trans.). In G. Simmel (Ed.), *Philosophische Kultur* (pp. 7–24). Kröner. http://www.osea-cite.org/tourismworkshop/resources/Simmel_The_Adventurer.pdf

Stewart, A. (2020). *Developing place-responsive pedagogy in outdoor environmental education: A rhizomatic curriculum autobiography* (Vol. 17). Springer.

Swarbrooke, J. (2003). *Adventure tourism: The new frontier.* Routledge.

Varley, P. (2006). Confecting adventure and playing with meaning: The adventure commodification continuum. *Journal of Sport & Tourism, 11*(2), 173–194.

Varley, P., & Semple, T. (2015). Nordic slow adventure: Explorations in time and nature. *Scandinavian Journal of Hospitality and Tourism, 15*(1–2), 73–90.

Varley, P. J. (2011). Sea kayakers at the margins: The liminoid character of contemporary adventures. *Leisure Studies, 30*(1), 85–98.

Vester, H. G. (1987). Adventure as a form of leisure. *Leisure Studies, 6*(3), 237–249.

Walle, A. H. (1997). Pursing risk or insight: Marketing adventures. *Annals of Tourism Research, 24*(2), 265–282.

Warren, K. (1996). Women's outdoor adventures: Myth and reality. In K. Warren (Ed.), *Women's voices in experiential education* (pp. 10–17). Kendall Hunt.

Weber, K. (2001). Outdoor adventure tourism: A review of research approaches. *Annals of Tourism Research, 28*(2), 360–377.

Zweig, P. (1974). *The adventurer. The fate of adventure in the Western world.* Princeton University Press.

The following case study introduces an adventure tourism operator which offers transformative experiences for clients, encouraging greater kin and care for the non-human elements of our planet.

Case study 3.1   Regenerative adventure tourism – TYF Adventure and B Corp Certification

Christopher Woodfield

TYF Adventure, Wales

TYF Adventure, based in St. David's, in Pembrokeshire Coast National Park, have been exploring a responsible business approach for over 35 years. Founded in 1986, TYF has been experimenting with the power of play and nature connection and was an early pioneer in the adventure and outdoor tourism sector. TYF founded the activity of coasteering, an immersive exploration of the coastscape, combining swimming, jumping, scrambling and walking, and now takes thousands of schoolchildren, young people, family groups and adults into the outdoors every year. As well as coasteering, TYF offers a range of outdoor and educational experiences including surfing, kayaking, wild swimming, coastal walks, paddleboarding and sea-cliff climbing.

TYF's mission is to help people fall in love with nature, grow fearless curiosity and build an unshakeable confidence in their ability to make a difference that counts. TYF is driven by a passion to connect people so deeply with nature that it changes the way they live. This can support customers, employees and the local community to rise to the climate and nature emergencies in an empowering way.

Central to this mission of connecting, inspiring and empowering others to be the best they can be, TYF realises and recognises that it can also facilitate positive change in how it operates as a business, putting purpose front and centre. In 2015, TYF was part of a group of Founding B Corp businesses in the UK and became the first in Wales. B Corp is an environmental and social certification standard for businesses and is a growing movement of positive change. Starting in the US, the B Corp certification has spread globally, and the UK is now the fastest-growing B Corp community in the world, with more than 1,000 B Corp–certified businesses across all sectors (B Lab UK, 2023).

B Corp businesses are held to strict standards of environmental and social criteria, with the certification covering all aspects of business operations. There are five main areas of focus: governance, employees, community, environment and customers. Businesses are scored against these five areas by undertaking a B impact assessment questionnaire in which they must score over 80 points. To be eligible for certification they must also complete a change in the company's articles of association to include the environment and community as key stakeholders. B Corp certification and the B impact assessment cover a wide range of aspects with more than 200 questions across the five impact areas, with businesses being rewarded for having progressive policies and procedures, fair pay structures, strong employee rights, environmental practices and carbon reduction schemes, spending within their local community, ethical supply chains and a whole host of other criteria.

To retain the B Corp certification, businesses are required to recertify every 3 years and demonstrate they are upholding or improving their environmental and

Figure 3.4 Sea kayaking with TYF Adventure, Pembrokeshire, Wales, UK. Photograph by TYF Adventure.

social standards. TYF has now recertified twice and is proud to have a strong reputation as a responsible and ethical business. TYF aligns its approach with supporting action on the UN Sustainable Development Goals and has a key focus on upskilling its staff, for example with team sustainability training, including carbon literacy, circular economy, doughnut economics, blue health and ocean literacy. Furthermore, in 2021, TYF pioneered a Planetary Repair Centre, offering a repair, upcycling and recycling service for customers and community members to ensure outdoor gear and clothing do not end up in landfills and are given a new lease of life. These are just some examples of TYFs approach to ensuring it delivers on its sustainability ambitions. Sustainability is a core component of the how, what and why of TYF, and the B Corp certification has been a useful mechanism to showcase this, as well as ensure continual improvement is embedded.

Adventure tourism and the wider outdoor community ecosystem have a key role to play in educating and inspiring positive change through immersive and awe-inspiring experiences in nature. There is a myriad of evidence highlighting consequential pro-environmental behaviour change linked to this. However, adventure tourism operators also have a role to play in not just advocating for sustainability and stewardship but also being role models and stewards and purposeful businesses themselves, combining positive impact on profit, people, planet and place. The B Corp certification should not be seen as a means to an end; rather, the start of a journey towards a destination of regenerative tourism where people and nature are thriving and flourishing.

B Lab UK (2023) 1000 UK B Corps [online]. https://bcorporation.uk/act-and-learn/campaigns/1000-uk-b-corps/ [Accessed on 06/02/2024]

TYF https://www.tyf.com/

4 Emerging and transformational trends in issues related to adventure tourism and recreation

Alan Ewert, Amy DiRenzo and Jon Frankel

Chapter learning outcomes

1 Identify three demographic trends and their corresponding implications for adventure recreation.
2 From an administrative perspective, describe three trends occurring in adventure recreation that are likely to impact organised/commercial programmes.
3 From an administrative perspective, identify three trends occurring in adventure recreation that are likely to impact or currently impact college/university programmes.

Introduction

There is a substantial and growing corpus of literature now available that describes both adventure tourism and outdoor adventure recreation. While the focus has generally concerned the consumer, this chapter approaches the phenomenon from the perspective of both the organisation and the staff and, more specifically, what trends and issues are impacting the people *designing, delivering, and managing* these experiences. It is reasonable to expect that the same issues impacting participants such as the COVID-19 pandemic, daily stress, and obligations to work, school, and family are also impacting both administrative and field-based staff, in addition to the organisations that they work for. These impacts have manifested into a variety of challenges for providers and programme administrators, including difficulty in hiring new staff and/or retaining experienced staff, lower morale among employees, staff departures prior to the end of their contract/season, and discrepancies between the stated mission of an organisation and the actual delivery of that mission.

For example, in a recent study, Owens (2022) indicates that potential staff members working in the adventure recreation industry have a different set of work-related expectations than has been typically seen in older and more established staff. Some of these differences include a desire for more flexibility in the work schedule, seeking unique experiences, asking the organisation to provide more transparency regarding the actual work they would be involved in, a value match between themselves and the organisation, and achieving a balance in the social pressures they experience on the job. To explore these concepts, this chapter includes a synopsis of both adventure tourism and outdoor adventure recreation and provides a brief history of the driving forces underlying both

DOI: 10.4324/9781003393153-6

endeavours. A review of the effects of climate change, the pandemic, and the economy are examined within the framework of how these issues are impacting the participant as well as the staff delivering the experience. The chapter then offers examples of anticipated trends that will impact the delivery of these experiences. It concludes with a synthesis of what these trends mean for the industry overall. We approach this issue from the perspectives of both organised programmes and programmes in higher or tertiary education. We represent a spectrum of backgrounds and experience in adventure tourism, outdoor adventure recreation, and outdoor recreation including organised programmes, commercial operations, and higher education.

The outdoor industry

For the purpose of this chapter, we have taken a broad and inclusive view of the outdoor industry. More specifically, we conceptualise the outdoor industry as being composed of three major components: (1) adventure tourism, (2) outdoor adventure recreation, and (3) outdoor recreation. There are nuanced definitional and conceptual differences among these terms, as we will briefly delineate. However, for the purpose of simplicity, we use the term *adventure recreation* throughout this chapter to encapsulate adventure tourism, outdoor adventure recreation, and outdoor recreation.

From a definitional perspective, we borrow from Buckley (2006) with his definition of adventure tourism as being '[g]uided commercial tours where the principal attraction is an outdoor activity that relies on features of natural terrain, generally requiring specialised sporting or similar equipment, and is exciting for the tour clients' (p. 1). In a similar fashion, we use Ewert and Sibthorp's (2014, p. 5) definition of outdoor adventure recreation: 'Activities and experiences usually done in a natural or outside environment that involves elements of challenge and either real or perceived risk, in which the outcome is uncertain but influenced by the skill and ability of the participant.' Finally, we take a broad view of outdoor recreation and rely on Jensen and Guthrie (2006), who suggest that this can be thought of as recreation engaged in out of doors, most commonly in natural settings. The activities that encompass outdoor recreation vary depending on the physical environment. They can include fishing, hunting, backpacking, and horseback riding, among others, and can be completed individually or collectively. Thus, under this definitional framework, outdoor recreation is a broad concept that encompasses a varying range of activities and landscapes.

We have included in this discussion all three concepts as a way of illustrating their overlap and to make the point that current and emerging trends often have pertinence in each area. For example, finding and retaining qualified field staff is a challenge whether the organisation is primarily involved with adventure tourism, outdoor adventure recreation, and/or outdoor recreation. Thus, horseback riding, trekking, or canoeing might be considered any of these activity types depending on the location, the intensity of the activity, or how the experience is structured and engaged in. Given the intent of this book, this chapter primarily focuses on adventure tourism and outdoor adventure recreation and, as previously stated, uses the term *adventure recreation* to capture all three concepts.

A brief history of growth and development

Perhaps not surprisingly, the literature is replete with documentation regarding the rapid growth of participation in adventure activities. Even with the pandemic, growth in the number of participants is impressive. For example, recent US data from the Bureau of Economic Analysis (BEA, 2022) show that the outdoor recreation economy accounted for 1.8% (US$689 billion in consumer spending) of current-dollar gross domestic product (GDP) for the nation in 2020. This was after a downturn because of COVID-19 pandemic impacts, such as lockdowns and stay-at-home orders.

Data for adventure recreation are often challenging to collect due to differences in the type of data collected and often remote locations for participation. Table 4.1 represents some of the most recent data regarding the number of participants in the US. Clearly, when aggregated, these activities represent a substantial and growing segment of the population across the globe.

One of the reasons for this rapid growth has been in the pursuit of personal health and well-being. For example, in 1987, the US President's Commission on Americans Outdoors proposed that the outdoors is like a Great Health Machine. Undoubtedly, the commission found that the public placed a high value on adventure recreation activities and programmes. By way of comparison, the Outdoor Recreation Resources Commission in 1959, found that 90% of the American public went outdoors for recreation. The President's Commission in 1987 found a similar level of involvement (89%) for a total of 188 million people. Fast-forward to 2021 and the data suggest more participants, different types of users, and a growing array of uses on wildlands and open spaces. More recently, and consistent with previous data in the US, a study released through the BEA in 2022 found that adventure recreation accounted for US$862 billion in gross economic output amounting to 1.9% of GDP and 4.5 million jobs in 2021.

Similar growth was seen in the UK with adventure and water sports spending rising from £1.5 billion to £1.8 billion between 2011 and 2019 (Office of National Statistics, 2021). Likewise, the estimated number of outdoor-related engagements within Great Britain annually peaked in 2016, with approximately 1.5 billion activities completed. The number of outdoor-related activities participated in annually increased by 30% between 2011 and 2016, up from 1.1 billion. In 2019, 1.4 billion activities were

Table 4.1 Participation rates of selected outdoor adventure activities 2010 and 2020[a,b]

Activity	2010		2020	
	Number in (000s)	*Percentage (%)*	*Number in (000s)*	*Percentage (%)*
Outdoor Backpacking	8,349	2.9	10,746	3.5
Mountain Biking (Non-Paved Surface)	7,142	2.5	8,998	5.0
Camping	30,996	12.2	36,082	11.9
Canoeing	10,553	3.6	9,595	3.2
Climbing (Traditional/Ice/ Mountain)	2,198	0.7	2,456	0.8
Whitewater Kayaking	1,842	0.6	2,605	0.9
Rafting	4,460	1.6	3,474	1.1
Scuba Diving	3,153	1.1	2,588	0.9
Cross-Country Skiing	4,157	1.5	4,768	1.6

[a] Data from 2021 USA Outdoor Participation Trends Report (Outdoor Foundation).
[b] Percentage of total number of users.

completed, a slight reduction since 2016. It should be noted, however, that adventure and water sports increased by 75%, from 57 million activities completed in 2011 to 100 million in 2019. Moreover, it is estimated that the market size of global adventure tourism reached nearly US$496 million in 2021. This was an increase from the US$374 million generated in 2020 (Statista, 2020). However, the increase in the number of participants and the complexities of issues such as the COVID-19 pandemic, demographic changes, technology, climate change, and economic impacts are not without their costs, and often, these are borne by both the organisation and the staff.

Effects of the COVID-19 pandemic

In the past several years, adventure recreation programmes have been impacted by several pandemic-related factors. The COVID-19 pandemic changed the world, as well as the patterns of use and participation in adventure recreation. Data on actual use and equipment sales present a mixed picture. While higher equipment sales for specific forms of recreational equipment have occurred, a report by the Outdoor Foundation (2021), indicated that 69% of the 20 national outdoor recreation trade associations in the US, representing more than 100,000 businesses, reported significant and negative impacts from the COVID-19 pandemic. Of the businesses responding to the survey, 79% reported laying off or furloughing a portion of their staff, and 11% indicated they closed or laid off most or all their staff. In addition, more than 89% of the outdoor industry businesses indicated a decrease in sales, with 39% indicating a decrease of 50–75%. A recent study by Leonard et al. (2021) examined the effect of the COVID-19 pandemic on organisations specifically aligned with adventure recreation. It was found that out of the 34 organisations studied, 40% reported as being closed down, 37% indicated staff reductions, and of the programmes that remained open, only 50% had been able to maintain their staffing levels. It is notable that in this study, all 14 of the outdoor behaviour health programmes represented in the survey reported staying open.

From a participant perspective, using a pre- and post-COVID-19 qualifier as a comparison measure for over 1,000 respondents, Rice et al. (2020) report that the number of outdoor activities commonly associated with adventure recreation has actually decreased in levels of participation as a result of the pandemic. A sample of these activities includes backpacking, camping, climbing, downhill and cross-country skiing, flatwater canoeing, kayaking, and rafting. On a positive note, findings indicate that more than 66% of the study respondents anticipated returning to levels and types of participation they previously engaged in after the pandemic subsided. Moreover, the COVID-19 pandemic has also impacted land management, with many agencies and parks implementing reservation systems to manage overuse and impact on the land, as well as enacting permitting restrictions that limit group size and overall capacity for commercial outfitters.

Demographic changes

Current data from the 2021 Outdoor Participation Trends Report (Outdoor Foundation, 2021) suggest that between 2010 and 2020, the number of participants has increased, with a few exceptions (canoeing, rafting, and scuba diving; see Table 4.1). Looking more deeply into the data, however, reveals a number of issues that impact adventure recreation (Outdoor Foundation, 2021) in the US and includes the following:

- Lack of diversity – In the US, nearly 75% of outdoor participants in 2020 were White, even though White, non-Hispanic, or non-Latino people comprise only approximately 60% of the population. A 7% annual decline was noted among Asians for the past 3 years, and among Blacks, participation rates were level. Participation rates among Hispanics and Latinos/Latinas (11% of total outdoor participants) grew to just under 48%, but their rates remained well below those of Whites.
- Relatively static female participation – Despite industry efforts to address gender disparities, since 2008, females have represented just 46% of outdoor participants, even though 51% of the population in the US are female.
- Declining intensity – In 2010, 24% of all participants reported partaking in outdoor activities more than twice a week, while in 2020, that number dropped to 20%. The same pattern occurred among children and young adults, suggesting that there is a reduced number of devoted outdoor participants but an increase in the number of casual ones.
- Fewer outings – The average number of annual outings per participant continued a steady, long-term decline, falling from 87 outdoor trips in 2012 to just 71 in 2020.
- Retaining post-COVID-19 participants – A report from the Outdoor Industry Association (2021) indicated that about one-quarter of new participants say they do not want to continue their new outdoor activities, a number that may grow sharply as consumers return to pre-pandemic habits. Using a national sample of more than 3,900 respondents, Taff et al. (2021) found that 36% of respondents indicated that they did not participate regularly in outdoor recreation prior to the pandemic or during the pandemic. Slightly more than 30% stated that they did participate regularly in outdoor recreation prior to the pandemic and continued to do so regularly during the pandemic. Furthermore, almost 14% indicated that they regularly participated in outdoor recreation prior to the pandemic and did not plan on doing so during the pandemic. Other data from this study reaffirm the prevalence of the issues raised above, with the findings that the majority of respondents in all categories, including those who were new to outdoor recreation amidst the pandemic, identified themselves as being White, and being the least ethnically diverse, compared to the other groups mentioned.

Along with demographic changes, a number of other factors are impacting the participation patterns inherent in adventure recreation. A sampling of these factors includes technology, climate change, and economic impacts.

Technology

Technology influences who participate and how they participate. In an earlier study, Ewert et al. (2000) contended that technology impacts participation patterns in adventure recreation in the areas of safety, comfort, access, and information. In addition, Cuthbertson et al. (2004) examined how modern technology has positively and negatively influenced adventure recreation. More recently, Van Kraalingen (2021), reviewing 33 articles and studies, looked at the impact mobile phones have on outdoor programmes and experiences. The data suggested that the portability and accessibility of mobile devices offer new opportunities while raising issues related to complexity, safety, and loss of desired experiential quality. The findings highlight three principal strategies that offer meaningful ways to manage the tensions between technology and outdoor learning:

mitigation, intentionality, and adaptation. Furthermore, the prevalence of social media platforms (e.g., TikTok, Instagram, Snapchat, and Twitter/X) is expected to be a contributor to growth in adventure recreation market segments as users are exposed to and learn about new activities and destinations (see Chapter 11). Consumers also tend to read reviews posted by previous participants that may influence, fairly or unfairly, which programme, experience, or operator they choose. From an administrative perspective, technology has also enhanced organisational risk management thanks to mapping capabilities, expanding cell service, and devices such as personal locator beacons.

Climate change

Projected rises in greenhouse gases imply issues such as changing minimum/maximum temperatures, potential evapotranspiration, and levels and types of precipitation (see Chapter 22 for an in-depth description of climate change issues). Adventure recreation programmes respond to these issues by modifying activities, locations, and timings. For example, Askew and Bowker (2018) employed a two-step approach using an estimation and simulation step for estimating models of adult participation rates and days per participant by activity at regional and national levels at 10-year intervals through to 2060. Their findings found snowmobiling and undeveloped skiing (cross-country skiing and snowshoeing) were the most negatively affected by climate change. Participation in both fishing and motorised water activities in the northern hemisphere will increase. However, climate change in the Rocky Mountain region of the US will negatively impact motorised activities, such as hunting and fishing. The Pacific Coast region for Canada and the US will show the most stability, with relatively small climate-induced shifts.

What these changes suggest for adventure recreation will be variable and specific for certain areas. For example, increasing dryness and higher temperatures will elevate the likelihood of more wildfires. While users in areas directly threatened by wildfires can usually move to less threatened areas, the principal negative factor is smoke and more specifically the air quality index. Poor air quality from smoke often extends over much larger areas, is often changeable over short periods, and is difficult to deal with. In addition, physical structures such as base camps can be threatened by fire or are unusable due to excess smoke. These and related factors add stress to both field-based and administrative staff, as well as participants.

Concomitant to higher temperatures and drier conditions will be the impact on available sources of water while in the field and water levels for lakes and rivers. In some cases, reduced river levels will affect activities such as rafting and canoeing. Reductions in lake levels will preclude access to certain areas and disrupt loading and unloading boats and rafts. These situations will often necessitate changes in itineraries and programme locations. For example, in wildfire-prone locations, programmes may need to be redesigned and/or not offered at certain times or locations. Record temperatures made headlines in 2022, documenting the effects extreme heat is having on global infrastructure as runways melted and railway tracks buckled under record temperatures (CNN, July 2022), often disrupting travel. Such extreme heat may also necessitate changes to programme itineraries as participants may not be able to move as far or as fast in an effort to avoid medical injury. As areas become too hot and too dry, course locations may also need to change. This would require additional organisational resources as staff would need to reconnoitre new course areas to determine their suitability, and new use permits would need to be obtained. A heightened awareness and concern for the effects of climate change may also lead participants to prioritise more environmentally sustainable travel and activities, although higher costs may result in reduced participation (Kumar & Deshmukh, 2021).

Economic impacts

Participation in adventure recreation often resides in the discretionary side of societal expenditures. As a result, it is often sensitive to economic trends and issues. For example, economic downturns, inflation, international crises, long-term weather conditions, and pandemics can all exert rapid and severe economic disruptions that can have immediate ramifications for these industries. In Australia, Spennemann and Whitsed (2021) report that the Australian outdoor recreation industry was severely affected by the COVID-19 pandemic. They found a pessimistic outlook for the industry, with extensive financial and job losses. They suggest an important point, that is, that the industry needs to adapt and protect itself against future disruptors and pandemics.

Clearly there have been some changes in the level of participation in adventure recreation, but other factors have remained. While interest in these types of activities has remained strong and even grown, the data also suggest that participation levels have varied depending on a number of factors, including age, type of activity, sex, racial background, and economic status.

In addition, the global adventure tourism market size is predicted to grow tenfold from US\$112 billion in 2020 to US\$1.1 trillion in 2028 (Kumar & Deshmukh, 2021). Contributors to this growth include increased social media networking, such as Instagram, Facebook and travel blogging; government tourism initiatives; and the attractiveness of varied natural landscapes. Moreover, going with a partner, family, or friend has generated the most growth in 2020 and is expected to be the highest growth segment through 2028, indicating more adventure travellers are interested in seeking adventure with a partner rather than solo. Adventure travellers are predominantly between the ages of 30 and 41 years, and as a result, potentially enjoy a high quality of health and physical fitness enabling them to engage in adventurous activity, and they are more financially stable.

As with climate change, economic changes will impact adventure recreation in both the types of programmes offered and how they are offered. These and other changes will often manifest in the areas of organised programmes and outdoor programmes in colleges and universities. The following is a synopsis of potential changes that these and related programmes will see based on a *current*, *possible*, and *probable* scale.

Organised programmes

In this chapter, we refer to organised programmes as including commercial or educational programmes often utilising a guide but generally outside the purview of a programme in a higher or tertiary institution. The following are examples of ongoing or expected trends:

- Competition for staff has increased with individuals moving from one opportunity to the next rather than staying with one operation for an extended time. A number of trends have occurred simultaneously or as a result of this competition. First, there is a push for better staff compensation, including higher wages and insurance benefits, and subsequent pushback from some operators. It is becoming more problematic to attract and retain highly qualified staff than more participants. In addition, there is a growing recognition of guiding as a profession as opposed to a hobby. Participants are searching for immersive experiences rather than just instruction (see Case study 4.1 for an example of this issue). **Current**

- Increased competition for qualified field staff within a landscape of shifting staff demographics. Increasingly field staff are asking for a more comprehensive and satisfying experience, including better living conditions, opportunities for advancement and progression in the industry, higher pay, health benefits, additional and extended time off, and shorter commitments. **Current**
- Increasingly, staff come from different backgrounds. For example, substantial numbers of employees come from college outdoor programmes and majors, suggesting that staff have debt and need higher rates of pay while, at the same time, often having less experience. Additionally, such employees arrive with training exclusively from a programme instead of personal experience, thus limiting their range of expertise (see also Chapter 18). **Current**
- Stress and mental health are becoming more acute and problematic. These impacts are affecting staff's ability to engage participants more fully. Moreover, because of the reduced numbers of new employees, current staff are being asked to fulfil different more complex schedules than what they were hired for. As a result, incoming staff are turning down jobs after going through the entire hiring process, often due to the belief that taking a job in adventure recreation would present too much stress in their lives. There is evidence that new employees may be unwilling, or unable, to navigate the inherent stressors of leading others in the field, which begs the question: Are new staff truly equipped to take care of other people in the outdoors? **Current/Probable**
- The shifting business model is based on what people want from an adventure recreation experience and what they can afford. This often includes shorter trips and less expensive experiences closer to home. There are also concerns regarding complications associated with international travel. In addition, participants often want to simply have a new experience with their friends (new or old) in a natural environment, with less focus on instruction and more on experience for the sake of experience. **Current/Probable**
- Organisations will increasingly offer year-round programming as a way to more efficiently use their available facilities and gear, provide year-round employment for their staff, and create more financially sustainable business models. **Probable**
- There is an increasing shortage of experienced field staff. With a corresponding increase in less experienced field staff, programme activities will be modified to offer lower levels of risk and technical skill. If activities are not modified, there is potential for an increased risk to participants as well as for an increase in incidents. In addition, less experienced staff will necessitate additional training and mentoring, thus increasing overall programme costs. Also, additional employees are often being hired mid-season to fill gaps created by injury or attrition, and then have to learn on the job without formal training. This results in increased risk potential as well as a higher potential for negative impact on the participants. For example, fewer staff with sufficient technical skills have resulted in a growing number of adventure recreation programmes changing experiences in a variety of ways, including portaging around previously run rapids, selecting less technical rock-climbing routes, and offering backpacking rather than mountaineering which requires a more technical and advanced skill set. **Current/Probable**

Case study 4.1   GIVE Foundation

When the COVID-19 pandemic ground most recreational travel to a halt, numerous organisations that rely on travellers to sustain their business were severely impacted, with some operators shutting down for good. Many staff were temporarily or permanently laid off. For the GIVE Foundation Volunteer Abroad Programs in Asia, Africa, and Central America with GIVE! (https://www.givevolunteers.org/), an organisation specialising in asset-based community development and with a focus on mobilising a community's own assets as the primary basis for development, the pandemic created a unique challenge. The organisation relies heavily on its established relationships with local leaders, guides, and community members to develop an action plan that benefits the local community based on what resources and assets (knowledge and skills) already exist within the host community. When the pandemic stopped global travel, programme administrators were forced to wrestle with how to continue to build their relationships, some of which took years to establish, and support the community development plan as well as support the organisation's staff. Ultimately, one option the organisation considered was keeping host-site staff employed and temporarily laying off US-based staff.

- Transportation costs, particularly for programmes that offer pickup at one location and transportation to an alternate site, will continue to increase. Cost will rise with unplanned evacuations, early withdrawals, and other situations for which additional transportation is needed. One result will be for organisations to modify the course start or end locations. **Probable**
- Locations burned by wildfires will reduce the number of available areas for course activities. This factor will necessitate finding new locations or put increasing pressure on current ones. **Current**
- In studying adventure recreation activities in locations such as Nepal, Wengel (2021) found that an emerging trend is a bottom-up approach to developing adventure sports, such as mountain biking and skiing. While promoted to both domestic and international visitors, one distinguishing feature is that locals who are passionate about these sports drive such initiatives. Thus, creating and expanding adventure tourism activities for locals not only provides economic benefits but also contributes to the well-being and recreation opportunities for locals. **Current/Probable**
- Whether a result of the pandemic, increased stress related to their home environment, or other factors, there is an increasing number of participants who enrol on adventure recreation programmes and have mental health needs requiring additional staff time and emotional energy. This may increase incidents as such participants are more likely to disregard directions from staff and inadvertently put themselves in situations with increased risk. **Current**
- Organisations are looking for ways to support staff mental health and stress management but are finding that tools and resources are neither organised nor readily available or fully understood. **Current/Probable**

The issue of staff stress levels is becoming more acute with the advent of many of the issues raised earlier, and as such, it warrants some future attention. Stress levels of staff often manifest themselves in a variety of ways and along somewhat predictable patterns. One way to assess the amount of stress and subsequent dysfunction an organisation is facing is through the use of a stress-continuum projection, modified from Responder Alliance (2021) as illustrated in Table 4.2.

Within this context, a range of stress level categories include normal, minimal, moderate, and severe. Within each category exists a number of issues that can predict the level of stress being experienced by field and administrative staff. For example, under normal levels of stress experienced by field-based and administrative staff, there exists a general acknowledgement of a relatively high level of orientation to the mission of the organisation. By comparison, a severe level of stress often results in a feeling among staff that the mission is less critical or even unknown. In another example, open communication between staff members, within both the field-based staff and the administrative staff, can often be seen within normal levels of stress. Conversely, under a severe level of stress, communication patterns often degrade into a culture of blame or a sense that "nobody is listening to me."

These categories can often be interchanged over time and situation. That is, an organisation may be experiencing moderate or severe levels of stress, which is situational or time-dependent, while at other times, the organisation may experience normal or minimal levels of stress. The important aspect represented by this model is the overall and ongoing pattern of stress levels. It can also be true that the organisation may be reporting

Table 4.2 Potential stress components seen in outdoor adventure recreation instructional and administrative teams

Normal	Minimal	Moderate	Severe
Mission Oriented	Mission Less Transparent	Mission Less Central	Mission Not Critical
Psychologically and Socially Healthy	Reduced Levels of Motivation	Resources Perceived as Less Available	Lack of Trust with Administration and Other Staff
Open Communication	Transactional Communication	Perceived Less Value as a Team Member	No One Is Listening to Me or My Needs
Emotionally Stable	Reduced Sense of Being Able to be Flexible in the Workplace	Communication Starting to Get Isolated and Strained	Culture of Blame
Client/ Customer-Focused	Reduced Focus on Client/Customer	A Growing Sense of Frustration and Not Being Heard	Sense of Isolation
High Energy Levels	Reduced Willingness to Ask for Help	Deterioration of Healthy Relationships	Reduction in Physical and Emotional Health
Capable of Flexibility in Decision-Making	Vulnerability Is Beginning to Be Questioned	Sense of Isolation	Sense of Hopelessness
Sense of Organisational Support	Requests for Help is Criticised	Suffering in Silence	Broken Relationships

Adapted from Combat and Operational Stress First Aid (Responderalliance.com).

normal or minimal levels of stress while the field-based staff or lower level administrative staff may be reporting moderate or severe levels of stress. In this case, clearly there is a disconnect between one organisational level and another (e.g. executive leadership and field staff).

Organisations attempting to address staff stress should consider implementing a holistic approach that includes focused training, cultural norm setting, regular structured communication, and systems designed to support staff in maintaining normal stress levels, as well as a tool to help identify different levels of stress and the subsequent impacts of those levels. When considering a predictive tool such as the Stress Level Continuum (Responder Alliance, 2021) as presented in Table 4.2, perhaps the most impactful approach for both staff and the organisation is to consider the trajectory of the stress responses experienced. That is, over time, do staff members individually or the organisation collectively find themselves working within a normal set of stress responses, or are they experiencing stress responses indicative of a moderate or severe level?

Programmes and offerings in colleges and universities

In this chapter, current and emerging trends in adventure recreation are examined due to their connection to preparing and training staff, both field-based and administrative. Colleges and universities often provide substantial numbers of personnel, both through direct programme training or for the development of a broad base of useful skills and knowledge in a variety of areas such as management, economics, human relations, programme design, and analytical skills. To date, Turner et al. (2022) report that within the US, there are approximately 128 organised and identifiable adventure recreation programmes in higher education, with 64% of these being public.

The following presents a number of current and emerging trends that pertain to colleges and universities and have ramifications for the adventure recreation field.

- Reduced risk tolerance – In an effort to reduce liability exposure and potential costs associated with offering courses and programmes, colleges and universities will increasingly become more risk-averse and rely more on outfitter and contracted types of services. As a result, there will be a movement toward externally contracted courses and away from faculty-led field courses. **Probable**
- Shorter outdoor experiences – To reduce transportation costs along with other logistical requirements, as well as more perceived controllability, colleges and universities will support shorter outdoor experiences, often utilising locations closer to campus. **Possible**
- Nearby nature settings – Given the limitations often associated with nearby nature settings such as municipal and regional parks, the presence of more people, and strict regulations on resource use (e.g. fires, water-based activities, and wilderness-type backpacking), colleges and universities will often be required to design different mixes of activities and skill-based endeavours. **Probable**
- Indoor facilities – In order to reduce perceived liability, transportation costs, and the utilisation of specialised equipment, colleges and universities will place a greater emphasis on the use of indoor facilities such as climbing walls and pools. Service staff will often replace academic faculty at these sites, with equipment such as kayaks and rock-climbing gear being provided by contracted or service staff. **Probable**
- Student health and college and university mission – Outdoor programmes will increasingly be tasked with identifying how these types of programmes contribute to student

welfare and academic issues such as grade point average, retention, and overall feelings of having a successful college and university experience. In addition, outdoor programmes will be asked to develop a rationale concerning how these programmes fit into the mission of the college or university. For example, are they essential to the academic growth of the student or simply a somewhat extraneous and "fun" experience? **Probable**

- Virtual nature experiences – These will be used in the classroom as a substitute or alternative to out-of-doors and away-from-campus learning experiences or field-based outings. Reasons for the growth of these experiences will include fewer environmental impacts, lower barriers to entry, such as skill base and level of physicality, higher volumes of people can be accommodated, transportability of the experience, and a lack of a need for transportation, equipment, and instructor skills. As virtual nature experiences improve in visualisation and other variables, the desirability of offering these electronic versions of the outdoor environment will increase. **Possible**

- Planned academic adventures – Some colleges and universities are offering short-term (approximately 2-week) study away/abroad courses that are faculty-led and locally contracted. These experiences cater to students who not only desire an adventure that comes with visiting an exotic location and having a meaningful experience but also want the comfort and security of having a planned itinerary, organised logistics, and a responsible point person, such as a professor, in the event something goes wrong. **Current**

Implications for the future

In this chapter, we identified a number of current and potential trends that are impacting the areas of adventure recreation. We examined these trends within the frameworks of organised programmes and those housed within the structure of higher or tertiary education. In addition, a review of the effects of climate change, the COVID-19 pandemic, and the economy were examined regarding how these issues can and are impacting these fields and the staff delivering experiences. While beyond the scope of this chapter, Cohen et al. (2014) posit that the influences of decision-making, values, motivations, self-identification and personality, expectations, attitudes, perceptions, satisfaction, and past experiences should also be considered in ascertaining emerging trends and issues for the sector.

From this analysis, the major factors that appear to be driving these emerging issues include demographics, responses to the pandemic, externalities such as economics and disposable income, legal concerns surrounding the offering of adventure-based activities, access to available and experienced staff, and locations that are accessible by adventure recreation programmes. In addition, the outcomes associated with climate change will present a varied picture for these programmes. Some areas and subsequent programmes will benefit from a warming climate. An example of this positive change from an adventure recreation perspective will be tours to areas where inclement weather often precludes successful trips. Conversely, warmer weather patterns will increase fire and smoke hazards, reduce water levels in specific rivers and other waterways, and increase the occurrence of heat-related illnesses. For example, at the time of this writing, the Rio Grande near Big Bend National Park in the US is now un-runnable due to low or non-existent water levels.

While forecasting the future is risky at best, it does seem clear that adventure recreation programmes and organisations have and will experience changes in a number of areas. These areas include programme design, the structure of the organisation offering the programmes, available staff and permits, and the number of participants. It may be true that courses become shorter in length, involve less travel and to less remote areas, and include a different set of activities or intensity of activities based on a reduction in the availability of experienced staff. How flexible and anticipatory adventure recreation organisations respond to the variety of extant influences such as changing economies, external threats, changes in natural resources, and other phenomena, will determine in large measure, the success and longevity of these programmes.

Review questions

1 What trends have impacted your personal adventure recreation experience(s)? Did they impact your experience positively or negatively? How so?
2 Do you anticipate any trends impacting the delivery of adventure recreation programmes not identified in this chapter? What might those be?
3 Brainstorm ways the adventure recreation industry can adapt to "future-proof" itself from future disruptors and/or pandemics and address some of the trends.
4 Imagine you are the programme administrator for the GIVE organisation discussed in the case study. How would you go about deciding who remains employed and who is impacted by a reduction in the workforce? What might the repercussions of your decision be?

Acknowledgements

A special thanks to Anne Morrison, Director of Safety for Northwest Outward Bound, Ken Gilbertson, PhD. Professor Emeritus, University of Minnesota – Duluth, and Curt Davidson, PhD., University of Wyoming, for their helpful and insightful comments in the preparation of this chapter.

Further reading

Ewert, A., & Sibthorp, J. (2014). *Outdoor adventure education: Foundations, theory, and research.* Human Kinetics.
Galloway, K. (2021, December 10). *Seasonal guides are speaking up about the stresses of the job.* Outside. https://www.outsideonline.com/health/wellness/seasonal-guide-mental-health-stress/
Owens, M. (2022, April 11). *Gen Z Staff: What exactly are they looking for?* American Camp Association. https://www.acacamps.org/blog/gen-z-staff-what-exactly-are-they-looking
Rice, W. L., Meyer, C., Lawhon, B., Taff, B. D., Mateer, T., & Reigner, N., Newman, P. (2020, April 18). *The COVID-19 pandemic is changing the way people recreate outdoors: Preliminary report on a national survey of outdoor enthusiasts amid the COVID-19 pandemic.* The Pennsylvania State University Department of Recreation, Park, and Tourism Management, and Leave No Trace Center for Outdoor Ethics. https://doi.org/10.31235/osf.io/prnz9

References

Askew, A. E., & Bowker, J. M. (2018). Impacts of climate change on outdoor recreation participation: Outlook to 2060. *Journal of Park and Recreation Administration, 36,* 97–120. https://doi.org/10.18666/JPRA-2018-V36-I2-83
Buckley, R. (2006). *Adventure tourism.*
Bureau of Economic Analysis. (2022). Outdoor Recreation Satellite Account, U.S. and States, 2021: New statistics for 2021; 2017-2020 Updated. BEA. https://www.bea.gov/data/special-topics/outdoor-recreation

Cohen, S. A., Prayag, G., & Moital, M. (2014). Consumer behaviour in tourism: Concepts, influences and opportunities. *Current Issues in Tourism, 17*(10), 872–909. https://doi.org/10.1080/1 3683500.2013.850064

Cuthbertson, B., Socha, T. L., & Potter, T. G. (2004). The double-edged sword: Critical reflections on traditional and modern technology in outdoor education. *Journal of Adventure Education and Outdoor Learning, 4*(2), 133–144.

Ewert, A., & Sibthorp, J. (2014). *Outdoor Adventure Education: Foundations, theory, and research*. Human Kinetics.

Ewert, A. W., Shultis, J., & Webb, C. (2000). Outdoor recreation and technologies: A Janus-faced relationship. *The Proceedings of the 2000 Social Aspects of Recreation Research Conference* (pp. 241–252). https://doi.org/10.1080/14729670485200491

Jensen, C. R., & Guthrie, S. (2006). *Outdoor recreation in America*. Human Kinetics.

Kumar, S., & Deshmukh B. (2021). Adventure tourism market by type, activity, type of travelers, age group, sales channel, and region: Global opportunity analysis and industry forecast 2021–2028. https://www.alliedmarketresearch.com/adventure-tourism-market

Leonard, A. M., Ewert, A. W., Lieberman-Raridon, K., Mitten, D., Rabinowitz, E., Deringer, S. A., & Anderson, I. (2021). Outdoor adventure and experiential education and COVID-19: What have we learned? *Journal of Experiential Education, 45*(3). https://doi.org/10.1177/ 10538259211050762

Office of National Statistics. (2021, April). *Tourism and outdoor leisure accounts, natural capital, UK: 2021*. Office of National Statistics. https://www.ons.gov.uk/releases/uknaturalcapital accountstourism

Outdoor Foundation. (2021). *2021 outdoor participation trends report*. OIA. https:// outdoorindustry.org/resource/2021-outdoor-participation-trends-report/

Outdoor Industry Association. (2021). *2021 special report: New outdoor participant (COVID and beyond)*. OIA. https://outdoorindustry.org/resource/2021-special-report-new-outdoor-participant-covid-beyond/

Owens, M. (2022, April 11). *Gen Z Staff: What exactly are they looking for?* American Camp Association. https://www.acacamps.org/blog/gen-z-staff-what-exactly-are-they-looking

Responder Alliance. (2021). The stress continuum. https://www.responderalliance.com/ stress-continuum

Rice, W. L., Meyer, C., Lawhon, B., Taff, B. D., Mateer, T., Reigner, N., Newman, P. (2020, April 18). The COVID-19 pandemic is changing the way people recreate outdoors: Preliminary report on a national survey of outdoor enthusiasts amid the COVID-19 pandemic. https://doi. org/10.31235/osf.io/prnz9

Spennemann, D. H., & Whitsed, R. (2021). The impact of COVID-19 on the Australian outdoor recreation industry from the perspective of practitioners. *Journal of Outdoor Recreation and Tourism, 41*. https://doi.org/10.1016/j.jort.2021.100445

Statista. (2020). Global-adventure-tourism-market-size. https://www.statista.com/statistics/ 1172869/global-adventure-tourism-market-size/

Taff, B. D., Rice, W. L., Lawhon, B., Newman, P. (2021). Who started, stopped, and continued participating in outdoor recreation during the COVID-19 Pandemic in the United States? Results from a national panel study. *Land, 10*, 1396. https://doi.org/10.3390/land10121396

Turner, J., Jostad, J., Andre, E., Bell, B., Collins, K. C. Gerbers, K., & Hobbs, W. (2022). Overview of the current Landscape of outdoor programs in higher education [Abstract] *Proceedings of the Coalition for Education in the Outdoors 15th Biennial Research Symposium*. https://doi. org/10.18666/JOREL-2022-11595

van Kraalingen, I. (2021). A systematized review of the use of mobile technology in outdoor learning. *Journal of Adventure Education and Outdoor Learning*. https://doi.org/10.1080/14729679 .2021.1984963

Wengel, Y. (2021). The micro-trends of emerging adventure tourism activities in Nepal. *Journal of Tourism Futures, 7*(2), 209–215. https://doi.org/10.1108/JTF-01-2020-0011

5 Quantitative research methods in adventure tourism

Tahir Albayrak and Meltem Caber

Chapter learning outcomes

1 Understand the philosophical bases of scientific paradigms and their main research methods.
2 Learn about quantitative research methods and their usage in adventure tourism-related studies.
3 Comprehend the methodological summary of the scientific articles that used quantitative methods in adventure tourism-based topics.
4 Gain a comprehensive perspective on adventure tourism and an understanding how scholars applied quantitative research methods in this context.

Introduction

Social science researchers have a range of scientific approaches to the study of social life, known as paradigms. Each paradigm has components of epistemology, ontology, and methodology that are defined in this chapter. Therefore, the adopted paradigm is a determinant of the research problem and how it is handled by the researchers. Relying on the epistemological, ontological, and methodological elements of a specific paradigm, researchers attempt to solve the problem and propose the theoretical or practical implications of their findings. In this chapter, social research paradigms and their components are first introduced to understand the current and most common research perspectives in the adventure tourism context. The introduction of the quantitative research methods and a review of studies that followed a quantitative methodology in adventure tourism literature in the period 1996–2022 provide the key themes of this chapter.

Research paradigms in social sciences

To understand the development of scientific approaches in certain tourism-related fields, such as adventure tourism, it is useful to first recall scientific research paradigms and methods that have been used to date. In principle, good research questions in social sciences can only be answered when the researchers determine the answers to 'what should be the focus of the research' and 'what data and methodology would be most helpful in answering the research questions' (Wright et al., 2004, p.748). Moreover, it would be incorrect to assume that scientific research is carried out without being affected

DOI: 10.4324/9781003393153-7

in any way by the researchers' point of view of the world and phenomena. This fact is conceptualised as a paradigm in science. A *paradigm* is defined as 'a way of viewing the world' (DeCarlo, 2018, p. 144) and 'a framework from which to understand the human experience' (Kuhn, 1970, p.152). In scientific studies, a paradigm is considered as 'a set of propositions that are held and shared by members of an academic area of study to be self-evident about the nature of truth and scientific knowledge in their discipline' (Boukezzoula, 2019, p. 453). Research paradigms are important since they show the researchers' perspective on an issue or phenomenon and play a significant role in underpinning their approaches to methodologies (Williams, 2020). The essential components of the paradigms are *ontology, epistemology, methodology, and methods.* Figure 5.1 illustrates the paradigm components and which broad question(s) each of these is trying to respond to.

Ontology means the study, theory, or science of being which is concerned with what is or the nature of reality. It is classified as objectivism and subjectivism. According to *objectivism*, social entities exist externally to the social actors who deal with them. On the contrary, *subjectivism* defends that there are social phenomena created from the perceptions and actions of the social actors who are concerned with their existence (Sheppard, 2020). *Epistemology* is concerned with 'the nature and forms of knowledge, how it can be acquired and how it can be communicated to other human beings' (Cohen, Manion, & Morrison, 2007, p. 7). This concept tries to describe a set of information and bridge the gap between perceived and actual reality. *Methodology* means how the researchers carry out a study by revealing the information and involves some comprehensive actions, such as deciding on research questions and methods (Fazlıoğulları, 2012). This refers to the research design, methods, approaches, and procedures (Keeves, 1997, cited in Kivunja & Kuyini, 2017). The methodological choices of the researchers are dependent on the 'strategy, plan of action, process or design' of the research (Crotty, 1998, p. 3). The *methods* 'produce data, which brings the information that will constitute the knowledge and would fit in the theory or would not' (Iofrida et al., 2018). The relationships between ontology, epistemology, methodology, and methods are visualised in Figure 5.2.

In social sciences, several predominant paradigms exist that own unique ontological and epistemological perspectives (Sheppard, 2020). In this chapter, the most predominant paradigms used in the adventure tourism context are presented with their supporting ontological and epistemological perspectives (e.g., Makombe, 2017; Scotland, 2012), which are namely (a) positivism/post-positivism, (b) an interpretivist approach, (c) a critical approach, and (d) a pragmatic approach.

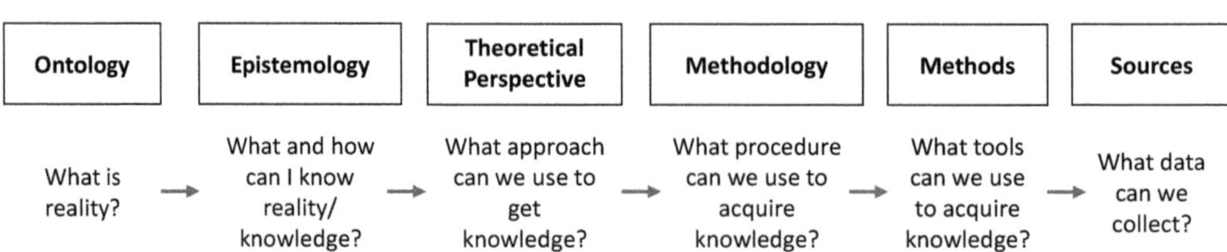

Figure 5.1 The components of a paradigm. Adapted from Patel (2017).

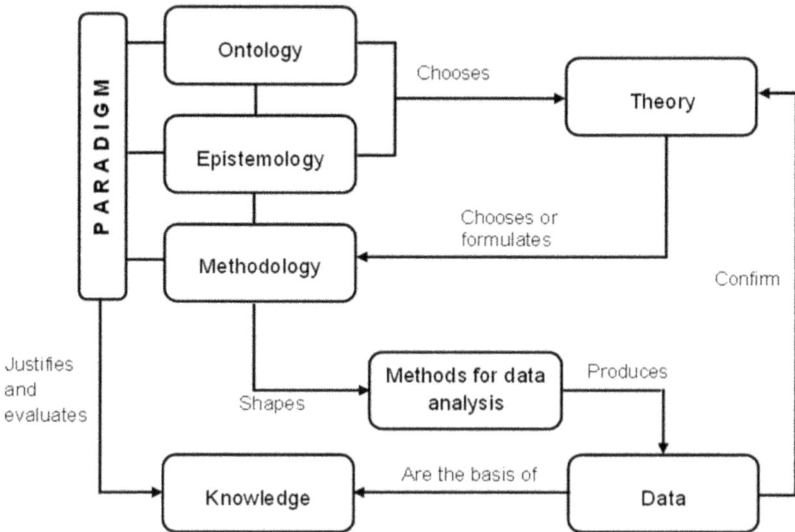

Figure 5.2 The relationships between ontology, epistemology, methodology, and methods. Adapted from Iofrida et al. (2018).

Positivism/post-positivism

Positivism, which is also known as the "scientific method" or "science research," is 'based on the rationalistic and empiricist philosophy that originated with Aristotle, Francis Bacon, John Locke, August Comte, and Emmanuel Kant' (Mackenzie & Knipe, 2006; Mertens, 2005, p. 8). Scientific studies which are in this paradigm own a deductive logic, and they use the formulation of hypotheses/tests, operational definitions, mathematical equations, statistical calculations, and extrapolations to obtain conclusions (Kivunja & Kuyini, 2017). Therefore, positivist paradigm–based studies use quantitative research methods. The ontology of positivism relies on realism, which presumes that reality is objective, quantifiable, and measurable. Supernaturalism is rejected and knowledge of a social phenomenon is accepted to be observable, recordable, and measurable, like the natural sciences. From the epistemological point of view, this approach treats researchers and the objects under study as different entities, mutually unaffected by each other (Ugwu et al., 2021). Whereas positivism is based on the verification of hypotheses, post-positivism deals with the falsification of hypotheses (Makombe, 2017). Post-positivism emerged at the beginning of the 20th century due to some critical reflections and amendments to positivism (Guba & Lincoln, 1994). O'Leary (2004, p.6) notes that post-positivists see the world as a multi-reality environment, adapting the view that 'what might be the truth for one person or cultural group may not be the truth for another'.

Interpretivist approach

The interpretive approach has the intention of understanding the research participants' views about the situation being studied since it recognises the impact of their backgrounds and experiences (Mackenzie & Knipe, 2006). The ontology of interpretivism is relativism, which suggests that scientific knowledge should be developed from the perspective of the subject people who are directly involved in a situation. In addition, it uses a subjectivist epistemology, by assuming that the researchers generate the meaning of their data from their own thinking informed by their interactions with the research participants (Kivunja & Kuyini, 2017). Qualitative research methods are considered interpretive, and the data collection and analysis can occur concurrently (Makombe, 2017). Since it is accepted that embedded reality can only be discovered by experiencing how concepts and theories are positioned in their natural environment, the most used study methods have been phenomenology, ethnography, action research, case study, participative observation, narrative inquiry, hermeneutics, content analysis, and field study (Demir, 2019; Kivunja & Kuyini, 2017). By its nature, this paradigm is seen as the opposite of positivism. You can read more about qualitative approaches to research in Chapter 6.

Critical approach

Unlike the positivist paradigm, the critical paradigm suggests that social sciences can never be totally objective since there are power, inequality, and social change issues in the world. It particularly uses a biased epistemology both for exploring the points of view of the oppressed, excluded, and exploited social groups and for defining social structures in society (Creswell, 2009). Researchers who adopt the critical paradigm usually prefer qualitative research methods (Akış Roney, 2011). Some of the study types based on this paradigm have been critical ethnography, feminist research, participatory action research, critical discourse analysis, cultural studies, participatory emancipation, and postcolonial/indigenous study (Çıvak & Sezerel, 2018; Kivunja & Kuyini, 2017).

Pragmatic approach

The pragmatic approach, which is also known as methodological eclecticism, ignores the differences in paradigms and proposes the use of mixed methods (Melosik, 2021). The researchers who adopt this approach may use both quantitative and qualitative methods (mixed methods) for understanding human behaviours. The ontology of the pragmatic paradigm rejects pure reality and advocates the search for individual-based interpretations. It uses relational epistemology by presuming that human behaviours and social relationships should be determined by case-based research methods which are selected as the best by the researchers (Ugwu et al., 2021). Commonly used study types include the naturalist study, narrative inquiry, case study, phenomenology, ethnography, action research, experimental study, quasi-experimental study, and causal-comparative study (Kivunja & Kuyini, 2017). Table 5.1 shows the main characteristics of these paradigms.

In the adventure tourism context, we can examine Beckman, Whaley and Kim's (2017) study ontologically, epistemologically, and methodologically as an example. This research aims to explore whitewater rafting tourists' behaviours, specifically investigating the relationships that adventure motivations have with adventure experiences (i.e. emotional response and place attachment) and behavioural outcomes (i.e. revisit intention and word of mouth). The researchers follow an objective and realistic ontology, and they propose a conceptual model relying on the edgework theory (Thompson, 1971). The epistemology is detached, and the research design is empirical since the research data were methodologically collected by a survey and the conceptual model was then tested by statistical analyses.

After presenting the philosophical and conceptual foundations of scientific research, we can move on to research methods. Research methods are categorised as qualitative,

Table 5.1 The paradigms and their characteristics (Makombe, 2017)

Research methodology	Paradigms				
		Positivist/ post-positivist	*Interpretivist*	*Pragmatic*	*Critical*
	Ontology	Realism/critical realism Objectivity	Relativism Subjective Historical Constructed reality (Pragmatism has some objectivity)		Relativism Subjective–Objective Constructed and historical reality
	Epistemology	Detached	Transactional Participatory	Mix detached and participatory in predetermined sequence	Transactional, Experiential
	Approach	Empirical	Normative Advocacy Activism (Pragmatism mixes empirical and normative in a predetermined sequence)		
	Research Method	Quantitative (with statistical representativeness a necessary condition for generalisation; Scientific method)	Qualitative (statistical representativeness not always a requirement)	Quantitative and Qualitative (statistical representativeness not always a requirement)	Qualitative Cooperative inquiry Collaborative/democratic dialogue
	Research Design(s)	Experimental Descriptive Case control Case study Causal Cross section Exploratory Longitudinal Observational	Descriptive Narrative Case study (Single/Multiple) Exploratory Historical (life/topical oral) Observational (participant/non-participant) Philosophical Dialectic (Pragmatism can have components of quantitative research designs)		Action Research Epistemic/political participation determines design

(*Continued*)

Table 5.1 (Continued)

Research methodology	Paradigms			
		Positivist/ post-positivist	Interpretivist	Pragmatic	Critical

		Positivist/ post-positivist	Interpretivist / Pragmatic	Critical
	Research Guide	Research questions and hypotheses Describe, control and predict Anti-Speculative ideals Caution, clarity, and precision	Sometimes research questions and hypotheses but mostly research questions only	Research questions with intended action
	Principle	Uncover the universal laws (which exist) governing social events	Describe, explain, and understand meanings, values, and beliefs of social phenomena from (sometimes with) participants (experiential, contextual, historical, local, specific) and researcher's perspectives	Co-creation of knowledge Subjects are participants and sometimes co-researchers
	Researcher's Posture	Objective Detachment or value freedom Measurement and testing Deterministic	Subjective Can be interactive Relativism/multi-perspectives Researcher can be immersed Integration of knowledge and values Insight and intuition (Objective detachment not necessary but still a possibility) Research subjects can become researchers/ co-researchers Blurry distinction between researcher and researched.	Blurry distinction between researcher and researched Participants are co-researchers

quantitative, and mixed methods (Creswell, 2009; Makombe, 2017). The "quantitative" term refers to only 'the activity of quantification; while the "qualitative" term describes what is examined in-depth without being linked to a research paradigm' (Symonds & Gorard, 2008, p. 17). "Mixed" research uses both quantitative and qualitative methods. This chapter specifically deals with quantitative methods used in the adventure tourism context, as qualitative methods are covered in the 'Chapter 6. Therefore, further details about quantitative research methods are presented in the next section.

Quantitative research methods

Bryman and Bell (2015, p. 160) define quantitative research methods as 'entailing the collection of numerical data and exhibiting the view of the relationship between theory and research as deductive, a predilection for natural science approach, and as having an objectivist conception of social reality'. Quantitative research methods in the social sciences are used to obtain generalisable knowledge for large populations, based on the assumption that social phenomena have an objective reality (outside the subjective perspectives of researchers; Szyjka, 2012). The seven key principles of quantitative research methods are given by Regoniel (2015): (1) contain measurable variables (such as age, the number of children, educational status, etc.); (2) use standardised research instruments (for ensuring the accuracy, reliability, and validity of data); (3) assume a normal population distribution (a normal data distribution curve is preferred over a non-normal, which requires a large population and random sampling to avoid researcher's bias in interpreting the results); (4) present data in tables, graphs, or figures; (5) use repeatable method (for reinforcing the validity of findings); (6) can predict outcomes (quantitative models or formula derived data analysis can predict outcome); and (7) use measuring devices (calibrated instruments ensure an objective and accurate collection of data). Accordingly, following the main principles of the positivist paradigm, the studies that use quantitative methods are usually validated by applying four criteria: internal validity, external validity, reliability, and objectivity (Lincoln & Guba, 1985). Although the literature reviews on research methods in tourism showed that there was a methodological superiority in favour of the positivist paradigm using quantitative methods, and this trend is still valid, the number of studies using the pragmatic paradigm based on qualitative and mixed methods has increased in the last decade (Nunkoo, 2020).

Study types in quantitative research methods

Quantitative research methods are divided into two groups: experimental and non-experimental. 'The main feature that distinguishes non-experimental studies from experimental studies is the lack of random assignment' (Rogers & Révész, 2020, p. 134). In experimental studies, random assignment is about placing participants from the research sample into different treatment groups by using randomisation. It means every member of the research sample has a known or equal chance of being placed in a control group or an experimental group (Bhandari, 2022). That helps improve the internal validity of an experiment and avoid biases. At that point, it should be noted that 'random assignment' and 'random sampling' are not the same things. Random assignment is dividing sample participants into control or experimental groups, while random sampling (also called probability sampling or random selection) is selecting people from a population of interest to include in research. Another difference between experimental and non-experimental studies is their use of research variables. Experimental studies can control and manipulate the independent variables, while non-experimental studies do not. In experiments, the researchers manipulate an independent variable for being able to measure its effect on a dependent variable, when controlling for other variables. In this process, the researchers mostly use different levels of an independent variable for different groups of participants, which is named as 'between-groups' or 'independent measures' design (Bhandari, 2022). In non-experimental studies, the researchers measure the variables in their natural state without any manipulation. These types of studies are appropriate when manipulating the independent variable is not possible or the researchers do not aim to test a causal relationship between different variables.

Experimental studies

True experimental studies 'involve the manipulation of one or more independent variables, where the dependent variables are measured in the form of pre- and post-testing. These studies are also performed on a control group and an experimental group' (Rogers & Révész, 2020, p. 133–134). In the adventure tourism literature, Su et al.'s (2022) research, which examined the effect of having a companion on adventure tourists' satisfaction and subjective well-being, is a good example of a true experimental study. The researchers conducted three situational experiments (Studies 1, 2, and 3) containing manipulation-based scenarios and an online survey on tourists visiting a national park in China. In Study 1, the researchers first examined whether the presence of a companion impacts tourist satisfaction and subjective well-being. This was a one-factor between-subjects experimental design (companion vs. no companion). In Study 2, a one-factor between-subjects (greater vs. comparable vs. lower relative ability) and a 2 (female vs male) × 3 (greater vs. comparable vs. lower relative ability) factorial between-subjects design were used. In Study 3, the same factorial between-subjects design was used as in Study 2. The companion's relative ability is the independent variable, satisfaction and subjective well-being are the dependent variables, and tourist gender is the mediator in these relationships. The authors showed that having a companion with greater/comparable relative ability increases tourist satisfaction and well-being. This effect is also higher for female tourists.

Quasi-experimental studies 'are used in natural settings when some control over the experimental conditions can be exerted, yet full control is either not possible or not desirable' (Hank & Wildemuth, 2017, p.91). In the adventure tourism context, Chhetri et al.'s research (2004) can be given as an example of quasi-experimental studies. In this study, the researchers divided the hiking university students into two groups and directed them to two different hiking routes in Grampians National Park, Australia. The authors empirically measured the participant experiences (consisting of various subjective meanings such as moods, emotions, and feelings) and compared the behavioural hiking experiences of the groups. By first using the multidimensional scaling (MDS) method, they identified two dimensions of experience: Dimension 1: "negative–positive experience" and Dimension 2: "intrinsic–extrinsic" related to geographic stimulus variables affected participant experience. Then a matrix has been prepared to show the differences between groups. Second, by using principal components analysis (PCA), they explored the experiential variables, such as attracting, relaxing, and exciting. Following this, the variation of experience variables by three types of landscapes (views, open spaces, forest stands) was examined by the analyses. In this study, the researchers controlled the group sizes and manipulated the walking routes.

Non-experimental studies

Some of the most used non-experimental studies are as follows (Khaldi, 2017):

Descriptive studies portray the characteristics of persons, situations, or groups and the frequency with which certain phenomena occur (Dulock, 1993). In a study (Valentine et al., 2004), the researchers investigated swim-with-whales operations based on the dwarf minke whales in the Great Barrier Reef of northern Queensland. They first examined the demographic characteristics of the tourists (e.g. mean of age, percentage of nationalities, percentage of experience levels). Following this, they calculated the frequencies and percentages of the most frequently mentioned elements of the respondents' expectations (e.g. get as close to whale[s], see whale[s]). Finally, tourist experiences were determined by grouping the closest distance approached by a minke whale (e.g. less than 2 m, between 2 and 4 m) and by grouping what the respondents were doing when the closest approach occurred (e.g. in-water holding rope, observing from on-board the boat).

Causal-comparative studies show the relations between the investigated variables to identify possible causal relationships. In Wei-Ching's (2022) study, the researcher performed an on-site survey to understand hikers' air pollution perceptions, stresses induced by air pollution, and coping behaviours. The authors compared the behaviours of the hikers from two cities (Xiangshan and Shoushan). The results showed that although the air quality of Xiangshan was better than Shoushan, Xiangshan hikers were more likely to use absolute substitution than Shoushan hikers when facing air pollution. The results of the regression analysis also indicated that the perception of air pollution and stress associated with air pollution were correlated with coping behaviours. Hikers tended to use behavioural coping strategies rather than cognitive coping strategies.

Correlational studies establish possible but not necessarily present relationships between variables. In Ong and Musa's study (2012), the researchers examined the scuba divers' behaviours. Their conceptual model proposed the direct and positive relationships between environmental concern and divers' underwater responsible behaviour and between specific scuba diving attitudes and divers' underwater responsible behaviour. Moreover, scuba diving attitudes were hypothesised to play a mediator role between environmental concern and divers' underwater responsible behaviour. Both correlation and regression analyses were used to examine the relationships among the variables. The results indicated that divers were mostly eco-centric and had highly responsible underwater attitudes. Divers' behaviour underwater had a direct relationship with both environmental concern and specific scuba diving attitudes. The mediator role of scuba diving attitudes was also confirmed in the relationship between environmental concern and responsible behaviour underwater.

Ex post facto studies are initiated without the intervention of the researcher and after the effect of an event/factor has occurred. For example, in adventure tourism literature, Zimmerhackel et al. (2018) investigated how shark conservation in the Maldives affected dive tourism demand. The results showed that increased shark abundance could increase tourism demand for diving, resulting in greater economic gains for the tourism industry. Moreover, study findings underlined the importance of shark diving for tourist perceptions and the future of the country's tourism industry.

Surveys describe the views of a large group of people at a given point in time. An example of this type of study from the adventure tourism literature belongs to Barbieri and Sotomayor (2013), who investigated surfer behaviours and preferences with a survey questionnaire. The researchers measured the perceptions of the seriousness of surfing, surf travel behaviours, and destination preferences of the participants on a 5-point scale (1 – *very unimportant*, 2 – *unimportant*, 3 – *neutral*, 4 – *important*, 5 – *very important*). Findings (based on arithmetic means and percentages) show that surfers give importance to various attributes when selecting a tourism destination and among those, the most important ones are related to the overall surfing appeal of the destination, the variety of waves, and the quality of the natural environment.

The widely used survey types in quantitative studies are (1) cross-sectional surveys which measure the exposure and the outcomes belonging to study participants at the same time, who are selected based on the inclusion and exclusion criteria set for the study (Setia, 2016), and (2) longitudinal surveys where quantitative data are collected continuously or repeatedly from particular individuals over prolonged periods (Caruana et al., 2015). Longitudinal surveys are also grouped under two categories: cohort surveys which 'follow the same group of individuals over time' and trend surveys which take 'repeated samples of different people each time but always use the same core questions' (Mathers et al., 2007, p. 5).

A systematic review of quantitative studies on adventure tourism

This section presents the findings of a non-experimental descriptive study, in which the scientific papers using quantitative research methods in adventure tourism were reviewed and the results were presented analytically with the aim of summarising the 'status quo' in the adventure tourism literature. Like previous literature reviews (Doran et al., 2022; Pickering & Byrne, 2014; Yang et al., 2017), a systematic review process consisting of the identification of research objectives, formulation of review protocol, literature search, extraction of the related literature, and synthesis of findings, was followed in this research. To formulate the review protocol, firstly search terms were identified based on the adventure tourism classification of Pomfret (2006). Moreover, the identified adventure activities in the literature review of Rantala et al. (2018; climbing, rafting, mountaineering, kayaking, diving, wildlife, hiking, skiing, biking, surfing, horse riding, safari, sailing, bungy jump) were used as the search terms on Web of Science. Based on the guideline of Preferred Reporting Items for Systematic Reviews and Meta-Analyses (PRISMA), the selection of the relevant manuscript was summarised in Figure 5.3. By focusing only on the "hospitality leisure tourism" category without a timeframe, 3188 results were identified. Then only English-language articles that were published in scientific journals indexed by Social Sciences Citation Index (SSCI) were considered. Meeting abstracts, book reviews, editorial materials, proceeding papers, letters, notes, corrections, and core sports journals (e.g. *Sports in Society*) were excluded. As a result of this process, 784 records were obtained.

Since the objective of this chapter is to identify quantitative data analyses used by adventure tourism-based articles, two researchers independently evaluated all records and selected only papers that used quantitative data analysis. Finally, 121 papers remained for the analyses.

For evaluating the quantitative data analyses performed in the selected papers, a codebook covering the research methods and statistical analyses was created. Then two evaluators independently scanned all the papers and recorded the information (such as

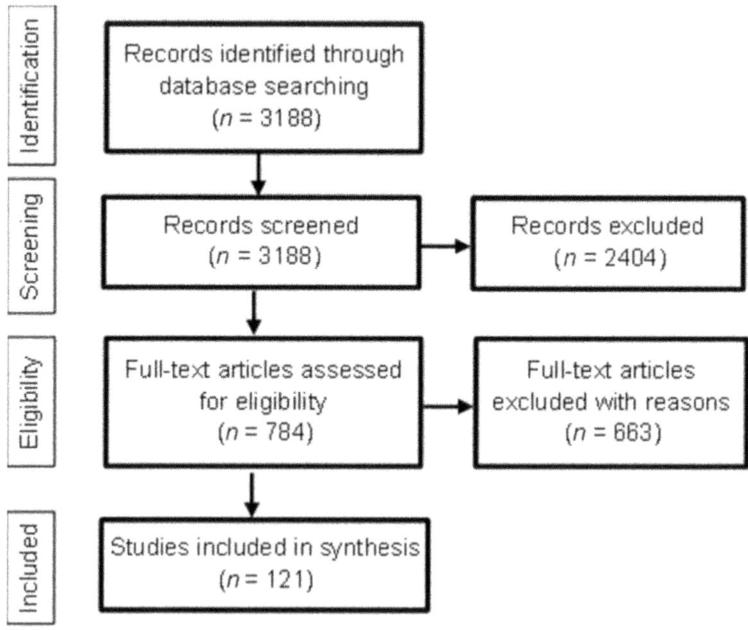

Figure 5.3 The PRISMA flowchart. Adapted from Moher et al. (2009).

methods and analyses) to the Excel sheet configured according to the codebook. Tables 5.2 and 5.3 reflect the general information about the selected papers. Accordingly, most of the reviewed papers were published between 2011 and 2020 in *Tourism Management* and *Current Issues in Tourism*. Skiing was the most investigated adventure activity by reviewed papers, followed by hiking, and scuba diving. Moreover, 13 papers investigated adventure tourism in a general manner without focusing on a specific activity. The majority of the studies were conducted in Australia, China, and the US, respectively.

Table 5.2 Adventure tourism papers using quantitative research methods in key journals and their publication years

Journals	1996–2000	2001–2010	2011–2020	2021–2022	Total
Tourism Management	2	8	18		28
Current Issues in Tourism		1	11	3	15
Journal of Outdoor Recreation and Tourism			5	7	12
Journal of Travel & Tourism Marketing			8	1	9
Scandinavian Journal of Hospitality and Tourism			6	1	7
Asia Pacific Journal of Tourism Research			6		6
Others	1	3	33	7	44
Total	3	12	87	19	121

Table 5.3 Literature information by adventure activities and research location

Adventure activity	f	Research location	f
Skiing	23	Australia	11
Hiking	14	China	10
Scuba diving	13	US	9
Adventure activities (general)	13	Taiwan	8
Mountaineering	9	Norway	7
Rafting	8	Austria	6
Surfing	6	Malaysia	6
Multiple activities	5	UK	6
Others	30	Multiple countries	7
		Others	51

Review findings

Paradigm used by reviewed papers

The literature review on adventure tourism presented in this chapter shows that most quantitative studies are non-experimental and based on the positivist paradigm (92 papers, 76%). The studies were mostly carried out with surveys in which questionnaire forms were used. The pragmatic approach was identified as another paradigm used by the reviewed papers (29 papers, 24%). Mixed research designs, combining qualitative and quantitative methods, are beneficial for pragmatist scientists as they can be adapted to exploratory-type studies. However, the ratio of experimental studies is currently few (7 papers) in all papers using mixed research methods, whether they are true experimental (1 paper) or quasi-experimental (6 papers).

Research methodologies

For answering the research questions, data should be collected from the members of a population (Taherdoost, 2016). There are two options for these: (1) probability (or random) sampling or (2) non-probability (or non-random) sampling. If the researchers decide to do probability sampling, they may select one of the sub-techniques such as simple random, stratified random, cluster sampling, systematic sampling, and multi-stage sampling. For the non-probability sampling, they may select to perform sub-techniques such as quota sampling, snowball sampling, judgement sampling, and convenience sampling. Among those, convenience sampling is one of the easiest since the participants are often readily available or reachable. The examination of the methods used by the selected papers in this study revealed that the most commonly used sampling method (83 papers, 68.7%) was convenience sampling (Table 5.4). However, nearly half of these (37 papers) only presented information about their sampling method and the name of the sampling method. Probability sampling methods were rarely (9%) applied by the researchers. Interestingly, 6.6% of the papers did not present any information about the sampling method that they used or any details about it.

Survey-based research can be done either in a paper-and-pencil style or online. Paper surveys are conducted face-to-face with voluntary participation of the targeted samples in the field of research (Figure 5.4), whereas online surveys are performed at online data collection platforms, such as Amazon Mechanical Turk (MTurk) and social media pages (e.g. Facebook). This systematic literature review's results show that most of the researchers

Table 5.4 Sampling methods and measurement scales

Sampling method	f	percentage	Measurement scale	f	percentage
Non-probability sampling			4-point scale	3	2.5
Convenience sampling	83	68.7	5-point scale	53	43.8
Purposive sampling	7	5.8	6-point scale	2	1.7
Snowball sampling	6	5.0	7-point scale	33	27.3
Quota sampling	2	1.6	9-point scale	2	1.7
Probability sampling			10-point scale	4	3.3
Simple random sampling	5	4.2	11-point scale	1	0.8
Systematic sampling	4	3.2	Multiple scale (e.g. 5 & 7-point)	9	7.4
Stratified sampling	2	1.6	Not applicable	13	10.7
No information	8	6.6	No information	1	0.8
Not applicable (e.g., students)	4	3.3			

Figure 5.4 A paper survey application with scuba divers in Türkiye. Photograph by Tahir Albayrak.

preferred paper surveys (111 articles, 91.8%), while online surveys were relatively few (7 articles, 5.8%), although this may now be changing, given the ease of online delivery. Just in two papers, both paper and online surveys were performed by the researchers.

In order to understand the general characteristics of quantitative research in the adventure tourism literature and reveal how the methodological issues were fulfilled, both the sample sizes and sample characteristics are examined in this study. To determine the sample size of a study, predetermined tables (prepared by Cochran, 1977;

Krejcie & Morgan, 1970) or formulas can be used which consider desired confidence and precision levels, as well as the estimated proportion of the investigated attribute in the population (Israel, 1992). If the sample size is between 30–500, parametric analyses techniques (that are based on assumptions about the distribution of the population from which the sample was taken); otherwise, non-parametric analysis techniques (that are not based on assumptions, therefore, the data can be collected from a sample that does not follow a specific distribution) can be used. The results of this study revealed that the average sample size in the reviewed papers ranged from 25 to 14,861 with a mean of 647 (*SD* = 1473). Since the presentation of sample characteristics is vital for the generalisability of the findings, the papers were also evaluated for this criterion. Typical sample characteristics to present in the research results may include gender, age, income level, and education level. These are important to mention since they show how representative the sample is of the population. There is an ongoing debate among scientists that the heavy use of college students in the social sciences as convenience samples limits the generalisability of the findings to the research population (Peterson & Merunka, 2014). Hence, two papers were excluded due to using students as the study sample. Most of the papers (91.7%) in this review were seen to present their sample characteristics. However, 6.6% of the evaluated papers did not provide any information about their sample characteristics.

Another facet of the analysis is related to the measurement of the scale items (Table 5.4). Each variable in a quantitative study can be empirically measured by a scale which contains the items. These items are one-by-one rated by the study participants relying on how close they are to their perspectives. The 5-point, 7-point, or 9-point Likert types of scales contain 5, 7 or 9 response options. In this literature review, 5-point Likert-type scales (43.8%) have been identified as more prevalent in the measurement of the scale items, followed by 7-point scales (27.3%). For coding of the scale, most researchers used values starting from 1 (e.g. 1 to 5, 1 to 7), while a researcher coded from –2 to +2 where 0 represents 'neither' option (Needham & Rollins, 2005). Surprisingly, in one study (Carnicelli-Filho et al., 2010), the authors mistakenly mentioned using a 5-point Likert scale, even though they coded a 6-point (i.e. 0, *no fear*, and 5, *extreme fear*). Some researchers (7.4%) preferred to use multiple scales in the same survey. For example, in den Breejen's (2007) study where day-by-day experiences in long-distance walking routes were examined, survey participants' motivation, experience, and focus attention were measured by a 5-point Likert scale (1 = *strongly disagree*, 2 = *disagree*, 3 = *neutral*, 4 = *agree*, 5 = *strongly agree*), whilst the participants' perceived identity was measured by a 7-point semantic differential scale (experienced/inexperienced; occasional/frequent; and serious/casual). A study by Stemmer et al. (2022) used 5-, 7-, 9-, and 11-point scales in the same study. However, none such articles provides information on whether they convert or equate these different scales into a single scale type.

Translation of the scale items from one language to another was also evaluated in this review study. In 21 papers, the researchers stated that they translated the scale items from English to another language, while only 8 of them performed a back-translation process as suggested by Brislin (1970). Moreover, 51 papers (42.2%) did not provide any information about the translation process. Reviewed studies were also evaluated whether they conducted a pilot study or not since it is an essential initial step of research. Because the pilot study, which is conducted on a small-scale basis, tests the feasibility of the proposed larger study. The results revealed that only 38% of the reviewed papers conducted a pilot study before the main survey.

Table 5.5 The scale item translations and response bias information

Translation of the items	f	%	Response bias	f	%
English	27	22.3	Response bias	20	16.5
Yes (no back translation)	13	10.7	Bias as a limitation	17	14.0
Yes (with back translation)	8	6.6	No information	79	65.3
No information	51	42.2	Not applicable	5	4.1
Not applicable (e.g. scale development)	22	18.2			

In quantitative research, response bias is a problem that reflects research participants' tendencies to respond in a way that is unrealistic, inaccurate, distorted or socially desirable mostly when they are asked to self-assess their behaviours. They may also prefer to give the same response or leave many of them blank if they do not understand the questions. This situation may occur when the measurement tool and survey design are weak (common method bias) and when the participants want to please the researcher or depending on their perception of the researcher (researcher bias). Since response bias is one of the main reasons for measurement errors (Yuksel, 2017) and is a negative fact against the reliability and validity of the findings, it was additionally investigated in this study. Surprisingly, only 20 papers (16.5%) presented information about how they handled different types of response biases (e.g. common method bias, non-response bias, researcher bias), while 14% of the articles acknowledged response bias as a limitation. In addition, almost two thirds of the papers (65.3%) did not provide any information about the potential response biases that may occur in the research (Table 5.5).

Data analysis procedures

Table 5.6 reflects the data analysis procedures used in the 121 papers, evaluated in the current literature review. Inspection of the analyses used by the papers showed the predominant use of descriptive statistics (72.7%). Researchers also widely conducted factor analysis (37.2%, EFA: exploratory factor analysis or PCA: principal component analysis) to obtain dimensions representing multivariate data. In addition, the correlation coefficient, as the measure of association, was the most common (30.6%) in the papers. Moreover, Pearson's chi-square statistic was used in 23.9% of the papers for testing the relationships between categorical variables.

While confirmatory factor analysis (CFA) and EFA are conducted to explore the underlying factors of a variable which is shown to have a multidimensional characteristic by scholars; other analyses are performed to test possible relationships among the research variables. If the data have a normal distribution curve, then the researchers use parametric analyses, such as one-sample *t*-test, independent-sample *t*-test, paired-sample *t*-test, analysis of variance (ANOVA), analysis of co-variance (ANCOVA), and multivariate ANOVA (MANOVA)–multivariate ANCOVA (MANCOVA; as shown in Table 5.6). In the case of non-normal distribution data, scholars use non-parametric analyses, such as Mann–Whitney *U* test and Kruskal–Wallis test.

For example, in Doran et al.'s study (2020), where the strategies of women in negotiating mountaineering participation constraints and the benefits that they obtained from participation were investigated, first EFA and then (for confirming the results of EFA) CFA analyses were performed by the researchers to find out the factorial structure of

Table 5.6 Data analysis procedures (N = 121)

Data analysis	f	Percentage
Descriptive statistics (*M, SD*, percentages etc.)	88	72.7
Measures of association		
Correlation coefficient	37	30.6
Tests of contingency tables and proportion comparison		
Pearson's chi-square	23	23.9
Cross-tabulation without a test	2	1.6
Fisher's coefficient	1	0.8
Non-parametric tests for comparison		
Mann–Whitney *U* test	1	0.8
Kruskal–Wallis test	4	3.3
Parametric tests for comparison		
One-sample *t*-test	4	3.3
Independent-sample *t*-test	10	8.3
Paired-sample *t*-test	3	2.5
ANOVA	24	19.8
ANCOVA	2	1.6
MANOVA–MANCOVA	2	1.6
Psychometric: Factor analysis		
Exploratory factor analysis (EFA)	23	19.0
Principal component analysis (PCA)	22	18.2
Models to analyse relationships among variables		
Linear regression	19	15.7
Logistic regression	7	5.8
Other regression models (Poisson, quantile, hurdle etc.)	9	7.4
Structural equation modelling (SEM)	31	25.6
Correlation coefficient	3	2.4
Asymmetry analyses (PRCA, AIPA, IRPA etc.) *	4	3.3
Hybrid choice model	1	0.8
Other analyses		
Cluster analysis	18	14.9
CHAID analysis	1	0.8
Multidimensional scaling	1	0.8

* PRCA: penalty reward contrast analysis; AIPA: asymmetric importance-performance analysis; IRPA: impact range performance analysis.

both constraint negotiation strategies and participation benefits. Findings showed that there are four dimensions of constraint negotiation ('time and prioritisation,' 'preparation and planning', 'confidence and adaptation', 'determination') and three dimensions of participation benefit ('fulfilment and achievement', 'freedom and self-interest', 'socialisation and bonding'). CFA is conducted using the 'structural equation modelling' (SEM) technique, which runs through AMOS or Smart PLS software, developed to test complex and multi-variable interactions in social sciences. Following EFA and CFA, structural relationships are tested (hypothesis testing) between the variables by structural modelling in SEM. Although both experimental and non-experimental data can be used in SEM, it requires a large size of the data set. It is an easy-to-use software that shows multiple statistical relationships simultaneously through visualisation and model validation (Dash & Paul, 2021). This technique is rather useful for confirmatory studies since it allows the analysis of the statistical relationships between observable (manifest) and non-observable (latent) variables (Fordellone & Vichi, 2020). Based on these statistical advantages, SEM is widely used by scholars working on adventure tourist behaviours. Another technique is called 'partial least squares structural equation modelling' (PLS-SEM) which is mostly used for exploratory studies that examine multi-variable relationships like SEM. Despite confirming the proposed interactions in a research model like in SEM, PLS-SEM aims to maximise explained variance in the dependent variables and to evaluate additionally the data quality depending on the measurement model characteristics (Hair et al., 2017). It also has the advantage of using smaller data sets than SEM.

Since participants in adventure tourism may have varying demographic and behavioural characteristics, examination of different market segments or participant groups provides valuable findings to researchers. To do that, researchers may employ cluster analysis so that they can define the groups and analyse the differences between separate clusters/groups. With the purpose of group comparisons, this literature review shows that parametric tests (e.g., tests and ANOVA) were mostly preferred by the researchers (37.1%), compared to non-parametric tests (4.1% in total). Among the parametric tests, the most widely conducted test was an ANOVA, with 19.8% of the papers. To analyse the relationships among study variables regression analysis was conducted in 35 papers (28.9%), while SEM was performed in 25.6% of the papers. Among the regression models, linear regression analysis was mostly preferred by the researchers (15.7%), followed by logistic regression (5.8%). The correlation coefficient was used in three papers (2.4%) to examine the relationships between study variables. Interestingly, four papers considered the asymmetric relationships between variables and used specific analyses (e.g. PRCA: penalty reward contrast analysis; AIPA: asymmetric importance-performance analysis) for that purpose. With 14.9%, cluster analysis was another popular procedure among the researchers.

To sum up, among the reviewed papers, in 27 papers, analyses that do not require normality assumption of the data (e.g., Kruskal–Wallis test) were conducted. Although normality assumption is a prerequisite for the analyses performed by the remaining papers ($n = 94$), 72.3% of them did not present any information related to this.

Since the employment of SEM requires rigorous steps, 31 papers in which SEM was performed were investigated in more detail. Among them, while five papers performed PLS-based SEM, 26 papers used covariance-based SEM. AMOS was the most common software used by 17 papers and followed by Smart PLS (4 papers). Among 31 papers, only 22 presented hypothesised relationships amongst study variables based on the related literature review. Before the SEM analysis, EFA was conducted by nine papers to find the underlying dimensions of the constructs and seven papers detected the normality of the data. While 28 papers conducted CFA to test the measurement model which is an initial step of SEM, only 20 of them presented the goodness of fit statistics. The correlation/covariance matrix among variables was presented in 21 papers. Although all papers using SEM provided the reliability of the variables, eight papers did not provide any information about discriminant validity. Furthermore, five papers did not present the AVE (average variance extracted) values of the constructs. The goodness of fit statistics, belonging to the structural model, were not provided by three out of 31 papers. Beta coefficients of the hypothesised relationships were presented in all papers. Moreover, mediation analysis was performed in six papers, while moderation analysis was conducted in seven papers.

Discussion

This chapter has two aims and subsections: (1) to introduce the terminology, paradigms, and research methods in the social sciences and (2) to present the results of a systematic literature review of research related to adventure tourism using quantitative research methods. In the first part, presenting the characteristics of social science research and quantitative research methods aims to provide readers with the necessary basic information. For this purpose, some of the concepts mentioned in the first subsection are reinforced with examples in the context of adventure tourism. Thanks to this information, it has been possible to convey how quantitative research methods are used in scientific articles on adventure tourism, what kind of approaches are adopted and what kind of methodological problems are observed. Besides learning the basics of quantitative studies in social sciences and their use in the adventure tourism context, understanding whether data analyses were performed properly by the researchers is crucial because a rigorous research process and data analysis are prerequisites to obtaining reliable and valid findings, and hence implications. Data collection is a crucial step of a quantitative research process. In the current research, reviewed manuscripts' methods were investigated in terms of the sampling process and measures.

In quantitative research, the aim is to generalise the findings obtained from the sample to the target population. In other words, the sample should be representative of the target population. Therefore, providing detailed information about the sampling process is needed. Among the reviewed papers, different sampling methods (e.g., convenience or probability) were used by the researchers, while convenience sampling was the most prevalent. In any case, as much detail as possible about the sampling must be provided. For example, in a study, the authors mentioned that they used simple random sampling (Mundet & Ribera, 2001). However, they did not provide any information about how each member of the population had an equal likelihood of being selected. Unfortunately, 6.6% of them did not provide any information about their data collection method. This makes interpreting their results difficult. Another requirement for the generalisability of the findings is the presentation of the sample characteristics. Unfortunately, 6.6% of the reviewed papers failed to present that information. Moreover, to assure generalisability, the researchers are advised to compare their sample characteristics with the target population. Furthermore, the target population must be defined and presented.

For the measurement of the scale items, researchers benefited from Likert-type scales. The 5-point Likert-type scale was mostly preferred by the reviewed manuscripts. However, researchers (Preston & Colman, 2000) argue that a 7-point scale has superiority to others in terms of reliability, validity, and discriminating power, although there is no agreement for the optimal number of response categories in the Likert scale. This is also in line with human information-processing capacity and short-term memory (Miller, 1956).

The use of the Likert-type scales leads to another discussion about whether it is an ordinal or an interval scale. Ordinal scales involve arranging information in a specific order. The data obtained by this type of scale can be named, grouped, and ranked. Since the ordinal numbers show only the sequence and the distances between the values are not equal, the data violates the assumptions of normal distribution. Thus, non-parametric statistics should be preferred for the analyses. In interval scales, variables have an order, the difference between two values is equal, and the presence of zero is arbitrary. Thus, parametric statistics can be applied to the data collected by interval scales. While some researchers argue that Likert-type scales should be considered as ordinal scales, an increasing number of researchers have treated them as interval scales (Wu & Leung, 2017). Considering that in most of the reviewed papers, researchers performed parametric statistics, it can be concluded they treated a Likert scale as an interval scale. However, performing parametric analyses instead of non-parametric tests may lead to incorrect results because of the scales' characteristics presented above. Although some scholars conclude that there is no difference between using parametric and non-parametric statistics (e.g., Mircioiu & Atkinson, 2017), this discussion is still ongoing in academia. A possible solution is to perform statistical analyses by using the summed scale score, which is interval type, instead of item level (Leung, 2011).

In any case, the sample size is directly related to the type of statistics to be used and the robustness of the analysis results. The precision level, the confidence level, and the population variability in the attribute being investigated are the three main criteria that directly influence sample size. In typical social science research, 95% confidence and ±5% precision levels are used for determining the sample size. Since, in most of the cases, the proportion of investigated attributes in the population is not known, the researchers consider it to be 0.5. Thus, a typical formula produces a sample size of 385 (at a 95% confidence level, ±5% precision, and 0.5 estimated proportion of attributes in the population). However, the type of analysis and sampling method should be also considered to decide the correct sample size (Israel, 1992). It is good that the sample size of the reviewed manuscripts is high enough (mean = 647; SD = 1473). The translation of the scale items from one language into a different one is another crucial part of the method. For the content equivalence of the translated scales, researchers agree that back-translation is required (Brislin, 1970). However, among the reviewed manuscripts, only seven papers performed and reported a back-translation process, while most of them did not provide any information about it, including the language of the questionnaire. Thus, future researchers are advised to present all information related to the language of the questionnaire and its translation process (if it applies).

As previously explained, the response bias has a negative effect on the validity of the research findings and therefore, also on the implications derived from them (Yuksel, 2017). However, in approximately two thirds of the reviewed manuscripts, the authors did not provide any information related to it. To avoid response bias, precautions must be planned before the survey (e.g. the wording of the items and obtaining research variables from different sources) and be presented in the manuscripts (Podsakoff et al., 2012). The researchers must also evaluate response bias before the analyses by using statistical procedures (e.g., Harman's single-factor test) and present it as a limitation, if necessary (Podsakoff & Organ, 1986).

Regarding data analysis used by the reviewed papers, results highlighted the use of very different statistical techniques, while parametric statistics are the most prevalent. The use of parametric analyses requires several assumptions, including the normal distribution of the data. Nevertheless, more than two thirds of the manuscripts did not provide any information on whether their data meets the requirement of normality or other assumptions. Therefore, future researchers are advised to provide that information to reflect the trustworthiness of the findings. Findings reflected that non-parametric tests for comparison were still infrequently (4.1%) applied by the researchers (Palmer et al., 2005).

On one hand, in correlational manuscripts, factor analysis (EFA or PCA), regression analysis, and SEM are three widely used statistics, respectively. In particular, the widespread use of factor and regression analyses is in line with other tourism manuscripts (Palmer et al., 2005). On the other hand, in comparative studies, parametric tests, especially ANOVA, are more prevalent. This finding is also consistent with previous findings in tourism (Palmer et al., 2005) and in other disciplines (Blanca Mena et al., 2018). With a prevalence of 14.9%, cluster analysis is another commonly used technique among the reviewed manuscripts. These results show that similar analysis techniques are used by researchers in the adventure tourism literature for relational and comparative studies. This is a finding that both supports the statistical reliability of the analyses and reflects the traditional views of scientists in choosing analysis techniques. As new analysis techniques, asymmetric analyses (e.g., PRCA, AIPA, IRPA) attract the interest of the researchers. In these types of analyses, researchers assume that there is an asymmetric relationship between the interested variables (e.g., tourism experience and overall satisfaction). However, in the literature review, these types of analyses are seen to be scarcely used by scholars yet. Although in previous literature reviews (Palmer et al., 2005) SEM was identified as one of the rarely used techniques (6.99% in 2005), the current research identified that its popularity has increased among the researchers (25.6%). To perform SEM, the covariance-based approach was most preferred (26 out of 31 papers). Like the previous literature reviews on SEM (Zhang et al., 2021), the current review revealed that the manuscripts using SEM did not clearly report each step of data analysis. For example, some papers failed to present the relationships between study variables deriving from the literature review (9 papers) and investigate distributional assumptions of the data (26 papers).

Conclusion

Since there is increasing criticism related to whether scientific research produces reliable knowledge, it is important to use robust methods and analyses in research. Therefore, in the current research, an overview related to the method and data analyses used in adventure tourism–related manuscripts was provided, as well as some practical suggestions for the researchers. It has been revealed that the studies in the literature generally adopt a positivist perspective and mostly provide causal or descriptive information about the variables. Although the pragmatist research framework using mixed research methods was adopted in only a few papers, this may assist the researchers who aim to conduct exploratory studies in understanding tourist behaviours participating in adventure tourism activities. Moreover, the results indicated the predominant use of parametric statistics. However, their basic assumptions (e.g., normality of the data) were mostly neglected by the researchers, which may negatively affect the reliability and validity of the findings. Furthermore, as a new perspective, asymmetric analysis techniques which consider the non-linear relationships between the variables can be used in future research. These may help to explore the complicated nature of human behaviours in adventure experiences. As in every research, the current study has some limitations that need to be acknowledged. For example, the investigation of only English-language manuscripts that were published in journals indexed by SSCI can be considered a limitation. Other indexes besides SSCI can be included in future research to expand the scope of the reviewed articles. In this review, the data analysis techniques of the examined articles were examined in a general manner. Future literature reviews may focus on a single analysis technique such as SEM for a more detailed investigation.

Review questions

1 Define *paradigm*. What are the main research paradigms in social sciences? Which do you think is best for studying adventure activities?
2 Name the types of research methods (designs) used in adventure tourism research.
3 What is the generalisability of the findings? How can it be ensured in a study on adventure tourism?

Further reading

Doran, A., Pomfret, G., & Adu-Ampong, E. A. (2022). Mind the gap: A systematic review of the knowledge contribution claims in adventure tourism research. *Journal of Hospitality and Tourism Management, 51*, 238–251. https://doi.org/10.1016/j.jhtm.2022.03.015

Okumus, F., & Rasoolimanesh, S. M. (2022). *Contemporary research methods in hospitality and tourism.* Emerald Publishing Limited. https://doi.org/10.1108/978-1-80117-546-320221021

Rantala, O., Rokenes, A., & Valkonen, J. (2018). Is adventure tourism a coherent concept? A review of research approaches on adventure tourism. *Annals of Leisure Research, 21*(5), 539–552. https://doi.org/10.1080/11745398.2016.1250647

Sheppard, V. (2020). Research methods for the social sciences: An introduction (open access). https://pressbooks.bccampus.ca/jibcresearchmethods/

Yuksel, A. (2017). A critique of "Response Bias" in the tourism, travel and hospitality research. *Tourism Management, 59*, 376–384. https://doi.org/10.1016/j.tourman.2016.08.003

Zhang, M. F., Dawson, J. F., & Kline, R. B. (2021). Evaluating the use of covariance-based structural equation modelling with reflective measurement in organizational and management research: A review and recommendations for best practice. *British Journal of Management, 32*(2), 257–272. https://doi.org/10.1111/1467-8551.12415

References

Akış Roney, S. (2011). What is scientific research? *Anatolia, 22*(2), 211–215.

Barbieri, C., & Sotomayor, S. (2013). Surf travel behavior and destination preferences: An application of the Serious Leisure Inventory and Measure. *Tourism Management, 35*, 111–121. https://doi.org/10.1016/j.tourman.2012.06.005

Beckman, E., Whaley, J. E., & Kim, Y.-K. (2017). Motivations and experiences of whitewater rafting tourists on the Ocoee River, USA. *International Journal of Tourism Research, 19*, 257–267. https://doi.org/10.1002/jtr.2109

Bhandari, P. (2022). *Random Assignment in Experiments: Introduction & Examples.* Scribbr. Retrieved from: https://www.scribbr.com/methodology/random-assignment/ on 24.12.2022.

Blanca Mena, M. J., Alarcón Postigo, R., & Bono Cabré, R. (2018). Current practices in data analysis procedures in psychology: What has changed? *Frontiers in Psychology, 9*, 2568. https://doi.org/10.3389/fpsyg.2018.02558

Boukezzoula, M. (2019). A plea for a focus on the contrasts between two paradigms and their implications for problem statement. *Arab World English Journal, 10*(3), 448–469. http://dx.doi.org/10.2139/ssrn.3466048

Brislin, R. W. (1970). Back-translation for cross-cultural research. *Journal of Cross-Cultural Psychology, 1*(3), 185–216. https://doi.org/10.1177/135910457000100301

Bryman, A., & Bell, E. (2015). *Business research methods* (4th ed.). Oxford University Press.

Carnicelli-Filho, S., Schwartz, G. M., & Tahara, A. K. (2010). Fear and adventure tourism in Brazil. *Tourism Management, 31*(6), 953–956. https://doi.org/10.1016/j.tourman.2009.07.013

Caruana, E. J., Roman, M., Hernández-Sánchez, J., & Solli, P. (2015). Longitudinal studies. *Journal of Thoracic Disease, 7*(11), 537–540. https://doi.org/10.3978/j.issn.2072-1439.2015.10.63

Chhetri, P., Arrowsmith, C., & Jackson, M. (2004). Determining hiking experiences in nature-based tourist destinations. *Tourism Management, 25*, 31–43. https://doi.org/10.1016/S0261-5177(03)00057-8

Çıvak, B., & Sezerel, H. (2018). Research paradigms and tourism literature, *Tourism Academic Journal*, 5(1), 1–14.

Cochran, W. G. (1977). *Sampling techniques* (3rd ed.). Wiley.

Cohen, L., Manion, L., & Morrison, K. (2007). *Research methods in education* (6th ed.). Routledge.

Creswell, J. W. (2013). *Choosing among five approaches* (3rd ed.). Sage.

Creswell, J. W. (2009). *Research design: Qualitative, quantitative and mixed methods approaches* (3rd ed.). Sage.

Crotty, M. (1998). *The foundations of social research: Meaning and perspective in the research process*. Sage.

Dash, G., & Paul, J. (2021). CB-SEM vs PLS-SEM methods for research in social sciences and technology forecasting. *Technological Forecasting and Social Change*, 173, 121092. https://doi.org/10.1016/j.techfore.2021.121092

DeCarlo, M. (2018). *Scientific inquiry in social work*. Open Social Work Education. Retrieved from: https://scientificinquiryinsocialwork.pressbooks.com/chapter/6-2-paradigms-theories-and-how-they-shape-a-researchers-approach/#footnote-220-1, on 31.07.2022.

Demir, E. (2019). *My notes: Scientific research methods*. Ankara University, Educational Sciences Institute, Measurement and Evaluation Department, Retrieved from: https://www.researchgate.net/profile/Ergul-Demir/publication/331223767_Bilimsel_Arastirma_Paradigmalari/links/5c6d2088299bf1e3a5b685d9/Bilimsel-Arastirma-Paradigmalari.pdf, on 01.08.2022.

den Breejen, L. (2007). The experiences of long distance walking: A case study of the West Highland Way in Scotland. *Tourism Management*, 28(6), 1417–1427. https://doi.org/10.1016/j.tourman.2006.12.004

Doran, A., Pomfret, G., & Adu-Ampong, E. A. (2022). Mind the gap: A systematic review of the knowledge contribution claims in adventure tourism research. *Journal of Hospitality and Tourism Management*, 51, 238–251. https://doi.org/10.1016/j.jhtm.2022.03.015

Doran, A., Schofield, P., & Low, T. (2020). Women's mountaineering: Accessing participation benefits through constraint negotiation strategies. *Leisure Studies*, 39(5), 721–735. https://doi.org/10.1080/02614367.2020.1763439

Dulock, H. L. (1993). Research design: Descriptive research. *Journal of Pediatric Oncology Nursing*, 10(4), 154–157. https://doi.org/10.1177/104345429301000406

Fazlıoğulları, O. (2012). Scientific research paradigms in social sciences. *International Journal of Educational Policies*, 6(1), 41–55.

Fordellone, M., & Vichi, M. (2020). Finding groups in structural equation modeling through the partial least squares algorithm. *Computational Statistics & Data Analysis*, 147, 106957. https://doi.org/10.1016/j.csda.2020.106957

Guba, E. G., & Lincoln, Y. S. (1994). Competing paradigms in qualitative research. In N. K. Denzin & Y. S. Lincoln *Handbook of qualitative research* (3rd ed., pp. 105–117). Sage.

Hair Jr, J. F., Matthews, L. M., Matthews, R. L., & Sarstedt, M. (2017). PLS-SEM or CB-SEM: updated guidelines on which method to use. *International Journal of Multivariate Data Analysis*, 1(2), 107–123. https://doi.org/10.1504/IJMDA.2017.087624

Hank, C., & Wildemuth, B. M. (2017). Quasi-experimental studies. In B. M. Wildemuth (Ed.), *Applications of social research methods to questions in information and library science* (2nd ed.). Libraries Unlimited.

Iofrida, N., De Luca, A. I., Strano, A., & Gulisano, G. (2018). Can social research paradigms justify the diversity of approaches to social life cycle assessment? *International Journal of Life Cycle Assess*, 23, 464–480. https://doi.org/10.1007/s11367-016-1206-6

Israel, G. D. (1992). *Determining sample size (Fact sheet PEOD-6). Program Evaluation and Organizational Development, Florida Co-operative Extension Service, Institute of Food and Agricultural Services*. University of Florida.

Keeves, J. P. (1997). *Educational research methodology and measurement*. Cambridge University Press.

Khaldi, K. (2017). Quantitative, qualitative or mixed research: Which research paradigm to use? *Journal of Educational and Social Research*, 7(2), 15–24. https://doi.org/10.5901/jesr.2017.v7n2p15

Kivunja, C., & Kuyini, A. B. (2017). Understanding and applying research paradigms in educational contexts. *International Journal of Higher Education*, 6(5), 26–41. https://doi.org/10.1177/001316447003000308

Krejcie, R. V., & Morgan, D. W. (1970). Determining sample size for research activities. *Educational & Psychological Measurement*, 30, 607–610.

Kuhn, T. S. (1970). The structure of scientific revolutions. In: *International encyclopaedia of unified science* (Vol. 2, Nr. 2, 2nd ed.). The University of Chicago.

Leung, S. O. (2011). A comparison of psychometric properties and normality in 4-, 5-, 6-, and 11-point Likert scales. *Journal of Social Service Research*, 37(4), 412–421. https://doi.org/10.1080/01488376.2011.580697

Lincoln, Y. S., & Guba, E. G. (1985). *Naturalistic inquiry*. Sage.

Mackenzie, N., & Knipe, S. (2006). Research dilemmas: Paradigms, methods and methodology. *Issues in Educational Research*, 16, 193–205.

Makombe, G. (2017). An expose of the relationship between paradigm, method and design in research. *The Qualitative Report*, 22(12), 3363–3382.

Mathers, N., Fox, N., & Hunn, A. (2007). Surveys and Questionnaires. The NIHR RDS for the East Midlands / Yorkshire & the Humber, Retrieved from: https://www.rds-yh.nihr.ac.uk/wp-content/uploads/2013/05/12_Surveys_and_Questionnaires_Revision_2009.pdf, on 04.08.2022.

Melosik, Z. (2021). There is no end of paradigms war in social sciences: A meta-analytical approach. *Educational Studies Review*, 1(32), 193–203. https://doi.org/10.12775/PBE.2021.011

Mertens, D. M. (2005). *Research methods in education and psychology: Integrating diversity with quantitative and qualitative approaches* (2nd ed.). Sage.

Miller, G. A. (1956). The magical number seven, plus or minus two: Some limits on our capacity for processing information. *Psychological Review*, 63(2), 81–97. https://doi.org/10.1037/h0043158

Mircioiu, C., & Atkinson, J. (2017). A comparison of parametric and non-parametric methods applied to a Likert scale. *Pharmacy*, 5(2), 26–38.

Moher, D., Liberati, A., Tetzlaff, J., & Altman, D. G. (2009). Preferred reporting items for systematic reviews and meta-analyses: The PRISMA statement. *Annals of Internal Medicine*, 151(4), 264–269. https://doi.org/10.7326/0003-4819-151-4-200908180-00135

Mundet, L., & Ribera, L. (2001). Characteristics of divers at a Spanish resort. *Tourism Management*, 22(5), 501–510. https://doi.org/10.1016/S0261-5177(01)00016-4

Needham, M. D., & Rollins, R. B. (2005). Interest group standards for recreation and tourism impacts at ski areas in the summer. *Tourism Management*, 26(1), 1–13. https://doi.org/10.1016/j.tourman.2003.08.015

Nunkoo, R. (2020). The state of research methods in tourism and hospitality. In R. Nunkoo (Ed.), *Handbook of research methods for tourism and hospitality management* (1st ed., pp. 3–23). Edward Elgar Publishing.

O'Leary, Z. (2004). *The essential guide to doing research*. Sage.

Ong, T. F. & Musa, G. (2012). SCUBA divers' underwater responsible behaviour: Can environmental concern and divers' attitude make a difference? *Current Issues in Tourism*, 15(4), 329–351. https://doi.org/10.1080/13683500.2011.604407

Palmer, A. L., Sese, A., & Montano, J. J. (2005). Tourism and statistics: Bibliometric study 1998–2002. *Annals of Tourism Research*, 32(1), 167–178. https://doi.org/10.1016/j.annals.2004.06.003

Patel, S. (2017). The research paradigm – methodology, epistemology and ontology – explained in simple language. Retrieved from: https://salmapatel.co.uk/academia/the-research-paradigm-methodology-epistemology-and-ontology-explained-in-simple-language, on 21.12.2022.

Peterson, R. A., & Merunka, D. R. (2014). Convenience samples of college students and research reproducibility. *Journal of Business Research*, 67(5), 1035–1041. https://doi.org/10.1016/j.jbusres.2013.08.010

Pickering, C. & Byrne, J. (2014). The benefits of publishing systematic quantitative literature reviews for PhD candidates and other early-career researchers. *Higher Education Research and Development*, 33(3), p. 534–548.

Podsakoff, P. M., MacKenzie, S. B., & Podsakoff, N. P. (2012). Sources of method bias in social science research and recommendations on how to control it. *Annual Review of Psychology*, 63(1), 539–569. https://doi.org/10.1146/annurev-psych-120710-100452

Podsakoff, P. M., & Organ, D. W. (1986). Self-reports in organizational research: Problems and prospects. *Journal of Management*, 12(4), 531–544. https://doi.org/10.1177/014920638601200408

Pomfret, G. (2006). Mountaineering adventure tourists: A conceptual framework for research. *Tourism Management*, 27, 113–123. https://doi.org/10.1016/j.tourman.2004.08.003

Preston, C. C., & Colman, A. M. (2000). Optimal number of response categories in rating scales: reliability, validity, discriminating power, and respondent preferences. *Acta Psychologica*, 104(1), 1–15. https://doi.org/10.1016/S0001-6918(99)00050-5

Rantala, O., Rokenes, A., & Valkonen, J. (2018). Is adventure tourism a coherent concept? A review of research approaches on adventure tourism. *Annals of Leisure Research, 21*(5), 539–552. https://doi.org/10.1080/11745398.2016.1250647

Regoniel, P. (2015, January 3). Quantitative Research Methods: Meaning and Characteristics [Blog Post]. In Research-based Articles. Retrieved from https://simplyeducate.me/2015/01/03/quantitative-methods-meaning-and-characteristics on 29.12.2022.

Rogers, J., & Révész, A. (2020). Experimental and quasi-experimental designs. In J. McKinley & H. Rose (Eds.), *The Routledge handbook of research methods in applied linguistics* (pp. 133–143). Routledge.

Scotland, J. (2012). Exploring the philosophical underpinnings of research: Relating ontology and epistemology to the methodology and method of scientific, interpretive and critical research paradigms. *English Language Teaching, 5*(9), 9–16.

Setia, M. S. (2016). Methodology series module 3: Cross-sectional studies. *Indian Journal of Dermatology, 61*(3), 261–264. https://doi.org/10.4103/0019-5154.182410

Sheppard, V. (2020). *Research methods for the social sciences: An introduction.* Retrieved from: https://pressbooks.bccampus.ca/jibcresearchmethods, on 31.07.2022.

Stemmer, K., Gjerald, O., & Øgaard, T. (2022). Crowding, emotions, visitor satisfaction and loyalty in a managed visitor attraction. *Leisure Sciences*, 1–23. https://doi.org/10.1080/01490400.2022.2028691

Su, L., Cheng, J., & Swanson, S. (2022). The companion effect on adventure tourists' satisfaction and subjective well-being: The moderating role of gender. *Tourism Review, 77*(3), 897–912. https://doi.org/10.1108/TR-02-2021-0063

Symonds, J. E., & Gorard, S. (2008). *The Death of Mixed Methods: Research Labels and their Casualties*, Heriot Watt University, Edinburgh: The British Educational Research Association Annual Conference. 3–6 September 2008.

Szyjka, S. (2012). Understanding research paradigms: Trends in science education research. *Problems of Education in the 21st Century, 43*, 110–118.

Taherdoost, H. (2016). Sampling methods in research methodology; how to choose a sampling technique for research. *International Journal of Academic Research in Management, 5*(2), 18–27. http://dx.doi.org/10.2139/ssrn.3205035

Thompson, H. S. (1971). *Fear and loathing in Las Vegas: A savage journey to the heart of the American Dream.* Random House, Inc..

Ugwu, C. I., Ekere, J. N. & Onoh, C. (2021). Research paradigms and methodological choices in the research process. *Journal of Applied Information Science and Technology, 14*(2), 116–124.

Valentine, P. S., Birtles, A., Curnock, M., Arnold, P., & Dunstan, A. (2004). Getting closer to whales—passenger expectations and experiences, and the management of swim with dwarf minke whale interactions in the Great Barrier Reef. *Tourism Management, 25*, 647–655. https://doi.org/10.1016/j.tourman.2003.09.001

Wei-Ching, W. (2022). Ambient air pollution perception and coping behaviour among hikers. *Current Issues in Tourism*, Early view, https://doi.org/10.1080/13683500.2022.2053075

Williams, R. T. (2020). The paradigm wars: Is MMR really a solution? *American Journal of Trade and Policy, 7*(3), 79–84. https://doi.org/10.18034/ajtp.v7i3.507

Wright, B. E., Manigault, L. J., & Black, T. R. (2004). Quantitative research measurement in public administration: An assessment of journal publications. *Administration & Society, 35*(6), 747–764. https://doi.org/10.1177/0095399703257266

Wu, H., & Leung, S. O. (2017). Can Likert scales be treated as interval scales?—A Simulation study. *Journal of Social Service Research, 43*(4), 527–532. https://doi.org/10.1080/01488376.2017.1329775

Yang, E. C. L., Khoo-Latimore, C. & Arcodia, C. (2017). A systematic literature review of risk and gender research in tourism. *Tourism Management, 58*, 89–100. https://doi.org/10.1016/j.tourman.2016.10.001

Yuksel, A. (2017). A critique of "Response Bias" in tourism, travel and hospitality research. *Tourism Management, 59*, 376–384. https://doi.org/10.1016/j.tourman.2016.08.003

Zhang, M. F., Dawson, J. F., & Kline, R. B. (2021). Evaluating the use of covariance-based structural equation modelling with reflective measurement in organizational and management research: A review and recommendations for best practice. *British Journal of Management, 32*(2), 257–272. https://doi.org/10.1111/1467-8551.12415

Zimmerhackel, J. S., Rogers, A. A., Meekan, M. G., Pannell, D. J., & Marit, E. K. (2018). How shark conservation in the Maldives affects demand for dive tourism. *Tourism Management, 69*, 263–271. https://doi.org/10.1016/j.tourman.2018.06.009

6 Qualitative methods in adventure tourism research

Tove I. Dahl, Tim Dassler and Arild Røkenes

Chapter learning outcomes

1 Become aware of the underlying philosophical foundations of research that frame the kinds of questions we pose and the answers we find.
2 Become familiar with the qualitative methods used in the most current adventure research from a selection of leading adventure research journals.
3 Understand a selection of established but less commonly used qualitative methods in adventure research.
4 See the strengths and challenges in current practice and the opportunities that lie ahead.

Introduction

Every time we ask a question, it reflects assumptions we carry about the world. Those assumptions intrinsically colour how we approach and make sense of our experiences. Our questions also frame our pursuit of answers and what we learn along the way. The same is true in research.

Phillimore and Goodson (2004) once asked if our approaches to qualitative research are more about recipes for practice or thinking tools. In this chapter, we regard them as thinking tools that are used to reason one's way to an informed question and then a reasoned pursuit of a trustworthy answer (see Figure 5.1 in Chapter 5). Methodological thinking tools help provide order to a discipline's foundation, identify relevant, interesting and urgent questions of the time, enable us to choose and clearly communicate our research approach, and better work collaboratively with others who think similarly. Such disciplinary order guides how we go about seeking, discovering and making claims about the answers we find. This chapter therefore complements this book's quantitative methods chapter (see Chapter 5) by focusing on the thinking tools that serve all these purposes with *qualitative* methods.

The foundations of the research enterprise

The overall essence of the research endeavour is to find reasoned and *trustworthy* answers to *informed* questions. The answers we find to our new questions are *trustworthy* based

DOI: 10.4324/9781003393153-8

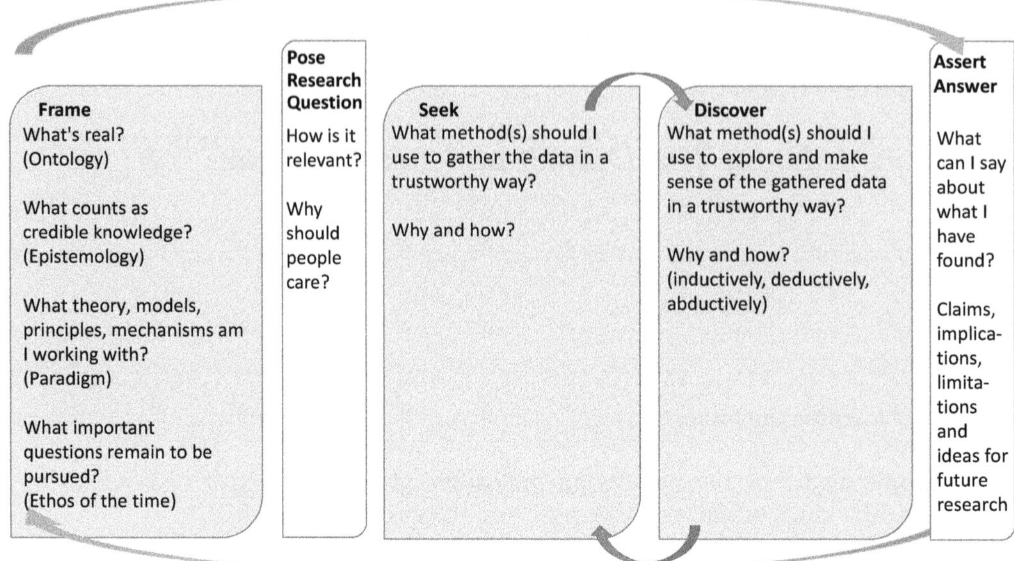

Figure 6.1 Guiding questions for the iterative qualitative research process.

on the subsequent appropriateness and quality of the approach we use to find them (Lincoln & Guba, 1986). The approach we adopt will involve methods (1) to seek out data from which to find an answer and (2) to organise and interpret them in a way that distils an answer that we can confidently support with evidence that fits our framework (see Figure 6.1 for an overview of the elements, like stepping stones, involved in the process; note how it complements Figure 5.1 in Chapter 5).

Our confidence in what we share in the end is a by-product of how well we have framed the question, chosen and carried out our approach, and how thoroughly and transparently we describe all these steps in our communication of the work. In the complex qualitative tradition, it is expected that we make all the steps of our work explicit so that others may better understand and evaluate it. Starting with the assumptions our work is based on, we commonly frame our work in terms of its ontology, epistemology and relevance to dominant research paradigms (Mackenzie & Knipe, 2006). Based on that, we select appropriate methodologies to carry it out.

Framing

Before choosing a method, we have to identify and frame our question (Carter & Little, 2007). Why are we asking what we are asking, why are we asking it this way, and why might the outcome matter? Røkenes et al. (2015) asked how guides contribute value to clients' adventure tourism experiences to better understand the complexities involved. To publish this work, however, they had to find a match between their approach and journal preferences.

Journals vary in the genres of research they publish, and therefore in how explicitly they want authors to describe their framework. The goals and backgrounds of the managing editors influence the expectations, aims and scope of their journals (Sánchez et al., 2019). Journals that specialise in research using particular methods (e.g. experimental designs) may require less explicit ontological, epistemological and paradigmatic framing since the choice of the journal itself already signals the kinds of work prioritised.

Less prescriptive journals, however, may publish articles from a variety of research traditions. For example, *Annals of Tourism Research* serves a broad range of fields in the social sciences and therefore publishes multiple research approaches. The *Journal of Hospitality and Tourism Management* encourages collaborative research between academia and the tourism industry, and *Tourism Management* encourages the use of innovative perspectives to challenge the field. More specifically, in a recent overview of articles published in the *Journal of Sustainable Tourism*, 21 different research methods were reported. The top three were questionnaires, interviews, and case studies, but methods also included action research, big data research, the Delphi method, and experiments (Xu, Nash & Whitmarsh, 2020). Readers of these kinds of journals come from varied backgrounds and therefore need more explicit framing to understand our question and the reasoning behind our research design. Røkenes et al., (2015) found a home for their question and approach in the *Scandinavian Journal of Hospitality and Tourism* that encourages interdisciplinary approaches. So, how should you frame your work?

Ontology

Let us start with what is "real." *Ontology* 'refers to fundamental, taken-for-granted assumptions about the ultimate reality of things' (Slife, 2004, p. 157). Different traditions vary on this score (Crotty, 1998). Being aware of our assumptions about reality can help us to position our research and be more reflective about our own role in the research process and what our results can say about the world.

Two of the most common ontologies are realism and idealism. If we assume that objects and value-free facts about the world are what is most real and possible to be studied in an objective way, we will ground our research in a *realist ontology* – commonly used in the natural sciences. If, on the one hand, we assume that ideas about the world are most real and possible to study through how people construct meaningful interpretations of the world, we will ground our research in an *idealist ontology* – commonly used in the humanities. Meanwhile, *abstractionism* assumes that 'abstractions, such as theories, techniques, and principles, capture and embody the fundamentally real' (Slife, 2004, p. 157) and knowledge is deduced from these abstractions. A *relational ontology*, on the other hand, considers relations as constitutive of knowledge and reality. We assume knowledge is premised on practices whereby what is real is created in nature–human and human–human relations (social and cultural) and even spatial-temporal relations, like how our own or others' history, experiences and possible futures affect who we are and what we do. Røkenes et al. (2015) acknowledged that self-reported and observed experiences of back-country skiers and mountain bikers are different (from the stance of either an outsider or participant). They therefore used multiple perspectives to describe the mountain experience. They also described the background of each of the researchers whose own perspectives of reality might influence their analyses.

Ontologies also guide assumptions about how researchers may *create* knowledge. From a phenomenological perspective, researchers discern their own and others' lived experiences in relation to a particular phenomenon to understand and describe the essence of that phenomenon. From a hermeneutic perspective, we discern knowledge in a participatory way where meaning emerges through dialogic communication. From a symbolic interactionist perspective, we discern knowledge through how people self-reflect on the symbolic meaning of their place in their environment (Cresswell, 2013; Crotty, 1998).

So, in qualitative research, we recognise that there may be multiple realities that can be classified as natural or logical (Toloie-Eshlaghy et al., 2011). Consequently, when trying to understand the world, our assumptions about what reality is, and how knowledge about reality is created will influence what we look for, and how we do so. It is our job as researchers to align our practices and communications with those understandings.

Epistemology

Epistemology is concerned with the 'nature of knowledge' (Schwandt, 2001, p. 71) and whether truth about the world can be discovered objectively (objectivism), is imposed by us on the world (subjectivism) or constructed in the interplay between human–nature and human–human relations (constructivism).

Objectivism assumes that a meaningful reality does objectively exist in and of itself, and, with appropriate methods, we as researchers can discover objective facts about the world free from any value judgements and subjective interpretations (James & Busher, 2009). In notable contrast, *subjectivism* assumes that there is no objective reality and that meaning is imposed by the subject on the object (Crotty, 1998). Then there is *constructivism* that assumes that 'all claims to knowledge take place within a particular conceptual framework' (James & Busher, 2009, p. 7) and that our theories, interpretations and explanations of the world are just that. They do not reflect an objective reality since no one can hold an unmediated version of an objective truth about the world. Thus, truth or meaning is neither objectively discovered nor subjectively imposed on the world but rather constructed in many ways that can be described conceptually.

Contrary to many quantitative approaches, in qualitative research, we recognise that research is, by nature, an interaction and that all participants and researchers come with values that affect (or introduce a bias in) what they "know" (Toloie-Eshlaghy et al., 2011). Hence, the need to be explicit about which epistemology our work is premised on. Røkenes et al. (2015), for example, acknowledged the subjectivism of back-country skier and mountain biker client/guide experiences and gathered complementary knowledge through participant observation and semi-structured interviews.

Paradigms

As described in Chapter 5, paradigms are embedded in ontologies and epistemologies. They provide the concrete framework that encodes theory, models, principles or mechanisms that may help us understand, explain or predict how things work (e.g. [post-]positivist, interpretive/constructivist, transformative, or pragmatic; see Mackenzie & Knipe, 2006). These, in turn, provide tools for confidently assessing what is currently known, where our knowledge needs to be refined or consolidated, where there are conflicting findings to be resolved, where there are entire gaps in our knowledge, and how we can use the answers to these considerations to guide us in making smart research decisions. Those decisions are the basis for formulating strong research questions and providing direction for smart answer-finding approaches.

The foundations of the research process

Before reading the qualitative methods in the practice case study (see Case study 6.1), consider the steps involved in constructing a trustworthy research design.

Posing a question

Qualitative research questions are often open-ended and exploratory, although they can vary in purpose (Cresswell, 2013). They may involve theory-building about the nature of play mechanisms (Zhu & Xu, 2022), developing an understanding of selected phenomena such as liminality (Goodnow & Bordoloi, 2017; Moriarty, 2011), or deeply exploring or reflecting on cases or individuals such as client–guide relations (Røkenes et al., 2015). They can also include piecing together rich understandings of cultural interactions (Cresswell, 2013), for example around the experience of challenging hiking experiences (Mykletun et al., 2021).

In 1990, Tesch created an overview of qualitative research questions and methods that identified 26 different methods used to pursue four main types of questions – those related to (1) the characteristics of language and its use, (2) the discovery of regularities, (3) searching for meaning, and the (4) process of reflection and wonder. The methods involve approaches of introspecting to purely observing to various degrees of mutual participation (Moriarty, 2011). Frequently based on small samples, these methods can nevertheless yield rich, thick data and descriptions (Ponterotto, 2006) that inform emergent understandings of the nature of a phenomenon, reveal patterns and explanations for how people think or behave, and can inspire new ways to classify and organise our understanding of the world,

Though qualitative questions can be quite open, the kinds of questions worthy of qualitative research approaches, suggests Tracy (2010), should distinguish themselves by being 'relevant, timely, significant, interesting or evocative' (p. 840) and not obvious or opportunistic. At the same time, by being open to different ontologies and epistemologies through the questions we pose, qualitative approaches lend themselves well to disrupting and challenging dominant thought and representations – thereby opening the door to more intellectual diversity. For example new kinds of questions can expose underrepresented or entirely new perspectives (Ivanova et al., 2021), such as interests relevant to people living in places of rapid climate change or diverse ethnicities, gender identities or physical abilities. You can read more about what matters to stakeholders in some of these groups in Part IV of this book.

Finding answers

When considering the path to answers, both how we seek out data (through data collection) and go about discovering what they hold (through data analysis and interpretation) are two separate, but related, methodological processes. For example, one set of data garnered in the seeking process, such as travellers' photographs, can be analysed in multiple ways in the discovery process. Likewise, all kinds of data collected in the seeking process can be analysed thematically in the discovery process. Therefore, when our research question invites choosing qualitative methods, we actually have to make distinct and linked design choices specific to the seeking and discovery processes of our approach.

Seeking

The kinds of data sought in qualitative research can vary considerably. What kind of knowledge are you looking for (e.g., confirmatory or exploratory), and how, when and where can you gather it?

Table 6.1 Commonly used qualitative methods for collecting data.

Method	Focus
Interviews	Structured, semi-structured or open conversations with individuals for the purpose of capturing deep and specific personal content (McGehee, 2012), e.g., Zhu and Xu (2022).
Focus groups	Researcher-facilitated structured, semi-structured or open conversations with groups of individuals. Suited to exploring social processes and how they create or change the self and its performances (Anvik et al., 2021). Together, they generate and construct shared meanings based on the totality of the individual's contributions (Cater & Low, 2012), e.g., Schlegelmilch & Ollenburg (2013).
Ethnography	Thick descriptions of experiences and underlying dynamics that capture 'how people conceptualize themselves vis-à-vis others' (Adams, 2012, p. 343) through the documentation and analysis of multiple forms and layers of meaning, e.g., Løland & Hällgren (2022).
Participatory methods	The use of inclusive methods such as action research, participatory action research, participatory rural appraisal and community-based participatory research enable participants to be co-participants in the research process with the goal of involving them in producing results they can use to their benefit (Dahl, 2013; Mackenzie & Knipe, 2006), e.g., Fitzpatrick & Handscomb (2017)

Some of the most common qualitative methods used for collecting data are summarised in Table 6.1. Other methods include accessing documents and archival materials, finding and analysing artefacts, observing and note taking/coding, conversations, diary-writing (Figure 6.2), participant observation, historical methods, action and applied research, mixed methods (Cresswell, 2013) mapping or drawing (Butler-Kisber & Poldma, 2010), or using combinations of methods to foster new kinds of thinking (e.g., see Prince, 2022). The useful but less commonly used methods of life histories, visual methods, relationship mapping, and Q-methodology are presented later.

Discovering

As Kuhn long ago asserted, research is essentially about puzzle-solving (Kuhn, 1970), and there are multiple ways to proceed. If we base our research questions on prior research, we are more likely to begin analysing our data deductively – examining our data from the stance of what we might expect based on our selected paradigm (theory, models, principles, or mechanisms) and comparing that to what we actually find in our data (Phillimore & Goodson, 2004).

If we pose our questions explicitly without using assumptions organised in an existing paradigm, we are more inclined to analyse data inductively – examining our data from a more open stance as to what they may reveal through our interpretation of them. From this stance, reflexivity is important. It involves recognising the subjectiveness of knowledge and accounting for that when describing our questions, approach and results to transparently address any preconceptions or biases we might bring to our work (Cresswell, 2013; Lincoln & Guba, 1986). Researchers may also blend their use of deductive and inductive methods, using both perspectives to inform the other to the extent that the question posed, and data collected allow for it (e.g., see Beames, Mackenzie & Raymond,

Figure 6.2 Diary writing in the field. Photograph by Adele Doran.

2022 for an informative example where they did a systematic review of sustainability guidance literature using a PRISMA approach, and then theoretically analysed the identified sources with a six-step reflexive thematic analysis).

In Tesch's (2013) detailed overview of qualitative methods used for analysing data, methods include phenomenology, content analysis, discourse analysis, symbolic interactionism, grounded theory, event structure analysis, action research, emancipatory research, case studies, life histories, educational connoisseurship, heuristic research, and many forms of ethnography (holistic, structural, educational, ethnoscience, and several other ethnographies 'of'). See the sampling section for some examples.

Asserting with trustworthiness and rigour

Once the data have been collected and analysed, we interpret what they mean through claims that position the findings in relation to our initial framework. How do our findings refine, consolidate, settle, bridge, or fill in what we know from before? How generalisable or transferable are they? What implications might they have for theoretical development or practical use? How confidently can we make these claims considering any limitations we might see in the way we formulated our question, collected our data or did our analyses? Finally, what do the findings invite of fruitful thought for future research based on the trustworthiness and rigour of our work?

Lincoln and Guba's (1986) four evaluative criteria for rigorous work have been used extensively for assessing the trustworthiness of naturalistic, or qualitative, inquiry. They include credibility (through efforts like triangulation and member checks), transferability (through thick descriptive data that make it possible for others to evaluate the degree of the data's fit to the problem), dependability (as related to the process), and confirmability (as related to the research 'products' such as data and claims made from them).

Complementing Lincoln and Guba's initial criteria, Tracy (2010, p.839) suggested eight 'big-tent' criteria for executing excellent qualitative research. In her quality measures for rigour, she included a worthy topic (already mentioned), rich rigour (in all phases of the work), sincerity (where the researcher is honest about their subjectiveness and transparent in their communication of the work), credibility (much like Lincoln and Guba's criterion, although also with the inclusion of multiple voices), resonance (presented in a way that moves the audience), significant contribution (e.g., to conceptual work, moral considerations and methodology), ethical work (procedurally, situationally, relationally, in how the work ends), and meaningful coherence (true to its question and design and meaningfully connected the parts with each other and to a greater whole). What all these criteria have in common is the goal of assuring robust research capable of advancing disciplinary knowledge.

Qualitative research methods used in adventure research today

To get an overview of qualitative methods currently being reported in mainstream adventure journals, in November 2022, we surveyed all the articles published from 2019 to 2022 in seven top journals that had "adventure" or "adventure tourism" in their title, abstract or among their key words and that used qualitative methods in their design; 21 relevant articles were found. Because we authors all work in Norway, where skiing is a notable part of the mountain adventure culture, and there were no examples of skiing research in the first sample we collected, we carried out an extra search for qualitative research on skiing, adding two additional articles from the same period to the sample.

The journals represented in this sample, and the number of articles that matched our criteria, include the *Annals of Tourism Research* (3), the *Journal of Hospitality and Tourism Management* (3), the *Journal of Outdoor Recreation and Tourism* (10), the *Journal of Sports Science & Medicine* (1), *Leisure Studies* (1), *Tourism Management* (1), and *Tourism Management Perspectives* (1). The selected articles were based on data collected in 21 different countries spanning Europe (12 articles), Asia (5), Australasia (5), North America (3), Central and South America (2), and Antarctica (1).

The articles used a range of seeking methods, with interviews used most frequently by far (16 studies). The interviews were conducted face-to-face and online, and the interview forms varied from open, to semi-structured, proposition-guided, probing, storytelling-focused, and multi-stage interviews. Other methods included observations (3), questionnaires/surveys with open-ended questions (2), and one each of case studies, mind maps, and drawings. Three of the included studies were also forms of literature reviews.

In terms of the analyses, thematic and content analyses were by far the most used, although some used deductive approaches, others inductive, and others a combination of both (see Case study 6.1: the Qualitative methods in practice section for an example). The tools used to conduct these analyses included software such as NVivo and MAXQDA, and approaches used included Cresswell's (2013) 6-step process; grounded theory's open coding, axial coding and selective coding (Charmaz, 2000); Scanlan et al.'s (2003) Scanlan Collaborative Interview Method (SCIM); and Ritchie and Spencer's (1994) five-step framework approach.

Case study 6.1   Qualitative methods in practice

What key factors seem to matter for the psychological well-being of land-, water- and air-based adventure guides, asked Mackenzie and Raymond (2020). They used a constructivist approach to gather and analyse adventure tour guide well-being experiences. The authors acknowledged the role their backgrounds would have on the design and analyses, varying in expertise from psychological theory, adventure guiding and work in sustainable tourism. The first author represented a strong background in the first two, and the second author represented a strong background in the second two. They used this to their advantage in their design.

Data collection involved semi-structured two-part interviews using the Scanlan Collaborative Interview Method (SCIM). In the first part, participants answered three open-ended inductive questions about their guiding and feelings of personal well-being. Participants then created a visual map of key concepts derived from the interview. In the second part, they were given cards with key concepts from personal well-being theory and asked to consider whether each concept was relevant to their well-being. If so, they were asked to include it where it fitted on their conceptual map.

To analyse and interpret the data, the interviews were transcribed and inductively coded by each author and then compared with the ideas in their visual maps. Final themes were derived inductively through author conversations and reflections around the themes' relations to the visual figures. The data were also interpreted deductively using personal well-being theory. Finally, three overarching themes emerged from the adventure guide experiences: the natural environment, the adventure activity, and people. A conceptual model was also derived from the analyses, complementing existing research and offering novel, reasoned ideas for how to better support adventure guide personal well-being.

A sampling of complementary qualitative research methods

Here are four examples of less-commonly used yet well-described methods from the literature. They include the kinds of questions pursued, how data are collected in the seeking phase, and how data are analysed in the discovery phase.

Life histories

The goal of life history work is to capture detailed, subjective biographical life or work histories to better understand the experiences and events that influence life trajectories (Ladkin, 2004). The pragmatic power of retrospection of long spans of time compensates for hard-to-gather longitudinal data, although the data are only as robust as the limits of recall and other documentation allow. Data can come in any form of documentation that captures life between birth and death, often through dialogue with a focal person and the researcher talking through meaningful episodes (Moen, 2006). It can also include self-produced content in the form of diaries, autobiographies and oral histories, or data gathered through interviews with memory prompts or biographical work using other reference tools. A well-blended example is the study of Canadian mountaineer Margaret Fleming

(Reichwein & Fox, 2001) using primary sources like interviews, diary entries, letters and photos, and secondary resources, such as books and articles that reflected the life and times of Fleming's mountaineering life. Epistemologically, however, researchers must be aware of and account for the distinction between what Bruner (2011) described as life imagined, lived, experienced, and told.

With the life history base established, discovery can proceed in multiple ways, like analysis by themes, story structure or language (Agar, 1980). Biases can colour participants' stories and researchers' understandings of them (Ladkin, 2004), although certain kinds of events may be more impervious to bias, for example those linked to participants' long-term goals or pleasant events. They can be checked through triangulation.

Visual methods

A substantial part of the tourist experience is encoded visually by the "tourist gaze" (Urry, 1992) and methods have been developed to capture these experiences (Park & Kim, 2018). The kinds of questions that visual data have been used to explore touch on topics as broad as adventure destination experience, management and development (Hunter, 2016); impression management, travel planning and travel activities (Park & Kim, 2018); as well as how local host residents experience adventure tourists (Brickell, 2012; Winter & Adu-Ampong, 2021).

Visual methods involve the procurement and/or use of visual images such as videos and photographs taken by visitors, tourism photographers or researchers-published by destination management organisations in brochures and other print media-destination websites or on social media such as Instagram and Facebook (Park & Kim, 2018). They can also include hand-made collages, maps or drawings (Figure 6.3) (Butler-Kisber & Poldma, 2010) and architectural drawings (Hillman et al., 2018). Furthermore, images are used in interviews as objects for guiding conversation. Wilson and Kumli (2019), for example, derived students' definitions of adventure by analysing the content of short videos produced by three student cohorts while on microadventures that they claimed represented their personal definition of adventure.

Depending on the research framework, question and kind of data collected, analyses can include, among other methods, semiotic analysis, content analysis, semantic analysis, compositional analysis, visual ethnography, and critical analysis, as well as analysis of people's experiences of images as expressed in interviews and diaries (Park & Kim, 2018).

Relationship mapping

Relationship mapping (a specific form of concept mapping) springs from the constructivist tradition (Wilson et al., 2016). In general, the rational-relational goal of mapping is to document the way we think, in a visual form, about relationships that are otherwise less accessible in other forms of expression (Butler-Kisber & Poldma, 2010). The process of creating a map is holistic and non-linear, and its form reveals implicit understandings of relationships in a clarifying way.

The kinds of questions that relational maps have been used to answer include how concepts interrelate, such as in how individual guides experience well-being (see Case study 6.1), or how the concepts of adventure in the field of tourism interrelate (Janowski

Could you please draw a map of the Queenstown area in the box below.

Please show any activities or locations that have been significant in your holiday.

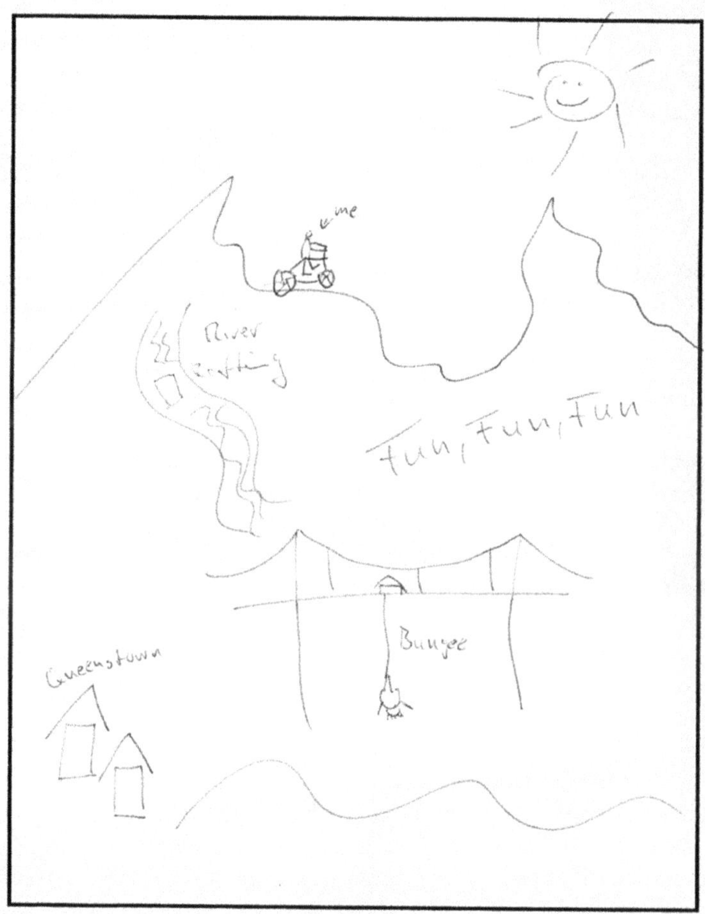

Thank You.

Figure 6.3 Tourist sketch of adventure activities in Queenstown, New Zealand. Carl Cater.

et al., 2021). Maps can be produced individually or in groups, and the final versions used in an analysis can be the original map or a map refined through reflection and discussion. Still, it is worth considering what kind of information can be gained or lost in the "smoothing out" aspect of the revision processes (Butler-Kisber & Poldma, 2010). Reflexivity is important here. Software can be also used to map relationships among other kinds of data, for example bibliometric tools that can be used to analyse relationships among data in published research databases (Molina-Collado et al., 2022).

As an object of analysis, maps produced by people are typically not sufficient by themselves. Some form of verbal information that links the map content to the ideas the maps express can provide the relational link to make it easier to analyse from the mapmaker's perspective (e.g., see Mackenzie & Raymond, 2020). With empathy, the researcher can further discern the humanness of the map content based on how it reflects the mapmaker's intellectual and emotional investments in the overall map message (Butler-Kisber & Poldma, 2010).

Q methodology

The Q methodology is grounded in the idea that there are multiple realities and that all knowledge is subjective (Stergiou & Airey, 2011). The Q methodology is an exploratory way to capture that in a structured form that makes what a person thinks or knows about something visible. The idea is not to extract a generalisable trait shared by an entire population through these sorts but to see how individuals construct unique meanings out of similar information. The degree of similarity among sorts can then be revealed through correlations and factor analyses. This method removes the influence of *a priori* expectations that a researcher may have about the topic, and it makes the data easier to analyse after data collection.

The data are based on a ready-made set of statements (the Q-sample) that include a representative sample of the range of key ideas relevant to the topic of the research question (Stergiou & Airey, 2011). A recent study (Lee et al., 2021) used this methodology to examine how tourists perceived the risk of COVID-19 as relevant to their travel choices. The researchers used 30 statements from a literature review and in-depth interviews with six tourists who had cancelled planned trips abroad.

The tourists were given 30 statements to evenly sort into seven categories (3 of them with labels). The left-most category (labelled 'most disagree') had space for three items, the next category for four items, the next for five items, the middle category (labelled 'neutral') for six items, the next for five items, the next for four items, and the right-most category (labelled 'most agree') had space for three items. After completing a Q-sort, participants can then comment and reflect on it to help with its interpretation (Stergiou & Airey, 2011). Based on the tourists' sorts and reflections, Lee et al. (2021) found that their participants could be classified into four types with four different reasons for perhaps cancelling their trip – those who (1) worried about their health, (2) worried about potential problems, (3) worried about tourism itself, or (4) worried about other issues.

Strengths, challenges and possibilities

The potential scope of qualitative research is vast. However, the actual practice captured in our small survey of mainstream journals shows a quite limited use of what is actually possible. Whether that is more a reflection of the nature of researchers' intellectual imagination, our time's burning questions, or gatekeeping through editorial decisions, is not clear. How can we use our qualitative thinking tools better?

Imagination

Imagination is our most important tool for seeing anew and advancing methodological variety and growth. Paradoxically, the expectation of originality in research is often met with the conservatism of tradition (Siler & Strang, 2017). Original research is always embedded in time, and state-of-the-art research at any given time is often characterised by a particular "flavour." This is, in part, because research often involves refining and/or expanding the scope of paradigms – work that takes time (Kuhn, 1970). Original work may therefore meet with disciplinary and conceptual resistance, conflict with the expertise of reviewers who do not understand the work or consist of ideas that are evaluated "before their time." This is evidenced by their acceptance later when the field has evolved in a way that recognises its value (Sánchez et al., 2019). Therefore, what counts as a burning question ripe for pursuit is somewhat moderated by the readiness of the research community to consider it. A lack of imagination is likely to limit nimble advancement and notable breakthroughs (Braun et al., 2017). If we truly embrace the idea of multiple ontologies and epistemologies, then we have so many more paradigms to develop and dimensions to tap from the multi-modality of human behaviour and experience.

Well-contextualised questions pursued with high trustworthiness and communicated well may stand a better chance of breaking through. Greater awareness of the range of ontologies and epistemologies we must work with, as well as continued work exploring more, will blow open the range of questions still to be asked. Our qualitative thinking tools can help us to formulate these questions in new ways. Still, this will demand going beyond current conventions and disciplinary ways of thinking (Kapiszewski & Wood, 2022), acknowledging common biases, and engaging in greater reflexivity.

For example, who are posing the questions, who are included in the research and how does that impact the way a field of study grows (Dahl, 2013; Mackenzie & Knipe, 2006)? Who assesses the quality of research and from which disciplinary background? Are we sufficiently tapping all adventure stakeholders, or are there potential stakeholders yet to come to the adventure table? How gendered or geographically limited are the questions (Ateljevic, Pritchard & Morgan, 2007)? How represented are the views of people with different abilities, socio-economic and ethnic backgrounds, family constellations, orientations or ages, for whom adventure opportunities or personal safety issues may vary wildly (see especially Part IV chapters in this book)?

What methods are being used? Visual methodologies that better capture visual experiences, or capture experiences visually, offer some examples of how to tap that. Likewise, when exploring the qualities of space and place, embodied approaches to adventure tourism experiences, as simple as walking with people *in situ* while talking about their experiences there, also open opportunities for reflection that might be lost in more traditional interview settings (Heijnen et al., 2022).

Access

Our 2019–2022 sample of selected journals was not a complete overview of all qualitative adventure research published in that time span. Niche journals, journals published in other languages, and research published in atypical adventure sources certainly held more. However, the included mainstream journals have long influenced research agendas by virtue of their broad readership, reading and citation metrics, editorship with preferred methodological and theoretical orientations, higher education reward systems for publishing in high-prestige journals, the ethos of what is considered original research, and the 'invisible college' of friends recognising friends' work, even in anonymous peer reviews (e.g., Crane, 1967; Sanches et al., 2019). These all pose ethical challenges for the wide dissemination of knowledge.

One positive development is the trend towards greater transparency and access through increased open science. More open reviewing and publishing processes are increasing (Wang et al., 2015). Journals and websites like PLOS and F1000 Research both offer models where there is significantly more transparency in the peer review processes and publishing. Moreover, in these types of journals, the soundness (or trustworthiness) of the work, more than the novelty of the results, drives what gets published.

As a bonus, openly available articles tend to get picked up sooner and are accessed over longer periods of time by more readers and, they tend to get cited more, and they tend to get picked up more in other media (Wang et al., 2015). With more research findings, and even data, freely available and widely read, we can anticipate a broader range of research methodologies to find their way into the literature and flourish. For qualitative adventure research, this could be a game-changer.

Conclusion

There will always be room for growth in adventure tourism research. A call to increase diverse and complementary research approaches would naturally warrant a greater presence of qualitative work in the balance (Gough and Lyons, 2016). By its very nature, the openness to multiple ontologies and epistemologies of qualitative thinking can bring considerable diversity to the questions we pose, the high-quality approaches we use to answer them and the contributions to the field that will emerge. It is up to all of us to know, imagine, and act on this.

Review questions

1 Choose two of the main adventure research journals from this chapter to explore the kinds of qualitative research approaches being published today. What are the aims and scope of these journals, what have they published in the last three years, and what does that reflect about their philosophical foundations (ontology, epistemology, relevant paradigms)? How is that relevant to how authors communicate their work to each journal's audience? What further clues might the composition of the editorial boards give you?
2 Take an interview-based study and suggest three other approaches one could use to address the heart of the research question (tweak the question as needed).
3 What might be necessary to successfully introduce, and get published, research that introduces an innovative qualitative approach?

Further Reading

Brinkman, S., & Kvale, S. (2018). *Doing interviews* (Vol. 2). Sage. ISBN: 978-1-47391-295-3
Cresswell, J. W. (2013). *Qualitative inquiry and research design: Choosing among five approaches* (3rd ed.). Sage.

Dwyer, L., Gill, A., & Seetaram, N. (Eds.). (2012). *Handbook of research methods in tourism: Quantitative and qualitative approaches*. Edward Elgar Publishing.

Phillimore, J., & Goodson, L. (2004). *Qualitative research in tourism: Ontologies, epistemologies and methodologies*. Routledge.

Tracy, S. J. (2010). Qualitative quality: Eight "big-tent" criteria for excellent qualitative research. *Qualitative Inquiry, 16*(10), 837–851.

References

Adams, K. M. (2012). Ethnographic methods. In L. Dwyer, A. Gill, & N. Seetaram (Eds.), *Handbook of research methods in tourism: Quantitative and qualitative approaches* (pp. 339–363). Edward Elgar Publishing.

Agar, M. (1980). Stories, background knowledge and themes: Problems in the analysis of life history narrative. *American Ethnologist, 7*(2), 223–239.

Anvik, B. T., & Olesen, E. S. B. (2021). Gruppen mener? *Tidsskrift for velferdsforskning, 24*(1), 21–34. https://doi.org/10.18261/issn.2464-3076-2021-01-03

Ateljevic, I., Pritchard, A., & Morgan, N. (Eds.). (2007). *The critical turn in tourism studies: Innovative research methodologies*. Elsevier.

Beames, S., Mackenzie, S. H., & Raymond, E. (2022). How can we adventure sustainably? A systematized review of sustainability guidance for adventure tourism operators. *Journal of Hospitality and Tourism Management, 50*, 223–231. https://doi.org/10.1016/j.jhtm.2022.01.002

Braun, V., Clarke, V., & Gray, D. (2017). Innovations in qualitative methods. In *The Palgrave handbook of critical social psychology* (pp. 243–266). Palgrave Macmillan. https://doi.org/10.1057/978-1-137-51018-1

Brickell, K. (2012). Visual critiques of tourist development: Host-employed photography in Vietnam. *Tourism Geographies, 14*(1), 98–116. https://doi.org/10.1080/14616688.2011.593187

Bruner, E. (2011). Afterword: The tour as imagined, lived, experienced, and told. In J. Skinner & D. Theodossopoulos (Eds.), *Great expectations: Imagination and anticipation in tourism* (Vol. 34, pp. 196–200). Berghahn Books. ISBN: 978-0-85745-278-8

Butler-Kisber, L., & Poldma, T. (2010). The power of visual approaches in qualitative inquiry: The use of collage making and concept mapping in experiential research. *Journal of Research Practice, 6*(2), Article M18. http://jrp.icaap.org/index.php/jrp/article/view/197/196

Carter, S. M., & Little, M. (2007). Justifying knowledge, justifying method, taking action: Epistemologies, methodologies, and methods in qualitative research. *Qualitative Health Research, 17*(10), 1316–1328. https://doi.org/10.1177/1049732307306927

Cater, C., & Low, T. (2012). Focus groups. In L. Dwyer, A. Gill, & N. Seetaram (Eds.), *Handbook of research methods in tourism: Quantitative and qualitative approaches* (pp. 352–364). Edward Elgar Publishing.

Charmaz, K. (2000). Grounded theory: Objectivist and constructivist methods. In K. Charmaz, N. K. Denzin, & Y. S. Lincoln (Eds.), *Handbook of qualitative research* (2nd ed., pp. 509–535). Sage. ISBN 0-7619-1512-5

Crane, D. (1967). The gatekeepers of science: Some factors affecting the selection of articles for scientific journals. *The American Sociologist*, 195–201.

Cresswell, J. W. (2013). *Qualitative inquiry and research design: Choosing among five approaches* (3rd ed.). SAGE. ISBN: 978-4129-9531-3

Crotty, M. J. (1998). *The foundations of social research: Meaning and perspective in the research process*. Sage.

Dahl, T. I. (2013). Children as researchers: We have a lot to learn. In G. Melton, A. Ben-Arieh, J. Cashmore, G.S. Goodman, & N. K. Worley (Eds.). *The SAGE handbook of child research* (pp. 593–618). Sage. https://doi.org/10.4135/9781446294758

Dwyer, L., Gill, A., & Seetaram, N. (Eds.). (2012). *Handbook of research methods in tourism: Quantitative and qualitative approaches*. Edward Elgar Publishing.

Fitzpatrick, J., & Handscomb, B. (2017). Co-creating spaces on an adventure playground: Using participatory action research as an approach to continuing professional development. In *Practice-based research in children's play* (pp. 147–166). Policy Press. https://doi.org/10.51952/9781447330059.ch009

Goodnow, J. M., & Bordoloi, S. (2017). Travel and insight on the limen: A content analysis of adventure travel narratives. *Tourism Review International, 21*(3), 223–239. https://doi.org/10.3727/154427217X15022104437701

Gough, B., & Lyons, A. (2016). The future of qualitative research in psychology: Accentuating the positive. *Integrative Psychological and Behavioral Science, 50*(2), 234–243. https://doi.org/10.1007/s12124-015-9320-8

Heijnen, I., Stewart, E., & Espiner, S. (2022). On the move: The theory and practice of the walking interview method in outdoor education research. *Annals of Leisure Research, 25*(4), 529–547. https://doi.org/10.1080/11745398.2021.1949734

Hillman, P., Moyle, B.D., & Weiler, B. (2018). Application of visual methods to perceptions of tourism development. *International Journal of Culture, Tourism and Hospitality Research, 12*(1), 124–129. ISBN: 978-1-138-49179-3

Hunter, W. C. (2016). The social construction of tourism online destination image: A comparative semiotic analysis of the visual representation of Seoul. *Tourism Management, 54,* 221–229. https://doi.org/10.1016/j.tourman.2015.11.012

Ivanova, M., Buda, D. M., & Burrai, E. (2021). Creative and disruptive methodologies in tourism studies. *Tourism Geographies,* 1–10. https://doi.org/10.1080/14616688.2020.1784992

James, N., & Busher, H. (2009). *Epistemological dimensions in qualitative research: The construction of knowledge online.* In Online Interviewing. SAGE Publications, Limited. https://doi.org/10.4135/9780857024503

Janowski, I., Gardiner, S., & Kwek, A. (2021). Dimensions of adventure tourism. *Tourism Management Perspectives, 37,* 100776. https://doi.org/10.1016/j.tmp.2020.100776

Kapiszewski, D., & Wood, E. (2022). Ethics, epistemology, and openness in research with human participants. *Perspectives on Politics, 20*(3), 948–964. doi:10.1017/S1537592720004703

Kuhn, T. S. (1970). *The structure of scientific revolutions* (2nd ed., enl., Vol. 2.2, pp. XII, 210). University of Chicago Press.

Ladkin, A. (2004). The life and work history methodology: A discussion of its potential use for tourism and hospitality research. In *Qualitative research in tourism* (pp. 254–261). Routledge. eBook ISBN: 9780203642986

Lee, W. S., Park, S., Jung, J., Mun, S., & Jung, J. (2021). A study on tourists' perceived risks from COVID-19 using Q-methodology. *Asia Pacific Journal of Tourism Research, 26*(10), 1057–1069. https://doi.org/10.1080/10941665.2021.1953087

Lincoln, Y. S., & Guba, E. G. (1986). But is it rigorous? Trustworthiness and authenticity in naturalistic evaluation. *New Directions for Program Evaluation, 1986*(30), 73–84.

Løland, S., & Hällgren, M. (2022). 'Where to ski?': An ethnography of how guides make sense while planning. *Leisure Studies,* 1–17. https://doi.org/10.1080/02614367.2022.2153905

Mackenzie, N., & Knipe, S. (2006). Research dilemmas: Paradigms, methods and methodology. *Issues in Educational Research, 16*(2), 193–205.

Mackenzie, S. H., & Raymond, E. (2020). A conceptual model of adventure tour guide well-being. *Annals of Tourism Research, 84,* 102977. https://doi.org/10.1016/j.annals.2020.102977

McGehee, N. G. (2012). Interview techniques. In L. Dwyer, A. Gill, & N. Seetaram (Eds.), *Handbook of research methods in tourism: Quantitative and qualitative approaches* (pp. 365–376). Edward Elgar Publishing.

Moen, T. (2006). Reflections on the narrative research approach. *International Journal of Qualitative Methods, 5*(4), 56–69. https://doi.org/10.1177/160940690600500405

Molina-Collado, A., Gómez-Rico, M., Sigala, M., Molina, M. V., Aranda, E., & Salinero, Y. (2022). Mapping tourism and hospitality research on information and communication technology: A bibliometric and scientific approach. *Information Technology & Tourism,* 1–42. https://doi.org/10.1007/s40558-022-00227-8

Moriarty, J. (2011). Qualitative methods overview. National Institute for Health Research School for Social Care. http://eprints.lse.ac.uk/41199/1/SSCR_Methods_Review_1-1.pdf 1 (kcl.ac.uk)

Mykletun, R. J., Oma, P. Ø., & Aas, Ø. (2021). When the hiking gets tough: "New adventurers" and the "extinction of experiences". *Journal of Outdoor Recreation and Tourism, 36,* 100450. https://doi.org/10.1016/j.jort.2021.100450

Park, E., & Kim, S. (2018). Are we doing enough for visual research in tourism? The past, present, and future of tourism studies using photographic images. *International Journal of Tourism Research, 20*(4), 433–441. https://doi.org/10.1002/jtr.2194

Phillimore, J., & Goodson, L. (2004). Progress in qualitative research in tourism: Epistemology, ontology and methodology. In L. Goodson & J. Phillimore (Eds.), *Qualitative research in tourism: Ontologies, epistemologies and methodologies* (pp. 21–23). Taylor & Francis Group. ISBN 9780415280877

Ponterotto, J. G. (2006). Brief note on the origins, evolution and meaning of the qualitative research concept thick description. *The Qualitative Report*, 11(3), 538–549. https://doi.org/10.46743/2160-3715/2006.1666

Prince, C. (2022). Experiments in methodology: Sensory and poetic threads of inquiry, resistance, and transformation. *Qualitative Inquiry*, 28(1), 94–107. https://doi.org/10.1177/10778004211014611

Reichwein, P., & Fox, K. (2001). Margaret Fleming and the Alpine Club of Canada: A woman's place in mountain leisure and literature, 1932–1952. *Journal of Canadian Studies*, 36(3), 35–60.

Ritchie, J., & Spencer, L. (1994). Qualitative data analysis for applied policy research. In A. Bryman & R.G. Burgess (Eds.), *Analysing qualitative data* (pp. 173–194). Routledge. ISBN 9780415060639

Røkenes, A., Schumann, S., & Rose, J. (2015). The art of guiding in nature-based adventure tourism–how guides can create client value and positive experiences on mountain bike and backcountry ski tours. *Scandinavian Journal of Hospitality and Tourism*, 15(1), 62–82. http://doi.org/10.1080/15022250.2015.1061733

Sánchez, I. R., Makkonen, T., & Williams, A. M. (2019). Peer review assessment of originality in tourism journals: Critical perspective of key gatekeepers. *Annals of Tourism Research*, 77, 1–11. https://doi.org/10.1016/j.annals.2019.04.003

Scanlan, T.K., Russell, D.G., Wilson, N.C., & Scanlan, L.A. (2003). Project on elite athlete commitment: I. introduction and methodology. *Journal of Sport and Exercise Psychology*, 25, 360–376. eISSN: 1543-2904

Schlegelmilch, F., & Ollenburg, C. (2013). Marketing the adventure: Utilizing the aspects of risk/fear/thrill to target the youth traveller segment, *Tourism Review*, 68(3), 44–54. https://doi.org/10.1108/TR-03-2013-0010

Schwandt, T. A. (2001). *The SAGE dictionary of qualitative inquiry* (2nd ed.). Sage. ISBN: 9781452217451

Siler, K., & Strang, D. (2017). Peer review and scholarly originality: Let 1,000 flowers bloom, but don't step on any. *Science, Technology, & Human Values*, 42(1), 29–61. https://doi.org/10.1177/0162243916656919

Slife. (2004). Taking practice seriously. *Journal of Theoretical and Philosophical Psychology*, 24(2), 157–178. https://doi.org/10.1037/h0091239

Stergiou, D., & Airey, D. (2011). Q-methodology and tourism research. *Current Issues in Tourism*, 14(4), 311–322. https://doi.org/10.1080/13683500.2010.537316

Tesch, R. (2013). *Qualitative research: Analysis types and software*. Routledge.

Toloie-Eshlaghy, A., Chitsaz, S., Karimian, L., & Charkhchi, R. (2011). A classification of qualitative research methods. *Research Journal of International Studies*, 20(20), 106–123.

Tracy, S. J. (2010). Qualitative quality: Eight "big-tent" criteria for excellent qualitative research. *Qualitative Inquiry*, 16(10), 837–851.

Urry, J. (1992). The tourist gaze and the environment. *Theory, Culture & Society*, 9(3), 1–26. https://doi.org/10.1177/026327692009003001

Wang, X., Liu, C., Mao, W., & Fang, Z. (2015). The open access advantage considering citation, article usage and social media attention. *Scientometrics*, 103(2), 555–564. https://doi.org/10.1007/s11192-015-1547-0

Wilson, J., & Kumli, B. (2019). College students' definition of adventure: Three years of student-generated videos. *Journal of Outdoor Recreation, Education, and Leadership*, 11(3). https://doi.org/10.18666/JOREL-2019-V11-I3-9880

Wilson, J., Mandich, A., & Magalhães, L. (2016). Concept mapping: A dynamic, individualized and qualitative method for eliciting meaning. *Qualitative Health Research*, 26(8), 1151–1161. https://doi.org/10.1177/1049732315616623

Winter, T., & Adu-Ampong, E. A. (2021). Residents with camera: Exploring tourism impacts through participant-generated images. *Annals of Tourism Research*, 87, 103112. https://doi.org/10.1016/j.annals.2020.103112

Xu, F., Nash, N., & Whitmarsh, L. (2020). Big data or small data? A methodological review of sustainable tourism. *Journal of Sustainable Tourism*, 28(2), 144–163. https://doi.org/10.1080/09669582.2019.1631318

Zhu, D., & Xu, H. (2022). Novice tourists' play experience in commercial outdoor adventure tourism: The perspective of reversal theory. *Journal of Outdoor Recreation and Tourism*, 39, 100529. https://doi.org/10.1016/j.jort.2022.100529

Part II

Consumers of adventure tourism

This section focuses on the consumer or the 'adventure tourist', an individual who has evolved far from the masculine *Übermensch* of times past. In Chapter 7, Gill Pomfret begins by explaining our propensity towards adventure, building on the theoretical and historical foundations of the previous section. By examining adventurous motivations, benefits and constraints (and negotiating these), the chapter contains some important case studies on both family and older adventure tourists, who continue to grow in importance. Consequently, as we have hinted in previous chapters, the extant understanding of adventure is now so broad in terms of both activities and participation that it is important to consider how we might segment the market. Therefore, in Chapter 8, Melissa Lötter introduces conceptual frameworks for segmenting adventure tourism markets,

DOI: 10.4324/9781003393153-9

examining these from the perspective of activities, environments and experiences. In many ways, demand for adventure tourism is emblematic of the fragmentation and democratisation of the entire tourism product away from the Fordist packaging of previous eras. It is again apparent how markets have contributed towards a close link between activities and destinations, creating some very powerful place myths for adventure tourism.

Nevertheless, the evolution of adventure tourism, and particularly its adoption by new markets, means that its cultural interpretation remains dynamic. In Chapter 9 Vinathe Sharma-Brymer, Eric Brymer and Soumya Mitra explore how the Western idea of adventure tourism explored in previous chapters can be disrupted by more localised, culturally diverse and inclusive experiences of adventure. Indeed, recent interest in so-called microadventures in the West may well have established analogues in other cultures, for example the aspects of challenge and sublime nature connection inherent in ancient religious pilgrimage. Developing this in Chapter 10, Young Sook Lee and Ming Feng Huang provide a historical overview of the development of adventure activities in Korea and Taiwan and their respective foundations in leisure and sport, as well as significant colonial influences on such practices. The concept of colonial modernity is invoked to explain how adventure tourism has notable characteristics in East Asia.

For many, technology has had a significant impact on the tourist experience of adventure, perhaps somewhat paradoxically in a practice that often seeks to get back to nature. Some of this is in the physical equipment used, from scuba tanks to ziplines, but arguably the greatest has been in the field of digital technology. In Chapter 11, Ingo Janowski, Sarah Gardiner and Anna Kwek delve into how Australian, German and Chinese cultures employ online channels differently in an adventure tourism context. This reminds us not to revert to technological determinism in the application of digital technologies, as it is societies that ultimately decide how adventure experiences are contextualised.

In the final chapter of this section, Susan Houge Mackenzie goes to the heart of contemporary adventure consumption, heralding a more recent turn to place well-being at its core. It would appear that a sector that once seemed to prioritise hedonism should place greater focus on eudaimonic goals, broadening and deepening the motivations we identified at the beginning of the section. Such an awareness of well-being has significant implications, not only for adventure tourists but also for the people and places that host them.

7 The adventure tourist

Gill Pomfret

Chapter learning outcomes

1 Present some of the key demand trends in adventure tourism and provide insights into adventure tourists.
2 Develop an awareness of the motives and benefits of adventure tourism.
3 Understand the constraints adventure tourists face and how they negotiate these.
4 Examine in more depth two growing markets: older adventure tourists and family adventure tourists.

Introduction

As the first chapter in Part II of this handbook, this chapter aims to provide insights into the consumers of adventure tourism experiences. Given the diversity of adventure tourism activities which are now on offer to tourists and the plethora of different markets that engage in these on holiday or in their home region, there is potential to present contemporary thinking on many consumer-related topics. However, these themes were considered fundamental to providing a foundation to facilitate understanding of adventure tourists: the demand for adventure tourism, the motives encouraging individuals to participate in adventure activities, the benefits gained from engagement, and the constraints negotiation process before and during the activity. This chapter takes each theme in turn and presents scholarly and industry research, which includes differing perspectives and demonstrates the complexities involved in gaining insights into the consumers of adventure tourism. Throughout the chapter, there are three illustrative case studies: Sheffield: The Outdoor City, older adventure tourists, and family adventure tourists. Case study 7.1 demonstrates how Sheffield (UK) encourages visitors to the city and its surrounding environs through The Outdoor City brand, its reputation based on its natural topography and a wealth of outdoor activity opportunities and adventure organisations. Case studies 7.2 and 7.3 focus on two growing markets for adventure tourism and aim to offer insights into these relatively under-researched groups of tourists. Notably, despite the lack of scholarly attention on older adults and families, there is a wealth of UK-based adventure tour operators that offer adventure holidays specifically for these markets.

DOI: 10.4324/9781003393153-10

Adventure tourism demand trends

As discussed in other chapters (e.g. Chapters 1, 2 and 3), there have been various iterations of adventure tourism since its initial stages of development. This continuous evolution alongside the plethora of activities offered means that adventure tourism has wide appeal to a diverse range of markets. It is, however, difficult to accurately estimate the demand for adventure tourism globally for several reasons. First, adventure tourism is perceived differently by different people. Therefore 'an adventure for one person, in a particular place, at a given time, may not be an adventure for another, or for the same person in a different place or time' (Priest, 1999, p.60). For instance, a tourist participating in windsurfing for the first time on holiday will view and experience the activity differently from an experienced and skilled windsurfer. Related to this, it would be rare for anyone to identify themselves as an adventure tourist. More often than not, they are just exploring a destination and seeking to engage in enjoyable and meaningful experiences. Second, because adventure tourism is eclectic in meaning and reflects a broad range of activities, it shares commonalities with other types of tourism. Originally, scholars suggested that adventure tourism was intricately linked to activity, nature-based, discovery, and expedition tourism (Swarbrooke et al., 2003). Although this is still the case, new notions of adventure have evolved post-pandemic (see Chapter 1) which comprise a much more extensive range of nature-based activities which take place both locally and further afield. Accordingly, walking through forested areas in your local neighbourhood, forest bathing (see Chapter 14), and camping in your back garden are now deemed adventurous. Third, some individuals organise their own adventure trips or recreational adventures independently without employing the services of commercial organisations. For instance, tourists may go on a hiking holiday which they have organised themselves, working out and navigating routes and trails, and deciding which destination(s) to visit and where to stay. The size of the independent adventure market is, therefore, difficult to measure. Fourth, the United Nations World Tourism Organisation (unwto.org), which provides statistics on inbound and outbound tourism globally, regionally, and nationally, can offer some intelligence on tourist arrivals to known adventure destinations (e.g., Queenstown, New Zealand). However, it is difficult to ascertain the exact purpose of the visit beyond travelling for leisure as the data do not offer these more granular insights. With these difficulties in mind, we now turn to an overview of current and growing adventure tourism demand trends.

The Adventure Travel Trade Association (ATTA, 2023a), an industry trade body and global network of adventure travel leaders, provides a useful snapshot of current and future key trends based on 110 survey responses from inbound and outbound adventure tour operators. Although this is a small sample size, it offers some useful indications about the industry's growth trajectory. These data are though exclusively concerned with commercial operations and do not cover those tourists who independently organise their own adventure holidays. Furthermore, the ATTA (2020) advocates a broad definition of adventure tourism which embraces at least two of the three key elements: being in nature, engaging with culture, and physical activity. Importantly, the adventure tourism industry seems to have recovered well from the financial impact of the pandemic. Survey data found that adventure trip bookings increased from 52% full in 2021 to 65% full in 2022. The average number of tourists using the respondents' tour operating services rose considerably by 213% in 2022 (4,243 tourists) compared with 2021 (1,355 tourists). Holidays in high demand include those which are custom-made, visits to remote destinations, specialist guided experiences, slow travel adventures, sustainable, low-impact tourism, and trips for solo travellers and family groups. In 2022, hiking/trekking/walking, safaris/wildlife viewing, cycling (road/paved surface), cultural, culinary, and overland 4×4 were the most popular adventure trips. According to Mintel (2022), cycling and walking holidays will continue to be popular, and snow sports and equestrian activity holidays also show strong growth potential amongst UK tourists.

The ATTA (2023b) survey (103 respondents drawn from ATTA membership organisations) reports that tourists are increasingly interested in enjoying sustainable adventure experiences, reflecting a broader societal trend for sustainable practices and green products. Travelling in small groups to more nearby destinations and the continued rise in demand for local micro-adventures post-pandemic, due, in part, to changing leisure trends during this era and the cost-of-living crisis, are becoming more popular. Adventure tourists are keen to enjoy authentic experiences with opportunities to interact with local communities, eat and drink locally produced offerings, and purchase locally made products. Coming out of the pandemic, tourists prioritised nature-based tourism and outdoor activities to enhance their physical and mental well-being (Adventure Travel Trade Association 2023). This has benefitted the adventure tourism industry as seen in, for example the strong growth in demand for wellness retreats where tourists engage in slow adventure activities, such as native plant and mushroom foraging, and forest bathing in remote destinations. Tourists are becoming more aware of climate change and trying to reduce their carbon footprint, and tour operators are responding to this challenge (see Chapter 21). For instance, Global Family Travels promotes climate action to its tourists in its carbon-neutral Icelandic adventure holiday. The industry is also focusing its efforts on diversity, equity and inclusion with more tour operators creating Indigenous experiences and encouraging tourists to engage with local communities. Moreover, TripAdvisor has made its platform more inclusive for the diverse communities that use its website by asking businesses to self-identify their attributes, for example disability-owned, LGBTQIA+-owned (Adventure Travel Trade Trends, 2023b).

Tourism motives and benefits

The relationship between motives and benefits

Motives are recognised as the driving force which underpins our behaviour, and benefits are the enjoyable outcomes of such behaviour (Brown, 2008). Both are essential facilitators of adventure tourism participation. This section of the chapter therefore examines the key motives and motivational theories which encourage tourists to engage in adventure activities during their holidays, and the benefits they gain from these experiences. Motives and benefits are inherently intertwined with one another (Brown, 2008). As noted by Lehto et al. (2002), scholars often use the terms *motives, perceived benefits*, and *benefits sought* interchangeably to describe the expected psychological outcomes from tourism participation. Similarly, they employ the terms *actual benefits* and *benefits realised* when referring to the positive outcomes of tourism. Often, therefore, the motives which initially encourage tourists to partake in adventure (or other) pursuits change into the "benefits realised," and, in turn, these benefits facilitate further participation (Ewert et al., 2013). Moreover, the decision to continue engagement in a particular adventure activity becomes stronger if the adventure experience has been enjoyable. For instance, an adventure tourist may be strongly motivated to go on a rock-climbing holiday for escapism, the novelty of a new activity or destination or both and skills development and enjoy the natural environment. Assuming such needs are fulfilled, these original motives transform into benefits and drive the tourist to further engage in climbing in their home region as well as on holiday.

Adventure tourism motives and benefits

There has been considerable research attention directed at understanding adventure tourist motives and its benefits (e.g. Buckley, 2012; Ewert et al., 2013; Pomfret & Bramwell, 2016) even though this is a challenging process as 'the [adventure] experience is essentially ineffable and can be fully understood only by actually participating in it' (Lyng, 1990, p. 862). Some scholars have addressed this by immersing themselves in the adventure setting and activity and conducting auto-ethnographic studies (e.g., Buckley, 2012; Doran, 2016). According to Buckley (2020, p. 94):

> [m]any different motivations have been proposed and demonstrated for people to take part in adventurous nature sports; either individually; as members of recreational associations; or as clients of adventure tour operators.

Research in the fields of adventure recreation (e.g., Ewert, 1985), outdoor education (e.g., Miles & Priest, 1990), psychology (e.g., Csikzentmihalyi, 1992), and sociology (e.g., Breivik, 1996) formed the basis of early scholarly work on adventure tourism motivation. It is recognised that tourism and recreation converge in an adventure setting, as these industries frequently share the same resources and facilities, and activity participation results in similar experiences and outcomes (Pomfret & Bramwell, 2016). For example a person who regularly engages in caving as a hobby in their home region is likely to be encouraged to explore new caving destinations when on holiday. They may be motivated by exploration, discovery and challenge when caving locally, and by the range of cave environments offered at a destination when on holiday.

Early adventure recreation studies examined a plethora of motives, primarily concerned with risk and danger, challenge, mastery and skills development, and control over one's environment (Breivik, 1996; Ewert, 1985; Loewenstein, 1999). Scholars have reviewed this extant research alongside tourism studies to establish the most prominent motive-related themes for adventure tourists. Buckley (2012) ascertained three motive categories and 14 distinct motives within extant literature. Notably, only 15 of the 50 studies reviewed investigated tourists, and the remainder focused on recreationists. The categories were internal motives based on activity performance (e.g., achievement), internal or external motives focused on participants' place in nature (e.g., spiritual), and external motives reflecting participants' social position (e.g., friends). Pomfret and Bramwell (2016) found that researchers had mostly examined motives for multi-activity participation, primarily within recreation rather than tourism settings. Furthermore, research was absent on certain activities, including snowboarding, horseback riding, surfing, and paragliding. Other work (Ewert et al., 2013) compared the motives of participants with different skill and experience levels in different activities: canoeists, rock climbers, white water kayakers, and sea kayakers. This study showed motivational differences, for example canoeists scored highly on social motives whereas rock climbers scored highly on sensation-seeking motives.

Industry research (Mintel, 2022) ascertained that UK tourists have different motives for different activities. Skills development and excitement are important to water sports and snow sports enthusiasts whereas mental wellbeing is particularly pertinent for those on mountaineering, rock climbing, yoga, and Pilates holidays. However, the overriding pull motive for most respondents is being outdoors in the natural environment. The aforementioned ATTA (2023a) study of tour operator perspectives on their clients found that the most popular adventure motives were concerned with enjoying novel and authentic experiences and getting off the "beaten track", while other motives included travelling "like a local", cultural encounters, and wellness/betterment goals. What this research highlights is that there are many diverse motives, not solely one, driving engagement. Some of these motives are common to most adventure tourism activities, whereas others are specific to the activity.

Motives, benefits, and well-being

The strong growth in adventure tourism supply and demand in recent years (ATTA, 2024) indicates that increasingly diverse markets enjoy this type of holiday. Reflecting these changing trends, more scholars are adopting an integrated approach to adventure tourism, focusing on the benefits that activity participation brings. This approach contrasts with the traditional view of adventure which focuses on masculinity, risk, danger, and deviant personalities (Houge Mackenzie & Brymer, 2020). See Chapter 12 for a detailed discussion of traditional and contemporary conceptualisations of adventure and well-being, well-being motives, and well-being benefits from activity participation. *Well-being* is a term which embraces several characteristics concerned with personal, community and societal happiness, quality of life, health, capability, engagement, and life satisfaction (Diener, 2006). One study (Pomfret, Sand & May, 2023) conducted a systematic literature review on adventure activity engagement and subjective well-being, the latter including diverse types of individual well-being, such as social, psychological, and emotional (Sirgy, 2010). Findings suggest that scholars explore the positive outcomes of adventure activity participation but do not always apply the well-being construct. Five well-being themes and 16 subthemes emerged from this review: personal development, immersion and transformation, physical and mental balance, extraordinary experiences, and community. See Chapter 1 for a more comprehensive overview of these findings.

Extrinsic and intrinsic motives and benefits

Some scholars (e.g., Ewert et al., 2013) believe that there are motivational differences between adventure tourists according to their experience and skill levels. They suggest that novice adventurers are more extrinsically motivated, therefore their motives develop from external environmental factors, such as socialising and camaraderie. Contrastingly, skilled and experienced adventures are driven more by intrinsic motives, such as challenging oneself and accomplishing self-set goals. However, this distinction is not always so clear-cut. Motives are likely to change during adventure activity engagement depending on their experiences. Experienced adventurers can be extrinsically motivated by the natural environment, and novice adventurers can be intrinsically motivated by the need for improved physical and mental health (Kerr & Houge Mackenzie, 2012). Accordingly, Ewert et al. (2013, p. 103) propose that adventure activity participants (canoeing, whitewater kayaking, sea kayaking, rock climbing) are motivated by three factors: social, sensation-seeking, and self-image:

> Thus, individuals may engage in rock climbing because of the sensations they feel while on a rock face but also because they like to engage in challenging and exciting recreation with friends or a like-minded group of people, or they value being known and recognized as a "rock climber".

Push and pull motivation theory

Push and pull motivation theory (Dann, 1981) is often used to understand different tourist groups and their motives. Push motives arise from the need to fulfil internal motives, such as the desire to escape from daily life, relaxation, skills development, and new experiences. Pull motives comprise those elements which influence the tourist's choice of destination and product, for example climate, mountains, bodies of water, adventure guides, and equipment hire. The promotion of adventure tourism products within a destination also acts as a pull motive for tourists. It is suggested that push motives are aligned to intrinsic motives whereas pull motives reflect extrinsic motives. Given that the natural environment is an essential element of adventure tourism, a destination's natural landscape will exert a strong pull on adventure tourist decision-making (see Case study 7.1). For instance, national parks are protected for their natural features, such as their wildlife and high ecological quality, and are carefully managed and developed to offer a plethora of adventure recreation and tourism activities for their visitors (Natural England, 2011). Unsurprisingly, motives to visit national parks include push as well as pull elements, although the latter are arguably the key draw. Motives include escapism, nature, personal development, strengthening relationships, experiencing places, the search for solitude, self-evaluation, challenge, contact with nature, and physical activity (Carrascosa- López, Carvache-Franco & Mondéjar Jiménez, 2021). During pandemic times, recreation, relaxation, health, fitness, and exploration were important motives driving national park visitation (Kruczek et al., 2023).

Researchers have applied push and pull theory to understand adventure tourists in different contexts. The key push motives of rock climbers on holiday in Antalya Türkiye were the physical settings (i.e., the natural environment) and challenges whereas the most important pull motives were climbing novelty seeking and climbing tourism infrastructure (Caber & Albayrak, 2016). In the same region of Türkiye, research on scuba divers (Albayrak, Caber & Cater, 2021) established that exploration and excitement, personal development, and socialisation and challenge were important push motives. Tourism and diving offerings, safety and accessibility, ancillary attributes, new areas and unspoiled diving destinations were key pull motives. Recent work on mountain bikers (Abernethy et al., 2022) found that destination attractiveness is determined by the interaction of push motives and destination pull factors. These findings indicate that push and pull factors do not operate in isolation; rather, they are inextricably linked with one another.

Case study 7.1   Sheffield: The Outdoor City

Sheffield (South Yorkshire, UK) has been successfully developing The Outdoor City brand since 2014, building on the rich natural assets it offers in and around the city. Sheffield is unique in that it is the only UK city with a national park, the Peak District National Park, within its boundary. One third of Sheffield is located within this park. Its natural topography offers a plethora of greenspaces, natural landscapes, internationally renowned gritstone, rivers, reservoirs, and rolling hills. These features, alongside the wealth of outdoor adventure activity opportunities available and the 'extensive linked pockets of green spaces throughout the city' (Sheffield City Council, 2015, p. 7), act as pull motives encouraging visitors to Sheffield. In 2021, Sheffield was voted as the UK's greenest city from a pool of 25 cities and plans to be a net-zero city by 2050. Welcome to Sheffield's website promotes the city's wide-ranging outdoor resources. These include mountain biking, rock climbing, hiking, road cycling, running, canoeing, kayaking, stand-up paddleboarding, orienteering, and caving. There are many outdoor activity providers which offer instruction and guidance in these pursuits. For instance, DC Outdoors has 20-plus years of running adventure pursuits and delivers a plethora of outdoor activities in Sheffield and the Peak District National Park, tailored to individual needs and experiences. Sheffield's Festival of the Outdoors is a month-long celebration of The Outdoor City, taking place in March every year. It includes many adventure-related events, such as Sheffield Urban CX, an urban hill climb event for road cyclists, and an urban stand-up paddleboarding experience. Sheffield is also home to Jagged Globe, a global climb, trek and ski adventure operator which was 'the first British company to organise a professionally-led expedition to Everest' in 1993. In sum, Sheffield is 'a hub for all kinds of adventures' https://www.welcome tosheffield.co.uk/visit/outdoors-nature/.

Other chapters explore the pull of adventure destinations and products in more depth. In Chapter 13, Sand and Gross examine the different sectors and types of organisations which make up the adventure tourism industry. Notably, the provision of incoming travel agencies, accommodation, suitable adventure products and natural environment resources within destinations function as pull motives for adventure tourists. How motivational these elements are is contingent on how supportive different governments are regarding adventure tourism development within their countries and the level of private investment in destination development. These factors are included in the ATTA's Adventure Tourism Development Index (ATDI) which ranks countries on their potential for sustainable adventure tourism. In Chapter 14, Farkić and Taylor highlight the need for destinations to continually focus their efforts on sustainable development and develop adventure products which are environmentally and socially sustainable. This increased focus on sustainability can act as a powerful pull element for adventure tourists. See Case studies 7.2 and 7.3 on older adventure tourists and family adventure tourists for insights into the motives driving participation by these markets, and the benefits that they gain.

Constraints and negotiation strategies for adventure tourists

It is evident that motives and anticipated benefits strongly encourage adventure tourism participation, yet tourists often encounter various constraints both pre- and during activity engagement. Such constraints can sometimes be negotiated, particularly if motivation levels are high, whereas others can be more difficult to overcome. Scholars have developed different frameworks and categorisations of constraints related to leisure. One renowned framework is the theory of hierarchical leisure constraints (Crawford, Jackson & Godbey, 1991), which proposes three categories of constraints that individuals may face when engaging in leisure pursuits, including adventure activities. Intrapersonal constraints are psychological and physical factors which influence participants' expectations, their personal beliefs about their skills and experiences of an activity, for example worrying about getting immersed in glacial water during whitewater rafting. Interpersonal constraints are social factors which influence our leisure choices, caused by others affecting individuals, for example not having anyone to share adventure experiences with or having different leisure preferences. Structural constraints are intervening factors that affect activity engagement and preference, for example a lack of money to purchase mountaineering equipment, or family-oriented factors, such as childcare or caring for older relatives.

As its name suggests, this framework is sequential and the effect of the three types of constraints on leisure participation is hierarchical. In 1993, Jackson, Crawford and Godbey introduced the term "constraint negotiation" to explain that leisure activity engagement relies on the sequential negotiation of constraints rather than an absence of such barriers. Initially, intrapersonal constraints shape people's leisure preferences. Then, they may encounter interpersonal constraints, depending on their leisure preferences and whether they are engaging in the activity with someone else or not. Once, and if, they overcome these constraints, individuals then need to deal with any structural constraints. If they successfully negotiate these, then they can participate in the leisure activity; if not, the result is non-participation (Crawford et al., 1991). This leisure constraints theory has been extensively applied to diverse types of leisure participants, for instance older tourists (Huber, Milne & Hyde, 2018), carers (Hunter-Jones et al., 2023), and millennials (Griggs & Lacey, 2022).

Scholars have applied the theory of hierarchical leisure constraints to adventure settings. Whether to participate in an activity, or continue to engage in it, depends on how important the three constraint types are, the interplay between different constraints, and the importance that individuals attribute to each one (Crawford et al., 1991). Additionally, negotiation efficacy influences individuals' ability to deal with the constraints they face. This is 'people's confidence in their ability to successfully use negotiation strategies to overcome constraints they encounter' (Loucks-Atkinson & Mannell, 2007, p.22). For instance, Fu, Liu and Yang (2022) found that although the lack of interest in ice and snow sports constraint dimension scored lowest amongst Chinese residents, suggesting strong levels of motivation, several other constraint dimensions also impacted their likelihood of engaging in these sports. These were personal feelings and attitudes, a lack of knowledge and skills, a lack of facilities and services, time, no companion, and a lack of interest. Other work (Alexandris et al., 2017) ascertained strong links between people's experience and loyalty levels in skiing, how psychologically connected they were to the sport, and the intensity of constraints they faced. Experienced, loyal skiers ("I live to ski") were less likely to encounter intense and high-level constraints compared to less experienced skiers. Newer skiers were not as psychologically connected to skiing as their more experienced counterparts, and therefore, they were more likely to not continue with the sport. Interestingly, structural constraints, particularly time and cost constraints, were important across all experience levels. However, more experienced skiers were strongly motivated to negotiate these constraints, which explains their continued engagement with skiing. Zarei et al. (2021) used the model to explore constraints which influence people's pro-environmental behaviour when engaging in mountain hiking. Findings revealed that constraints strongly influence pro-environmental behaviour. Structural constraints can be reduced by providing proper restrooms and parking spaces and controlling the carrying capacity of an area by limiting the number of people who visit. Intrapersonal and interpersonal constraints can be reduced by, for example, implementing volunteer programmes on hiking routes to provide visitors with information about the local natural environment and environmentally friendly practices.

Other adventure scholars (e.g., Doran, 2016; Fendt & Wilson, 2012) have opted to classify constraints broadly into practical, socio-cultural, and personal constraints. This approach has been used to examine constraints faced by women before and during adventure activity participation. For instance, Doran's (2016) content analysis of literature on women's adventure experiences revealed fewer personal constraints compared to practical and socio-cultural ones. Prominent personal constraint subcategories were self-doubt, fear, a sense of guilt, and a perception of not being adventurous. There was a wider range of socio-cultural sub categories: a lack of companions, gender stereotypes, social expectations, unwanted male attention, masculine portrayal of adventure, and experiencing localism. Similarly, practical constraints had several subcategories: a lack of money, time, familiarity with the destination and role models, need for specific equipment, expedition mentality, and poor promotion of adventure opportunities for women. Women use wide-ranging and complex strategies to negotiate these constraints at different points of their adventure experience journey. These are grouped into three categories: determination (e.g., strong motivation), planning and preparing (e.g., activity training to develop skills, fitness and knowledge), and the need to prioritise participation and make compromises (e.g., adjusting their chosen adventure activity). Successfully negotiating constraints and, consequently, participating in the adventure activity can lead to feelings of empowerment for women.

Other work (Doran, Schofield & Low, 2018), which explores women's participation in a range of mountaineering tourism activities, returns to the hierarchical leisure constraints model. Aside from the presence of intrapersonal, interpersonal, and structural constraints, these researchers established a new and distinct constraint dimension of family constraints. This involves family and/or friends not understanding their participation in mountaineering tourism and household duties/family commitments. Their later study (Doran, Schofield & Low, 2020), which examines women's negotiation of mountaineering constraints, found four negotiation dimensions. Three of these are listed earlier (Doran, 2016) whereas the fourth new dimension is confidence and adaptation. Although the latter is reflected in determination strategies, that is women feeling confident enough in their own skills to participate, and planning and preparing strategies, that is improving confidence through training beforehand, in this study confidence and adaptation is a distinct dimension. The previously mentioned studies clearly indicate that constraints and their negotiation influence participation in adventure activities, alongside motives and other influences. For instance, for mountaineer tourists, these include hardiness, self-efficacy (belief in one's capabilities) and negotiation-efficacy (Doran & Pomfret, 2019). See Case studies 7.2 and 7.3 on older adventure tourists and family adventure tourists for insights into the constraints that these markets potentially face and how they negotiate these.

Case study 7.2   Older adventure tourists

Almost 39% of the UK population will be aged 50 years or over by 2030, rising to 42% by 2050 (Office for National Statistics, 2024). Relatedly, there is plentiful evidence to suggest that older adults will form an increasingly significant market for the adventure tourism industry. Being outdoors in the natural environment and feeling physically rejuvenated – spurred by the desire to stay active and fit – are key drivers for UK tourists aged 45 years and older (Mintel, 2022). The Adventure Travel Trade Association (ATTA, 2024) notes that Booking.com anticipates more women over 50 years of age will travel alone in future years, empowered by social and financial shifts and feelings of freedom. Table 7.1 lists some operators which offer adventure holidays for older adults. Mostly, these organisations target the 50-plus age group. Despite this growing market, few scholars have explored outdoor adventure activity participation amongst older adults (Smith & Dalmer, 2023). Moreover, researchers often frame engagement in such pursuits in later life in a 'potentially condescending way' (p. 3). They adopt the active ageing or successful ageing approaches, which homogenise older adults and view ageing as a negative process.

The benefits of any type of physical activity (inclusive of adventure pursuits) for older adults include improved self-esteem (Devereux-Fitzgerald et al., 2016), enhanced mental health (Chen et al., 2012), opportunities to continue with lifelong leisure patterns, and enhanced social interaction (Barnett, Guell & Ogilvie, 2012). Adventure scholars (e.g., Buckley, 2020) suggest that 'by keeping older people healthy and happy, adventurous outdoor nature sports thus make large contributions to national economies' (p. 93). Participation in these activities, therefore, improves older adults' physical, social, and mental health; develops their resilience; leads to productive working lives; and results in positive contributions to their community. For those with reduced capabilities,

Table 7.1 Adventure holidays for older adults

Adventure tour operator	*Website*
Silversurfers	https://www.silversurfers.com/
Silver Travel Advisor	https://silvertraveladvisor.com/
Explore	https://www.explore.co.uk/
Responsible Travel	https://www.responsibletravel.com/
On The Go Tours	https://www.onthegotours.com/uk
Wild Frontiers	https://www.wildfrontierstravel.com/en_GB
Intrepid Travel	https://www.intrepidtravel.com/uk
Solos	https://www.solosholidays.co.uk/

engagement can bring additional benefits, such as euphoria, which can temporarily help reduce the stress of ageing and feelings of chronic pain (Buckley, 2020; Gregory & Dimmock, 2019). Smith and Dalmer's (2023) scoping review of extant research on older adults and outdoor adventure activities ascertained that many scholars focused on the physical (e.g., improved physical fitness) and mental health and well-being (e.g., improved self-esteem and confidence) benefits from participation.

Other investigations examine the benefits of doing specific adventure activities. For instance, older and experienced women divers in Australia (Gregory & Dimmock, 2019; Gregory & Edney, 2019) reaped many benefits from diving including improved quality of life, well-being, and confidence, meeting new people, learning about other cultures and countries, feeling physically fitter, and personally challenging themselves. Wheaton's (2017) study on active older adults established that 'silver surfers' (p. 96) enjoyed the activity of surfing as well as the lifestyle. They felt youthful, active and connected to their surfing community, and embraced a "living for the moment" attitude.

Although older adults gain many benefits from adventure activity participation, they can also face numerous constraints which influence their decisions around recreation and tourism participation generally. Structural constraints are the most influential barriers and include a lack of information, time, and financial resources and perceptions of travel risks, such as safety fears or worrying that something may go wrong during the activity or trip. Intrapersonal constraints include the physiological effects of ageing and their impact on health and psychological factors, such as feeling too old to engage in adventure activities or losing interest in travel. Interpersonal constraints include a lack of social resources, such as not having a companion (spouse or friend) to do activities with or not sharing the same interests (Moal, 2020). Although these constraints potentially affect older adults' engagement in adventure activities, their strong motivation to participate can often result in successful negotiation (Chen et al., 2021) using a range of strategies. These include taking organised trips (see Table 7.1), joining activity clubs and associations, and modifying the activity to accommodate their needs (Huber, Milne & Hyde, 2018).

Little is known about the constraints that older adults face in adventure settings. Yet it is important to understand the effects of these barriers on participation and explore ways to negotiate these to facilitate healthy and active ageing. Scholars (Currie, Colley & Irvine, 2021) have established that this demographic is less likely to access outdoor environments for recreational purposes. Through interviews, they identified numerous constraints. These were intrapersonal, for example health and mobility (chronic and acute conditions); interpersonal, for example safety concerns (being attacked by dogs); and structural, for example weather and daylight and access to outdoor spaces (poor paths, cost to get there).

Case study 7.3   Family adventure tourists

Families are important to the adventure tourism industry and organisations are increasingly diversifying their markets and focusing on the family adventure market, primarily in response to changing trends post-COVID-19 (ATTA, 2023a) and the need to spend quality time with family and friends on holiday. More tourists are prioritising time with their families' post-pandemic, which is driving the demand for multigenerational holidays (Mintel, 2024). Operators offer commodified, soft multi-activity adventure experiences to cater for the varied skill levels, motives, and ages of different family members. Itineraries are geared towards fulfilling everyone's needs through designing a carefully balanced set of activities for the whole family as well as parent- and children-only pursuits (Pomfret & Varley, 2019). Table 7.2 lists a selection of UK adventure tour operators which offer family adventure holidays. Most of these are operators with product offerings for several different markets inclusive of families. For instance, 50% of The Adventure People's clients are solo travellers. G Adventures offers its G Adventure Land collection for people aged 18 to 30-something, and separate land-based trips for families, whereas its Geluxe holidays are aimed at the premium market. A smaller number of operators specifically cater for families and specialise in adventure and activity holidays, such as Families Worldwide and Take the Family.

Table 7.2 Tour operators and family adventure holidays

Adventure tour operator	*Website*
Stubborn Mule Travel	https://www.stubbornmuletravel.com/
Families Worldwide	https://www.familiesworldwide.co.uk/
Intrepid Travel	https://www.intrepidtravel.com/uk
G Adventures	https://www.gadventures.com/
Exodus Adventure Travels	https://www.exodus.co.uk/
Tour Dust	https://www.tourdust.com/
The Adventure People	https://theadventurepeople.com/
PGL	https://www.pgl.co.uk/
Explore	https://www.explore.co.uk/
KE Adventure Travel	https://www.keadventure.com/
On The Go Tours	https://www.onthegotours.com/
Take the Family	https://www.takethefamily.com/
Neilson Holidays	https://www.neilson.co.uk/
Trailfinders	https://www.trailfinders.com/
World Expeditions	https://worldexpeditions.com/

Although it is evident that there is a plentiful supply of operators offering adventure holidays to families, it should be noted that this market also takes independently organised trips. This is particularly the case when parents or guardians are experienced and competent adventurers who are keen to develop a family legacy of shared traditions, wisdom, values, and history, known as generativity (Hebblethwaite & Norris, 2011). They can achieve this by passing on their adventure skills, for example rock climbing or orienteering, to their children.

There are many studies which provide insights into family tourists, as indicated in a bibliometric analysis of family tourism research (Li, Lehto & Li, 2020). The success of family holidays is contingent on 'a harmonic balanced set of different individual pleasures' (Larsen, 2013, p. 171). For instance, young children thrive from fun activities and play, whereas teenagers enjoy challenges, socialising, and personal development on holiday. However, families are at their happiest when participating in activities together (Schänzel, 2010). Less is known about those families who engage in adventure activities on holiday. This is compounded by the difficulties involved in conducting research on families, such as ensuring the perspectives of all family members, particularly young children, are reflected during data collection (Pomfret, 2021). Individual family members have differing needs, and, therefore, it is difficult to fully comprehend their holiday motives, experiences, and benefits enjoyed.

Unsurprisingly, therefore, the motives and benefits of family adventure tourism are wide ranging and reflect not only generic holiday motives but also the fulfilment of specific adventure needs. Furthermore, the motives which drive adventure activity participation differ between parents and children and between different age groups of children (Jamal, Aminudin & Kausar, 2019; Pomfret, 2021). They include shared fun and novelty, exciting experiences, bonding and togetherness, enhanced family communication, trust and family relationships, generativity, skills development, and family well-being and health (Obrador, 2012; Pomfret, 2021). Family members can influence each other's subjective well-being. For instance, parents can help younger children to feel safe and secure during adventure activity engagement. Parental decisions to get their children involved in adventure provide personal development opportunities and family bonding (Pomfret, 2021). The subjective well-being of individual members, in turn, affects family well-being. This is the collective mental, social, and physiological health of each member, and it reflects enhanced family resilience, functioning, and cohesion (Noor et al., 2014).

Constraints negotiation research on families is limited, with most studies in this area focusing on individuals. Yet, adventure tourism activities can be challenging for families as they usually necessitate specific skills, experience, and knowledge to safely participate (Reis, Thompson-Carr & Lovelock, 2012; Pomfret, 2021). Other considerations are the composition of families (single parent or both parents or guardians), the number of children, and their ages and genders. For instance, gender can influence what type of outdoor activities children are exposed to during their childhood, with parents encouraging more male-oriented activities for boys, such as fishing or hunting. Outdoor-oriented parents, who introduce their children to adventure at an early age, are more likely to continue to collectively participate in adventurous pursuits as a family unit (Lovelock et al., 2016). Young children can be constrained by their age in their level of prowess in a particular adventure activity because of the skills and experience required (Pomfret, 2021). To become skilled in such a pursuit, they need time combined with regular exposure to, and practice in, the activity. Therefore, it is important to understand constraints and how these can be successfully negotiated by the families themselves as well as by adventure organisations.

Constraints for families include a lack of free time and financial resources, the ages of children and the stage in their life cycle, parenting arrangements, access to the natural environment and information about this and activity opportunities, and appropriate infrastructure (Agate, Williams & Barrett, 2011). Women are often further constrained in their individual leisure pursuits when they become mothers (Reis et al., 2012). Intrapersonal constraints, which reflect individual preferences, can discourage family participation in adventure activities. Weather conditions, time commitments, for example school or work, and the stage of childhood development – which affects preferences and the growing need for independence – influence mothers and daughters' engagement in such pursuits (Izenstark & Ebata, 2022). Relatedly, interpersonal constraints, such as peer pressure and the need to conform, can affect older children and adolescents' leisure activity choices (Lovelock et al., 2016). Structural constraints affect family participation in adventure activities. Reis et al. (2012) found that New Zealand's national government organisation, the Department of Conservation, and Dunedin City Council did not focus its efforts on families in providing outdoor recreation events and opportunities. However, Wellington Regional Council did run a series of outdoor events and activities for families. Although this is an old study, it signals the importance of planned outdoor activities and opportunities as a key negotiation strategy to overcome such structural constraints. In summary, there are numerous constraints that families may face in adventure activity participation. However, if strong motivational levels and successful negotiation strategies combine, families have a high propensity to engage in adventure activities either within their home region or when on holiday.

Conclusion

This chapter provided some useful insights into the demand for adventure tourism, who adventure tourists are, how they are motivated, the benefits they experience from adventure activity engagement, the constraints they potentially face before and during their adventure pursuits, and the strategies they negotiate to overcome these. It is evident that adventure tourism is enjoying a strong growth trajectory, and it is becoming more accessible to almost everyone as adventure organisations, and destinations continuously expand their product offerings to cater for the needs of an increasingly diverse profile of adventure tourists. The following chapters (Chapters 8–12) investigate several important themes concerning adventure tourists, focusing on their segmentation, differing perspectives of adventure experiences which are more localised and inclusive, the development of adventure tourism in Korea and Taiwan, technology and adventure experiences, and the increasingly important topic of well-being through adventure.

Review questions

1 Choose three different adventure activities and research the motives and benefits of these. What are the key similarities and differences between the three activities regarding motives and benefits?
2 Choose three adventure tourist markets (e.g. older adults, families, women, couples). Make a table of the key intrapersonal, interpersonal, and structural constraints that these adventure tourists face. In a table, develop a list of the most appropriate and effective negotiation strategies to overcome these constraints.
3 Investigate a known adventure tourism destination and develop a list of the key pull motives which entice adventure tourists to visit there and engage in adventure activities.

Further reading

Doran, A., Schofield, P., & Low, T. (2020). Women's mountaineering: Accessing participation benefits through constraint negotiation strategies. *Leisure Studies*, *39*(5), 721–735. https://doi.org/1 0.1080/02614367.2020.1763439
Jamal, S. A., Aminudin, N., & Kausar, D. R. (2019). Family adventure tourism motives and decision-making: A case of whitewater rafting. *Journal of Outdoor Recreation & Tourism*, *25*, 10–15. https://doi.org/10.1016/j.jort.2018.11.005
Mintel (2022). *Physical activity holidays UK*. Mintel Marketing Intelligence
Pomfret, G., & Bramwell, B. (2016). The characteristics and motivational decisions of outdoor adventure tourists: A review and analysis. *Current Issues in Tourism*, *19*(14), 1447–1478. https:// doi.org/10.1080/13683500.2014.925430
Smith, E. S., & Dalmer, N. K. (2023). Understanding older adults' participation in outdoor adventure activities: A scoping review. *Journal of Adventure Education and Outdoor Learning*. https:// doi.org/10.1080/14729679.2023.2295842

References

Abernethy, B., Dixon, A. W., Holladay, P. J., & Koo, W. G. (2022). Determinants of Canadian and US mountain bike tourists' site preferences: Examining the push-pull relationship. *Journal of Sport and Tourism*, *26*(3), 249–268. https://doi.org/10.1080/14775085.2022.2069147
Adventure Travel Trade Association. (2020). *Adventure tourism development index*. ATTA.
Adventure Travel Trade Association. (2023a). Adventure travel industry snapshot. June 2023. 2023 Adventure Travel Industry Snapshot | Adventure Travel Trade Association
Adventure Travel Trade Association. (2024). Industry outlook 2024: The key trends, challenges, and issues shaping adventure travel. Industry Outlook 2024: The Key Trends, Challenges, and Issues Shaping Adventure Travel | Adventure Travel Trade Association adventuretravelnews.com
Adventure Travel Trade Association. (2023b). Creating sustainable experiences in adventure travel. Creating Sustainable Experiences in Adventure Travel | Adventure Travel Trade Association
Agate, S., Williams, J. E., & Barrett, N. (2011). From Mickey Mouse to Max and Cheese: Enhancing user experience for the family market. *International Journal of Business Innovation and Research*, *5*(4), 338–358. https://doi.org/10.1504/IJBIR.2011.041055
Albayrak, T., Caber, M., & Cater, C. (2021). Mass tourism underwater: A segmentation approach to motivations of scuba diving holiday tourists. *Tourism Geographies*, *23*(5–6), 985–1000. https://doi.org/10.1080/14616688.2019.1696884
Alexandris, K., Du, J., Funk, D., & Theodorakis, N. D. (2017). Leisure constraints and the psychological continuum model: A study among recreational mountain skiers. *Leisure Studies*, *36*(5), 670–683. https://doi.org/10.1080/02614367.2016.1263871
Barnett, I., Guell, C., & Ogilvie, D. (2012). The experience of physical activity and the transition to retirement: A systematic review and integrative synthesis of qualitative and quantitative evidence. *International Journal of Behavioral Nutrition and Physical Activity 9*(1), 97–107. http:// www.ijbnpa.org/content/9/1/97

Breivik, G. (1996). Personality, sensation seeking and risk taking among Everest climbers. *International Journal of Sport Psychology, 27*(3), 308–320.

Brown, S. (2008). Travelling with a purpose: Understanding the motives and benefits of volunteer vacations. *Current Issues in Tourism, 8*(6), 479–496.

Buckley, R. (2012). Rush as a key motivation in skilled adventure tourism: Resolving the risk recreation paradox. *Tourism Management, 33*(4), 961–970. http://dx.doi.org/10.1016/j.tourman.2011.10.002

Buckley, R. (2020). Nature tourism and mental health: Parks, happiness and causation. *Journal of Sustainable Tourism, 28*(9), 1409–1424. https://doi.org/10.1080/09669582.2020.1742725

Caber, M., & Albayrak, T. (2016). Push or pull? Identifying rock climbing tourists' motivations. *Tourism Management, 55*, 74–84. https://doi.org/10.1016/j.tourman.2016.02.003

Carrascosa-López, C., Carvache-Franco, M., & Mondéjar Jiménez, J. (2021). Caravache-Franco 2021. Understanding motivations and segmentation in ecotourism destinations. Application to natural parks in Spanish Mediterranean area. *Sustainability 13*, 4802. https://doi.org/10.3390/su13094802

Chen, F., Dai, S., Xu, H., & Abliz, A. (2021). Senior's travel constraint, negotiation strategy and travel intention: Examining the role of social support. *The International Journal of Tourism Research, 23*(3), 363–377. https://doi.org/10.1002/jtr.2412

Chen, L. J., Stevinson, C., Ku, P. W., Chang, Y. K., & Chu, D. C. (2012). Relationships of leisure-time and non-leisure-time physical activity with depressive symptoms: A population based study of Taiwanese older adults. *International Journal of Behavioral Nutrition and Physical Activity 9*(28), 1–10. http://www.ijbnpa.org/content/9/1/28

Crawford D., Jackson E., & Godbey G. (1991). A hierarchical model of leisure constraints. *Leisure Sciences, 13*(4), 309–320.

Csikzentmihalyi, M. (1992). *Flow*. Harper & Row.

Currie, M., Colley, K., & Irvine, K. N. (2021). Outdoor recreation for older adults in Scotland: Qualitatively exploring the multiplicity of constraints to participation. *International Journal of Environmental Research and Public Health, 18*(18). https://doi.org/10.3390/ijerph18147705

Dann, G. (1981). Tourist motivation: An appraisal. Annals of Tourism Research, 8(2), 187–219. https://doi.org/10.1016/0160-7383(81)90082-7

Devereux-Fitzgerald, A., Powell, R., Dewhurst, A., & French, D. P. (2016). The acceptability of physical activity interventions to older adults: A systematic review and meta-synthesis. *Social Science & Medicine, 158*, 14–23. https://doi.org/10.1016/j.socscimed.2016.04.006

Diener, E. (2006). Guidelines for national indicators of subjective well-being and ill-being. *Journal of Happiness Studies, 7*(4), 397–404.

Doran, A. (2016). Empowerment and women in adventure tourism: A negotiated journey. *Journal of Sport & Tourism, 20*(21), 57–80. https://www.tandfonline.com/action/showCitFormats?doi=10.1080/14775085.2016.1176594

Doran, A., & Pomfret, G. (2019). Exploring efficacy in personal constraint negotiation: An ethnography of mountaineering tourists. *Tourist Studies, 19*(4), 475–495. https://doi.org/10.1177/1468797619837965

Doran, A., Schofield, P., & Low, T. (2018). Women's mountaineering tourism: An empirical investigation of its theoretical constraint dimensions. *Leisure Studies, 37*(4), 396–410. https://doi.org/10.1080/02614367.2018.1452283

Doran, A., Schofield, P., & Low, T. (2020). Women's mountaineering: Accessing participation benefits through constraint negotiation strategies. *Leisure Studies, 39*(5), 721–735. https://doi.org/10.1080/02614367.2020.1763439

Ewert, A. (1985). Why people climb: The relationship of participant motives and experience level to mountaineering. *Journal of Leisure Research, 17*(3), 241–250.

Ewert, A., Gilbertson, K., Luo, Y. C., & Voight, A. (2013). Beyond "Because it's there": Motivations for pursuing adventure recreational activities. *Journal of Leisure Research, 45*(1), 91–111. https://www.tandfonline.com/action/showCitFormats?doi=10.18666/jlr-2013-v45-i1-2944

Fendt, L. S., & Wilson, E. (2012). 'I just push through the barriers because I live for surfing': How women negotiate their constraints in surf tourism. *Annals of Leisure Research, 15*(1), 4–18. https://doi.org/10.1080/11745398.2012.670960

Fu, L., Liu, Y., & Yang, Z. (2022). Influence of ice and snow sports participation experience on participation constraints among residents in southern China: A quantitative analysis based on propensity score matching. *Journal of Resources & Ecology*, *13*(4), 624–634. https://doi.org/10.5814/j.issn.1674-764x.2022.04.008

Gregory, S. F., & Dimmock, K. (2019). Alive and kicking: The benefits of scuba diving leisure for older Australian women. *Annals of Leisure Research*, *22*(4), 550–574. https://doi.org/10.1080/11745398.2019.1605914

Gregory, S. F., & Edney, J. (2019). Divers or divas? A market analysis of the mature aged female diver: An Australian perspective. *Tourism in Marine Environments*, *14*(3), 143–162. https://doi.org/10.3727/154427319X15635387000925

Griggs, N., & Lacey, G. T. (2022). Barriers and limitations to national park visitation by millennials: Perceptions from second-generation Australians. *Annals of Tourism Research Empirical Insights*, *3*. https://doi.org/10.1016/j.annale.2022.100074

Hebblethwaite, S., & Norris, J. (2011). Expressions of generativity through family leisure: Experiences of grandparents and adult grandchildren. *Family Relations*, *60*(1), 121–133.

Houge Mackenzie, S., & Brymer, E. (2020). Conceptualizing adventurous nature sport: A positive psychology perspective. *Annals of Leisure Research*, *23*(1), 79–91. https://doi.org/10.1080/11745398.2018.1483733

Huber, D., Milne, S., & Hyde, K. (2018). Constraints and facilitators for senior tourism. *Tourism Management Perspectives*, *27*. https://doi.org/10.1016/j.tmp.2018.04.003

Hunter-Jones, P., Sudbury-Riley, L., Chan, J., & Al-Abdin, A. (2023). Barriers to participation in tourism linked respite care. *Annals of Tourism Research*, *98*. https://doi.org/10.1016/j.annals.2022.103508

Izenstark, D., & Ebata, A. T. (2022). Why families go outside: An exploration of mothers' and daughters' family-based nature activities. *Leisure Sciences*, *44*(5), 559–577. https://doi.org/10.1080/01490400.2019.1625293

Jackson, E. L., Crawford, D. W., & Godbey, G. (1993). Negotiation of leisure constraints. *Leisure Sciences*, *15*(1), 1–11. https://psycnet.apa.org/doi/10.1080/01490409309513182

Jamal, S. A., Aminudin, N., & Kausar, D. R. (2019). Family adventure tourism motives and decision-making: A case of whitewater rafting. *Journal of Outdoor Recreation & Tourism*, *25*, 10–15. https://doi.org/10.1016/j.jort.2018.11.005

Kerr, J. H., & Houge Mackenzie, S. (2012). Multiple motives for participating in adventure sports. *Psychology of Sport & Exercise*, *13*(5), 649–657. http://dx.doi.org/10.1016/j.psychsport.2012.04.002

Kruczek, Z., Szromek, A. R., Jodłowski, M., Gymrek, K., & Nowak, K. (2023). Visiting national parks during the COVID-19 pandemic–An example of social adaptation of tourists in the perspective of creating social innovations. *Journal of Open Innovation: Technology, Market, and Complexity*, *9*, 100062. https://doi.org/10.1016/j.joitmc.2023.100062

Larsen, J. (2013). Family flow: The pleasures of "being together" in a holiday home. *Scandinavian Journal of Hospitality & Tourism*, *13*(3), 153–174. https://doi.org/10.1080/15022250.2013.808523

Lehto, X., O'Leary, J., & Lee, G. (2002). Mature international travellers: An examination of gender and benefits. *Journal of Hospitality and Leisure Marketing*, *9*(1/2), 53–72. https://doi.org/10.1300/J150v09n01_05

Li, M., Lehto, X., & Li, H. (2020). 40 years of family tourism research: Bibliometric analysis and remaining issues. *Journal of China Tourism Research*, *16*(1), 1–22. https://doi.org/10.1080/19388160.2020.1733337

Loewenstein, G. (1999). Because it's there: The challenge of mountaineering … for utility theory. *Kyklos*, *52*(3), 315–343.

Loucks-Atkinson, A., & Mannell, R. (2007) Role of self-efficacy in the constraints negotiation process: The case of individuals with fibromyalgia syndrome. *Leisure Sciences 29*(1), 19–36. https://doi.org/10.1080/01490400600983313

Lovelock, B., Walters, T., Jellum, C., & Thompson-Carr, A. (2016). The participation of children, adolescents, and young adults in nature-based recreation. *Leisure Sciences*, *38*(5), 441–460. https://doi.org/10.1080/01490400.2016.1151388

Lyng, S. (1990). Edgework: A social psychological analysis of voluntary risk taking. *The American Journal of Sociology*, *95*(4), 851–886.

Miles, J. C., & Priest, S. (Eds.) (1990). *Adventure education*. Venture Publishing Inc.

Mintel. (2022). *Physical activity holidays UK*. Mintel Marketing Intelligence.

Mintel. (2024). *Holiday review UK*. Mintel Marketing Intelligence.

Moal, G. M. (2020). The travel constraints faced by retired travelers in the 21st. *Journal of Consumer Marketing, 38*(1), 69–77. http://dx.doi.org/10.1108/JCM-02-2020-3665

Natural England. (2011). *Guidance for Assessing Landscapes for Designation as National Park or Area of Outstanding Natural Beauty, England.*

Noor, N., Ghandi, A., Ismahalil, I., & Saodah, W. (2014). Development of indicators for family well-being in Malaysia. *Social Indicators Research, 115*(1), 279–318.

Obrador, P. (2012). The place of family in tourism research: Domesticity and thick sociality by the pool. *Annals of Tourism Research, 39*(1), 401–420. http://dx.doi.org/10.1016/j.annals.2011.07.006

Office for National Statistics. (2024). *Principal projection – UK population in age groups*. Home - Office for National Statistics (ons.gov.uk)

Pomfret, G. (2021). Family adventure tourism: Towards hedonic and eudaimonic wellbeing. *Tourism Management Perspectives, 39*,100852. https://doi.org/10.1016/j.tmp.2021.100852

Pomfret, G., & Bramwell, B. (2016). The characteristics and motivational decisions of outdoor adventure tourists: A review and analysis. *Current Issues in Tourism, 19*(14), 1447–1478. https://doi.org/10.1080/13683500.2014.925430

Pomfret, G., Sand, M., & May, C. (2023). Conceptualising the power of outdoor adventure activities for subjective well-being: A systematic literature review. *Journal of Outdoor Recreation & Tourism, 42*, 100641. https://doi.org/10.1016/j.jort.2023.100641

Pomfret, G., & Varley, P. (2019). Families at leisure outdoors: Well-being through adventure. *Leisure Studies, 38*(4), 494–508. https://doi.org/10.1080/02614367.2019.1600574

Priest, S. (1999). The adventure experience paradigm. In Miles, J. C., & Priest, S. (Eds.). *Adventure programming*. Venture Publishing, Inc.

Reis, A. C., Thompson-Carr, A., & Lovelock, B. (2012). Parks and families: Addressing management facilitators and constraints to outdoor recreation participation. *Annals of Leisure Research, 15*(4), 315–334.

Schänzel, H. A. (2010). Whole family research: Towards a methodology in tourism for encompassing generation, gender, and group dynamic perspectives. *Tourism Analysis, 15*(5), 555–569.

Sheffield City Council. (2015). *The outdoor city economic strategy*. Sheffield City Council.

Sirgy, M. J. (2010). Toward a quality-of-life theory of leisure travel satisfaction. *Journal of Travel Research, 49*(2), 246–260. https://doi.org/10.1177/0047287509337416

Smith, E. S., & Dalmer, N. K. (2023). Understanding older adults' participation in outdoor adventure activities: A scoping review. *Journal of Adventure Education and Outdoor Learning*. https://doi.org/10.1080/14729679.2023.2295842

Swarbrooke, J., Beard, C., Leckie, S., & Pomfret, G. (Eds.) (2003). *Adventure tourism: The new frontier*. Butterworth Heinemann.

Wheaton, B. (2017). Surfing through the life-course: Silver surfers' negotiation of ageing. *Annals of Leisure Research, 20*(1), 96–116. https://doi.org/10.1080/11745398.2016.1167610

Zarei, I., Ehsani, M., Moghimehfar, F., & Aroufzad, S. (2021). Predicting mountain hikers' pro-environmental behavioral intention: An extension to the theory of planned behavior. *Journal of Park and Recreation Administration, 39*(3), 70–90.

8 Adventure tourism segmentation

Melissa Jeanette Lötter

Chapter learning outcomes

1 Discuss the post hoc approach to divide the adventure tourism market into groups based on variables that define the adventure target group and/or market segment.
2 Identify specific bases for segmenting the adventure tourism market using the conceptual framework.
3 Relate the factors associated with the activity segmentation bases to collect data about the specific characteristics of the adventure tourist market.
4 Understand the importance of applying an effective market strategy.

Introduction

Tourism in developed countries evolved from Fordist to post-Fordist to contemporary tourism. During the twentieth century, Fordist patterns (mass production and consumption of non-differentiated products) were applied to the tourism industry (Denegri-Knott & Zwick, 2012; Hsu & Gartner, 2012). Fordist tourism included the following:

1 Production strictness: highly standardised, large-scale, dependent on scale economies
2 Low prices
3 Undifferentiated products: similar facilities and/or experiences
4 Substantial number of tourists to a route of mass production
5 Collective observation of tourists: focused on descriptors designed to concentrate tourists' seasonally differentiated consumption
6 Collective consumption by undifferentiated tourists
7 Demand for familiarity by tourists

By the end of the twentieth century, people had more free time, around five to six weeks of annual paid holidays, more disposable income influenced by the globalisation of financial markets, and more information through access to modern technologies. As a result, society changed to more singles, later marriages, fewer children, and people with varied tastes and demands (Denegri-Knott & Zwick, 2012; Hsu & Gartner, 2012). Arva and

DOI: 10.4324/9781003393153-11

Deli-Gray (2011) and Prideaux et al. (2012) noted that post-Fordist tourism ultimately led to the following:

1 Rejection of certain forms of mass tourism
2 De-differentiation of tourism
3 More information about alternative holidays and attractions through the media
4 Increase of holiday types
5 Visitor attractions based on lifestyle
6 Spread of alternative sights and attractions
7 More personalised tourism products
8 Fast turnover of tourist sites and experiences
9 Fewer repeat visits
10 Increased variety of preferences

However, it is argued that Fordist and post-Fordist tourism coexist side by side in different segments of a society or as individuals move through different life stages (Arva & Kuruvilla, 2013; Sharpley & Stone, 2014). For example, modern mass tourism includes tourism of poorer countries and poorer people for whom money is more important than fantasy, self-realisation, or virtuality. Modern niche tourism entails tourism of the post-Fordist rich countries and well-off middle-class people. Furthermore, one person can follow efficiency in their buying today and be completely post-Fordist yet seek aesthetic values and fantasy tomorrow. As a result, the twenty-first century has progressed from Fordist to post-Fordist tourism to what is now known as contemporary tourism (Arva & Deli-Gray, 2011; Arva & Kuruvilla, 2013; Novelli, 2007; Robinson et al., 2011; Urry & Larsen, 2011).

Contemporary tourism exhibits a unique set of principles that govern supply and demand. This situation results from tourism, a service industry characterised by intangibility and heterogeneity. For example, contemporary tourism includes two major trajectories, namely (1) mass and (2) alternative or niche tourism. Mass tourism consists of large-scale, low-priced, and non-sustainable high-impact tourism. By contrast, niche tourism consists of small-scale, high-priced, and sustainable tourism.

However, as noted by Hanlon (2015), the difference between Fordist, post-Fordist, and contemporary tourism is that tourism marketers have come to realise that there are seemingly contradictory tendencies and developed products that exist together with post-Fordist values. For example, fantasy, virtual reality, or personalisation are consistent if these aspects are correctly mixed and created. In other words, as noted by Ritzer (2014), mass tourism has been influenced by the inclusion of contemporary tourism activities, which have become forms of niche tourism products and markets in themselves. Tourism in this sense is evolving, where leisure and recreation develop into something more substantial. The next section looks at the differences between macro and niche tourism markets.

Macro- and micro-niche tourism markets

Niche tourism is regarded as a form of tourism that seeks to avoid adverse impacts and instead wants to enhance positive socio-cultural and environmental development. According to Novelli (as cited by Nieminen, 2012) and Robinson et al. (2016), niche tourism is usually characterised by:

1 Small-scale, individual, independent, or small group activity
2 It tends to be slow, controlled, and regulated
3 It demonstrates an emphasis on travel as an experience of host cultures
4 It focuses on the maintenance of traditional societal values and the host environment

Stated differently, these authors regard niche tourism as a small-scale phenomenon, developing and selling active differentiated products at varying, usually expensive, prices that cater to more responsible individuals or small groups seeking quality sustainable activities, environments, and/or experiences. Robinson et al. (2011) and Deloitte (2015) claim that as demand grows in areas of popularity that can be initially described as niche tourism, supply continues to grow. Consequently, there is a geographical element whereby locations with highly specific offers can establish themselves as niche tourism destinations (Ali-Knight, 2010; Benur & Bramwell, 2015; Nelson, 2013). In addition, mass tourism is increasingly incorporating characteristics associated with niche tourism as companies and/or destinations offer active, differentiated, sustainable, and high-quality products accompanied by personalised marketing approaches (Vainikka, 2013; Weaver, 2014).

As a result, in comparison to mass tourism, the only factor that truly differentiates mass tourism and niche tourism is quantity, that is scale of production and consumption. This is because both contemporary tourism trajectories could result in large-scale, low-cost production and consumption, which cater to tourists' tastes and demands, with accompanying impacts. This notion, as stated by Poon (as cited by Vainikka, 2013), can signify the end of mass tourism or mass tourism changing into multiple forms (Hernandez et al., 2015; Torres as cited by Vainikka, 2013). This is because the latter does not need to be homogeneous, as different types of tourism create mass tourism in various combinations.

With the preceding in mind, contemporary tourism should instead be viewed as niche tourism with macro niche tourism, that is large-scale tourism segments like natural area tourism, and micro niche tourism, that is small-scale tourism segments like adventure tourism, trajectories. In other words, Novelli (2007), Ali-Knight (2010), and Robinson et al. (2011) argue that niche tourism at one end of the spectrum should be viewed as a relatively large market sector (macro niches), capable of further segmentation. At the other end of the spectrum, niche tourism should focus on specific small markets that would be difficult to split further (micro niches).

For example, Novelli (2007) proposes that niche tourism consists of five macro- and 30 micro-niche tourism markets. The five macro niches consist of cultural, environmental, rural, urban, and others, each with its own micro-niche tourism markets. Swarbrooke et al. (2003) propose that niche tourism consists of four macro niche tourism markets: activity, nature-based, discovery/cultural, and expedition tourism. However, these authors further claim that macro- and micro-niche tourism markets could include products that are seen as belonging to other macro and micro niches. Finally, Newsome et al. (2012) argue that niche tourism consists of four macro- and 10 micro-niche tourism markets. The four macro niches consist of natural, cultural, event, and other – each with its own micro-niche tourism markets.

Comparing the research of Novelli (2007), Swarbrooke et al. (2003), and Newsome et al. (2012), it is evident that there are typically three primary macro-niche markets, namely cultural, event, and natural, and seven secondary micro-niche tourism markets, namely adventure tourism (see Chapter 2 for a history of adventure tourism), ecotourism, festival tourism, heritage tourism, nature-based tourism, religious tourism, and sports tourism, that constitute niche tourism. Therefore, when referring to niche tourism, reference is only made to these three primary macro niches and seven secondary micro niches, as these emerged over time as the main trajectories of the niche tourism market.

Segmenting niche tourism markets and adventure tourism

Differentiated product development and personalised marketing approaches are essential to attract different niche tourist groups (see Chapter 14 for further insights into sustainable adventure tourism products). Differentiated by the scale of sustainable production and consumption patterns, niche tourism marketing should move away from contemporary mass tourism marketing strategies toward those associated with target marketing. Target marketing, which, according to Tsiotsou and Goldsmith (2012), includes segmentation, targeting, and positioning, should be used by adventure tourism marketers to discover marketing opportunities by indicating their preferred products. In other words, adventure tourism marketers should target groups with greater purchase interest rather than scattering their marketing efforts in an unfocused manner.

There are two basic market segmentation approaches: (1) a priori and (2) post hoc (Pearce & Butler, 2010; Wedel & Kamakura, 2012). The a priori approach involves dividing a market into groups without the benefit of primary market research. For example, intuition, experience, and the analysis of secondary data sources are used to group tourists into segments. By comparison, the post hoc approach uses primary market research to divide a market into groups based on the selected variables that define a specific target group and/or market segment (Pearce & Butler, 2010; Wedel & Kamakura, 2012). Adventure tourism marketers typically follow a post hoc approach by (1) identifying bases for segmenting the market and (2) developing profiles of resulting segments (Kotler & Armstrong, 2015; 2016).

Step 1 of this procedure typically involves exploratory interviews and focus groups to gain insight into adventure tourists' motivations, attitudes, and behaviours (Kotler & Armstrong, 2015; 2016). A formal questionnaire is then formulated to collect data about the broad geographic, demographic, psychographic, and/or behavioural characteristics of the adventure tourists constituting the market. After that, factor analysis is applied to the data obtained from the questionnaire to remove highly correlated variables before using cluster analysis (see Chapter 5). This is done to create a specified number of different segments. Step 2 of this procedure requires that every cluster be profiled in distinguishing geographic, demographic, psychographic, and/or behavioural patterns named after a dominant distinguishing characteristic. But, as market segments change, this procedure should be repeated periodically (Kotler & Armstrong, 2015; 2016). Moreover, instead of solely looking at broad segmentation bases, adventure tourism marketers should identify the specific preference segments of activities, environments, and/or experiences to develop profiles of the resulting micro-niche segment.

Key factors that differentiate macro- and micro-niche tourism markets

Cultural, event, and natural area tourism are considered macro-niche tourism markets in their own right (see Figure 8.1). Their differences are discerned in terms of the degree and emphasis rather than the presence or absence of unique characteristics. For instance,

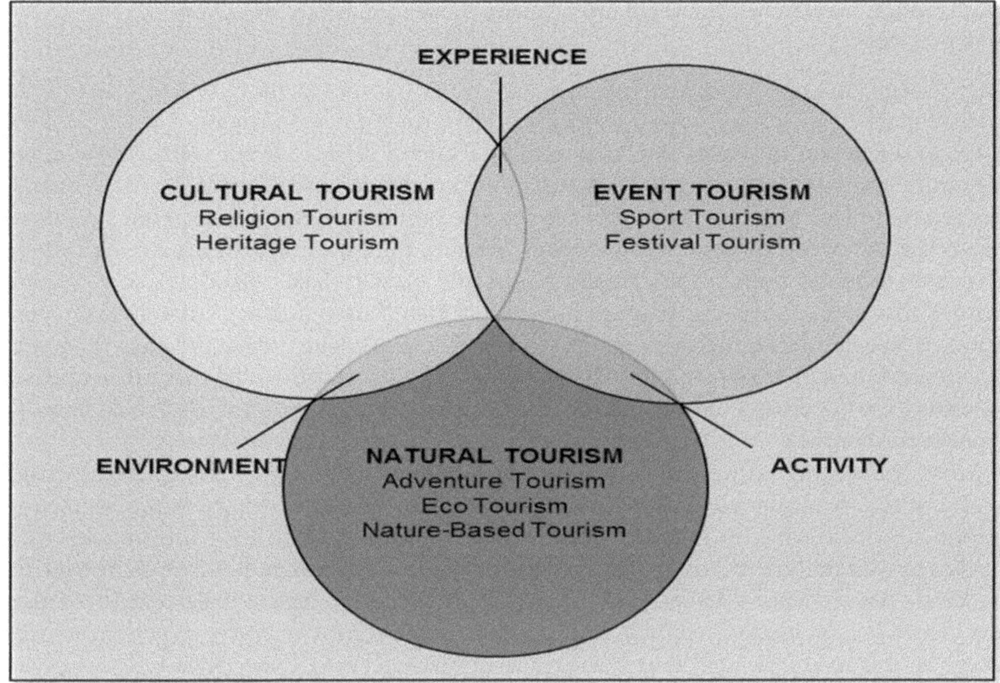

Figure 8.1 Key factors that differentiate niche tourism markets. Adapted from Sung (2000).

cultural tourism is perceived as experience-orientated, event tourism as activity-orientated, and natural area tourism – including adventure tourism – as environment-orientated.

However, as illustrated in Figure 8.1, there is a clear overlap between macro niche tourist groups. Thus, Sung (2000) suggests the key differentiating factors of cultural tourism are the environment and the tourist experience, the key differentiating factors of event tourism are the tourist experience and the physical activity, and the key differentiating factors of natural area tourism (including adventure tourism) are the physical activity and the environment, with tourist experiences becoming a more prevalent factor.

According to Sung (2000), micro-niche tourism markets are also distinguished by activities, experiences, and environments. For example, sport and adventure tourism emphasise the physical stimulation of heritage, ecotourism focuses more on environmental stimulation, and religious-, festival-, and nature-based tourism stresses mental stimulation.

Viewing the key factors that differentiate macro and micro niche tourism markets as specific products, it is therefore believed that in addition to the broad segmentation bases, niche tourism markets should be segmented according to their preference for activities, environments, and/or experiences. The three segments of niche tourism – activities, environments, and/or experiences – reflect the Adventure Travel Trade Association's (2022) definition of adventure travel, which is 'any tourist activity, including two of the following three components: a physical activity, a cultural exchange or interaction and engagement with nature'. This highlights the usefulness of applying this segmentation method to the specific niche tourism of adventure tourism.

A conceptual framework for segmenting niche tourism markets based on their preferences

Market segments can be drawn up in many ways. However, macro- and micro-niche tourism marketers should focus on specific preference segments instead of solely looking at broad segmentation bases. As illustrated in Figure 8.2, since event tourists (including adventure tourists) prefer activity-based and experience-based products, it is recommended that niche tourism marketers use activity-based and experience-based segmentation as specific bases to segment this market. Second, since natural area tourists (including adventure tourists) prefer environment-based and activity-based products, it is recommended that niche tourism marketers use environment-based and activity-based segmentation as specific bases to segment this market. Finally, since cultural tourists prefer experience-based and environment-based products, it is recommended that niche tourism marketers use experience-based and environment-based segmentation as specific bases to segment this market.

More specifically, since adventure and sports tourists mostly prefer activity-based products, it is recommended that niche tourism marketers use activity-based segmentation as a specific basis to segment these markets. Second, since eco- and nature-based tourists mostly prefer environment-based products, it is recommended that niche tourism marketers use environment-based segmentation as a specific basis to segment these markets. Third, since religion, festival, and heritage tourists mostly prefer experience-based products, it is recommended that niche tourism marketers use experience-based segmentation as a specific basis to segment these markets.

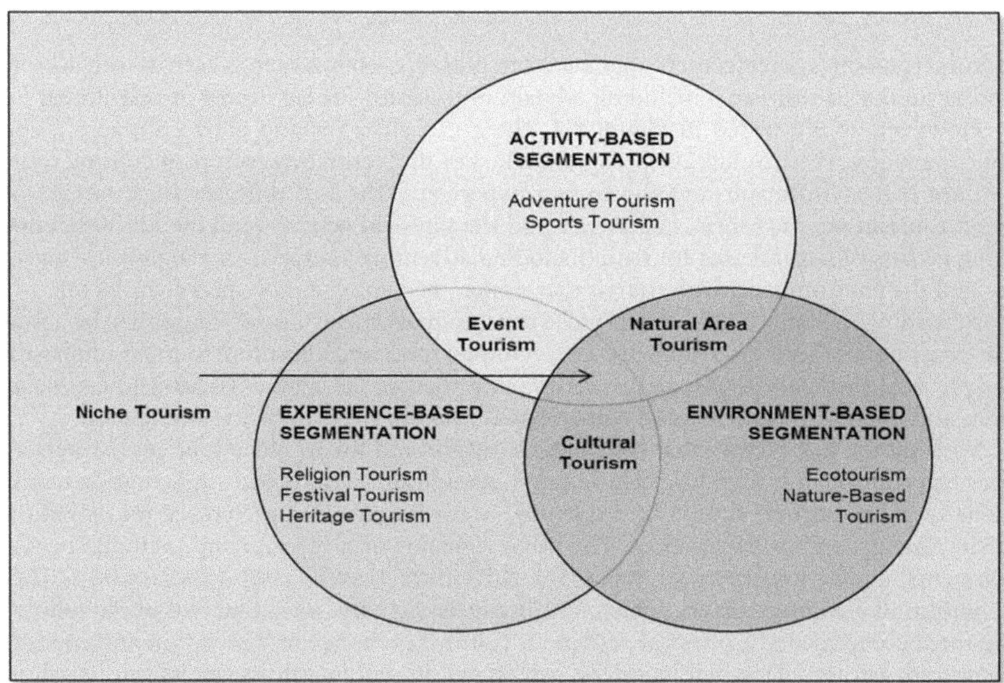

Figure 8.2 A conceptual framework for segmenting niche tourism markets.

It is important to note that although environment-based products are preferred mainly by nature-based tourists and experience-based products are typically preferred by heritage tourists, nature-based tourists prefer experience-based products, and heritage tourists also prefer environment-based products. Thus, it is recommended that niche tourism marketers use experience-based segmentation to segment nature-based tourists and environment-based segmentation to segment heritage tourists.

Specific bases for segmenting niche tourism markets

Once adventure tourism marketers used the conceptual framework (see Figure 8.2) to identify which specific segmentation bases to use when segmenting a macro- or micro-niche tourism market, they could use the following factors as the foundation for segmenting adventure tourism markets according to their preference for activities, environments, and/or experiences. For example, once an adventure tourism marketer identifies activity-based segmentation to segment the adventure tourism market, this section provides them with the specific activity factors of (1) soft nature, (2) risk equipped, (3) hard challenge, (4) rugged nature, (5) winter snow, and (6) question marks as the foundation for segmenting adventure tourism markets.

Activity-based segmentation

Physical activity refers to any bodily movement that works one's muscles. It requires more energy than resting (Stevenson & Waite, 2011). In the words of Budeanu, Miller and Moscardo (2016, p.285), 'physical activities have been identified as a critical link between travel motivation and destination choice'. The image of a particular tourist destination is based on the activities offered. For example, the Great Barrier Reef is regarded as a water-based adventure activity mecca primarily associated with scuba diving. Hence, activities play a vital role in destination marketing, and destination marketers will benefit from knowledge about the potential activity preferences of tourist groups (see Case study 8.1).

Case study 8.1   Active and adventure tourism in the planning of local destination management in Slovenia

In the southeast of Slovenia, the Heritage Trails initiative commenced when the UK and the Slovene team conducted a 'Tourist Resource Inventorisation & Selection' study based on natural, built, and living cultural heritage resources in the selected region. Approximately 150 sites were included by the partners in the Heritage Trail, with 28 sites chosen for networking in a trail system for the area.

The idea was to develop a tourism product capable of providing opportunities for stays of up to seven days in the region, unlocking opportunities for new jobs and economic diversification in the rural parts of Dolenjska and Bela Krajina, the Southeast region of Slovenia. Some individuals and groups of partners developed new products with past traditions, giving them a fresh and new outlook, as required for modern EU tourism markets. However, six years after this initial phase, the demand side was identified by marketing managers of the well-accepted Heritage Trails. This led to the development of specific products for active tourism. The thinking was led by the following facts:

- More than 75% of tourists from foreign markets were seeking active holidays
- More than 50% of the reservations were made on the internet
- More tourists wanted to change destinations every couple of days

So, the tourism marketers determined that they would need to create an adventure product that would do the following:

- Be used by the individual traveller in the same manner as by a tour operator
- Connect to the actual tourist offer in the region
- Be supported by all new standards and used technologies
- Support active holidays
- Be differentiated from other products in the field of active holidays

This was the reason for opening the EU-funded project "Thematic Routes of SE Slovenia," consisting of 1,500 km of walking, cycling, horseback riding, and water routes. Visitors could enjoy multi-day itineraries, either already prepared by regional tour providers or self-prepared and designed online by potential visitors. These routes connected the region's natural and cultural heritage with other tourist offers, such as accommodation, activities, information, and services.

In this case, the idea was that active tourism requires and employs active physical and intellectual participation of the visitor, regardless of the local destination in the region. Wholly digitised and located by GPS, routes are now presented in the renewed portal and the newly built mobile portal. The product is also offered on Facebook and YouTube. Biking and horseback riding routes can also be visualised. Potential tourists can look at and plan their holidays online and in detail from their homes. Once in the region, they can use mobile and GPS devices (and printouts) to navigate the area. For those who need more time to create holidays by themselves, the active tourist packages are pre-prepared and shown on the web as well.

Source: Koščak & O'Rourke (2018).

Coupled with sociodemographic characteristics and differences in vacation travel preferences, attitudes and motives, activity-based segmentation has been well-researched in the tourism industry (Mumuni & Mansour, 2014). Numerous authors, as indicated in Table 8.1, have segmented contemporary tourism based on tourists' preferences for activities: destination orientation (sightseeing), outdoor activities, beach holidays (leisure), and sports activities. These are perceived to be the most suited characteristics to differentiate contemporary tourist groups. Nevertheless, all other aspects, namely general tourism, entertainment, active relaxation (recreation), culture, and visiting family and friends – except activity combos – are also significant.

Table 8.1 Categories in a sample of activity-based segmentation studies

	General tourism	Destination orientation (sightseeing)	Activity combo	Entertainment	Outdoor activity	Beach holiday (leisure)	Active relaxation (recreation)	Cultural	Sports	Family and friends
Hsieh, O'Leary & Morrison (1992)		X	X	X	X				X	X
Lang, O'Leary & Morrison (1994)		X	X		X				X	
Choi & Tsang(2000)		X		X	X				X	X
Sung, Morrison & O'Leary (2000)		X				X	X			
Kim & Jogaratnam (2003)		X				X		X	X	
Beritelli & Boksbrger (2005)	X	X				X	X			X
Yang (2009)	X				X			X		
Choi, Murray & Kwana (2011)					X		X	X		
Mumuni & Mansour (2014)	X			X		X				
Total	3	6	2	3	5	4	3	3	4	3

Activity-based segmentation refers to segregating groups of niche tourists by activity preferences (Manthiou et al., 2011). A more comprehensive segmentation approach was undertaken by Sung (2000) who performed the only segmentation analysis using recreational activities as the key variable. As the variety and availability of recreational activities to satisfy a wide range of interests and abilities are limitless, the study by Sung (2000) suggested that activity should be one of the specific bases for analysing sports and adventure tourism segments. Market segmentation should be based on what these adventure tourist groups do on holidays and what is marketable.

Although the exact size of the global adventure tourism market is unknown, due to definitional problems and a lack of sufficient data and research to date, it is agreed that some recreational activities belong within the sports and adventure tourism category (Adventure Travel Trade Association, 2022; see Chapters 1, 2, and 3 for more information about defining and conceptualising adventure tourism). Lists of recreational activities have been compiled and provide a theoretical foundation regarding the various activities available. Still, there needs to be a classification of recreational activities that can be used as a conceptual framework for identifying specific bases for segmenting these adventure tourism markets.

The study conducted by Sung (2000) is the only one that included 48 recreational activities that provided six effective activity groupings (activity factors) based on relationships among recreational activities representing different market segments. This analysis linked products, that is recreational activities, and market segments. It is important to note that these activity factors include various recreational activities related to sports and adventure tourism and do not specifically focus on one adventure or sports tourism activity. Six distinct factors were identified and labelled as (1) soft nature, (2) risk equipped, (3) hard challenge, (4) rugged nature, (5) winter snow, and (6) question marks. However, Sung (2000, p. 68) advocates that there are definite opportunities for comparing the provider-based results with comparable results through niche tourist-based research analyses.

For the present chapter, the six recreational activity factors outlined in Table 8.2 serve as a foundation for segmenting event (sports) and natural area (adventure) tourism markets according to activity preferences. Further research should include other adventure and/or sports activities and/or develop recreational activity factors that differentiate sports and adventure tourism. In this context, when segmenting adventure tourism markets, activity factors should be examined with distinct tourist groups with varying geographic, demographic, psychographic, and/or behavioural characteristics (Hinch & Higham, 2011). The profile of adventure tourist groups might vary according to tourist characteristics and responses, providing several ways to link activities more effectively to a potential adventure tourism target market (Sotiriadis et al., 2014).

Environment-based segmentation

The environment denotes the total of all surroundings of a living organism, including natural forces and other living things, which provide conditions for development and growth as well as danger and damage (Stevenson & Waite, 2011). To identify distinct types of tourist groups and understand differences between segments regarding demographic, attitudinal, and behavioural characteristics, environment-based segmentation divides a market into specific groups of tourists based on their environmental attitudes (Kim & Weiler, 2013).

Table 8.2 A categorisation of activity groupings

Soft nature	• Hiking & walking tours
	• Nature trips
	• Bird watching
	• Bicycling & mountain biking
	• River canoeing
	• Backpacking
	• Camping
	• Horseback riding
	• Fishing
	• Orienteering
Risk equipped	• Paragliding & hang gliding
	• Soaring
	• Skydiving
	• Spelunking/caving
	• Bungee jumping
	• Windsurfing
	• Rappelling
	• Ballooning
	• Survival training
	• Wilderness training
	• Sailing
Hard challenge	• Ice climbing
	• Rock climbing
	• Mountain climbing
	• Sea canoeing
	• Dog sledding
	• River kayaking
	• Sea kayaking
Rugged nature	• Jungle exploring safaris
	• Snorkelling
	• Arctic trips
	• Trekking
	• Rafting
Winter snow	• Cross country skiing
	• Alpine skiing
	• Downhill skiing
	• Nordic skiing
	• Snowshoeing
Question marks	• Snowmobiling
	• Four-wheel drive trips
	• Motorcycling
	• Water skiing
	• Hunting
	• Heli-skiing
	• Rogaining (long-distance orienteering)
	• Scuba diving

Source: Sung (2000).

Because niche tourism has the potential to impact the environment positively and negatively, sustainable and responsible practices are essential for the preservation of natural areas (see Chapter 14). These practices are components of existing product segments such as cultural and natural area tourism. Sustainable or responsible tourism entails conducting business in which responsibility towards the destination's ecology, culture, and communities take precedence (Mody et al., 2014).

Two responsible tourism segments, three distinct responsible tourism segments, three segments of green tourism, and three segments of sustainable tourism are identified. These display various levels of commitment, attitude, knowledge, and/or behaviour about sustainability, namely the following:

1 Responsible tourism segments: high and low environmental attitudes (Kim & Weiler, 2013)
2 Distinct responsible tourism segments: responsibles, novelty seekers, and socialisers (Mody et al., 2014)
3 Green tourism segments: the uncommitted, the green activist, and the undefined (do Paço et al., 2009)
4 Sustainable tourism segments: the reflective, the unconcerned, and the pro-sustainable tourist (López-Sánchez & Pulido-Fernández, 2016)

Considering the preceding points, it is suggested that environment-based segmentation, in addition to general contemporary tourism characteristics, includes (1) conservation (environmental involvement), (2) participation (social involvement), (3) spiritual setting (personal development), (4) observation (natural areas), and (5) education. Because these factors have not yet been evaluated empirically, caution should be applied when using these as a possible base for segmenting niche tourism markets.

Nevertheless, the five environmental factors are suggested to segment cultural and natural area tourism markets, specifically eco and heritage tourism. Because adventure tourism also falls within the natural area tourism market, these environmental factors could provide further insights when profiling adventure tourist groups. This is evident in the work of Cao et al. (2022), who propose that adventure tourists' pro-environmental behaviour is one of the key factors for the sustainable development of natural scenic spots.

Experience-based segmentation

Experience is based on an event or occurrence which leaves an impression on someone (Stevenson & Waite, 2011). Viewing experiences as the specific products which cultural and event tourists prefer, underlying dimensions connected to how these tourists experience specific activities seem appropriate. Although no universal definition exists for experience-based segmentation, several common characteristics of tourist experiences can be identified. For example, from the demand-side perspective, tourist experiences should (1) be personal and exceptional, (2) involve tourists' perception and participation, (3) engage tourists emotionally, (4) be shared with others, and (5) be remembered for some time (Ladeira et al., 2012). From a supply-side perspective, the service experience composition should focus on (1) the physical environment (setting), (2) interaction with employees (actors), and (3) interaction with other tourists (audience) as significant dimensions of service experiences (Ali et al., 2015).

The importance of tourist experiences is increasingly being recognised for segmentation purposes. Previous studies on tourist experiences have all suggested the multidimensionality of affairs and the centrality of tourists' perceptions (Chen & Lamberti, 2013). This indicates how the importance attached to the different experiential stimuli might represent an exciting segmentation variable. Other conceptual frameworks for classifying tourist experiences have also been provided over time. However, for the most part, these fall into two main streams that are conceptually entwined.

On one hand, the experiences are analysed from a tourist's perspective (Pine & Gillmore, 2011). This approach identifies four realms of tourist experiences according to the degree of tourist involvement, that is passive versus active participation, and the desire with which the tourist connects or engages with the event performance, that is absorption versus immersion. As a result, four types of experiences are generated, namely (1) entertainment (passive/absorption), (2) educational (active/absorption), (3) escapist (active/immersion), and (4) aesthetic (passive/immersion). This conceptualisation has characterised studies in marketing and, interestingly, in the specific setting of niche tourism.

On the other hand, Schmitt (2000) suggests a brand- or company-side viewpoint which defines five primary areas of experiential stimulation, namely (1) sensory, (2) emotion, (3) cognition, (4) behavioural, and (5) relational. Further refinements of this approach have led to different scales and conceptualisations, including neglecting the relational dimension in defining a brand personality (Schmitt et al., 2009) and subdividing behavioural stimuli into two subgroups (Gentile et al., 2007). These are (1) pragmatic stimuli, aimed at stimulating the tourist to do something, and (2) lifestyle stimuli, aimed at showing the tourist a series of cultural traits, for example values and beliefs characterising a specific way of living somehow considered to be desirable by tourists themselves.

In this context, considering the evident overlaps between the two streams, Chen and Lamberti (2013) endorse Gentile et al.'s (2007) approach to examine the kinds of experiences looked for by tourists and the constituting elements that tourists value the most. This provides a more appropriate perspective on market segmentation. Chen and Noci (2014) advise that the experiential expectations of niche tourist groups be characterised according to the importance attributed to four main dimensions:

1 Sensory experience: elements that can be seen, heard, touched, smelled, or tasted
2 Emotional experience: the affective self and the feelings of the tourist
3 Cognitive experience: the tourist's conscious self, including acquiring new knowledge
4 Pragmatic experience: active engagement in the representation/experience

As a result, four clusters are identified to depict the critical characteristics of each dimension:

1 Involved cluster: attributes with relatively high importance to all the experiential dimensions, which suggests a high involvement in the experience.
2 Indifferent cluster: attributes with relatively low importance to all the experiential dimensions, which suggests not being particularly interested in the experience.
3 Introverted cluster: attributes with higher importance to the experiential dimensions of emotion and cognition, which suggests a more significant concern for personal feelings and thoughts during the experience.
4 Extroverted cluster: attributes with relatively high importance to the experiential dimensions of sensory and action, which suggests a more significant concern for the physical and social environment during the experience.

However, experiences with various products are expected to differ in perceived value, which can affect the overall evaluation of the experience (Prebensen & Rosengren, 2016). For example, adventure tourists value various elements differently, depending on the place and situation. Even though the niche tourism industry is an excellent example of the experience industry, activity products are still important aspects of the adventure tourists' experience (Schmitt, 2000).

Why should adventure tourism and other niche tourism markets be segmented?

To guide the formulation of successful marketing strategies, adventure tourism marketers must understand the advantages and disadvantages and the requirements of effective market segmentation. As indicated in Table 8.3, there are numerous advantages and disadvantages in dividing a broad market, usually consisting of existing and potential customers, into subsets of adventure tourists exhibiting shared characteristics.

To capitalise on the advantages and offset the disadvantages of market segmentation, Kotler and Armstrong (2016) advocate that effective market segmentation requires segments to be the following:

- Measurable: The degree to which the segment's size and purchasing power can be measured
- Accessible: The extent to which segments can be accessed and served
- Sustainable: The degree to which segments are large or profitable enough to serve as markets
- Actionable: The extent to which effective programmes can be designed to attract and serve segments

However, with the increase of online distribution channels, particularly with the rapid development of e-commerce, practical plans that can fulfil the requirements of adventure tourists buying products online are required (Du et al., 2016). More than

Table 8.3 Advantages and disadvantages of market segmentation

Advantages	*Disadvantages*
• Identifies tourists' needs and wants, as well as the extent the current product is satisfying these needs • Finds and compares niches, locating unserved or under-served markets • Easily develops and modifies marketing approaches as it is simpler to address the needs of smaller groups of tourists if they have many characteristics in common • Efficiently sets targets and priorities creating a more efficient use of marketing resources by focusing on the segments with the best potential • Effectively develops marketing mix strategies • Effectively develops market positioning strategies, including the expansion and retention of tourists • Increases company performance	• Expensive • Tendency to appeal to unviable markets • Difficult to distinguish how accurately or broadly to segment the market • Increased competition in certain market segments

Source: Blythe (2013); Dibb & Simkin (2013); Pride & Ferrel (2014).

62% of adventure tourists plan their trips on the internet (Adventure Travel Trade Association, 2022).

Incorporating innovative technologies in an adventure tourism company has become more essential than ever, as explored in Chapters 11 and 20. For example, upcoming technological trends include virtual reality tours, professional photography, videos, social media, operator websites, online adventure travel agents, finding new insights with big data, blockchain, and implementing chatbots (Peak Flow Online Business Management, 2022). As such, more focus is placed on accessibility, sustainability and actionability when segmenting online. Nevertheless, technology should be an aid to help improve efficiency, achieve goals, and improve the adventure tourist experience. It is not supposed to take away human intervention. It should elevate its roles to focus on creating, communicating, and connecting to improve the adventure tourist experience.

Towards an effective marketing strategy

Once adventure tourism marketers have identified their specific segmentation bases and developed associated profiles, the next step is establishing segment attractiveness measures and selecting the target market. Because adventure tourism companies vary widely in terms of their ability to serve different market segments (Pesonen, 2013), they should focus on those parts of the market that can best be suited (differentiated marketing) rather than trying to compete in an entire market (undifferentiated marketing).

After evaluating the different market segments according to their attractiveness measured in size, expected growth, structural attractiveness, and the company's objectives and available resources, it should be decided which and how many segments to serve (Dibb & Simkin, 2013). Segmented (differentiated) marketing, niche (concentrated) marketing, and/or individual (local/micro) marketing, as opposed to undifferentiated (mass) marketing, should be implemented to increase targeting precision and rank potential market segments by profit opportunity and risk.

Segmented marketing refers to isolating some broad segments that make up a market. At the same time, *niche marketing* (micromarketing) refers to separating some broad segments that make up a market into more narrowly defined groups. Individual marketing (customised marketing) isolates individuals within narrowly defined groups. Moving away from segment marketing must focus on niche and/or individual marketing. However, this crucial decision affects and often directly determines the adventure tourism company's marketing mix (Boone & Kurtz, 2014).

After segmenting and targeting an adventure tourism market, a value proposition should be decided on (Kotler & Armstrong, 2015; 2016). This is done by differentiating the market offering to create superior consumer value and arranging for a market offering to occupy a clear, distinctive, and desirable place relative to competing products in the minds of the targeted adventure tourists. In this sense, products should be positioned in the minds of adventure tourists by identifying possible competitive advantages, selecting suitable competitive advantages, and developing a marketing mix to communicate and deliver the chosen position (Dibb & Simkin, 2013). Furthermore, when selecting a positioning strategy, the strengths and weaknesses of an adventure tourism company must be reviewed. A chosen position will place the company in a superior position against competitors (Kotler & Armstrong, 2015; 2016).

Unfortunately, today, the advertising of touristic products and destinations faces a significant challenge: saturation. It is becoming harder to position a destination at a tourist's top of mind due to the vast array of social media travel bloggers, influencers, and globalised advertising campaigns marketing various destinations worldwide in traditional and digital media. In the case of adventure tourism, despite not being a massive product like sun and beach, the same saturation challenge is becoming more evident. More and more destinations tap into the narrative of adventure tourism (see Chapter 2). For example, Queenstown is known as the world's adventure capital, Chamonix is Europe's adventure capital, and Moab is the adventure capital of America. They either attempt to position themselves within this growing segment or differentiate themselves from other destinations by using a description that connects with more tourists.

For example, Iceland claims to be "for the offbeat traveller." In contrast, South Africa tells tourists to "escape the mundane and truly Live Again." Adding a platform like Instagram to the mix, unfortunately, ends with all places that tend to look the same. As such, Del Palacio (2019) suggests that adventure tourism marketers should follow five basic principles when promoting their adventure destination or product, namely the following:

1 Know your product
2 Talk to your actual consumer
3 Micro-targeting or focalised segmentation is key
4 Be coherent about how you communicate your product and its reality
5 Capitalise your content and the content generated by your clients

Building and maintaining a consistent positioning strategy is complex, as many counterforces exist. For example, contracted advertising agencies may not support and work against a chosen position; new management may need to understand and change the positioning strategy. Budgets may be reduced for critical support programmes. Nevertheless, developing a positioning strategy must coincide with continuous support from management, employees, and vendors because adventure tourism products, prices, places, and promotions should be blended to produce the response it wants in the selected adventure tourism market.

Conclusion

Marketers focusing on niche tourism marketing are confronted with new and emerging challenges significantly influenced by external factors such as innovative technologies, business practices, and tourists' demands, which are deeply interrelated with demographic and socio-economic profile changes. As such, this chapter aimed to assist adventure tourism marketers in identifying the specific bases for segmenting adventure tourism markets by proposing a conceptual framework. In this context, it is recommended that specific segmentation bases, in conjunction with broad geographic, demographic, psychographic, and/or other behavioural segmentation bases, be used to identify and group adventure tourists according to the presence or absence of those factors that affect their purchase decisions so that marketing strategies can be adjusted to meet the specific needs and wants of every selected adventure tourist group or groups.

For example, instead of using assumptions and perceptions to identify which specific segmentation bases to use when segmenting adventure tourism markets, adventure tourism marketers can use the conceptual framework (see Figure 8.2) to identify that activity-based segmentation is the most appropriate basis for segmenting adventure tourism markets. In conjunction with broad geographic, demographic, psychographic, and/or other behavioural segmentation bases, the six recreational activity factors outlined in Table 8.2 should serve as the foundation for segmenting adventure tourism markets according to their activity preferences.

However, before adventure tourism marketers segment, target, and position their markets, they should clearly understand their micro, market, and macro environments. After that, once their markets are segmented, targeted, and positioned, adventure tourism marketers should implement sustainability and relationship marketing to attract and retain their markets to enhance their satisfaction (Berry as cited by Hudson & Hudson, 2013).

Review questions

1 Discuss the two-step process that adventure tourism marketers should follow when dividing their market into groups.
2 Using the conceptual framework for segmenting niche tourism markets, identify the specific bases for segmenting:

- Adventure tourism
- Ecotourism
- Natural area tourism
- Nature-based tourism
- Sports tourism

3 Discuss the factors that *adventure* tourism marketers should use to collect data pertaining to the specific characteristics of their identified niche tourist group.
4 Differentiate between market targeting and market positioning by providing a practical example of each.
5 Research practical examples of an adventure tourism company implementing effective target marketing.

Further reading

Armstrong, G., & Kotler, P. (2020). *Principles of marketing*. Pearson Education.
Morrison, A. M. (2022). *Tourism marketing: In the age of the consumer*. Taylor & Francis.
Newsome, D., Moore, S. A., & Dowling, R. K. (2012). *Natural area tourism: Ecology, impacts and management*. Channel View Publications.
Novelli, M. (2007). *Niche tourism*. Taylor & Francis.
Sung, H. H. (2000). *An analysis of the adventure travel market: From conceptual development to market segmentation*. [Doctoral dissertation, Purdue University].
Swarbrooke, J., Beard, C., Leckie, S., & Pomfret, G. (2003). *Adventure tourism: The new frontier*. Butterworth-Heinemann.

References

Adventure Travel Trade Association. (2022). *Adventure Travel Industry Snapshot*. https://learn. adventuretravel.biz/research/2022-adventure-travel-industry-snapshot
Ali, F., Amin, M., & Cobanoglu, C. (2015). An integrated model of service experience, emotions, satisfaction, and price acceptance: An empirical analysis in the Chinese hospitality industry. *Journal of Hospitality Marketing & Management*, Advanced online publication.
Ali-Knight, J. (2010). *The role of niche tourism products in destination development*. [PhD Tourism Management thesis, Edinburgh Napier University].

Arva, L., & Deli-Gray, Z. (2011). New types of tourism and tourism marketing in the post-industrial world. *Applied Studies in Agribusiness and Commerce, 5*, 33–37.

Arva, L., & Kuruvilla, S. J. (2013). A global perspective on the development of tourism marketing. *Indore Management Journal, 3*(4), 31–41.

Benur, A. M., & Bramwell, B. (2015). Tourism product development and product diversification in destinations. *Journal of Tourism Management, 50*, 213–224.

Beritelli, P., & Boksbrger, P. E. (2005). Activity-based market segmentation: A behavioural approach. *Tourism, 53*(3), 259–266.

Blythe, J. (2013). *Consumer behaviour*. SAGE.

Boone, L., & Kurtz, D. (2014). *Contemporary marketing*. Cengage Learning.

Budeanu, A., Miller, G., Moscardo, G., & Ooi, C. (2016). Sustainable tourism, progress, challenges, and opportunities: An introduction. *Journal of Cleaner Production, 111*(B), 285–294.

Cao, X., Qiu, J., Wang, L., & Zhou, G. (2022). An Integrative Model of Tourists' Pro-Environmental Behaviour Based on the Dual Path of Rational Planning and Embodied Emotion. *International Journal of Environmental Research and Public Health, 19*(13), 7910. https://doi.org/10.3390/ijerph19137910

Chen, S., & Lamberti, L. (2013). Segmenting Chinese tourists by the expected experience at theme parks. *International Journal of Engineering Business Management, 5*(22), 1–9.

Chen, S., & Noci, G. (2014). Segmenting Chinese tourists by the expected experience at theme parks. *Procedia - Social and Behavioural Sciences, 109*, 1230–1236.

Choi, H. S. C., Murray, I., & Kwana, K. (2011). Activity based segmentation of Canadian domestic pleasure travelers to New Brunswick. *International Journal of Hospitality and Tourism Administration, 12*(3), 202–224. https://doi.org/10.1080/15256480.2011.590736

Choi, W. M., & Tsang, C. K. L. (2000). Activity based segmentation on pleasure travel market of Hong Kong private housing residents. *Journal of Travel & Tourism Marketing, 8*(2), 75–97. https://doi.org/10.1300/J07.v08n02_06

Del Palacio, E. (2019, June 26). *Promotion of nature and adventure tourism: 5 basic principles for an integral strategy*. ATMEX. https://atmex.org/en/promotion-of-nature-and-adventure-tourism-5-basic-principles-for-an-integral-strategy/

Deloitte. (2015). *Global powers of luxury goods: Engaging the future luxury consumer*. https://www2.deloitte.com/content/dam/Deloitte/global/Documents/Consumer-Business/gx-cb-global-power-of-luxury-web.pdf

Denegri-Knott, J., & Zwick, D. (2012). Tracking presumption work on eBay reproduction of desire and the challenge of slow re-mcdonaldization. *American Behavioural Scientist, 56*(4): 439–458.

Dibb, S., & Simkin, L. (2013). *Market segmentation success: Making it happen*. Routledge.

Do Paço, A. M. F., Raposo, M. L. B., & Filho, W. L. (2009). Identifying the green consumer: A segmentation study. *Journal of Targeting, Measurement and Analysis for Marketing, 17*(1), 17–25.

Du, F., Yang, F., Liang, L., & Yang, M. (2016). Do service providers adopting market segmentation need cooperation with third parties? An application to hotels. *International Journal of Contemporary Hospitality Management, 28*(1), 136–155.

Gentile, C., Spiller, N., & Noci, G. (2007). How to sustain the customer experience: An overview of experience components that co-create value with the customer. *European Management Journal, 25*(5), 395–410.

Hanlon, G. (2015). *The dark side of management: A secret history of management theory*. Routledge.

Hernandez, J. M., Suarez-Vega, R., & Santana-Jimenez, Y. (2015). The inter-relationship between rural and mass tourism: The case of Catalonia, Spain. *Journal of Tourism Management, 54*: 43–57.

Hinch, T., & Higham, J. (2011). *Sport tourism development*. Channel View.

Hsieh, S., O'Leary, J. T., & Morrison, A. M. (1992). Segmenting the international travel market by activity. *Tourism Management, 13*(22), 209–223. https://doi.org/10.1016/0261-5177(92)90062-C

Hsu, C. H. C., & Gartner, W. C. (2012). *The Routledge handbook of tourism research*. Routledge.

Hudson, S., & Hudson. L. (2013). *Customer service for hospitality and tourism*. Goodfellow.

Kim, A. K., & Weiler, B. (2013). Visitors' attitudes towards responsible fossil collecting behaviour: An environmental attitude-based segmentation approach. *Journal of Tourism Management, 36*, 602–612.

Kim, G., & Jogaratnam, G. (2003). Activity preferences of Asian international and domestic American university students: An alternate basis for segmentation. *Journal of Vacation Marketing, 9*(3), 260–270.

Koščak, M., & O'Rourke, T. (2018). *Active & adventure tourism in the planning of local destination management - with case studies from Slovenia and Scotland.* https://doi.org/10.18690/978-961-286-226-8.6.

Kotler, P., & Armstrong, G. (2015). *Principles of marketing* (16th ed.). Pearson.

Kotler, P., & Armstrong, G. (2016). *Principles of marketing* (17th ed.). Pearson.

Ladeira, W. J., Costa, G., & Santini, F. D. O. (2012). Background and dimensions of consumption experience in Brazilian hotels. *Tourism and Hospitality Research, 12*(4): 215–227.

Lang, C., O'Leary, J. T., & Morrison, A. M. (1994). Activity segmentation of Japanese female overseas travelers. *Journal of Travel & Tourism Marketing, 2*(4), 1–22. https://doi.org/10.1300/J073v02n04_01

López-Sánchez, Y., & Pulido-Fernández, J. I. (2016). In search of the pro-sustainable tourist: A segmentation based on the tourist's "sustainable intelligence". *Journal of Tourism Management, 17,* 59–71.

Manthiou, A., Liang, R., Morrison, A. M., & Shin, J. (2011). *Segmenting International visitors in Shanghai by activity preference.* http://scholarworks.umass.edu/gradconf_hospitality/2011/Presentation/106/

Mody, M., Day, J., Sydnor, S., Jaffe, W., & Lehto, X. (2014). The different shades of responsibility: Examining domestic and international travellers' motivations for responsible tourism in India. *Journal of Tourism Management Perspectives, 12,* 113–124.

Mumuni, A. G., & Mansour, M. (2014). Activity-based segmentation of the outbound leisure tourism market of Saudi Arabia. *Journal of Vacation Marketing, 20*(3), 239–252.

Nelson, V. (2013). *An introduction to the geography of tourism.* Rowman & Littlefield.

Newsome, D., Moore, S. A., & Dowling, R. K. (2012). *Natural area tourism: Ecology, impacts and management.* Channel View Publications.

Nieminen, K. 2012. *Religious tourism: A Finnish perspective.* [Master of Hospitality Management thesis, Haaga-Heila university of applied sciences].

Novelli, M. (2007). *Niche tourism.* Taylor & Francis.

Peak Flow Online Business Management. (2022). *The future of travel technology.* https://peakflowobm.com/the-future-of-travel-technology/

Pearce, D. G., & Butler, R. (2010). *Tourism research: A 20-20 vision.* Goodfellow.

Pesonen, J. (2013). *Developing market segmentation in tourism: Insights from a Finnish rural tourism study.* [Social sciences and business studies dissertation, University of Eastern Finland].

Pine, B. J., & Gillmore, J. H. (2011). *The experience economy.* Harvard Business.

Prebensen, N. K., & Rosengren, S. (2016). Experience value as a function of hedonic and utilitarian dominant services. *International Journal of Contemporary Hospitality Management, 28*(1): 113–135.

Pride, W., & Ferrell, O. C. (2014). *Foundations of marketing.* Cengage Learning.

Prideaux, B., Fyall, A., Leask, A., Hall, C. M., Boyd, S. W., Voase, R., Christadoulakis, S., Robinson, K., Middleton, V. T. C., Stevens, T., Kazasis, F., Anestis, G., Wanhill, S., Wall, G., Henderson, J. C., Braun, B. M., Mccracken, M., Robbins, D., Goulding, P., & Shackley, M. (2012). *Managing visitor attractions: New directions.* Taylor & Francis.

Ritzer, G. (2014). *The McDonaldization of society.* SAGE.

Robinson, P., Fallon, P., Cameron, H., & Crotts, J. C. (2016). *Operations management in the travel industry.* (2nd ed.). CABI.

Robinson, P., Heitmann, S., & Dieke, P. (2011). *Research themes for tourism.* CABI.

Schmitt, B. H. (2000). *Experiential marketing: How to get customers to sense, feel, think, act, relate to your company and brands.* The Free Press.

Schmitt, B. H., Zarantonello, L., & Brakus, J. J. (2009). Brand experience: What is it? How is it measured? Does it affect loyalty? *Journal of Marketing, 73*(3), 52–68.

Sharpley, R., & Stone, P. (2014). *Contemporary tourist experiences: Concepts and consequences.* Routledge.

Sotiriadis, M., Van Zyl, C., & Nduna, L. (2014, June 25-27). *Evaluating the benefit/activity- based segmentation of the tourism market in the context of nature-based attractions.* [Conference proceeding]. *4th Advances in Hospitality and Tourism Marketing and Management Conference,* Mauritius.

Stevenson, A & Waite, M. (2011). *Concise Oxford English dictionary: Luxury edition*. Oxford University.

Sung, H. H. (2000). *An analysis of the adventure travel market: From conceptual development to market segmentation*. [Doctoral dissertation, Purdue University].

Sung, H. Y., Morrison, A. M., & O'Leary, J. T. (2000). Segmenting the adventure travel market by activities: From the North American industry providers' perspective. *Journal of Travel & Tourism Marketing, 9*(4), 1–20. https://doi.org/10.1300/J073v09n04_01

Tsiotsou, R. H., & Goldsmith, R. E. (2012). *Strategic marketing in tourism services*. Emerald Group.

Urry, J., & Larsen, J. (2011). *The tourist gaze 3.0.*. SAGE.

Vainikka, V. (2013). Rethinking mass tourism. *Journal of Tourist Studies, 13*(3): 268–286.

Weaver, D. B. (2014). Asymmetrical dialectics of sustainable tourism: Toward enlightened mass tourism. *Journal of Travel Research, 53*(2): 131–140.

Wedel, M., & Kamakura, W. A. (2012). *Market segmentation: Conceptual and methodological foundations*. Springer Science & Business Media.

Yang, N. K. (2009). Activity-based market segmentation in a rural tourism destination: A case study of west-central Michigan. PhD dissertation, Michigan State University.

9 Adventure tourism and tourist experience
Revisiting diversity and inclusion

Vinathe Sharma-Brymer, Eric Brymer and Soumya Mitra

Chapter learning outcomes

1 Understand the multidimensional nature of adventure tourism.
2 Reflect on person–environment relationship in adventure tourist experience.
3 Analyse sociocultural diversities of adventure tourist experience.
4 Apply a critical understanding of diversities to work with adventure tourists by enhancing affordances for participants, organisers, hosts, and environments.

Figure 9.1 Religious nature/human tourism site in India. Photograph by Abdul Rasheed.

DOI: 10.4324/9781003393153-12

Vignette 9.1 Spiritual adventure tourism

I am Kamala. I am a 59-year-old Indian woman and a retired schoolteacher. I practise the Hindu religion with values of compassion, respect, and care. Since my late forties, I have taken up religious tourism with friends, colleagues and relatives, deepening my understanding and experience of spirituality. As a small group of five to eight people, we travel to temples, holy rivers, mountains, lakes and other such places which hold reverence and respect for spirituality that is found in our religious practices and in the natural world. I have found strength, peace, happiness, and life satisfaction in my religious/spiritual tourism across India. As a tourist, I think I have lived adventure in its multiple meanings as every step of our travels has involved risks, challenges, and obstacles. We have learnt to face these with determination, skill building, and surrendering to the natural world and its elements. For example, in 1994, I and five others travelled as a group from Bangalore city to the remote Ganga Sagar in Eastern India, where the holy River Ganga joins the Bengal Sea. Our adventurous experience was a mixture of apprehension, anxiety, fear of the unknown, peace, fulfilment, accomplishment and satisfaction. For a traditional Indian woman such as myself, all my travels have been adventurous in seeking more-than-human experiences and gaining newer meanings of life. My adventurous spiritual tourist experiences have influenced me in deepening my community connections with kindness and peace.

Vignette 9.2 Humanist approaches to adventure tourism

I am Anand, an Indian male in my early 30s. My personal connection to outdoor adventure tourism has evolved from the journeys I have taken throughout my life. As a child, I grew up in the foothills of the Indian Himalaya and spent a great deal of time outdoors, which shaped my relationship with the natural world and motivated me to associate with adventure tourism. Coming from the dominant Hindu religious socio-cultural context, I realised my privileges at an early age with good opportunities for a formal education. However, the thought-provoking discussions I had had with my father as a teenager encouraged me to embrace Humanism as my value in life over Hinduism as a religion. This motivated me later in life to embrace outdoor adventure education as a career choice. My lived experiences while working with different adventure tourism organisations in India, Tanzania, the US, Vietnam, and Australia helped me understand and respect individual perspectives due to differences in cultures, languages, and beliefs. I think my lived experience of my relationship with nature has taught me to see the adventure tourist experience from an ecological perspective, meaning that all beings are valued. This perspective has enabled me to promote humanistic values of humility and reciprocity.

Introduction

From a tourist experience perspective, adventure tourism is an embodied experience different to the everyday, mundane activities of life. Participant characteristics and sociocultural identities, the environment in which adventure is experienced, and the structure of adventure activity impact on this embodied experience. From this context, adventure tourism experiences are cultural experiences and result in long-term positive transformational benefits for participants and communities (Clarke et al., 2022) at individual and collective levels. For example, spiritual tourists travelling to remote places feel a sense of adventure, empowerment, and fulfilment. As Kamala's description (Vignette 9.1) shows, this could reflect in personal and collective values and behaviours for improving community connections with enhanced intra-cultural harmony.

In this chapter, we discuss themes that help us better understand the adventure tourist experience in relation to adventure activity or programme choices and level of engagement for optimum benefit. In recent times, Western-oriented notions of adventure tourism are being challenged as participants are increasingly varied, representing diverse genders (see Chapter 27), socio-cultural and racial identities (Mukherjee et al., 2020), age groups and ability levels (Bauer, 2012; Pomfret, 2019) and a keen interest in microadventure experiences in post-COVID-19 times (Houge Mackenzie & Goodnow, 2021). In this light, we offer reflections from both Western and non-Western adventure tourism engagement, such as spiritual tourism and recreational tourism, involving non-formal, place-based adventure (Sharma-Brymer, 2022). We also explore localised microadventure tourism, which is regarded as mobility-related adventure in non-Western societies (Cohen & Cohen, 2015). We recognise that the adventure tourist is a person who travels farther than 50 miles for an experience that may involve diverse levels of adventure (Jordan et al., 2016) and offer a comparative analysis of the documented lived experience of adventure tourism participants from diverse sociocultural contexts.

Traditional perspectives on adventure tourism

Traditionally, adventure tourism has been associated with purposeful, risky, commercialised outdoor activities that take place away from the participant's home and in the natural environment (Hall, 1992). However, in recent years, other types of religious, spiritual, and wildlife activities have also been considered as examples of adventure tourism, further diversifying the adventure tourist experience (Buckley, 2006; Swarbrooke et al., 2003). This creates ambiguity for researchers and theorists alike on what constitutes the adventure tourist experience (Cater, 2006; Swarbrooke et al., 2003; Varley, 2006).

At one level, the adventure tourist might be an independent highly skilled professional or amateur with a deep knowledge of the activity and environment, for example whitewater kayakers, skiers in the Alps region, or four-wheel driving enthusiasts travelling along the Himalayan terrain (Brymer, 2013; Brymer & Houge MacKenzie, 2015; Sharma-Brymer, 2018). Such trips might or might not require a specialist tour guide/operator/expert facilitator. At another level, the adventure tourist might be totally inexperienced, requiring commercial adventure tourism companies to guide and orient their clients' level of activity engagement, thereby influencing the tourist experience. Examples include rafting and trekking where the participants may purely focus on the activity or enjoy being immersed in an adventurous environment. Adventure tourism also provides opportunities for varying levels of activity difficulty, time and financial costs. Adventure tourists can choose an extended experience, such as a mountaineering expedition, to completely live a "be in the moment" experience, or a short-lived experience, such as whitewater rafting.

Understanding the concept of adventure tourism and its participants is important when examining the tourist experience. For example, if we understand participation as opportunities for individuals with specific personality characteristics, such as risk-takers searching out the "adrenaline rush," then this might suggest that their activities are all about thrills and are not appropriate for a broader population (Brymer & Schweitzer, 2017). Adventure businesses might target a small niche of participants, with low-level skills, and researchers might focus on explicating their personality characteristics. A framework that focuses on risk-taking might also suggest hedonistic experiences. Participant groups for this framework might stereotypically be young, male, and predominantly Western (Brymer & Schweitzer, 2017). In the following sections, the authors explore some key characteristics of adventure tourist experiences based on traditional and contemporary conceptual understanding of the notion of adventure tourism and the adventure tourist.

Adventure tourist experience as thrill-seeking experience

As noted earlier, adventure tourism is synonymous with outdoor activities where participants search for adventurous experiences involving varying levels of risk, challenge, and thrill (Muller & Cleaver, 2000). Activities can range from guided or staged adventure tourism activities, often called soft adventure (e.g., balloon rides, bungee jumps, safari tours), where activity and outcome-based experiences are described as predictable, safe, and reliable, to hard adventures which involve greater levels of participant commitment, skills, and responsibility (Swarbrooke et al., 2003). While risk and uncertainty are collectively considered fundamental characteristics of adventure tourism, the emphasis on these can challenge the traditional definition of adventure tourism (McKay, 2014; Pomfret & Bramwell, 2016). For example, what might be perceived as risky by one individual may be perceived as boring and mundane by another. At the same time, tour operators need to minimise real risk and offer predictable activities (Beedie, 2003; Taylor et al., 2013). This traditional approach relies on experiences characterised by risk (real or perceived), thrills, and excitement (Cater, 2006; Pizam et al., 2004), suggesting that tourists are less likely to return to activities previously undertaken. However, critics have pointed out that the tourist experience is becoming more diverse (Pomfret & Bramwell, 2016) because motivations and experiences are varied (Naidoo et al., 2015), intergenerational participation is growing rapidly (Mintel, 2010), and the gender gap is less pronounced (Bentley et al., 2010; Tourism Alliance, 2015). For example, approximately half of all adventure travellers and tour operators were women in the beginning decade of the 21st century (Adventure Travel Trade Association, 2022). In summary, the emphasis on risk and thrill as fundamental to the adventure tourist experience has been critiqued as limited (Kerr & Houge Mackenzie, 2012; Pomfret & Bramwell, 2016). Reasons for participating and searching out adventure tourism activities and identifying adventure within normalised tourism activities might be more nuanced, which has implications for management initiatives and how we understand the adventure tourist.

Adventure tourist experience as extraordinary human experience

Beyond the ideas of risk and thrills, researchers have proposed that the adventure tourist experience is more positive and profound. For example, Buckley (2006) found that the experience was characterised by factors such as overcoming fear, achievement and developing relationships with like-minded people. Naidoo et al. (2015) noted that the experience was developmental, where participants enjoyed enhanced self-image and a new appreciation of their capabilities. Adventure tourist experiences could also include personal insights and enlightenment (Sharma-Brymer, 2020). Some experiences often trigger deep personal changes that are positive, permanent, instant, and often unexpected (Brymer, 2013), although sometimes deliberately sought. Such experiences have been linked to the rediscovery of the nature–human relationship, which acts as a facilitator to reveal inner strengths. For example, the climber Lynn Hill wrote:

> It's all about learning to adapt totally to the environment you're in. I think it provides the perfect opportunity for learning about what makes you tick. When you're that involved in the external world, you can really explore your inner nature.
>
> (cited in Olsen, 2001, p. 66)

The adventure tourist experience reflects a therapeutic experience, where ideas of self are transcended (Brymer & Schweitzer, 2017), or feelings of well-being through a connection to nature. Being at one with the natural world is spoken about in various related terms such as being a part of the natural world, merging with, or being open to, the environment (Ryff, 2021). This experience often initiates a better understanding of personal well-being and a desire to treat the natural world in a more positive manner (Sharma-Brymer, 2022). In summary, the adventure tourist experience can be profound, allowing participants to explore forgotten aspects of being human that open up experiences which are hard to attain in other ways. Experiences are believed to be essential to what it means to be fully aware human beings. From this perspective, adventure tourist offerings are less about thrills and excitement and more about the continued journey of self-discovery and connection, where initiatives can be designed as a series of developmental programmes. At its extreme, this also suggests the possibility of a partnership between tourism and health such as seen in green social prescribing or nature prescription.

The adventure tourist experience as transformational

Adventure tourism is transformational in nature as tourists go through physical, emotional, and spiritual journeys (Brymer & Schweitzer, 2017). Reisinger (2013) defined transformational experiences as rich, emotional, and sensual, whereby the tourist can reach their full authentic potential. Nandasena et al. (2022) asserted that adventure tourism can result in 'transformative experiences through risk-taking, overcoming personal fears, self-affirmation, teamwork, and tourists realizing their true potential' (p. 288). Brymer (2013) argued that transformational outcomes lived in everyday life, not just during the adventure experience, include 'increased humility, enhanced spirituality and personal growth' (p. 116). Other studies show that participants actively change behaviours as well as experience increased empowerment, self-esteem, confidence, and a better appreciation of risk and culture. More profoundly, the transformational experience from adventure tourism can facilitate a sense of belonging in the world (Brymer & Hough Mackenzie, 2015). This affirms personal values of trust, reciprocity and relationality with others. The two vignettes (Vignettes 9.1 and 9.2) presented in this chapter further support these research findings. Kamala (Vignette 9.1) revealed how her spiritual tourism involved adventure and connection with the natural world. These experiences solidified her values of kindness and peace whilst deepening connections with her community.

For adventure tourists, some changes experienced during activity participation may translate over into their daily life for improving quality of life. For example, new insights learnt in perceiving and managing risks with an acute orientation to the environment may enrich their ordinary everyday life events and experiences (Clinch & Filimonau, 2017). Improved understanding of risk management is directly related to how the safety and well-being of others are managed. At times, physical engagement in adventure activities enhances self-care behaviours in everyday life and facilitates greater physical and emotional well-being for an individual (Pontea et al., 2021). This could translate to practices of enhanced self-awareness of others and the environment, firming up relationship building, and valuing self and others. Overall, these practices individually and collectively contribute to eudaimonic well-being (see also Chapters 1 and 12).

Embedding opportunities in adventure tourism activities to facilitate participant experience of risks and risk management, teamwork, and facing challenges should account for socio-cultural diversities. For example, tourists from different socio-cultural backgrounds might experience the activity and outcome differently, becoming aware of their own capacity and abilities. This self-awareness could translate into newer understandings of gendered roles in personal and community life and suggest specific roles for adventure tourism operators in community initiatives.

In summary, the adventure tourist experience can facilitate positive changes in individual tourist's values, attitudes, and behaviours (Lindberg & Jensen, 2021). For example, self-reliance, independence, increased self-esteem, confidence, and more positive relationships after experiencing risks, challenges and their outcomes. Adventure tourism can influence individual and collective experiences to the extent of further helping to see the values of "doing good" in the world. The experiences are transformative in that they encourage and facilitate the sustenance of such qualities and behaviours in everyday life, thereby influencing the individual's relationship with others, the community, and the environment (see Vignette 9.1).

Adventure tourism as a skill development experience

The adventure tourism experience can lead to the development of personal and professional skills that contribute to the growth of the adventure tourism industry (Beedie, 2008; McKay, 2014) and lead to profound life-changing personal experiences for the tourist. As a multidimensional industry, adventure tourism offers several career options (Adventure Travel Trade Association, n.d.), such as activity instructors, trip leaders, facilitators, guides, managers, and others. Furthermore, participation in adventure tourism can also influence young children to learn and develop the necessary skills for diverse career choices involving adventure such as sports, education, coaching, allied health, and business. Beedie (2008) suggested that the aspirations triggered by adventure tourism experiences open opportunities for personal and professional skill development. In some parts of the world, the lure of adventure activities and the scope to travel (McKay, 2014) can afford both adventure tourist experiences as well as opportunities for becoming a professional in the adventure tourism industry. For example, kayakers, skiers, and mountaineers have been able to practise their passionate adventure activities whilst also working as professionals in specific adventure areas (McKay, 2014). In such instances, the adventure tourist can transfer their skill sets to become the adventurer, professional and mentor of other adventure tourists. This promotes and influences adventure tourist experiences. In Vignette 9.2, Anand highlighted his self-growth as an enthusiastic adventurer, influenced by the outdoors as well as his religion. He chose outdoor and adventure education as a career path in his life, sharing humanistic values of reciprocity and humility. Please see Chapter 19 for a fuller examination of adventure tourism professionals and careers.

The adventure tourism sector is geographically diverse, requiring open-mindedness for learning, adapting, and applying different sets of skills to operate in particular contexts (Buckley, 2010; Shephard & Evans, 2005). This may result in the adventure tourist learning to appreciate and relating to the local demographics, flora, fauna and other features. This may be necessary for the tourist to immerse themselves in the experience through, for example a safari trip in Tanzania or a multi-day rafting trip down an interesting river with ample ecological features.

Adventure travel companies are increasing worldwide, actively expanding in activities that operators may present to a diverse range of tourists (Williams & Soutar, 2005). For those not necessarily drawn to becoming an adventurer themselves, this presents opportunities for microadventures, catering to a different clientele. Adventure tourists might also use their new skills to facilitate growth in others through team development programmes or performance coaching. Nevertheless, adventure tourism has positive affordances for both the participant and the organiser. The organiser/operator also experiences adventure tourism at a different level by learning skills related to technical competencies, first aid, customer service, group management, natural and cultural history interpretation, and sustainability. Developing new skills and competencies necessitates a self-examination of values, attitudes and behaviours that are pro-environment, pro-humanity, and pro-sustainability. Both professionals and/or operators and participants can be encouraged to perceive the interconnectedness between the key characteristics of adventure such as autonomy, self-awareness, self-determination and motivation. The realisation of affordance within this interconnectedness can profoundly influence the adventure tourist experience and those who contribute to its construction.

To summarise, the adventure tourist experience creates a platform that opens up opportunities for participants as well as the operators to learn new skills. At one level, this includes activity-focused professional skills such as technical competencies required to lead or be involved in teams providing tourism experiences in particular activities, for example guides and porters on Mount Everest. At another level, this experience leads to associated skills such as knowledge of first aid, group management skills, and the skill to interpret and connect with others to value place-based experiences. Martínez and Houge Mackenzie (2022) noted that the guides who felt more linked to nature also felt more responsible for the environment and life in all its forms (Figure 9.2). Through adventure tourism activities, tourists can also develop a deeper connection to nature that can help create sustainable outdoor practices.

Adventure tourist experience of the human–nature relationship

As noted above, the adventure tourist experience can involve a refreshed human–nature relationship (see Chapter 1). For many tourists, the significance of humans' relationship with the environment is profound and it is a crucial element embedded within their participation in adventure tourism (Figure 9.3). This can be the perception of dependency on the environment, their efforts to appreciate the place/environment, and/or developing the attitude of "being in the moment" while living the adventure in a specific environment. However, the tourist experience is not always positive for the environment, and as such,

Figure 9.2 A local guide leading an adventure tourist experience in India. Photograph by Leo Saldanha.

Figure 9.3 Tourists revere at a river confluence in Karnataka, India. Photograph by Abdul Rasheed.

the adventure tourist experience might create a greater perceived separation between the tourist and the environment or reinforce the environment as a resource or playing field. For example, Pomfret (2006) argues for a push and pull explanation of adventure tourism participation where human motivation factors "push" a person towards participation in adventure and the natural qualities of the activity construction "pull" individuals towards a specific destination and activity. This association can be negative for the environment where human activities impact on biodiversity loss, and climate change (Martínez & Houge Mackenzie, 2022). Hall (1992) noted that environmental deterioration happened as people engaged in adventure activities in diverse natural places, including national parks and wilderness areas, which were typically difficult to reach. Following that observation, Olivier (2006) argued that adventure tourist experiences focusing on risky and dangerous leisure activities may have more detrimental impacts on the environment. Even if the intention of adventure tourism opportunities is to create positive human–nature relationships, the outcomes may not always be positive. Please see Chapter 22 for a thorough discussion of the environmental impacts of adventure tourism, and their management.

Wilderness locations frequently provide opportunities that are unavailable in mundane everyday life contexts (Palmberg & Kuru, 2000). This could be a strong motivation for adventure tourists seeking more-than-human experience through relating to the natural world as closely as possible. According to Ewert and Hollenhorst (1989) and Hall (1992), adventure tourists want to understand and/or re-establish a connection with nature. Research has shown that tourists engage in adventurous activities to comprehend better or learn about nature with authentic personal experiences (Pennington-Gray & Kerstetter, 2002). Living their experience in nature in its truest sense with transcendental feelings could encourage pro-nature values and attitudes in adventure tourists (Sharma-Brymer, 2022).

In summary, adventure tourism brings a mix of negative and positive experiences for participants who consider newer insights into human–nature relationships. It is important to understand that adventure tourism, when regulated or legislated to keep a check on its impact on the environment, has positive affordances for tourists' experiences to make sense of their relationship with nature. These experiences could be significant for human well-being, developing pro-environmental behaviours and sustainability.

Religious and spiritual tourism as an adventurous experience

The lived tourist experience of feeling "adventurous" can also be associated with 'a tempo-spatial interruption from daily routine' (Cheer et al., 2017, p. 252). From this perspective some spiritual tourism experiences are characterised by travelling to challenging environments, collective excitement, anticipation and anxiety, and experiencing personal life satisfaction. They can, therefore, be included as forms of adventure tourism. In traditional non-Western societies, such as in Asia, groups of families and friends undertake trips to religious places and spiritual retreats seeking self-discovery experiences that are personally and collectively meaningful (Figure 9.4). An example of a traditional spiritual adventure experience involves transcontinental tourists travelling to circumvent Mount Kailash in the Himalayan region. This trip is an adventurous, immersive trekking experience through challenging environments. Although considered religious and deeply spiritual for people of many religions, participating in this activity entails elevated levels of adventure. Similar spiritual adventure tourist examples can be found across the globe, such as in Japan, China, and Nepal.

In many respects, religious and spiritual tourism overlap. Shinde (2018, p. 58) refers to religious tourism as 'patterns of travel where visitors fulfil religious and recreational needs by visiting places of religious importance and pilgrimage sites'. Mukherjee et al. (2020, p. 428) define spiritual tourism as travels 'where a pilgrim or tourist can achieve their self-personal transformation and self-discovery'. Following their traditional beliefs and practices, pilgrims as tourists are a mix of spiritual tourists, seeking more-than-human, self-realisation experiences, and traditional religious pilgrims, who are content with the completion of their pilgrimage dictated by tradition (Cheer et al., 2017). For other tourists, recreation can be filled with spiritual activity experiences (Haintzman, 2010). Although religion and spirituality can be distinctly separated from each other, they overlap in tourism where pilgrims are also spiritual tourists (Willson et al., 2013) taking part in adventure. Examples are when pilgrims travel to the Gangotri shrine in the Himalayan mountains, where River Ganga is worshipped in a goddess form, and Goumukh, the river's origin situated in a glacier. Travelling to Gangotri is termed as religious whereas taking up the arduous journey on foot to the mouth of the River Ganga reflects spirituality combined with feelings of oneness with nature. The experience of adventure embedded within these two activities reflects the pilgrim's association with the risks present in the unknown socio-cultural environments of the pilgrimage, as well as in the natural world. The adventure tourist, in this example, lives a deep sense of commitment, self-determination and choice. The lived experience can equally be transformational and transcending (see Vignette 9.1).

Religious and spiritual tourists travelling to transcontinental destinations is indicative of this type of tourism's worth and popularity. The tourist is adventurous in seeking

Figure 9.4 A Hindu shrine in the Himalayan mountains. Photograph by Leo Saldanha.

opportunities to discover new insights into the meaning of life and life satisfaction. They find interconnectedness between the person, activity and environment, which influences the holistic meaning of their lived experience. Some other examples are the passionate and deeply spiritual Western tourists seeking spiritual teachers and spiritual immersion in ancient religions, such as Buddhism (Shinde, 2021). It can be argued that adventure is inherently built into the experiences of these spiritual tourists. They move away from their mundane everyday life, travel to new and unfamiliar socio-cultural and geographical environments, and face the challenge of understanding newer ways of life. Another example is immersion in the natural world and its spirit as practised by the First Nations people of Australia (Tacey, 2016). Spiritual tourists are increasingly seeking such experiences in cultural and adventure tourism along with learning and connecting to Aboriginal knowledge, worldviews, and perspectives. One such deeply spiritual experience is Dadirri offered by Aboriginal Elder and scholar, Dr Miriam-Rose Ungunmerr (Ungunmerr, n.d.). This example calls for newer perspectives on how adventure tourism is conceptualised and processed.

Pilgrims as adventure tourists are influenced by a range of juxtaposing social and cultural factors intersecting faith, collective social beliefs, values and norms, and cultural identities. Cultural, religious, and spiritual immersion is the end goal of these tourists. However, their experiences involve the adventure of travelling to distant, remote places, interacting with other linguistic groups and their cultures, and being challenged by the inherent risks of travel, health, food, and safety. For example, groups of urban Indian families normally visit places of river confluence during auspicious seasons, not only for religious purposes but also to feel adventurous as they are meant to take a holy dip in the river. Members of faith organisations embark on spiritual tourism to temples, churches, gurudwaras (Sikh holy worshipping centres), and mosques during holy seasons. Caste-based groups of Hindus may undertake a fortnight-long temple tourism trip as part of fulfilling their religious vows. Within these experiences, the tourists accept challenges and risks, living the excitement and enjoying the life satisfaction which characterise the adventurous experience.

Whether organised at the group level or through a commercial travel agency, adventurous spiritual tourism promotes participation from a wide range of age, gender, ability and socio-economic population groups. Invisible within their tourism experiences are micro and meso levels of adventure. For example, for a group of low-income and lower caste people in Indian society travelling to holy shrines may be a unique adventurous experience. It may entail different levels of access and participation due to sociocultural and religious norms, power, wealth and control. Access to resources and effective management of risks impact the tourist's experience in determining their level of adventure. While the experience may not be adventurous for men, the women may feel very adventurous in stepping out of their heteronormative traditional home environments. The pilgrimage may be liberating for them and may feel out of the ordinary, adventurous, and thrilling. There is also the most important accomplishment of a lifelong dream, which feels adventurous for women of traditional societies such as India, cutting across religion, caste, class, financial status, and language.

Similarly, participants of spirituality tourism through packaged programmes, for example yoga, nature–nurture–nutrition, meditation, and Ayurvedic healing, feel transformation through culture and the adventurous feeling of accomplishment. Living an out-of-the-ordinary experience, they are filled with excitement, fulfilment, anxiety, fear, and a transformational and heightened level of self-exploration and life satisfaction. These experiences enhance social, emotional and collective well-being (Ryff, 2021). Their experiences could also help in cementing relationships, affirming humane values of empathy and compassion, and "doing good", which supports the sustainable development of communities (Carvache-Franco et al., 2022; UNWTO, 2014). Notably, taking part in the activity and change in their everyday mundane environment may enhance adventurous feelings for many people. Kamala's experiences (see Vignette 9.1) as a religious and spiritual tourist encompass adventures lived outside of her mundane ordinary life with transformation.

In summary, studying adventurous experiences within religious and spiritual tourism is an emerging research area with the potential for direct and profound impact. Researchers could consider socio-cultural diversities within those participant experiences that expand the meaning of adventure along with its scope.

Diverse sociocultural factors influencing adventure tourist experience

Visiting a place of recreation and adventure with or without activity involvement is an invitation for urbanites to get away from daily mundane routines. Driven by this focus, participants prefer spending time outdoors, at nature resorts and doing leisurely activities as a group. Adventure tourism increasingly involves escape from daily routines, social surveillance – which is inherent in traditional non-Western societies – and seeking opportunities for improving personal well-being (Akhoundogli & Buckley, 2021). Socio-cultural factors such as class, caste, faith and gender seemingly have a profound impact on people's choice, participation, and interest in micro/local/easy tourism. This is influenced by community affiliation, belongingness, and within-group experience, for example young corporate-working adults engaging in micro-level local adventure tourism, such as climbing a hill in the local region, or undertaking organised paragliding in a city.

For these urbanites, the criteria for choice and participation in adventure activities are collective experiences, financial affordability, health benefits, and like-mindedness. These are in addition to one's religion, caste, class, language, or gender. For a group of Indian women representing the Dalit caste (the lowest of the four castes), visiting a local dam and reservoir in their regional area may be a unique adventure filled with excitement, curiosity, joy, and fulfilment. This could be encouraged and organised by a non-government organisation working for Dalit women's empowerment. The women come together as an organised collective group without fearing patriarchal control and/or invoking indignation from the community. This microadventure experience may fill them with self-confidence and social mobility, which they can transfer to their daughters. Their understanding of adventure tourism may be different as they are not seeking religion or spirituality-based tourist experiences in remote places or thrill-based activities. Their family and community influence their choice, decision, access, and participation in these adventure activities where their overall participation may be more collectively valued than valued at a personal level.

However, there are varied outcomes of adventure activity participation at the individual level. A noteworthy example is an outdoor adventure and education organisation operating in the backwaters of River Sharavathi in Karnataka state of southern India (Gunari, 2021). Tourists to the region participate in kayaking, swimming, learning to ride coracles (small boats), nature walks, and hill-trekking activities. They are mostly middle-income groups of families, friends, schoolchildren, people with disabilities, women-only groups, and college students. Besides environmental and outdoor education lessons, they experience adventure at their own level as well as collectively (Sharma-Brymer, 2018). For example, after their adventurous experience, younger women bring other female members of their family, relatives, and friends bring their children with disabilities, and college students bring their families and friends. These newcomers might never have experienced water sports or nature walks by the river. They encounter these novel experiences with curiosity, excitement, anxiety, fear and satisfaction, and building personal confidence and self-esteem. This enhances individual agency and collective values of caring for the environment, compassion, and humility. The person–environment relationship is well-recognised with self-awareness and proactivity.

At a micro level, adventure tourism has been in practice across population groups that have diverse economic, social and cultural identities (income, education, class, caste, family-level of conservative outlook, community-driven norms) in many developing countries (e.g., Indian subcontinent, South-East Asian countries). Adventure tourists from these geographical parts are also influenced by accessible global-level outdoor adventure pursuits such as mountaineering. Importantly, adventure tourists from such countries acknowledge opportunities for expressing agency and personal capabilities (Sharma-Brymer, 2022). For example, in contemporary times, younger generations of employees working for multinational companies and earning high incomes value adventure tourism experiences for personal development, leisure, rejuvenation, and well-being. At the same time, their mothers may still be expected to seek approval from their husbands to take part in nature retreats with friends. With gender inequalities still operating strongly in patriarchal societies, women have many socio-cultural barriers to overcome for a deeper experience of adventure tourism, although their societies are witnessing an exponential growth in adventure tourism. A noteworthy point here is that this new genre of adventure tourists approach adventure tourism with an exploratory attitude for self-discovery rather than for the more traditional notions of adventure, such as risk-taking and outcome-driven activities (middle class, urban, women, employees of global companies, differently abled, and others).

In summary, the profile of adventure tourist experiences in developing countries, especially traditional countries such as the Indian subcontinent, has seen a steady transformation in recent years owing to technological advancements and wide-spread effects of marketisation. It has given rise to different types of adventure tourists with wider participation and promotion. In this sense, adventure tourist experiences are helping break down barriers across class, caste, gender and ability, promoting equality, equity and participant agency. From this perspective, it is also important to recognise that one size does not fit all. An adventurous activity can facilitate varied experiences, facilitating further interest in various activities. This suggests that future initiatives might be more about developing relationships with tourist diversities as differentiated from standard activities.

Conclusion

Adventure tourist experiences may not always be activity-oriented but can also be diversified for enhancing community connections, personal development and confidence. This aligns with research studies reporting on the positive benefits of adventure tourist experiences long after the adventure event ends. Through this comparative analysis, we have attempted to clarify the types of adventure tourists, their lived experiences, and adventure tourist experiences influenced by socio-cultural, religious, and economic factors. From a tourist experience perspective, the variety of experiences suggests the need for a very different conceptual appreciation of the tourist diversities and theoretical frameworks to understand the tourist as well as support the design of adventure tourism experiences. Adventure tourism has evolved beyond risk and thrill activities designed to churn through tourists towards more powerful relational opportunities for the growth of future participant communities. To realise this potential, future research should consider wide-ranging factors that impact opportunities for enhancing both global and local adventure tourist experiences. This will facilitate better diversity and inclusion in adventure tourism activities and outcomes. Future research needs to recognise the relationship between the tourist, the social, cultural, economic, and physical environment and the activity and focus more on the development of ecological frameworks that strengthen this relationship.

Review questions

1 Compare and contrast traditional perspectives of adventure tourist experiences with transcendental adventure tourist experiences. Give examples.
2 What factors can be considered influential in shaping the experiences of a microadventure tourist? Discuss these by constructing a case study from your own country-specific context.
3 How do you identify the person–environment relationship in an adventure tourist's experience undertaking a pilgrimage to a remote holy place? What role do socio-cultural factors play in creating opportunities for adventure during their tour?
4 Review gendered experiences of adventure tourism drawing from your own country-specific examples.

Further reading

Beedie, P. (2008). Adventure tourism as a 'New Frontier' in leisure. *World Leisure Journal*, 50, 173–183. https://doi.org/10.1080/04419057.2008.9674551
Brymer, E., & Houge Mackenzie, S. (2015). The impact of extreme sports on host communities' psychological growth and development. In Y. Reisinger (Ed.), *Transformational tourism: Host benefits* (pp. 129–140). CABI.
Cheer, J., Yaniv Belhassen, Y., & Kujawa, J. (2017). The search for spirituality in tourism: Toward a conceptual framework for spiritual tourism. *Tourism Management Perspectives*, 24, 252–256. https://doi.org/10.1016/j.tmp.2017.07.018
Pomfret, G. (2019). Conceptualising family adventure tourist motives, experiences and benefits. *Journal of Outdoor Recreation and Tourism*, 28(100193), 1–8.

References

Adventure Travel Trade Association. (2022). Adventure Travel Industry Snapshot 2022. https://www.adventuretravel.biz/store/industry-snapshot-2022/
Adventure Trade Travel Association. (n.d.) Adventure travel guide qualifications & performance standard – IV. *Adventure Travel Guiding Core Competencies*. https://www.adventuretravel.biz/education/adventure-edu/
Akhoundogli, M., & Buckley, B. (2021). Outdoor tourism to escape social surveillance: Health gains but sustainability costs. *Journal of Ecotourism*, 22, 1–21.
Bauer, I. (2012). Australian senior adventure travellers to Peru: Maximising older tourists' travel health experience. *Travel Medicine and Infectious Disease*, 10, 59–68.

Beedie, P. (2003). Adventure tourism. In S. Hudson (Ed.), *Sport and adventure tourism* (pp. 203–239). Haworth Hospitality Press.

Beedie, P. (2008). Adventure tourism as a 'New Frontier' in leisure. *World Leisure Journal, 50*, 173–183. https://doi.org/10.1080/04419057.2008.9674551.

Bentley, T. A., Cater, C., & Page, S. J. (2010). Adventure and ecotourism safety in Queensland: Operator experiences and practice. *Tourism Management, 31*(5), 563–571.

Brymer, E. (2013). Extreme sports as transformational experiences. In Y. Reisinger (Ed.), *Transformational tourism.* CABI.

Brymer, E., & Houge Mackenzie, S. (2015). The impact of extreme sports on host communities' psychological growth and development. In Y. Reisinger (Ed.), *Transformational tourism: Host benefits* (pp. 129–140). CABI.

Brymer, E., & Schweitzer, R. (2017). *Phenomenology and the extreme sports experience.* Routledge.

Buckley, R. (2006). *Adventure tourism.* CABI.

Buckley, R. (2010). *Adventure tourism management.* Butterworth-Heinemann.

Cater, C. I. (2006). Playing with risk? Participant perceptions of risk and management implications in adventure tourism. *Tourism Management, 27*, 317–325.

Carvache-Franco, M., Contreras-Moscol, D., Orden-Mejía, M., Carvache-Franco, W., Vera-Holguin, H., & Carvache-Franco, O. (2022). Motivations and loyalty of the demand for adventure tourism as sustainable travel. *Sustainability, 14*(8472), 1–17. https://doi.org/10.3390/su14148472

Cheer, J., Yaniv Belhassen, Y., & Kujawa, J. (2017). The search for spirituality in tourism: Toward a conceptual framework for spiritual tourism. *Tourism Management Perspectives, 24*, 252–256. https://doi.org/10.1016/j.tmp.2017.07.018

Clarke, J., Previte, J., & Chien, M. (2022). Adventurous femininities: The value of adventure for women travelers. *Journal of Vacation Marketing, 28*(2), 171–187. https://doi.org/10.1177/13567667211038952

Clinch, H., & Filimonau, V. (2017). Instructors' perspectives on risk management within adventure tourism. *Tourism Planning & Development, 14*(2), 220–239.

Cohen, E., & Cohen, S. A. (2015). A mobilities approach to tourism from emerging world regions. *Current Issues in Tourism, 18*(1), 11–43. https://doi.org/10.1080/13683500.2014.898617.

Ewert, A. W., & Hollenhorst, S. J. (1989). Testing the adventure model: Empirical support for a model of risk recreation participation. *Journal of Leisure Research, 21*(2), 124–139.

Gunari, R. V. (January 31, 2021). *Venture out for adventure: Honnemaradu in Shivamogga district is a paradise for fun and thrills.* https://www.newindianexpress.com/cities/bengaluru/2021/jan/31/venture-out-for-adventure-2257460.html

Hall, C. M. (1992). Adventure, sport and health tourism. In B. Weiler & C. M. Hall (Eds.), *Special interest tourism* (pp. 141–158). Belhaven.

Haintzman, P. (2010). Nature-based recreation and spirituality: A complex relationship. *Leisure Sciences, 32*, 72–89. https://doi.org/10.1080/01490400903430897

Houge Mackenzie, S., & Goodnow, J. (2021). Adventure in the age of COVID-19: Embracing Microadventures and locavism in a post-pandemic world. *Leisure Sciences, 43*(1–2), 62–69. https://doi.org/10.1080/01490400.2020.1773984

Kerr, J., & Houge Mackenzie, S. (2012). Multiple motives for participating in adventure sports. *Recreation, Parks and Tourism Administration, 13*(5), 649–657.

Jordan, E., Chancellor, C., & Norman, W. C. (2016). Who is a tourist? A comparison of straight line and driving distances. *Travel and Tourism Research Association: Advancing Tourism Research Globally.* 19.

Lindberg, F., & Jensen, Ø. (2021). Adventure regime of tourism experiences. *Current Issues in Tourism, 24*(20), 2905–2920.

Martínez, E. V., & Houge Mackenzie, S. (2022). Climate change and adventure guiding: The role of nature connection in guide wellbeing. *Frontiers in Public Health, 10.* https://doi.org/10.3389/fpubh.2022.946093

McKay, T. J. M. (2014). White water adventure tourism on the Ash River, South Africa. *African Journal for Physical, Health Education, Recreation and Dance, 20*(1), 52–75.

Mintel. (2010). Activity holidays-UK-February 2010. https://academic.mintel.com/sinatra/oxygen/display/id=479866

Mukherjee, S., Bhattacharjee, S., & Singha, S. (2020). Religious to spiritual Tourism: An era of paradigm shift in India. *Journal of Xi'an University of Architecture & Technology, XII*(II), 427–441.

Muller, T., & Cleaver, M. (2000). Targeting the CANZUS baby boomer explorer and adventurer segments. *Journal of Vacation Marketing, 62*(2), 154–169.

Nandasena, R., Morrison, A. M. and Coca-Stefaniak, J. A. (2022). Transformational tourism: A systematic literature review and research agenda. *Journal of Tourism Futures, 8*(3), 282–297.

Naidoo, P., Ramseook-Munhurrun, P., Seebaluck, N. V., & Janvier, S. (2015). Investigating the motivation of baby boomers for adventure tourism. *Procedia - Social and Behavioral Sciences, 175*, 244–251.

Olivier, S. (2006). Moral dilemmas of participation in dangerous leisure activities. *Leisure Studies, 25*(1), 95–109.

Olsen, M. (2001). *Women who risk: Profiles of women in extreme sports.* Hatherleigh Press.

Palmberg, I. E., & Kuru, J. (2000). Outdoor activities as a basis for environmental responsibility. *The Journal of Environmental Education, 31*(4), 32–36.

Pennington-Gray, L. A., & Kerstetter, D. L. (2002). Testing a constraints model within the context of nature-based tourism. *Journal of Travel Research, 40*(4), 416–423.

Pizam, A., Jeong, G., Reichel, A., Van Boemmel, H., Lusson, J., Steynberg. L., State-Costache, O., Volo, S., Kroesbacher, C., Kucerova, J., & Montmany, N. (2004). The relationship between risk-taking, sensation-seeking, and the tourist behavior of young adults: A cross-cultural study. *Journal of Travel Research, 42*, 251–260.

Pomfret, G. (2006). Mountaineering adventure tourists: A conceptual framework for research. *Tourism Management, 27*, 113–123.

Pomfret, G., & Bramwell, B. (2016). The characteristics and motivational decisions of outdoor adventure tourists: A review and analysis. *Current Issues in Tourism, 19*(14), 1447–1478. https://doi.org/10.1080/13683500.2014.925430

Pomfret, G. (2019). Conceptualising family adventure tourist motives, experiences and benefits. *Journal of Outdoor Recreation and Tourism, 28*(100193), 1–8.

Pontea, J., Coutoa, G., Sousab, A., Pimentela, P., & Oliveirac, A. (2021). Idealizing adventure tourism experiences: Tourists' self-assessment and expectations. *Journal of Outdoor Recreation and Tourism, 35*, 100379.

Reisinger, Y. (2013). *Transformational tourism: Tourist perspectives.* CABI.

Ryff, C. D. (2021). Spirituality and well-being: Theory, science, and the nature connection. *Religions, 12*(914), 9–15.

Sharma-Brymer, V. (2022). Giving back to nature: An autoethnographic analysis of adventure as transformational. In E. Brymer & P. Reid (Eds.). *Adventure Psychology: Going knowingly into the unknown.* Routledge Series.

Sharma-Brymer, V., & Brymer, E. (2020). The elephant in the room: An autoethnographic approach. In E. Laws., N. Scott., X. Font, & J. Koldowski (Eds.) *The Elephant Tourism Business – International Problems and Progress.* CABI.

Sharma-Brymer, V. (2018). Locations of resistance and agency: The actionable space of Indian women's connection to the outdoors. In T. Gray & D. Mitten (Eds.), *The Palgrave Macmillan international handbook of women and outdoor learning* (pp. 307–319). Palgrave Macmillan.

Shephard, G., & Evans, S. (2005). Adventure tourism: Hard decisions, soft options and home for tea: Adventure on the hoof. In M. Novelli (Ed.), *Niche tourism: Contemporary issues, trends and cases* (pp. 201–209). Butterworth-Heinemann Ltd.

Shinde, K. (2021). "Imported Buddhism" or "Co-Creation"? Buddhist cultural heritage and sustainability of tourism at the world heritage site of Lumbini, Nepal. *Sustainability, 13*(11), 5820. https://doi.org/10.3390/su13115820

Shinde, K. (2018). Governance and management of religious tourism in India. International *Journal of Religious Tourism and Pilgrimage, 6*(1), 58–71.

Taylor, S., Varley, P., & Johnston, T. (2013). *Adventure tourism: Meanings, experience and learning.* Routledge.

Swarbrooke, J., Beard, C., Leckie, S., & Pomfret, G. (2003). *Adventure tourism: The new frontier.* Elsevier Butterworth-Heinemann.

Tacey, D. (2016). Spirituality and healing. *Ata: Journal of Psychotherapy Aotearoa New Zealand, 20*(1), 19–33.

Tourism Alliance. (2015). *UK Tourism Statistics 2015.* https://www.tourismalliance.com/downloads/TA_369_395.pdf

Ungunmerr, M-R. (n.d.). *Dadirri: Inner deep listening and quiet still awareness.* https://www.miriamrosefoundation.org.au/dadirri/

United Nations World Tourism Organization. (2014). *Affiliate members global report, Volume 9 –Global report on adventure tourism.* UNWTO.

Varley, P. (2006). Confecting adventure and playing with meaning: The adventure commodification continuum. *Journal of Sport & Tourism, 11*(2), 173–194.

Williams, P., & Soutar, G. (2005). Close to the "edge": Critical issues for adventure tourism operators. *Asia Pacific Journal of Tourism Research, 10*(3), 247–261.

Willson, G. B., McIntosh, A. J., & Zahra, A. L. (2013). Tourism and spirituality: A phenomenological analysis. *Annals of Tourism Research, 42,* 150–168. https://doi.org/10.1016/j.annals.2013.01.016

10 Adventure tourism in South Korea and Taiwan

Through the lens of colonial modernity

Young Sook Lee and Ming Feng Huang

Chapter learning outcomes

1 Understand the South Korean adventure tourism-relevant concept of Leiports (레포츠).
2 Identify similarities and differences between Taiwanese mountain trekking adventure activities and western adventure trekking activities.
3 Recognise the applicability of colonial modernity theory to explain the East Asian adventure tourism phenomenon.

Introduction

To a certain extent, the Western ideals of adventure such as risk and challenge outlined earlier in this book (see Chapters 2 and 3) have been transported all over the world. However, the assumption that these ideals are conceptualised and put into practice the same way all over the globe is oversimplifying the adventure tourism sector. Risk and challenge may be conceptualised and practised in a more nuanced way in different societies. We aim to introduce and argue that the notion of colonial modernity may prove to be an adequate notion to explain the contemporary East Asian adventure tourism phenomenon.

Reflecting certain characteristics of colonialism and modernism, colonial modernity suggests that there are different forms of modernities. For example, to illustrate postmodern forms of modernities, Bauman (2000) suggested liquid modernity in relation to solid modernity. In distinguishing liquid versus solid modernities, Bardhi and Eckhardt (2017) argued that changing and adapting modernities can be termed as liquid modernity, while fixed and rigid modernity can be termed solid modernity. They provided practices of liquid versus solid modernities (e.g., types of ownership, ways of consumption, length of experience in consumption) in consumption contexts. Applying the liquid modernity concept to the tourism sector, Vogel and Oschmann (2013) argued that liquid modernity has provided the conditions for the development of the cruise industry as it is today. That is, global regulations and practices are adapted based on immediate conditions of local economic, political, social, and environmental conditions. As a comparative application of multiple modernities in the sports sector, Giulianotti and Robertson (2012) proposed multiple modernities as a way to advance our understanding of 'glocalization' (p. 449)

DOI: 10.4324/9781003393153-13

and Asian sports. Acknowledging these multiple modernities, Barlow (1997) outlined how the concept can be understood as the East Asian pathway to modernity: '"Colonial modernity" can be grasped as a speculative frame for investigating the infinitely pervasive discursive powers that increasingly connect at key points to the globalizing impulses of capitalism' (p. 6).

Colonial modernity is chosen in this chapter as an alternative approach to colonialism/post-colonialism and/or modernism/post-modernism theories in trying to understand the East Asian adventure tourism phenomenon in a more nuanced manner. Indeed, even from the very beginning, Taiwanese and South Korea's tourism and adventure development did not exactly emulate that of the West in either concept or practice. In the South Korean context, tourism focused more on how to rebuild the war-torn society after the Korean War rather than on people's quality of life *per se* (Lee, 2006). Although a tourism-related application of colonial modernity in Taiwan appears to be lacking to date in the form of an academic output, Taiwanese architecture and urbanisation have been conceptualised as a manifestation of colonial modernity (Hsia, 2002). The very lack of colonial modernity investigated in the Taiwanese tourism context calls for such an attempt. We suggest in this chapter that colonial modernity can help us understand contemporary forms of East Asian adventure tourism. Focusing on South Korea's Leiports (레포츠) and Taiwanese mountain-trekking adventure activities, this chapter proposes that adventure tourism in these countries emerged as part of the societies' modernisation in the post-colonial era and, accordingly, reflect their culture- and history-specific experiences, in addition to the imported Western ideals of adventure tourism.

The remainder of the chapter is as follows: Tourism research in modernism and colonialism is reviewed. The review highlights the need for a more nuanced approach (i.e. colonial modernity) to better understand East Asian adventure tourism phenomenon. Two cases of East Asian adventure tourism (South Korean Leiports and Taiwanese mountain-trekking adventure activities) are presented subsequently. This is followed by discussions that conceptualise the presented cases through the lenses of colonial modernity.

Modernism and colonialism approaches in tourism studies

Modernism has often been understood as having a close relationship with the growth and development of a society (Gidwani, 2002). While acknowledging these two features of modernisation to a certain level, Taylor (1999) firmly warned against homogenising the notion. That is, accepting growth and development as a unilateral global process risks over-simplifying the concept of modernity. Nevertheless, while the entire notion of modernity has been increasingly challenged and scholars have found it hard to agree on a single definition (Anderson, 1998), researchers in tourism studies have found the concept useful. As tourism is inherently based on the modern way of ordering the world (Franklin, 2003) and the division between work and non-work time and space (Roberts, 1999), researchers considered the modernity concept helpful to conceptualise tourism phenomena.

However, the modernity framework as the theoretical base in tourism studies has been critiqued as being Eurocentric (Franklin, 2003; Wang, 2000), if not primarily British (Ryan, 2003). This criticism gains more attention when we consider the multiple forms of modernities, as argued by various researchers (see Eisenstadt, 2000; Lau, 2003;

Mukhia, 2010 for detailed descriptions). Within the context of tourism studies, the bias in the epistemological position that subsequently places the Euro/Western hegemonic structure as superior still prevails (Cho, 2012).

Positioning the East Asian society as the active body that chooses and determines its political, economic and social elements within limits is an alternative to the Euro/Western hegemonic standpoint. Research on East Asian tourism phenomena shows that specific social and political conditions exist, but these issues have not been fully considered at a fundamental level (Oakes, 2021). Within adventure tourism research, the Euro/West-centric issue has been recognised. For example, Weber (2001) argued a need to go beyond the western construct of our understanding. Cheng et al. (2018) identified a lack of non-Western understanding in the study area. Recently, Martin and Ren (2020) demonstrated Chinese adventure tourism in the Arctic setting and alluded to the meaning of adventure from the Chinese perspective. Likewise, Buckley et al. (2014) identified particular characteristics of whitewater rafting or piaoliu in China, which is more passive and relies on significant human-made alterations to watercourses (see Figure 10.1). Also, Wengel (2021) has provided some specific micro-level trends in Nepal where domestic adventure tourists are on the rise. The current chapter adds to the increasing attention given to non-Euro/West-centric research on adventure tourism. Conceptualising South Korea's Leiports (레포츠) and Taiwanese mountain trekking from a colonial modernity perspective may prove to be a viable solution to recognising the multiple modernities aspect in adventure tourism studies.

Similarly, colonialism and post-colonialism approaches in tourism studies have focused primarily on contested identities of represented residents from ex-colonisers' nostalgia (Hall & Tucker, 2004; Nash, 1989), sometimes characterising the ex-colonisee as the "noble savage" (Deutschlander & Miller, 2003; Pomering & White, 2011). Thus, identity, representation and commodification have been the central issues in the development and practice of tourism from the colonialism perspective (Bandyopadhyay, 2011; Burns, 2008). However, this approach does not fully explain colonisees' view of their travel and tourism, as opposed to how they are viewed by the ex-colonisers and how they respond to being "othered" (Callon & Latour, 1981). Therefore, the need for cultural, political, and developmental sensitivities in tourism research is required (Alneng, 2002; Hazbun, 2010; Winter, 2009).

These efforts highlight the commodification of the places as objects to be viewed by ex-colonisers with little emphasis on the residents' perspective as ex-colonisees. Investigating colonialism's influence on ex-colonisees' perceptions and practices has enriched tourism scholars' understanding of the colonialism approach. The identified research themes, however, of representation, commodification, and contested identities, by "othering" the ex-colonisees in the tourism context, testify to the privileging of the ex-colonisers' worldview above that of the ex-colonisees. Conceivably, privileging the ex-colonisers' viewpoint may result from the economic imperatives in the tourism trade, where the expectations of tourists, many of whom originate from ex-coloniser countries, need to be met. Changing patterns in international tourism and tourist flows, however, suggest that expectations and perceptions of tourists from ex-colonisee countries need to be understood not just in comparison to the ex-colonisers' view but also from an independent perspective. We suggest that the colonial modernity approach may narrow the existing enquiry gaps in the modernism, post-modernism, colonialism, and post-colonialism perspectives. Before moving on to our analysis of the East Asian adventure from the colonial modernity perspective, we present cases (see Case studies 10.1 and 10.2) from South Korea and Taiwan in the following section.

Figure 10.1 Whitewater rafting or *Piaoliu* (漂流) in China. Photograph by Carl Cater.

Case study 10.1   A South Korean modern adventure: Leiports (레포츠)

The word "레포츠" (*Leiports*) is a Korean adoption of English words that combine leisure and sports, and in this sense, it shares a great deal with both outdoor recreation and adventure tourism. At the time of writing this chapter, the word "레포츠" is listed in the Korean standard dictionary as a Korean word. The range of activities in Leiports (레포츠) is wide, and the most popular ones include climbing, hiking, golf, fishing, and whitewater rafting. The exact time when the term *Leiports* (레포츠) appeared in South Korea is uncertain, but by the mid- to late 1990s, several popular books had been published on this subject in Korean in South Korea. Definitional attempts for this newly coined word began in 2000. To date, only one refereed journal publication on Leiports (레포츠) by Lee and Jennings (2010) appears in the English language, a sharp contrast to more than 3,000 refereed journal articles on this topic written in Korean and published in South Korea. In this chapter we accept the definition of Leiports (레포츠) as: 'Sports performed and travel and tourism experiences engaged in during leisure time; that is, sport and travel and tourism activities that may or may not be undertaken for their own sakes free from the need of making a living' (Lee & Jennings, 2010, p. 131).

According to a study by Lee and Jennings (2010), more than 50 universities and junior colleges in South Korea offered full degree programmes on Leiports management, potentially providing graduates for the adventure tourism sector. The contents of the programmes show that adventure tourism, as conceptualised in Euro/Western settings, is met with Korea-specific historic and cultural experiences that produced Leiports (레포츠). For example, the education programmes consist of multidisciplinary courses, ranging from sports-leisure-relevant, environmental, sociological, psychological, biological, nutritional, historical, cultural, economic, and marketing aspects (Cho, 2009).

While the broad historical and social contexts are intertwined with the significance of sports in the quality of human life, the programmes also have strong practical education components. For example, as part of the marine Leiports programme, Hanseo University provides practical training programmes. These include training sessions in marine rescue, marine sports, marine safety, marine leisure experiences, boat/yacht driving licences, and marine sports for a specialist trainer. Thus, the contents of the Leiports programmes in South Korea appear to be closer to the Western notion of adventure tourism than tourism generally or sports tourism.

Case study 10.2   A Taiwanese modern adventure: Mountain hiking, trekking, and climbing

Despite the rich and long history of mountaineering, mountain climbing and hiking in Europe described in Chapter 2, such activities are more recent developments in Taiwan. The country's geographical characteristics and the influence of Japanese mountaineering culture (Hsu, 2022) presented an opportunity to promote these activities as forms of adventure tourism. The following section elaborates on how Japanese colonial rule influenced the development of mountain hiking in Taiwan.

Taiwan is a long and narrow island forced upward by the collision of the Eurasian Plate and the Philippine Sea Plate on which the highest peak is Yu Shan (Mountain Jade 玉山), reaching 3,952 metres in height. Two thirds of the main island is covered by forested mountains and contains more than 250 peaks over 3,000 metres. The varied terrain provides a favourable environment for adventure tourism activities and has facilitated their growth in popularity in Taiwan (Taiwanese Tourism Bureau, 2022a). Taiwanese mountain hiking and climbing development can be divided into four time periods: (1) pre-1895, (2) 1895–1945, (3) 1945–1990s, and (4) post-2000. These follow the development sequence of "adventure," "sports," "exercise," and "leisure" as detailed in the following discussion.

Since the late Ming dynasty (mid- to late 1600s), many Han Chinese people emigrated from south-east mainland China to lowland and coastal parts of Taiwan. At this time, Indigenous people occupied mountain areas with a ferocious reputation of head-hunting those who invaded their territory. To avoid such unnecessary conflicts between Han Chinese and the Indigenous people, the Qing dynasty government prohibited the former from entering the mountains. However, there were several attempts at mountainous exploration because of global trade expansion in this area. For instance, records show that explorations were made by John Dodd, a Scottish tea merchant, Dr George Leslie Mackay, a Scottish missionary, and Robert Swinhoe, the first British consular representative (Lin, 2002).

After Japan invaded and colonised Taiwan, military operations continued to suppress the local armed forces, in order not to allow the locals time in the mountains. However, since 1896 Japanese scholars have conducted surveys and scientific research to investigate and explore the mountain areas. About two decades later, due to the pacification of the Indigenous population, the Japanese government started to encourage Taiwanese civilians to get involved in fitness and sports. This led to the Taiwanese Mountaineering Association (台灣山岳會) holding its inaugural meeting at the top of Guanyin Mountain (觀音山). Following this, various organisations, groups, and schools set up climbing clubs and contributed to the early development of mountaineering in Taiwan. A well-known example was in 1927 when 12 girls, including Taiwanese student Lin Tsai Wan from Taipei No. 3 Girls' High School, who climbed Yu Shan (Mountain Jade 玉山; Li, 2005). However, most of those involved in the committee of the Taiwanese Mountaineering Association were administrative officers and were mainly Japanese, not Taiwanese residents. This situation did not really change until the end of World War II.

In the 1970s, with the liberalisation of Taiwan's Organisation Law, mountaineering groups, clubs, and associations were established in close succession to each other. In the 1980s, Taiwan's economy took off, the media flourished, and the government opened up overseas travel and introduced foreign climbing techniques. Therefore, climbing organisations started to organise different mountaineering activities. The most active period for mountaineering activities in Taiwan was between the 1970s and the 1990s. Although it was very difficult to access the mountains, due to several different regulations set up by the government, and the control of access being very strict, mountain hikers and climbers were very enthusiastic and devoted to their activities. The list of the "100 Mountains" (Baiyue 百岳) was also finalised around this time, in 1972. This list can be seen as parallel to the Munroes, which is a list of peaks over 3,000 feet in Scotland, and the Nuttals which are mountains over 2,000 feet in Wales and England. Table 10.1 shows the 18 categories of Taiwanese mountains and how they have been classified according to aesthetic, emotive, and historical reasons.

According to Ye (2012), the prior Minister of the Department of Health of Taiwan, there were about 5 million hikers in Taiwan at the time of his publication, but most of them were hikers in the suburban mountain area. There were about 500,000 people who had hiked or climbed mountains over 3,000 metres, but only 30,000 to 50,000 were regular hikers and climbers. To better promote mountain hiking and climbing activities in more recent years, the Taiwan Outdoors (2022) has integrated the national hiking trail system and divided this into five levels of difficulty.

Table 10.1 Description of the Taiwanese 100 Mountains (Baiyue 百岳)

Category	Description
Five Mountain	These five mountains have a unique shape, a very good uninhibited view from the summit and stand out from their surroundings.
Three Sharp	These are very tall and steep on both sides, like a pen's tip.
Ten Dangerous	These are dangerously steep, the rock cliff is sharp, but the target is obvious.
Ten High	These mountains are high and huge with ups and downs, although the summit is wide, and the slope is gentle.
Nine Lofty	The mountain is like a tower, standing out from other mountains, and the shape is beautiful.
Nine Barrier	The peak looks like a wall, blocking the view to see through it like a barrier
Eight Beauty	The shape of the mountain is pretty, with gentle slopes and no massive jumbles of boulders.
Ten Smooth	Smooth indicates there are no steep uphill or downhill slopes, and the hiker does not need to climb rocks.
Ten Green	The peaks have a dense covering of trees and plants, so hikers have to shuttle back and forth in the bamboo undergrowth
Ten Rock	This means hikers have to climb the rocks to reach the summit.
Seven Precipitous	Steep, precipitous mountain terrain with scattered rocks and many sheer cliff faces.
Eight Cute	Cute refers to the summit as sharp with cliffs and a steep slope.
Eight Lean	Lean means the ridge is narrow and long, the hillside is thin and has a steep slope or cliff on both edges.
Nine Flat	The peak is wide and flat, there are no huge rocks or trees, the grass is short, and the hiker can walk easily.
Nine Secluded	Secluded means the mountain is located in a remote place which is located away from the main range.
Eight Hill	This represents a gently sloping hill with an easy ascent to the summit. This type of hill can easily be hiked up en route to a bigger mountain.
Six Easy	Easy refers to a mountain which is close to the path and convenient to ascend.
Six Shoulder Edge	This refers to peaks at the end of a ridge connected to a taller mountain.

In the early days (after World War II to the late 1990s), mountain hiking and climbing activities in Taiwan were mainly on high altitude mountains, for example hiking to finish the 100 Mountains list. Gradually, however, with the popularisation of sports, and the government becoming more devoted to planning and development, mountain hiking, and climbing activities in the suburban area became mainstream. Concurrently, the concept of Taiwanese mountaineers has also changed, from hiking and climbing high mountains to more emphasis on the relationship between people and mountains and forests. These trends are well illustrated by contemporary mountaineering figures. According to the annual survey conducted by the Taiwanese Tourism Bureau (2022b), hiking, mountain climbing, and camping were some of the most popular activities which Taiwanese people did in their leisure time. In 2005, about 15.2% of the population went hiking, mountain climbing, or camping during their leisure time, but this showed a significant increase over the following decade so that, by 2018, this number was almost 50% (see Figure 10.2). Even during and after the COVID-19 pandemic period, from early 2020 until the present day, there were slightly over 40% of the population participating in these activities. Enthusiasm for mountain hiking may be because of not only the geographic characteristics but also the mentality of pursuing a healthy life. Different research (Kim, et al., 2015; Mokras-Grabowski, 2016) revealed similar motivations in mountain hiking in which health is one of the most important

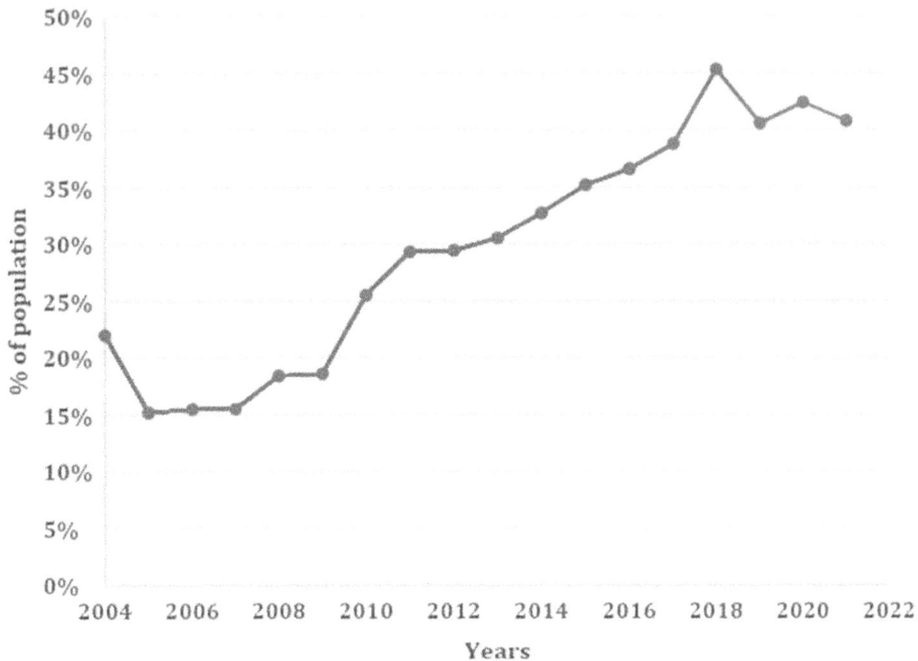

Figure 10.2 Trend of Taiwanese people choosing hiking, mountain climbing and camping in their leisure time 2004–2021 (Taiwanese Tourism Bureau, 2022b).

driving forces for mountain hiking participation. The trend is in line with the claim by Buckley et al. (2008) that in Chinese culture, health is one of the important aspects encouraging people's involvement in nature-based recreation.

As we explore the world of mountaineering, the contrast between European heritage and Taiwanese experiences becomes apparent. That is, the European heritage adventurers appear to perceive mountaineering as a conquering experience, while Taiwanese mountain hikers regard it as concerned with becoming one with nature (Figure 10.3). For example, Mr. Zhang, a mountaineering activity organiser, emphasised that we are meant to visit mountains rather than always reach the summit. He claimed that it is better to get to know one mountain well than to climb 100 Mountains. He believed that mountaineers need to approach nature with a humble attitude of understanding and experiencing, rather than an attitude of conquering it (Li, 2017).

Figure 10.3 Taiwanese mountain hiker adventure tourists on BeiDaWu Shan. Photograph by Ming Feng Huang.

Conceptualising South Korean and Taiwanese adventure through a colonial modernity lens

We suggest that colonial modernity is a better theoretical framework than others in understanding East Asian adventure tourism phenomena. As we extend our view on the usefulness of the theoretical framework, we highlight other sectors in East Asian societies where the colonial modernity notion has been adopted. In South Korea, colonial modernity – under the Japanese imperial rule of 1910–1945 – can be traced back to various sectors of society. The establishment and operation of the Kyongsong Broadcast Corporation (Robinson, 1999) is an example which illustrates the struggle for Korean identity under colonial modernity. In Taiwan, similar evidence of colonial modernity could be found in different places, for example, the land survey, police, and baojia system and public hygiene (Chang, 2014). Wang (1999) also emphasised how colonial rule policies and local society's ancient customs impacted the implementation of the modern law system when he investigated how Taiwan inherited and embraced Western law. Mackie (1998) has described Japan's place in contemporary East Asia within the context of colonial modernity. Specifically, the study suggests that while Japan's colonial rule is in history, contemporary post-colonial East Asian societies witness a clear presence of colonial modernity.

Critically evaluating the East Asian form of modernity, scholars (Barlow, 1997; Lee & Cho, 2012) suggest three main characteristics of colonial modernity:

(a) Japan as the sole non-Western coloniser in East Asia
(b) Issues arising from this colonisation are a national matter rather than a racial one
(c) Dual colonisation by Japan and the West, represented by America

South Korean Leiports in particular (레포츠) reflect these three characteristics of colonial modernity. The first characteristic can be affirmed by the very historic fact that Japan was the sole non-Western coloniser in East Asia and its historic influence is still felt in the region (Mackie, 1998). The second point – that colonial modernity as a national issue rather than a racial one – may not be found directly in Leiports (레포츠) but can be found in two areas. The first to affirm the colonial modernity characteristic is the phenomenon of the more recent Korean Wave (Hallyu in Korean and Hanliu in Chinese). Evaluating the contemporary East Asian popular culture of the Korean Wave (Hallyu in Korean and Hanliu in Chinese), Huang (2011) argues that the government-led popular industry developments in East Asia are closely related to 'branding nations' (p. 3) of the region. In the case of South Korea, the country branded itself as a 'cool, happening and modern' country, as argued by Huang (2011, p. 7). In this sense, South Korea may be branded as a nation of popular culture in the contemporary East Asian region. Huang's study, indeed, emphasises that the unit of comparison in contemporary East Asian popular culture is a "nation" rather than a race. The second example to be found is in traditional martial arts. For a long time, Korea has had a historic and cultural tradition for physical activities that are practised through traditional martial arts. Korean traditional martial arts such as Hapkkido or the Way of Coordinated Power, Kumdo or Korean Swordsmanship, Ssireum or Korean grappling, and Taekwondo strongly emphasise physical and spiritual strengths—an emphasis that has been significantly influenced by the East Asian cultural philosophies of Confucianism and Buddhism (Green & Svinth, 2010). This historically embedded practice of employing sport or physical activities invoked nationalism and confirms that the issue at hand in the South Korean forms of adventure-related activities is based on nationality rather than race.

The third characteristic, that Japan as a non-Western coloniser constantly seeks verification from the superior Western colonial power, largely represented by America, can be verified. That is, when Kim (2000) researched the rate of increasing consumption in adventure tourism pursuits such as scuba diving related to the increase of income in South Korea, the American experience was the benchmark to be compared to. The creation of the word Leiports (레포츠) is another example to affirm the third characteristic. In this instance, the presence of the West extends to language, and this is through America and English. Therefore, when adopting English words into the Korean language, the use of English-originating words connotes something positive and even prestigious (Lee, 1991). When English words are adopted through the phonetic creation of a new term in the Korean language, such as 레저 (*leisure*), the use represents more positive sentiment than when the Korean word is used. This is because it implicitly signifies modernisation, advancement, and a contemporary outlook. Such a practice is termed 'prestige borrowing' in Korean linguistics, to explain the social choice of foreign words adopted by contemporary South Korean society (Lee, 1991, p. 10). The South Korean societal choice for "ports" as the suffix in the term *Leiports* (레포츠) signifies the activities – in either singular or plural forms – that represent the South Korean meaning of travel, tourism, and sports (corresponding broadly to adventure tourism), which has an additional embedded nationalist component for cohesiveness and pride.

Just like South Korea, the impact of Japanese colonial rule can still be seen in Taiwan today. Taking the Taiwanese 100 Mountains as an example, it is very difficult to mention 100 mountains without mentioning the history of mountaineering in Japan. The establishment of 100 Mountains was influenced and inspired by Japanese mountaineering (Taiwan Outdoor, 2022). Wang (2013) mentioned that in 1964, the well-known Japanese mountaineer and writer Hisaya Fukada used the character of the mountain, history, personality of the mountain, and elevation of more than 1,500 metres as the selection criteria to select 100 Japanese mountains. A book titled *100 Famous Japanese Mountains* was hence published. This book describes mountain travel in Japan in a systematic way, which not only strongly influenced Japanese mountaineering but also indirectly gave birth to the emergence of Taiwanese 100 Mountains.

In the 1970s, under the leadership of veteran mountaineer Lin Wenan, together with three other mountaineers with similar experience Cai Jingzhang, Ding Tongsan and Xing Tianzheng began to plan Taiwanese 100 Mountain (Baiyue). 100 Famous Japanese Mountains was used as the model, hoping to improve the atmosphere of mountaineering/ mountain hiking in Taiwan with clear routes and specific goals. In 1971, Taiwan's mountain community launched the first Great Vertical Walk in the Central Mountain Range, which lasted for a month, and brought mountaineering activities to the public's attention for the first time. The following year, four people including Lin Wen'an established the "100 Mountain Club" on the top of Yangtou Mountain, and the 100 Mountain era of mountaineering in Taiwan officially started (Huang, n.d.). Thus, the Taiwanese modern mountaineering development certainly demonstrates aspects of colonial modernity. In the case of Taiwanese mountain trekking, only the first characteristic of colonial modernity can be identified at the time of writing this chapter. We argue that further research seeking the colonial modernity notion in Taiwanese adventure tourism experiences is an essential next step.

Conclusion

This chapter examined the colonial modernity approach in adventure tourism in conceptualising South Korean Leiports (레포츠) and Taiwanese mountain-trekking activities. We argue that the colonial modernity approach has much to contribute to conceptualising the contemporary East Asian adventure tourism phenomenon. Leiports (레포츠) epitomises South Korean cultural and historical enunciations of travel, tourism, leisure, and sports. Leisure, or *yoe ga* (여가), and related travel and tourism, was largely viewed as something Western and subsequently related to over-consumption and unnecessary waste, while Leiports (레포츠) is a concept that South Korean society created through its own modernising, industrialising, and democratising experiences, thus having less negative implications and denotes the society-specified pleasure, advancement, and new lifestyle. Similarly, although modern mountaineering can be traced back to Europe in the mid-18th century, Taiwan has not simply followed the West as Taiwan's society has evolved to incorporate its own circumstances and cultural and historical developments. Leiports (레포츠) and mountain-trekking participants in Taiwan should not be viewed as performing Euro/West-defined adventure tourists. The participants should be understood within the South Korean and Taiwanese society-specific cultural and historical framework that is dynamic and discursive with its interactions with Euro/West.

East Asian adventure tourism can be further researched related to modern concerns with wellbeing and slow trends. In the globally connected world of today, the boundaries between the shapers and those who get shaped are increasingly blurred. While not a delayed replication of Euro/Western pathways, East Asian adventure tourism cannot be claimed to have no initiating influence from the Euro/West, but this has been heavily filtered through colonial modernity. As Asian markets engage with adventure tourism globally, we should seek to further understand key differences (Martin & Ren, 2020). We, therefore, recommend more research into East Asian adventure tourism with the colonial modernity notion as its theoretical underpinning. The emergence and acceptance of Leiports (레포츠) in South Korea and mountain trekking activities in Taiwan affirm colonial modernity as the East Asian pathway to understanding modern adventure tourism.

Review questions

1 What are the main elements of modernity?
2 What is the main difference between modernity and colonial modernity?
3 Can you present a colonial modernity example in your region/country, related to adventure tourism development?
4 What are the similarities and differences in the South Korean and Taiwanese experiences in terms of adventure tourism developments?

Further reading

Bui, H. T., Jones, T. E., & Apollo, M. (2021). Reflections for trans-regional mountain tourism. In *Nature-Based Tourism in Asia's Mountainous Protected Areas: A Trans-regional Review of Peaks and Parks* (pp. 293–316). Springer International Publishing.
Martin, N., & Ren, C. (2020). Adventurous arctic encounters? Exploring Chinese adventure tourism. *Scandinavian Journal of Hospitality and Tourism*, 20(2), 126–143.
Zhu, D., & Xu, H. (2022). Novice tourists' play experience in commercial outdoor adventure tourism: The perspective of reversal theory. *Journal of Outdoor Recreation and Tourism*, 39, 100529.

References

Alneng, V. (2002). The modern does not cater for natives: Travel ethnography and the conventions of form. *Tourist Studies*, 2(2), 119–142.
Anderson, P. (1998). *The origins of postmodernity*. London: Verso.
Bandyopadhyay, R. (2011). A photo ethnography of tourism as neo-colonialism. *Annals of Tourism Research*, 38(2), 714–718.

Bardhi, F., & Eckhardt, G. M. (2017). Liquid consumption. *Journal of Consumer Research, 44*(3), 582–597.

Barlow, T. (1997). Introduction: On colonial modernity. In T. E. Barlow (Ed.), *Formations of colonial modernity in East Asia* (pp. 1–20). Durham & London: Duke University Press.

Bauman, Z. (2000). *Liquid modernity.* Cambridge: Polity Press.

Buckley, R., Cater, C., Zhong, L., & Chen, T. (2008). Shengtai Luyou: Cross-cultural comparison in ecotourism. *Annals of Tourism Research, 35*(4), 945–968.

Buckley, R., McDonald, K., Duan, L., Sun, L., & Chen, L. X. (2014). Chinese model for mass adventure tourism. *Tourism Management, 44,* 5–13.

Burns, P. (2008). Tourism, political discourse, and post- colonialism. *Tourism and Hospitality Planning & Development, 5*(1), 61–71.

Callon, M., & Latour, B. (1981). Unscrewing the Big Leviathan: How Actors Macro Structure Reality and How Sociologists Help Them Do So, pp. 277– A Study on Curriculum of Leisure Sports Related Departments at College and University in Gwangju & Jeonnam and Methodology: Towards an Integration of Micro- and Macro-Sociologies. London: Routledge and Kegan Paul.

Chang, L. C. (2014). Colonialism and modernity in Taiwan: Reflections on contemporary Taiwanese Historiography. In S. Weigelin-Schwiedrzik (Ed.), *Broken narratives* (pp. 133–164). Brill: Boston.

Cheng, M., Edwards, D., Darcy, S., & Redfern, K. (2018). A tri-method approach to a review of adventure tourism literature: Bibliometric analysis, content analysis, and a quantitative systematic literature review. *Journal of Hospitality & Tourism Research, 42*(6), 997–1020.

Cho, C. H. (2009). *A Study on Curriculum of Leisure Sports Related Departments at College and University in Gwangju & Jeonnam.* [MA dissertation, Chonman National University, South Korea]. Department of Physical Education. Graduate School of Education. Chonnam National University, South Korea.

Cho, Y. (2012). Colonial modernity matters? Debates on colonial past in South Korea. *Cultural Studies, 26*(5), 645–669.

Deutschlander, S., & Miller, L. J. (2003). Politicizing aboriginal cultural tourism: The discourse of primitivism in the tourist encounter. *The Canadian Review of Sociology and Anthropology, 40*(1), 27–44.

Eisenstadt, S. N. (2000). Multiple modernities. *Daedalus, 129,* 1–29.

Franklin, A. (2003). *Tourism: An introduction.* London, Thousant Oaks and Delhi: Sage Publications.

Gidwani, V. (2002). The unbearable modernity of 'development'?: Canal irrigation and development planning in Western India. *Progress in Planning, 58*(2002), 1–80.

Giulianotti, R., & Robertson, R. (2012). Glocalization and Sport in Asia: Diverse Perspectives and Future Possibilities. *Sociology of Sport Journal, 29*(4), 433–454.

Green, T. A., & Svinth, J. R. (2010). *Martial arts of the world: An encyclopedia of history and innovation* (Vol. One: Regions and individual arts). Oxford: ABC-Clio.

Hall, C. M., & Tucker, H. (2004). Tourism and postcolonialism: An introduction. In C. M. Hall & H. Tucker (Eds.), *Tourism and postcolonialism: Contested discourses, identities and representations* (pp. 1–24). London: Routledge.

Hazbun, W. (2010). Modernity on the beach: A postcolonial reading from southern shores. *Tourist Studies, 9*(3), 203–222.

Hsia, C. J. (2002). Theorizing colonial architecture and urbanism: Building colonial modernity in Taiwan. *Inter-Asia Cultural Studies, 3*(1), 7–23.

Hsu, M. C. (2022). Across the administrative boundaries—the first mile of the long distance trail collaboration in Taiwan. *Tourism Planning & Development, 19*(3), 235–244. https://doi.org/10 .1080/21568316.2021.1931425

Huang, F. S. (n.d.). The story of Taiwanese 100 Mountains in those years. Retrieved from http:// www.alpineclub.org.tw/front/bin/ptdetail.phtml?Part=yw_177-1&Rcg=35

Huang, S. (2011). Nation-branding and transnational consumption: Japan-mania and the Korean wave in Taiwan. *Media Culture & Society, 33*(1), 3–18.

Kim, H., Lee, S., Uysal, M., Kim, J., & Ahn, K. (2015). Nature-based tourism: Motivation and subjective well-being. *Journal of Travel and Tourism Marketing, 32,* 76–96. https://doi.org/10.1 080/10548408.2014.997958

Kim, S. K. (2000). Changing lifestyles and consumption patterns of the South Korean middle class and new generations. In C. Beng-Huat (Ed.), *Consumption in Asia* (pp. 61–81). London: Routledge.

Lau, J. K. W. (2003). *Multiple modernities: Cinema and popular media in transcultural Asia.* Philadelphia, PA: Temple University Press.

Lee, D.-G. (1991). A study on the loan words of English. *Theses (South Korea), 25*(1), 9–43.

Lee, H., & Cho, Y. (2012). Introduction: Colonial modernity and beyond in East Asian contexts. *Cultural Studies, 26*(5), 601–616.

Lee, Y. S. (2006). Korean war and tourism: Legacy of the war on the development of tourism industry in South Korea. *International Journal of Tourism Research, 8*, 157–170.

Lee, Y. S., & Jennings, G. (2010). The development of leiports (leisure and sports) studies and programs in tertiary education in South Korea. *Journal of Teaching in Travel and Tourism, 10*(2), 125–142.

Li, C. L. (2017, November 10). *Mountainous Protected Areas of East Asia: Governance Conservation & Mountain Tourism.* Paper presented at the *pre-conference workshop in the Asia Pacific Conference*, Ritsumeikan Asia Pacific University, Beppu, Oita, Japan.

Li, X. S. (2005). *Taiwan mountaineering history: A process of struggle and progress.* LI XISHENG: Taipei.

Lin, J. Y. (2002). *The leaf in the wing.* Taipei: Classic Press.

Mackie, V. (1998). Dialogue, distance and difference: Feminism in contemporary Japan. *Women's Studies International Forum, 21*(6), 599–615.

Martin, N., & Ren, C. (2020). Adventurous arctic encounters? Exploring Chinese adventure tourism. *Scandinavian Journal of Hospitality and Tourism, 20*(2), 126–143.

Mokras-Grabowski, J. (2016). Mountain hiking in Tatra National Park. *Tourism, 26*(1), 71–78.

Mukhia, H. (2010). Alternative modernities, alternatives to modernity or multiple modernities – what else? *Pakistan Perspectives, 15*(2), 1–11.

Nash, D. (1989). Tourism as a form of imperialism. In V. Smith (Ed.), *Host and guest* (2nd ed., pp. 33–48). Philadelphia: University of Pennsylvania Press.

Oakes, T. (2021). Afterword: A critical reckoning with the 'Asian Century' in the shadow of the anthropocene. *Tourism Geographies, 23*(4), 937–943.

Pomering, A., & White, L. (2011). The portrayal of Indigenous identity in Australian tourism brand advertising: Engendering an image of extraordinary reality or staged authenticity? *Place Branding and Public Diplomacy, 7*(3), 165–174.

Roberts, K. (1999). *Leisure in contemporary society.* Guildford and King's Lynn: Biddles Ltd.

Robinson, M. (1999). Broadcasting, cultural hegemony, and colonial modernity in Korea, 1924–1945. In G.-W. Shin & M. Robinson (Eds.), *Colonial Modernity in Korea* (pp. 52–69). Cambridge, MA: Harvard University Press.

Ryan, C. (2003). *A history of tourism in the English-speaking world Aspect of Tourism: Recreational Tourism.* Church Point, NSW: Footprint Books.

Taiwan Outdoors (2022). History of Mountaineering in Taiwan. Retrieved from https://taiwanoutdoors.com/history-of-mountaineering-in-taiwan/

Taiwanese Tourism Bureau. (2022b). Survey of Travel by R. O. C. Citizens. Retrieved November 11, 2022, from https://admin.taiwan.net.tw/English/FileUploadCategoryListE003140.aspx?CategoryID=bf74751a-2aed-4b50-87ce-471505630e60&appname=FileUploadCategoryListE003140

Taylor, P. (1999). *Modernities: A geohistorical interpretation.* Minneapolis: The University of Minnesota Press.

Tourism Bureau. (2022a). General Information. Retrieved November 11, 2022, from https://eng.taiwan.net.tw/m1.aspx?sNo=0002004

Vogel, M. P., & Oschmann, C. (2013). Cruising through liquid modernity. *Tourist Studies, 13*(1), 62–80.

Wang, B. (2013). [Little Dictionary] What is 100 Mountain. https://hiking.biji.co/index.php?q=news&act=info&id=916

Wang, N. (2000). *Tourism and modernity: A sociological analysis.* Oxford: Elsevier Science Ltd.

Wang, T. S. 王泰升 (1999). *Taiwan rizhi shiqi de falü gaige* 臺灣日治時期的法律改革 (Legal Reforms in Taiwan during the Period of Japanese Rule). Taibei: Lianjing.

Weber, K. (2001). Outdoor adventure tourism: A review of research approaches. *Annals of Tourism Research, 28*(2), 360–377.

Wengel, Y. (2021). The micro-trends of emerging adventure tourism activities in Nepal. *Journal of Tourism Futures, 7*(2), 209–215.

Winter, T. (2009). Asian tourism and the retreat of anglo- western centrism in tourism theory. *Current Issues in Tourism, 12*(1), 21–31.

Ye, J. C. (2012). Taiwanese mountaineering activities. Retrieved November 11, 2022, from http://blog.udn.com/yestaipei/7023646

11 Adventure tourists and technology

Ingo Janowski, Sarah Gardiner and Anna Kwek

Chapter learning outcomes

1 Understand the many ways in which technology impacts the adventure tourism experience and consumer journey.
2 Consider national and cultural differences in the embracement of technology in adventure tourism.
3 Reflect on the critical view of technology taking away from the original sense of adventure.
4 Understand practical ways to utilise technology in adventure tourism operations.

Introduction

This chapter discusses technological applications interwoven with contemporary adventure tourism participation from a consumer perspective. Adventure tourists are increasingly reliant on technology. They not only expect technology to support and enhance the adventure activity itself but also rely on technology in their decision-making process (pre-adventure), to capture their experience (during adventure), and share their adventure (during and post-adventure). These numerous touch points with technology along the consumers' adventure tourism journey are discussed in this chapter.

The chapter first provides a general overview of the increasing relevance of technology for adventure tourists. It highlights online channels and electronic word of mouth (eWOM) as essential tools and resources for adventure tourists. A cross-cultural case study (see Case study 11.1) overviews and contrasts Australian, Chinese, and German youth consumers' usage of online channels pre- and post-adventure and their preferences towards the digital recording and sharing of their experience using social media and other digital technology communication channels. This Case study highlights culturally diverse needs and wants regarding the use of technology in adventure tourism. Practical implications for adventure tourism providers, a discussion on the future of technology in adventure tourism, and suggested areas for future research are also presented.

DOI: 10.4324/9781003393153-14

Adventure tourists' interaction with digital technology: An overview

Digital technology shapes how adventure tourism is managed, marketed, and consumed (Adventure Travel Trade Association [ATTA], 2021; Beames et al., 2019). Despite its increasing influence on adventure tourism, a lack of respective attention and knowledge remains (Cheng et al., 2018; Ewert & Sibthorp, 2014). While the impact of technology on the supply (industry) side of adventure tourism is discussed in Part III of this handbook (see Chapter 20), this chapter focuses on the demand (consumer) side of the industry. It shows that technology is not a peripheral element in the adventure tourist journey anymore but a vital touchpoint across all stages of the adventure experience (ATTA, 2021; [UNWTO], 2014) in both hard, for example equipment, and soft, for example digital, forms.

Technological advancement has meant that it has become embedded and often essential to our everyday lives. It is, therefore, not surprising that adventure tourists are frequently aided by, and often reliant on, technology to plan for and engage in the experience. Technology continues to evolve across many visitor experience areas, including information and communication, access and transportation, safety, equipment, clothing, comfort, artificial environments, and the virtual consumption of adventure tourism. Examples of hard advancements that have emerged over time and influenced the consumer experience include artificial surfing wave pools (Cardwell & Higgins, 2016), indoor skydiving venues (Wheal, 2017; Figure 11.1), and artificially created snow and simulated climbing walls that mimic the activity in the natural environment (Swarbrooke et al., 2003). Technology-enabled devices to support adventure tourism, such as mountaineering clothing, are also available (Berger & Greenspan, 2008). Moreover, adventurers can access

Figure 11.1 Indoor skydiving, Queenstown, New Zealand. Photographer Carl Cater.

live video streams to monitor weather conditions (Beames et al., 2019) and GPS tracking devices (Thorpe, 2017), as well as a wide range of online channels, smartphone apps, increased cellular coverage, signalling devices, dive computers, snowmobiles, and other off-road vehicles (Ewert & Sibthorp, 2014).

The importance of technology for adventure tourists has been exemplified early on by Berger and Greenspan (2008). Their study explored how adventure tourists benefit from technology in a mountaineering context and explains how technology influences the mountaineer's self-identity, behaviours, and interactions. It found that mountaineers use several spheres of technology to aid their safety and success, to express who they are, and to facilitate meaning from the adventure experience. These spheres include (1) gear and climbing tools, for example functional clothing, tents, and poles; (2) research equipment, such as laptops; (3) health and safety, for instance insulin injections and satellite phones; and (4) comfort, recreation, and enjoyment, such as mobile showers, movie DVDs, and heated tent floors (Berger & Greenspan, 2008). These and other technological tools aid the accessibility and consumption of adventure tourism because they can "cushion" an outdoor adventure experience for tourists who are accustomed to an urban lifestyle devoid of immersion in, and interaction with, nature (Beedie & Hudson, 2003). Having the right top-quality equipment can trigger a sense of pride and achievement within the mountaineers, positively contributing to their overall experience (Ewert & Sibthorp, 2014).

Creating and distributing photos and videos of adventurous experiences via social media, blogs, and vlogs has contributed to the globalisation and rising demand for adventure tourism (ATTA, 2021; Berger & Greenspan, 2008). Today, dedicated mirrorless cameras, action cameras, consumer drones and modern smartphones that utilise advanced computational photography are widely used. This, combined with increased internet connectivity, means that adventure tourists can digitally capture and share their experience with a potentially global online audience, even from remote areas. The audience can thus virtually take part in the adventure and interact with adventure tourists' shared content, such as through comments, direct messages, or email (Berger & Greenspan, 2008). The COVID-19 pandemic has further accelerated and expanded the tourists' use of technology for communication, information exchange, decision-making, and consumption (Sorooshian, 2021). The significant interrelation between tourism and technology is expected to increase in the aftermath of the pandemic (Gretzel et al., 2020).

As illustrated, technology has a widespread influence on adventure tourists. Table 11.1 overviews the various functions of technology discussed in this chapter and at what stage in the adventure tourism consumer journey they predominantly apply. Amongst the many areas of technological advancements, the internet stands out as having arguably the most significant impact, especially as a tool to inspire and inform the consumer and as a platform to share adventure experiences with one's network and, more broadly, across the internet with strangers. The tourism industry is one of the most heavily impacted industries by the internet, particularly with the development of online portals and user-generated reviews (UNWTO, 2014). For example, TripAdvisor, which brands itself as 'the world's largest travel platform' features over 859 million reviews and opinions and is available in 49 markets and 28 languages (Tripadvisor, 2022). As early as 2014, the UNWTO declared that TripAdvisor and other online booking systems offering peer reviews, easy price comparison, and even virtual experience tours are also increasingly popular with adventure tourists. Since then, social media channels have grown in popularity. These platforms are frequently used for researching and planning tourism activities, sharing travel tips and experiences, and finding travel companions (Hasan & Neela,

Table 11.1 Main functions of technology in adventure tourism

	Pre-Adventure	In-Adventure	Post-Adventure
Communication	✓	✓	✓
Entertainment	✓	✓	✓
Sharing	✓	✓	✓
Information	✓	✓	
Access & Transport	✓	✓	
Inspiration	✓		
Booking & Payment	✓		
Facilitation		✓	
Health & Safety		✓	
Equipment		✓	
Clothing & Comfort		✓	
Documentation		✓	
Artificial Environment		✓	
Virtual adventure		✓	
Recall & Reflection			✓
Reviewing			✓

2021). As such, user-generated content and eWOM on review websites and social media provide impartial and valuable information and inspiration, influencing tourists' perceptions, expectations, and actions (Liu et al., 2019; Narangajavana Kaosiri et al., 2019).

Today, a broad range of online channels are at the disposal of constantly connected consumers (Beames et al., 2019). Adventure travellers increasingly use smartphones and other mobile devices to search for and book adventure tourism experiences (UNWTO, 2014). Research by Think with Google (2017) showed that 48 per cent of smartphone users in the United States are comfortable researching, planning and booking their entire trip solely using their smartphone. This connectiveness to the internet has only intensified during the COVID-19 pandemic, which caused an increase in remote workplace environments and digital nomadism (Kelly, 2022). Digital nomads are those individuals who work remotely while travelling the world (Thompson, 2019). They usually stay longer in a destination than the average tourist, shift their location every few months, and live a minimalistic lifestyle with few possessions. Nowadays, some countries even offer a specialised "digital nomad visa", which is alluring to adventure travellers seeking an immersive, in-depth cultural experience (Kelly, 2022).

Because online channels are highly popular with adventure tourists, it is increasingly critical for industry providers to utilise them for marketing purposes (UNWTO, 2014). According to research by the ATTA (2020), social media is the most popular direct marketing channel for adventure tourism operators, and 38 per cent of businesses rated social media advertising as one of the most helpful tools for acquiring new guests. In addition, 51 per cent of all bookings were made online, whether through the company website or online booking platforms.

Interestingly, the extent to which adventure tourists use and rely on online channels may differ not only between people of different age groups but also between national cultures. For example, frequent use of social media for information search and the sharing of travel experiences is a phenomenon associated with youth travellers generally (Zeng & Gerritsen, 2014) and Chinese youth travellers in particular (Cheng & Foley, 2018; Martin & Ren, 2020; Wu & Pearce, 2016). Cai et al. (2019) found that

Chinese backpackers are very intensive users of online channels across their greater backpacking experience, meaning before, during, and after their trip. They are very active across travel forums, searching for information, and they increasingly use online channels to find travel companions (Cai et al., 2019; Xiang, 2013). The lack of siblings for most Chinese youth consumers of this generation, a result of the one-child policy, underlies the strong popularity of, and reliance on, the internet, generally, and social media, specifically (Cheng, 2017; Chu & Choi, 2010). Accordingly, they seek a sense of belonging in the online environment, which allows them to gain social capital and refine their identities through sharing their unique, exciting, and courageous adventure tourism experiences (Cheng, 2017). Case study 11.1 dives deeper into the Chinese youth market and cross-cultural differences in how youth consumers utilise online channels in the context of adventure tourism.

While technology may empower and benefit consumers in their pursuit of adventure, as this chapter exemplifies, there are also downsides to adventure tourists' reliance on technology. One possible danger lies in adventure tourists' increasing use and reliance on technology. For instance, being well equipped but lacking experience in the adventure activity and/or environment may lead to an illusion of competence, overconfidence, and misjudgement of risk (Beames, 2017; Johnston & Edwards, 1994). Furthermore, Cuthbertson et al. (2004) propose that technology-enabled commodification of adventure tourism accelerates the separation of consumers away from nature. This is concerning because immersion in and interaction with the natural environment are one of the key elements of adventure tourism (Janowski et al., 2021). Beedie and Hudson (2003) similarly argue that technology supports the commodification of adventure tourism, meaning that it contributes to the accessibility and softening of the adventure experience. This trend arguably leads to the distancing of commercial adventure tourism offerings away from the original meaning of adventure which necessitates elements of risk and uncertainty (Kane & Tucker, 2004; Weber, 2001). The adventure tourism terminology has therefore been described as paradoxical (Fletcher, 2010; Imboden, 2012; Varley, 2006). Other secondary negative effects of technology in adventure tourism include negative environmental impacts, such as through the resource extraction to produce technological products and the limiting of adventure tourists' creativeness, learning opportunities, and engagement with people and places (Beames, 2017; Beames et al., 2019). As such, technology should be viewed and utilised in a way that supports or facilitates certain aspects of the adventure tourist journey and not as a replacement of the vital human components of adventure tourism (ATTA, 2021).

Adventure tourists' interaction with online channels: Pre-adventure

Technology is essential for global tourism and the tourist's information and booking process via search engines, social media, websites, and other online channels, even more so in the aftermath of the COVID-19 pandemic (ATTA, 2021). TripAdvisor proclaims that the internet is the second-most important inspiration source for leisure travellers (word of mouth from family and friends being the most important) and the number one resource for trip planning. Accordingly, 58 per cent of all leisure travellers use a search engine to start their planning process (UNWTO, 2014). Social media platforms, such as Facebook, Twitter, and Instagram, are also popular with adventure tourists, who search for captivating and inspiring stories that appeal to their values of sustainable and responsible travel (UNWTO, 2014). In addition to social media, travel blogs are found to stimulate and inform prospective travellers, particularly in China (Martin, 2019; Martin &

Ren, 2020). User-generated information and personal tales of adventure tourism experiences on social media platforms and blogs are thus aiding prospective adventure tourists' decision-making, leading to eWOM rapidly replacing traditional WOM (Hasan & Neela, 2021).

The internet is not only a critical pre-adventure source of inspiration and information but also frequently used for booking adventure experiences. Research by the Adventure Travel Trade Association (2021) shows that 34 per cent of all direct bookings are made through the adventure tourism operator's website and that 35 per cent are made through social media, email, and phone, while 8 per cent of all bookings are made through online travel agencies (OTAs). Today's high prevalence of smartphones and other mobile devices has further accelerated the importance of online channels for adventure tourists (UNWTO, 2014).

Adventure tourists' use of technology during the adventure

As mentioned in the introduction to this chapter, technological advancements directly impact the tourist's adventure experience itself, such as in the form of access, comfort, equipment, safety, and clothing. However, as illustrated by the case study (see Case study 11.1) in this chapter, one of the significant impacts of technology during the adventure is the digital capturing of the experience in the form of still (photo) and motion (video) imagery. This gives tourists a way to reflect and reminisce on the experience themselves post-experience and to share it with others. Over the past two decades, digital photography and videography technology rapidly advanced and turned from a niche to a mass phenomenon. This is due to smartphones' wide availability with impressive image-capture capabilities, enabled by computational photography that overcomes small image sensor limitations (Hasan & Neela, 2021; Zubarev, 2020). Also, dedicated action cameras, such as the "GoPro Hero", a compact, lightweight, water- and shockproof digital camera and, increasingly, sophisticated consumer-oriented camera drones that are relatively affordable are further enabling factors of digital adventure tourism "mass" recording (Thorpe, 2017). Camera drones enable image capture from a unique bird's-eye view, thus allowing a novel perspective and record of the adventure. Today, it is common for soft and hard adventure tourists of all ages to capture themselves during the activity, whether for personal memory or to share their experience online and/or in person. Accordingly, digital recording and sharing of one's adventure tourism experience have developed into an integral part of the experience for many (Thorpe, 2017).

Findings by the ATTA (2019; 2020) further consolidate that photography is important for adventure tourists. The photography of nature and wildlife was found to be one of the "hot" adventure travel activity trends, meaning that it is in high demand. Looking at the top trending activities by region, it is evident that photography is essential in Asia, where it is the second top trending adventure travel activity, while it ranks fourth in Europe, North America, and South America (ATTA, 2019). Taking a remarkable photo or selfie is a critical motivational factor for Chinese adventure tourists. The China Adventure Tourism Market Study (Martin, 2019) furthermore states that photos, videos, and blogs are commonly shared by Chinese travellers both during and post-adventure. This study identifies WeChat (more than 1 billion monthly users) as the most popular sharing platform for Chinese, followed by the micro-blogging site Weibo (450 million active users). These insights align with findings from the cross-cultural case study that identified adventure tourism photography as highly desired across cultures.

Case study 11.1   Cross-cultural consumers' usage of online channels in adventure tourism

Social media and eWOM increasingly affect the everyday life and tourism decision-making of young people (Bizirgianni & Dionysopoulou, 2013). This case study compares youth consumers' (ages 18–29 years) perceptions, needs, and wants towards adventure tourism across the three national cultures of Australia ($n = 37$), mainland China ($n = 34$), and Germany ($n = 20$). Data were collected via five focus group discussions ($N = 24$) and individual online interviews ($N = 67$). The discussions and interviews were voice recorded, transcribed, and thematically analysed via the three stages of coding – open coding, axial coding, and selective coding (Neumann, 2014). The findings reveal noteworthy differences. Pre-experience, Chinese youth consumers are particularly reliant on various online channels that stimulate their adventure tourism interest and serve as research tools compared to their Australian and German counterparts. As stated by one Chinese interviewee, 'the internet will influence the Chinese young people a lot, because in our free time we always use our mobile phones and surf the internet'. Chinese participants named a wide variety of Chinese online channels, including the online travel agency C-trip, the question-and-answer website Zhihu, the microblogging site Weibo, the video-sharing website Bilibili, the travel platform Mafengwo, the search engine Baidu, and the community platform Little Red Book. Western platforms such as Google, YouTube, and TripAdvisor are also frequented when travelling internationally. Overall, it was evident that intensive online research and evaluation of adventure activities precede Chinese youth consumers' commitment to an adventure activity. Online channels are less relevant for Australian and German youth consumers who seek information from personal contacts. They are also comparably more independent and spontaneous in their adventure tourism decision-making. However, Australians and Germans would still look up adventure tourism options and gain information online pre-departure. For example, a German interviewee stated, '[I]f you don't have any real references, so like maybe friends who did it already or something, then the only thing you can actually do is look at [online] references and how other people took the experience'. Google (search and reviews) is their most common starting point, while TripAdvisor is also frequented and valued by some.

Whereas online channels are highly important pre-adventure, particularly for the Chinese market, they play a much less significant role during the adventure activity. However, the case study unearthed another technological aspect of relevance during participation, that of the digital capture of the adventure. Youth consumers conveyed a desire to document their experience, mainly in the form of photography and, to a lesser degree, videography. Many youth consumers across cultures communicated a passion for taking photos of their adventure experience themselves, either with their smartphone or a dedicated digital camera. Yet, some would want to outsource the photo-taking to the operator/guide to focus on the activity at hand. Cultural differences are again apparent in the tourists' motivations behind their photo-taking. Australians and Germans primarily seek photos for in-person sharing and personal memory. In contrast, the Chinese principally seek online sharing of their footage.

Post-adventure, Chinese youth consumers share their adventure tourism experience online via social media posts (e.g., via WeChat Moments), blogs, and vlogs. This online sharing is driven by a longing for their peers' attention, admiration, and praise. For example, they expressed: 'If I do something incredible, crazy, I will share to everybody I know. ... Yeah, look at me. ... Show how fun the activity is. How good am I? [laughter]', and 'They will try to like my post and they will leave a lot of comments to say wow, it's exciting, wow, you are so great, and this gives me a sense of fulfilment from the outside world. ... Chinese young girls, they like to show off on the social media'.

Online sharing thus provides young Chinese adventure tourists with a sense of fulfilment and increased self-esteem. Their adventure is seen as something challenging and extraordinary that few others have experienced, distinguishing them from others and elevating their social status. This identified Chinese motivation for self-promotion through adventure tourism echoes previous research, including Jin et al. (2019) and Cheng (2017). While the increasing importance of social media for youth consumers as an extension or partial replacement of traditional friendship groups has been highlighted (Décieux et al., 2019; Manago & Vaughn, 2015), this case study accentuates cultural differences in that regard. Chinese youth consumers are much more deeply involved with and reliant on social media than Australian and German youth consumers. For example, one Australian interviewee commented that taking photos is 'more for myself to remember than anything', while a German stated, 'I would share it in person. I'm not really the type who shares things on social media'. As adventure tourists' post-experience online sharing stimulates others' adventure tourism interest, along with the previously discussed review websites and industry-owned content (e.g., operators' websites), pre-adventure, during the adventure, and post-adventure usages of technology are interlinked.

Adventure tourists' interaction with online channels: Post-adventure

In the post-adventure stage, tourists evaluate their experience and form opinions about the re-experiencing and/or recommending their adventure activity (Triantafillidou & Petala, 2016). This is where technology comes into play yet again. Adventure tourists predominantly utilise technology to share their experiences verbally and visually. While some use the internet to provide feedback in the form of reviews, such as via TripAdvisor, the online sharing of the adventure via personal social media channels, blogs, and vlogs has become ubiquitous (Liu et al., 2019). This sharing process is the sixth stage of an adventure tourist's consumer journey after dreaming, considering, planning, booking, and experiencing (UNWTO, 2014), as depicted in Figure 11.2.

Adventure tourists often seek to show others what exciting experiences they engaged in to gain "insider" status and to increase their social value (Beames et al., 2019; Giddy, 2018). Over the past decade, the sharing and rating of flattering, boastful, and attention-seeking photos, particularly selfies, has become much more frequent and

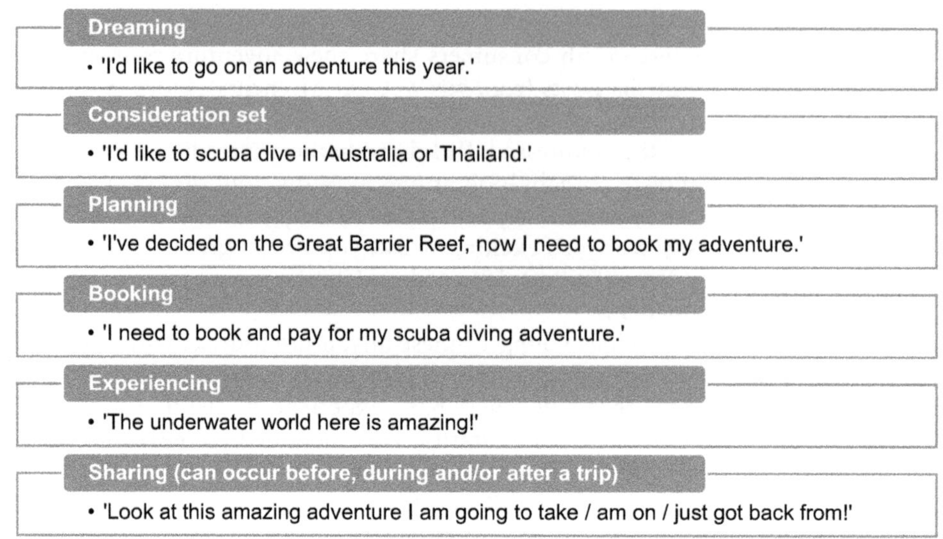

Figure 11.2 Six stages of the adventure tourist journey. (adapted from the UNWTO, 2014).

constitutes an integral part of many people's everyday lives (Hasan & Neela, 2021; Kang et al., 2018). This has also been described as the 'increasingly social media-driven "capture and share" culture in adventure' (Beames et al., 2019, p. 94). Such online sharing of one's adventure is thus a means to distinguish oneself and develop one's self-identity (Puchan, 2004), particularly given that adventure tourism activities are often novel (Gardiner & Kwek, 2017) and incorporate risk and challenge (Janowski et al., 2021). As such, they give adventure tourists a chance to 'show off their achievements, skills, status, and happiness to impress others as a means of positive self-presentation' (Hasan & Neela, 2021, p. 3).

Besides satisfying the identity-related consumer-focused motivation, online sharing of one's adventure inherently has a second function: generating eWOM in the form of marketing messages about the adventure experience and location (Hasan & Neela, 2021; Liu et al., 2019). Sharing personal tourism experiences, imagery, related sentiments, and information aids others' decision-making (Kang & Schuett, 2013; Zhang et al., 2010). From an industry perspective, this is a desirable no-cost promotion (Kim & Tussyadiah, 2013), especially since the user-generated nature of eWOM is perceived as unbiased and reliable in the eyes of prospective adventure tourists. As such, adventure tourism marketers seek ways to stimulate their customers' eWOM and to identify and encourage social media influencers to disseminate content favourable to their destination/business success (Hasan & Neela, 2021; Kim et al., 2015).

While some social media platforms are prevalent with adventure tourists, Instagram stands out as a popular channel for sharing photos and videos with family, friends, and the wider online community (Thorpe, 2017). Facebook, Twitter, YouTube, and other social networking platforms are also commonly used (Hasan & Neela, 2021; Thorpe, 2017). As indicated in this chapter's case study (see Case study 11.1), the choice of online channels varies between cultures. For instance, Chinese adventure travellers

> **Vignette 11.1   Controversy – Adventure tourism as an avenue for digital detox?**
>
> In contrast to the increasing demand for digital technology in adventure tourism discussed throughout this chapter, there is also evidence that adventure tourism can be desired for the opposing notion of digital detox. That is, adventure tourism provides an avenue to distance oneself from the use of technology and to reconnect with the natural world. The Adventure Travel Trade Association (ATTA, 2019) identified 'digital detox' as the number one 'warm trending consumer motivation', indicating that it is a strong motivator for adventure travellers. One year later, this motivation ranked second, whereas 'adventure travel as a status symbol' was the number one warm consumer motivation (ATTA, 2020). Ironically, this number one motivation leads back to the importance of online sharing of the adventure tourism experience and is thus contradictory to the idea of digital detox. Yet, given that online sharing is mainly done post-experience, the two trends are not necessarily in conflict with each other.

predominantly use WeChat, while the video-sharing platform YouKu and travel blogs like Mafengwo are also well-liked for sharing and researching adventure tourism experiences. These online channels thus help other Chinese users to save time and energy (Martin, 2019; Martin & Ren, 2020).

Practical implications for industry providers

Although this chapter focuses on the demand side of adventure tourism and technology, several practical implications for the industry are evident (see also Chapter 20). Because online channels are influencing adventure travellers' decision-making pre-adventure, industry providers must have a strong online presence. In particular, they should concentrate on information and design of their business website to make it user-friendly. Ensuring a presence on external platforms, such as a business's Facebook page, TripAdvisor listing, or updated Google business profile, is also important (ATTA, 2021). Consumers' high rate of smartphone usage furthermore necessitates responsive (i.e., mobile-friendly) websites and convenient online booking and payment options (UNWTO, 2014). Leveraging the user-generated eWOM space (i.e., social media, blogs, vlogs, review websites) is also critical to arouse interest and provide information, especially to consumers from uncertainty/risk-avoidant cultures that habitually rely on online research. Besides ensuring good search engine optimisation (SEO) for one's own content, it is also necessary to be present on specific online channels relevant to one's target market. For instance, a WeChat account is decisive when targeting Chinese adventure travellers. This app is increasingly accessible to international businesses as it has an English-language version (Martin, 2019). C-Trip, Mafengwo, Bilibili, Zhihu, Weibo, Little Red Book, and Baidu are other channels worth considering for building a presence on to reach this market that is increasingly interested in (predominantly soft) adventure tourism.

Influencer marketing can also effectively capture consumer interest online. This involves identifying suitable individuals with a significant following in the target audience, inviting them to experience one's adventure tourism product and having them share that experience via their online channel(s). If unreasonable for small adventure tourism businesses, collaboration with other industry providers and/or destination management organisations to engage influencers might be an option. Other measures to drive eWOM should also be taken. Where possible, adventure guides could document their clients' adventures in the form of photography and/or videography and make the footage available at low or no extra cost. This should result in high-quality content online sharing by clients, promoting the adventure tourism business (i.e., free advertisement; see Figure 11.3).

While some consumers appreciate guides taking photographs of them, many are interested in taking photographs themselves. Operators may thus incorporate photo-taking opportunities throughout the adventure experience, promote such opportunities in their marketing material, and/or even design photography-centred adventure products. Operators could also provide additional incentives to online sharing, such as by offering a small souvenir or discount voucher when clients share and tag the provider in their post, conducting a photo competition, handing out certificates of achievement to clients, or providing photo props (i.e., objects to be added to a scene for a photograph). For

Figure 11.3 Photo booth at Zip World, North Wales. Photograph by Carl Cater.

Figure 11.4 Photo prop at Zip World, North Wales. Photograph by Carl Cater.

instance, this could be a physical sign that states, "I survived [name of experience] with [name of company] at [location]", or similar, which guests can pose in front of after completing their adventure (Figure 11.4). The in-trip sharing of the experience in areas with limited internet connectivity could be enabled by the guide carrying a remote hotspot that clients can connect to, or other connecting opportunities, such as during a break in a local café (ATTA, 2021).

Finally, industry providers should encourage their clients to follow the business on social media and leave an online review of their adventure experience. This could be done through a follow-up email post-experience (ATTA, 2021). Regularly monitoring and responding to online reviews is also recommended to keep in touch with existing customers and positively influence their and prospective future clients' perceptions of the adventure product.

The future

Technology will continue to evolve and likely become an even more significant factor in our everyday lives and adventure tourism. This might be in the form of a more digitalised decision-making and booking journey for consumers and/or increasing artificial and simulated experiences, such as through virtual reality (VR) or augmented reality (AR). This would increase accessibility to adventure tourism experiences and lead to an even broader industry scope that already encompasses a vast array of tourism experiences (Swarbrooke et al., 2003). For example, the Oculus Quest 2 VR headset, in partnership with National Geographic, allows virtual expeditions to Antarctica where users can kayak around icebergs, climb ice walls, and even capture digital photos of what they can see in the virtual world (Joseph, 2020). Whether such digitally enhanced/enabled experiences can still be considered adventure tourism, given the fundamental elements of the natural environment, including physical activity, risk, and challenge (Janowski et al., 2021), ultimately comes down to an individual consumer's perception.

The consumers' growing use of and dependence on technology in combination with other socio-cultural trends, such as increasing urbanisation, may also lead to a higher longing for digital detox and escapism. This can be facilitated by traditional forms of adventure tourism that are based around the previously mentioned key elements rather than technologically supported experiences. Thus, it will be interesting to see how adventure tourism demand and supply will evolve.

The vast array of technological applications and developments in adventure tourism offers broad opportunities for future research. For example, this chapter's cross-cultural case study has identified youth consumers' desire to document their adventure through photography for various reasons, including personal keepsakes and online and offline sharing. Future investigations could explore what kind of imagery different youth consumers seek from their adventure, their preferences towards self-taken and organiser-taken photographs, and how such preferences may differ between cultures. The role of online channels in the consumer decision-making process could also be explored further, particularly how and at what stage of the customer journey the various online channels come into effect.

Conclusion

This chapter outlined the vital role of technology for adventure tourists. The broad application spectrum of technology in adventure tourism was highlighted, and the role of online channels (particularly pre- and post-adventure) and the digital capture of consumers' adventure tourism experiences (during the adventure) were discussed in-depth. The cross-cultural case study showed that demand and motivations for using online channels in an adventure tourism context vary between cultures. This necessitates industry providers' consideration of socio-cultural influencing factors and allows the deduction of future trends across different adventure tourist segments regarding their embracement of technology. For example, the emerging Chinese youth market shows a comparably high reliance on online channels. In contrast, the more traditional adventure tourism youth markets of Australia and Germany are less inclined to use technology along their customer journey. Such contemporary technological preferences may change as emerging markets develop into core markets. Adventure tourism is an exciting, fast-changing industry that is in high demand. It will be interesting to observe how technology will continue to impact and shape adventure tourists' motivations and decision-making, and their adventure experience, evaluation, and revisit intentions.

Review questions

1 What are the advantages and disadvantages of technological developments in adventure tourism?
2 How would you utilise technology pre-, in-, and post-adventure?

3 What does adventure tourism mean to you? Would you label a virtual experience adventure tourism? Would you consider the natural environment essential, or could artificial environments facilitate adventure tourism?
4 Would you want to document your adventure tourism experience? If so, how and why?

Further reading

Adventure Travel Trade Association. (2021). *Creating, Communicating, and Connecting: Technology in Adventure Travel*. https://cdn-research.adventuretravel.biz/research/87905718971204128923571 2018924/Technology-Research-Report-2021.pdf
Gardiner, S., & Kwek, A. (2017). Chinese participation in adventure tourism: A study of generation Y international students' perceptions. *Journal of Travel Research*, 56(4), 496–506.
Hasan, M. K., & Neela, N. M. (2021). Adventure tourists' electronic word-of-mouth (e-WOM) intention: The effect of water-based adventure experience, grandiose narcissism, and self-presentation. *Tourism and Hospitality Research*, 14673584211026326.
Thorpe, H. (2017). Action sports, social media, and new technologies: Towards a research agenda. *Communication & Sport*, 5(5), 554–578.

References

Adventure Travel Trade Association. (2019). *Adventure Travel Trends Snapshot - Part I.* https://www.adventuretravel.biz/research/2019-industry-snapshot
Adventure Travel Trade Association. (2020). Adventure Travel Trends Snapshot 2020 Report. https://cdn-research.adventuretravel.biz/research/5bbf8f4a1010a8.2358546546/Snapshot-Trends-2020-Report.pdf
Adventure Travel Trade Association. (2021). *Creating, Communicating, and Connecting: Technology in Adventure Travel*. https://cdn-research.adventuretravel.biz/research/87905718971204128923571 2018924/Technology-Research-Report-2021.pdf
Beames, S. (2017). Innovation and outdoor education. *Journal of Outdoor Recreation and Tourism*, 20(1), 2–6.
Beames, S., Mackie, C., & Atencio, M. (2019). *Adventure and Society*. Springer. https://doi.org/10.1007/978-3-319-96062-3
Beedie, P., & Hudson, S. (2003). Emergence of mountain-based adventure tourism. *Annals of Tourism Research*, 30(3), 625–643.
Berger, I. E., & Greenspan, I. (2008). High (on) technology: Producing tourist identities through technologized adventure. *Journal of Sport & Tourism*, 13(2), 89–114.
Bizirgianni, I., & Dionysopoulou, P. (2013). The influence of tourist trends of youth tourism through social media (SM) & information and communication technologies (ICTs). *Procedia-Social and Behavioral Sciences*, 73, 652–660.
Cai, W., Cohen, S. A., & Tribe, J. (2019). Harmony rules in Chinese backpacker groups. *Annals of Tourism Research*, 75, 120–130.
Cardwell, D., & Higgins, M. (2016). *Surf's Up, and the Ocean Is Nowhere in Sight*. The New York Times. Retrieved May 23, 2022, from https://www.nytimes.com/2016/09/03/business/energy-environment/surfs-up-and-the-ocean-is-nowhere-in-sight.html
Cheng, M. (2017). Understanding Chinese post-80s' outbound adventure tourism experience. PhD thesis. University of Technology Sydney, Australia. OPUS at UTS: Understanding Chinese post-80s' outbound adventure tourism experience - Open Publications of UTS Scholars.
Cheng, M., Edwards, D., Darcy, S., & Redfern, K. (2018). A Tri-Method Approach to a Review of Adventure Tourism Literature: Bibliometric Analysis, Content Analysis, and a Quantitative Systematic Literature Review. *Journal of Hospitality & Tourism Research*, 42(6), 997–1020. https://doi.org/10.1177/1096348016640588
Cheng, M., & Foley, C. (2018). Understanding the distinctiveness of Chinese post-80s tourists through an exploration of their formative experiences. *Current Issues in Tourism*, 21(110), 1312–1328.
Chu, S. C., & Choi, S. M. (2010). Social capital and self-preservation on social networking sites: A comparative study of Chinese and American young generations. *Chinese Journal of Communication*, 3(4), 402–420. https://doi.org/10.1080/17544750.2010.516575

Cuthbertson, B., Socha, T. L., & Potter, T. G. (2004). The double-edged sword: Critical reflections on traditional and modern technology in outdoor education. *Journal of Adventure Education and Outdoor Learning, 4*(2), 133–144.

Décieux, J. P., Heinen, A., & Willems, H. (2019). Social media and its role in friendship-driven interactions among young people: A mixed methods study. *Young, 27*(1), 18–31. https://doi.org/10.1177/1103308818755516

Ewert, A. W., & Sibthorp, J. (2014). *Outdoor adventure education: Foundations, theory, and research*. Human Kinetics.

Fletcher, R. (2010). The emperor's new adventure: Public secrecy and the paradox of adventure tourism. *Journal of Contemporary Ethnography, 39*(1), 6–33.

Gardiner, S., & Kwek, A. (2017). Chinese participation in adventure tourism: A study of generation Y international students' perceptions. *Journal of Travel Research, 56*(4), 496–506.

Giddy, J. K. (2018). A profile of commercial adventure tourism participants in South Africa. *Anatolia, 29*(1), 40–51. https://doi.org/10.1080/13032917.2017.1366346

Gretzel, U., Fuchs, M., Baggio, R., Hoepken, W., Law, R., Neidhardt, J., Pesonen, J., Zanker, M., & Xiang, Z. (2020). e-Tourism beyond COVID-19: A call for transformative research. *Information Technology & Tourism, 22*(2), 187–203.

Hasan, M. K., & Neela, N. M. (2021). Adventure tourists' electronic word-of-mouth (e-WOM) intention: The effect of water-based adventure experience, grandiose narcissism, and self-presentation. *Tourism and Hospitality Research*, 14673584211026326.

Imboden, A. (2012). Between risk and comfort: Representations of adventure tourism in Sweden and Switzerland. *Scandinavian Journal of Hospitality and Tourism, 12*(4), 310–323. https://doi.org/10.1080/15022250.2012.752624

Janowski, I., Gardiner, S., & Kwek, A. (2021). Dimensions of adventure tourism. *Tourism Management Perspectives, 37*. https://doi.org/10.1016/j.tmp.2020.100776

Jin, X., Xiang, Y., Weber, K., & Liu, Y. (2019). Motivation and involvement in adventure tourism activities: A Chinese tourists' perspective. *Asia Pacific Journal of Tourism Research, 24*(11), 1066–1078.

Johnston, B. R., & Edwards, T. (1994). The commodification of mountaineering. *Annals of Tourism Research, 21*(3), 459–478.

Joseph, P. (2020). "10 of the best virtual reality travel experiences." Retrieved May 1, 2023, from https://www.travelmag.com/articles/virtual-reality-travel-experiences

Kane, M. J., & Tucker, H. (2004). Adventure tourism: The freedom to play with reality. *Tourist Studies, 4*(3), 217–234. https://doi.org/10.1177/1468797604057323

Kang, J.-W., Lee, H., & Namkung, Y. (2018). The impact of restaurant patrons' flow experience on SNS satisfaction and offline purchase intentions. *International Journal of Contemporary Hospitality Management*.

Kang, M., & Schuett, M. (2013). Determinants of sharing travel experiences in social media. *Journal of Travel & Tourism Marketing, 30*(1–2), 93–107.

Kelly, H. (2022). *Adventure Travel Trends 2022*. Adventure Travel Trade Association. https://www.adventuretravelnews.com/adventure-travel-trends-2022

Kim, D., Jang, S. S., & Adler, H. (2015). What drives café customers to spread eWOM? Examining self-relevant value, quality value, and opinion leadership. *International Journal of Contemporary Hospitality Management, 27*(2), 261–282.

Kim, J., & Tussyadiah, I. P. (2013). Social networking and social support in tourism experience: The moderating role of online self-presentation strategies. *Journal of Travel & Tourism Marketing, 30*(1–2), 78–92.

Liu, H., Wu, L., & Li, X. (2019). Social media envy: How experience sharing on social networking sites drives millennials' aspirational tourism consumption. *Journal of Travel Research, 58*(3), 355–369.

Manago, A., & Vaughn, L. (2015). Social media, friendship, and happiness in the Millennial generation. In M. Demir (Ed.), *Friendship and happiness* (pp. 187–206). Springer. https://doi.org/10.1007/978-94-017-9603-3_11

Martin, N. (2019). China Adventure Tourism Market Study. Retrieved January 17, 2022, from https://bannikin.com/wp-content/uploads/2019/06/Bannikin-China-Adventure-Tourism-Report.pdf

Martin, N., & Ren, C. (2020). Adventurous Arctic encounters? Exploring Chinese adventure tourism. *Scandinavian Journal of Hospitality and Tourism, 20*, 1–18.

Narangajavana Kaosiri, Y., Callarisa Fiol, L. J., Moliner Tena, M. A., Rodríguez Artola, R. M., & Sanchez Garcia, J. (2019). User-generated content sources in social media: A new approach to explore tourist satisfaction. *Journal of Travel Research, 58*(2), 253–265.

Neumann, W. L. (2014). *Social research methods: Qualitative and quantitative approaches neumann* (7th ed.). Pearson.

Puchan, H. (2004). Living "extreme": Adventure sports, media and commercialisation. *Journal of Communication Management, 27*(2), 261–282.

Sorooshian, S. (2021). Implementation of an expanded decision-making technique to comment on Sweden readiness for digital tourism. *Systems, 9*(3), 50.

Swarbrooke, J., Beard, C., Leckie, S., & Pomfret, G. (2003). *Adventure tourism: The new frontier.* Routledge.

Think with Google. (2017). *Smartphone travel booking statistics.* Retrieved August 26, 2022, from https://www.thinkwithgoogle.com/marketing-strategies/app-and-mobile/smartphone-travel-booking-statistics/

Thompson, B. Y. (2019). The digital nomad lifestyle:(remote) work/leisure balance, privilege, and constructed community. *International Journal of the Sociology of Leisure, 2*(1), 27–42.

Thorpe, H. (2017). Action sports, social media, and new technologies: Towards a research agenda. *Communication & Sport, 5*(5), 554–578.

Triantafillidou, A., & Petala, Z. (2016). The role of sea-based adventure experiences in tourists' satisfaction and behavioral intentions. *Journal of Travel & Tourism Marketing, 33*(sup. 1), 67–87.

Tripadvisor. (2022). *Media Centre - About us.* Tripadvisor. Retrieved November 06, 2022, from https://tripadvisor.mediaroom.com/au-about-us

UNWTO. (2014). *Global Report on Adventure Tourism.* https://www.e-unwto.org/doi/pdf/10.18111/9789284416622

Varley, P. (2006). Confecting Adventure and Playing with Meaning: The Adventure Commodification Continuum. *Journal of Sport & Tourism, 11*(2), 173–194. https://doi.org/10.1080/14775080601155217

Weber, K. (2001). Outdoor adventure tourism: A review of research approaches. *Annals of Tourism Research, 28*(2), 360–377.

Wheal, J. K. S. (2017). *The High-Tech Race to Make Deadly Adventure Sports Safe for Anyone.* Outside. Retrieved May 23, 2022, from shttps://www.outsideonline.com/culture/books-media/ben-mcgrath-riverman-interview/

Wu, M. Y., & Pearce, P. (2016). Tourism blogging motivations: Why do Chinese tourists create little "Lonely Planets"? *Journal of Travel Research, 55*(4), 537–549.

Xiang, Y. (2013). The characteristics of independent Chinese outbound tourists. *Tourism Planning, 10*(2), 134–148.

Zeng, G., & Gerritsen, R. (2014). What do we know about social media in tourism? A review. *Tourism Management Perspectives, 10*, 27–36.

Zhang, Z., Ye, Q., Law, R., & Li, Y. (2010). The impact of e-word-of-mouth on the online popularity of restaurants: A comparison of consumer reviews and editor reviews. *International Journal of Hospitality Management, 29*(4), 694–700.

Zubarev, V. (2020). *Computational photography part I: What is computational photography?* Digital Photography Review. Retrieved August 4, 2022, from https://www.dpreview.com/articles/9828658229/computational-photography-part-i-what-is-computational-photography

12 Psychological well-being and adventure tourism

Implications for people and places

Susan Houge Mackenzie

Chapter learning outcomes

1 Analyse emerging and contemporary conceptualisations of adventure and well-being.
2 Evaluate current tensions in adventure tourism in relation to the well-being of people and places.
3 Identify implications and applications of contemporary well-being approaches for adventure tourism.
4 Reflect on how adventure tourism and well-being can become more integrated into future research and practice.

Introduction

Research and public policy have long supported links between traditional sporting pursuits and well-being. However, adventure recreation and tourism literature have primarily focused on the primacy of risk or sensation-seeking in adventure motivations and outcomes. This standpoint is limited by assumptions that participation is (a) dependent on specific personality structures, (b) solely motivated by risk-taking and hedonism, (c) only attractive or accessible to a narrow demographic, and (d) may be perceived as dysfunctional, deviant, or even damaging to people or places. In contrast, recent research suggests that adventure pursuits may provide unique psychological benefits due to their context. This chapter critically evaluates traditional perspectives of adventure in relation to emerging literature which suggests adventure tourism pursuits should be (re)conceptualised in terms of how they foster both hedonic and eudaimonic well-being[1] for participants, guides and destination communities. The significance of this perspective is that it (a) reframes adventure tourism in terms of the broader benefits it can foster for people and places and (b) offers more nuanced perspectives on *how* and *when* adventure tourism is likely to foster these well-being benefits. See also Chapter 1 for an overview of the inextricable links between adventure tourism and well-being.

DOI: 10.4324/9781003393153-15

Traditional and contemporary conceptualisations of adventure and well-being

Traditional frameworks for understanding adventure motivations and outcomes have focused primarily on the primacy of risk and thrill-seeking motives, and hedonistic rewards, such as excitement or pleasurable sensations (Kar & Tarafder, 2020; Llewellyn & Sanchez, 2008). See Chapter 7 for an overview of the motives and benefits associated with adventure tourism. These limited models of adventure experiences and outcomes have often engendered a focus on the importance of "pushing" participants out of their "comfort zone" to achieve hedonistic benefits in adventure tourism and recreation. For instance, risk-taking, danger, and control were central to Ewert and Hollenhorst's (1989) early model of adventure experiences: 'It is [the] positive valuation of risk and danger that makes adventure recreation fundamentally different from other recreation experiences' (p. 127). Despite challenges to this view, many researchers and practitioners maintain that confronting risk, fear, or danger produces optimal stress and discomfort, which in turn promotes outcomes such as improved self-esteem, "character building", and psychological resilience (e.g., Ewert & Yoshino, 2011; MacGregor et al., 2014). Although fear, risk and associated sensation-seeking have long been conceptualised as integral to adventure experiences (e.g., Zuckerman, 2007), the pivotal importance of these elements to adventure motivations and well-being outcomes remains contested.

An emerging body of research has challenged many of these traditional assumptions by suggesting that adventure motives, experiences and benefits extend beyond hedonistic rewards associated with risk and sensation-seeking. For instance, multiple studies counter these traditional narratives by illustrating how adventure recreation participants purposefully use these activities to facilitate psychological well-being (e.g., Buckley & Brough, 2017; Brymer & Schweitzer, 2013) and transcendence, even in activities with high serious injury rates, such as mountaineering and BASE jumping (e.g., Monasterio & Cloninger, 2019). Qualitative studies in particular suggest that adventure motives are more diverse than previously thought and are often linked to eudaimonic well-being outcomes, rather than hedonic or competitive outcomes (e.g., Ewert et al., 2020). For instance, Houge Mackenzie et al. (2021) proposed that adventure may enhance eudaimonic subjective well-being through the mechanisms of nature contact and satisfaction of basic psychological needs.

Several studies across multiple adventure domains have identified a range of motivations beyond thrill-seeking, such as personal control, courage, attention restoration, personal growth, self-actualisation, achievement, and mastery (e.g., Berman et al., 2008; Brymer & Oades, 2009; Buckley & Brough, 2017). In addition to challenging sensation-seeking models of adventure, a series of studies by Barlow et al. (2013) concluded that different adventure activities involve distinct motivations, such as emotional regulation and agency. Notwithstanding, many of these studies (e.g., Barlow et al., 2013) are still rooted in deficit models, which frame adventure pursuits as compensatory activities through which to counteract difficulties with emotional regulation, personal agency and anxiety. Varley and Semple (2015, p. 77) eloquently summarised how many current adventure conceptualisations ignore the 'holistic social nature of the [adventure tourism] experience':

Many theories of adventure ... encapsulate the adventure motive as a desire for borderline experiences occupying the threshold between catastrophe and adventure. Such representations, with their almost fatalistic proximity to disaster seem essentialist and elitist, and intuitively are at odds with the motives of many contemporary adventure travelers.

The role of eudaimonic well-being in emerging adventure frameworks

Well-being has been theorised in various ways across disciplines. Public policy approaches typically view well-being in terms of economics (e.g., enabling "preference satisfaction" by increasing incomes) and "objective" human development dimensions (e.g., life expectancy, education, living standards). Increasingly, subjective well-being frameworks, which incorporate both immediate emotions and overall life satisfaction (e.g., Waterman et al., 2008) are being applied to operationalise psychological well-being (PWB) across a range of domains. Two distinct approaches to PWB dominate the literature: hedonia (e.g., short-term positive affective experiences) and eudaimonia (e.g., longer term fulfilment, life satisfaction, meaning; e.g., Ryff, 2013; Ryan & Martela, 2016). As there is empirical support for both hedonic and eudaimonic well-being theories, the relative importance of these approaches remains contested. Nevertheless, some scholars (e.g., Huta & Waterman, 2014; Ryan & Martela, 2016) highlight the primacy of eudaimonic well-being approaches on the basis that eudaimonia also fosters more stable and enduring hedonic happiness.

Eudaimonic well-being encompasses meaning, purpose, optimal functioning, self-realisation, and flourishing, whereas hedonic well-being consists of pleasure, positive emotions, and avoidance of pain (Waterman et al., 2008). Hedonic motivations are often momentary and linked to shorter term goals. The hedonic approach aligns with traditional models of adventure tourism and recreation experiences focused on sensation or thrill-seeking and associated affective responses during or immediately following an adventure activity (Castanier et al., 2010). In contrast, eudaimonic motives and outcomes are more complex and less widely explored in adventure research.

An emerging body of knowledge highlighting the role of eudaimonic motives and well-being outcomes in adventure, such as meaning, connection to others, self-transcendence, and self-development, suggests that these aspects play an important role in adventure pursuits and might be more pivotal to long-term subjective well-being than hedonic motives. For instance, Ewert et al. (2020) investigated hedonic and eudaimonic motives in rock climbing, whitewater boating, and mountain biking and identified the following eudaimonic motives: social connections, personal growth, a sense of achievement, identity work, and escape. Social connections and identity in particular have emerged across a range of exploratory studies of adventure motives in rock and mountain climbing, backcountry skiing and snowboarding, mountain biking, and kayaking (Frühauf et al., 2017; Kerr & Houge Mackenzie, 2012), alongside eudaimonic experiences, such as self-transcendence in mountaineering and BASE jumping (Monasterio & Cloninger, 2019).

It is noteworthy that, while the eudaimonic motives and well-being outcomes discussed earlier are worthy of further attention in frameworks of adventure experiences, this does not imply hedonic motives and well-being outcomes are unimportant in adventure contexts. Research suggests that hedonic and eudaimonic aspects of adventure may function synergistically to foster a range of PWB outcomes in diverse adventure contexts. For example, in addition to eudiamonic motives, Ewert et al. (2020) identified the following as hedonic motives for rock climbing, whitewater boating, and mountain biking: flow experiences, sense of mastery, challenge, connection to nature, autonomy, and risk or danger. Although these authors framed the connection to nature as hedonistic, previous research has associated nature connectedness with eudaimonic well-being, characterised by an enhanced sense of meaning, purpose, or connection to other living beings (Zelenski & Nisbet, 2014). Connection to nature that extends beyond hedonistic rewards has been noted in multiple contemporary adventure studies (Frühauf et al., 2017; Kerr & Houge

Mackenzie, 2012; MacIntyre et al., 2019). These tensions highlight the dangers of attempting to privilege some elements of adventure experiences over others or assuming that hedonic and eudaimonic elements of adventure can be disentangled. Rather, these elements likely interact in a synergistic and holistic manner to produce a range of PWB outcomes.

Current tensions: How does adventure tourism impact the well-being of people and places?

Well-being outcomes for adventure participants

The developments identified earlier reflect the emergence of the Positive Tourism movement, an offshoot of the relatively new positive psychology discipline, which essentially seeks to understand what makes life worth living (Seligman & Csikszentmihalyi, 2000). Positive psychology approaches stemmed from dissatisfaction with "disease models" of psychology that focused almost exclusively on pathology and ill-being and largely ignored optimal functioning in individuals and communities. Over the past decade, Positive Tourism approaches emerged for similar reasons. Tourism literature has focused primarily on how tourism creates ill-being for individuals and societies, a view that often overlooks the well-being potential of tourism. Filep and Laing (2019, p. 350) argue that 'to focus on the negative is to ignore the transcendental qualities of tourism, and to overlook its potential for improving the human condition, both collectively and individually'.

A number of recent adventure tourism studies illustrate how these activities can foster personal transformations and well-being outcomes not only for adventure participants but also for adventure guides and adventure tourism communities. Research suggests that adventure participants experience well-being benefits such as emotional regulation, improved quality of life, positive life transformations, social connections, pleasure, increased resilience, self-esteem, autonomy, competence, physical activity levels, nature connections, and flow (e.g., Brymer & Oades, 2009; Brymer & Schweitzer, 2013; Ewert & Yoshino, 2011; Willig, 2008). Studies of rock climbing, whitewater paddling, skiing, paragliding, and skydiving provide strong evidence that these activities support key constructs associated with PWB, for example, agentic emotion regulation, self-esteem, and flow (Castanier et al., 2010; Willig, 2008). Mykletun and Mazza (2016) found that the psychosocial benefits of competitive adventure tourism events included positive social interactions, paratelic (play) states, flow, exploration, self-insight, and personal transformations. In addition, Lynch and Dibben (2016) identified enjoyment, challenge, progressively improving performance, opportunities for self-responsibility, and balancing challenge and skill as key intrinsic motivations for competitive adventure tourists. These authors further identified competition, social interaction, personal identity, aesthetics of place, and novelty as key extrinsic motivations.

Experiences of autonomy and social connection have also been linked to well-being outcomes across multiple adventure contexts. For instance, MacGregor et al. (2014, p. 175) found that 'regular climbing provides participants with an agentic emotional experience that then benefits their everyday functioning; such benefits are not derived from other (low-risk) activities'. The unique challenges posed by adventure tourism activities have also been linked to PWB outcomes via fostering feelings of competence and accomplishment (e.g., Tsaur et al., 2015). This literature suggests that the satisfaction of basic psychological needs for competence, relatedness, and autonomy may be important pathways through which adventure tourism facilitates PWB, as discussed further on.

Well-being outcomes for adventure guides

Research on tour guiding has progressed significantly in recent decades with an increasing focus on how guides fulfil roles beyond simply delivering information or ensuring tourists enjoy their holiday. Weiler, Black, and colleagues have produced a range of critical texts identifying the diversity and complexity of guide roles. These roles include fostering sustainable tourism and brokering tourism experiences across multiple domains, such as accessing local culture (e.g., Weiler & Black, 2015). Guides also broker cultural understanding and empathy by "choreographing" travel activities (Beedie, 2003). It has even been suggested that tour guides may facilitate self-development and lasting personal transformations in spiritual tourism contexts (Parsons et al., 2019). These increasingly complex roles reflect an evolution towards co-created tourism experiences, characterised by shared control between guides and clients and active collaboration to create customised tours (Weiler & Black, 2015). This literature reflects the critical roles that guide play in facilitating personally meaningful and environmentally sustainable tourist experiences.

Co-creating tours entails cultivating unique experiences that are both enjoyable and meaningful for clients and guides (Weiler & Black, 2015). This conceptualisation of symbiotic, shared experiences aligns with Positive Tourism approaches in that it suggests how adventure guiding can provide mutual benefit and meaning for clients and guides. Notwithstanding, research has predominantly highlighted the negative impacts of tour guiding and similar front-line roles, for example emotional labour (Zapf & Holz, 2006). The negative well-being impacts of emotional labour in particular have been documented in various cultural contexts. For instance, Wong and Wang (2009) reported that Taiwanese guides experienced significant ill-being outcomes due to high emotional labour demands, which was exacerbated by pay rates and group size. However, these authors also identified potentially positive consequences of emotional labour 'when it is experienced as self-enhancing or when workers are in control of their emotions' (p. 256).

Adventure guiding literature reflects the tensions highlighted by Filep et al. (2019), namely the dominance of dystopian accounts of guiding work contrasted with emerging literature, which suggests more hopeful potentials of tour guiding. Conceptually, Weiler and Black (2015) identified these potentials as embedded in co-created tours that support both hedonic (e.g., positive emotions) and eudaimonic (e.g., meaning, life satisfaction) well-being for clients and guides. Empirically, studies have reported how adventure guides often engage in "deep acting" (i.e., aligning one's true feelings with emotions required by a job) and other techniques that reduce emotional labour and increase authenticity and meaningfulness. In Arctic adventures, Mathisen (2019) illustrated how storytelling changed guides' focus from delivering a "product" to meaningful interactions with individual clients. In whitewater contexts, Houge Mackenzie and Kerr (2013) identified a range of inter and intrapersonal factors that influenced guide experiences. They found, for example, that guiding elicited negative emotions and stress due to interactions with the natural environment, clients and co-workers, while also identifying how developing a psychological protective frame reduced stress and facilitated positive emotions for guides. In extended adventure encounters, Sharpe (2005) reported how adventure guides created "authentic" and rewarding experiences through emotion regulation and deep acting, which resulted in feelings of achievement, whereas surface acting was eschewed as it led to suppressed emotions and produced "canned" trips. Likewise, Torland (2011) attributed the equally high job satisfaction levels she found across male and female Australian adventure guides to successful deep acting. Torland concluded that, as guides often worked 'in close proximity to their clients, in unpredictable natural environments and sometimes over extended periods of time … [they] will typically make "friends" with their clients rather than taking on the role as a more "superficial" service provider' (p. 381).

Well-being outcomes for adventure tourism host communities

Although less widely studied, there is initial evidence that adventure tourism experiences and events may enhance PWB for destination communities in multiple ways, such as fostering self-determination, competence, community belonging, interpersonal relationships, and meaning and purpose (e.g., Brymer & Houge Mackenzie, 2015). While these are established pillars in the PWB literature (e.g., Ryan & Deci, 2017), individual PWB benefits are often overlooked in theoretical frameworks of community-based tourism. For instance, witnessing new adventure tourism activities in local places can arouse curiosity for host community members, which develops into participation and the associated PWB benefits discussed earlier. Mountaineering tourism in Nepal has been directly linked to increased self-determination in host communities, in addition to cultural pride and safety (Sharma, 2009). There is also anecdotal evidence that host–tourist interactions associated with BASE jumping events enhance well-being by fostering new ways of thinking amongst both locals and visitors (Brymer & Houge Mackenzie, 2015). Beal and Smith (2010) argue that even watching "extreme" adventure activities can facilitate psychological transformations in host communities as individuals benefit from the spectacle, the vicarious realisation of individuality, re-enchantment, and unpredictability inherent in these activities. This proposition is reflected by community PWB benefits associated with Mavericks, the iconic California big wave surfing competition. In addition to the immediate PWB benefits of viewing awe-inspiring big wave surfing, the associated Jay Moriarity Foundation fosters a range of enduring PWB benefits by providing opportunities for young people to find meaning and purpose (i.e., eudaimonic well-being) in their lives through adventure.

On a philosophical level, it has also been suggested that hosts can experience spiritual transformations via interactions with adventure tourists. According to Steiner and Reisinger's (2006) Heideggerian framework for wellness through tourism, adventure tourism hosts may be transformed through the "fourfold" impact of exposure to these activities. Specifically, these activities may foster well-being via opportunities to experience *mortality* (via exposure to the reality of death), contact with *divinities* and *earth* (e.g., via exposure to spectacular natural events or places), and transcending sensual experiences either physically or mentally via encounters with *sky* (e.g., mountaineering, skydiving). They argue that these experiences bring us into authentic encounters not only with earth and sky, that is grounding and freeing nature, but also divinities and mortality, which collectively foster wellness. As adventure tourism relies on intimate relationships with nature and larger forces, exposure to these activities, whether directly or vicariously, may instigate positive personal transformations. These examples highlight (a) the potential of adventure tourism to positively transform community PWB on various levels through various mechanisms and (b) how examining adventure tourism using psychological models can yield novel insights in the relationships between adventure tourism and community PWB.

How does adventure enhance well-being? Unravelling the mechanisms

Despite emerging literature highlighting potential adventure benefits across diverse populations, our theoretical models of *how* adventure facilitates PWB require development. Inherent in this need for additional research to inform practice should be the recognition that *adventure is not inherently good*. While appearing obvious to some, this premise is particularly important for adventure researchers and practitioners to bear in mind, as we are those most likely to make the (erroneous) assumption that adventure is inherently beneficial. There are many situations in which adventure tourism experiences can inflict physical or emotional harm or simply lack meaningful PWB impacts, if they are not appropriately facilitated. Thus, contemporary, nuanced understandings of adventure should include a recognition that adventure is not inherently beneficial or egalitarian. To the contrary, adventure tourism participation and guiding has traditionally been a leisure activity available to a niche group of privileged individuals (e.g., Western, affluent, young males). Although this demographic is changing, the well-being potential of adventure tourism will remain unrealised for many populations if it cannot be made more inclusive. Thus, contemporary models of adventure tourism and well-being must reflect a nuanced awareness that (1) *adventure is not an inherent or irrefutable "good" for participants or guides – or the planet* and (2) we need to develop *genuinely inclusive adventure experiences* if well-being benefits are to be fully and broadly realised. The following section examines some key mechanisms that can inform these models (i.e., nature connection and fulfilment of basic psychological needs) and how guides and operators can create the appropriate conditions to support PWB for people and places.

Nature: A key well-being mechanism in adventure tourism

Adventure tourism inherently involves natural environments, therefore considering nature as an important mechanism via which adventure can facilitate PWB is an important avenue for developing research and practice. The role of nature in well-being has received increasing attention and support across health, sport, and leisure literature, and facilitating nature contact appears to be one way that adventure activities enhance well-being (e.g., Mutz & Müller, 2016). Systematic reviews (e.g., Thompson et al., 2011) have concluded that being physically active outdoors, a hallmark of adventure tourism, provides synergistic benefits beyond those attained from physical activities performed in non-natural environments. For example, "green exercise" studies demonstrate the additive benefits of physical activity in natural settings compared to non-natural environments (e.g., Pretty et al., 2005). In an experiment involving the same physical activity across indoor, urban, and natural settings, Ryan et al. (2010) found that natural contexts significantly enhanced subjective well-being. Eigenschenk et al.'s (2019) systematic review of nature-based physical activities identified a range of benefits including enhanced general well-being, psychological stability, life satisfaction, emotional intelligence, intellectual flexibility, mindfulness, empathy, self-esteem, self-actualisation, social capital, educational performance, and intrinsic motivation.

Nature connection has been correlated with a range of well-being outcomes in both adventure and non-adventure contexts, such as positive affect, vitality, autonomy, personal growth, and life satisfaction (e.g., Capaldi et al., 2015; Hanna et al., 2019). Both the frequency of nature contact, and the degree of nature present appear positively correlated with subjective well-being. Based on considerable evidence that nature provides an array of direct well-being benefits, both for the general population and adventure participants, the construct of nature connectedness (i.e., a sense of psychological connection with nature) has been proposed as a central mediator of hedonic and eudaimonic well-being outcomes in adventure tourism (Houge Mackenzie et al., 2021).

The pivotal role of natural environments in adventure experiences has been identified across a range of "extreme" to "lower risk" adventure activities. For instance, Buckley and Brough (2017) explored how a range of nature-based tourism activities may be used for therapeutic purposes, while Kerr and Houge Mackenzie (2020) found that connecting with nature was a primary motivation for BASE jumping, which refuted traditional framings of thrill-seeking as a core motive. In a systematic review of tourism and PWB, Garcês et al. (2018) highlighted the unique association between nature and well-being and called for further research on mechanisms underpinning this relationship. Even at the "extreme" end of the adventure spectrum, evidence suggests that nature plays a critical role in facilitating well-being benefits (Brymer & Schweitzer, 2013). Studies by Brymer and colleagues have repeatedly demonstrated how connecting with the natural world is central to "extreme" adventure motivations and well-being outcomes in both recreational and tourism contexts.

Recent research also identified the key role of nature in adventure guide well-being (Houge Mackenzie & Raymond, 2020; Houge Mackenzie et al., 2021). However, guides' valuing of the natural environment may be a double-edged sword due to the climate crisis. Valdivielso-Martínez and Houge Mackenzie (2022) found that guide well-being was negatively impacted if guides perceived that their work was directly contributing to the deterioration of specific natural environments, they highly valued and/or larger climate change impacts. Thus, reaping the nature-based benefits of adventure is predicated on (a) the preservation of the natural environments in which those adventures unfold and (b) ensuring participants and guides have opportunities to be mindful and psychologically immersed in those environments rather than overly focused on physical tasks or overwhelming emotions (e.g., fear, anxiety). Accordingly, building in dedicated, undistracted time for adventure tourism guides and participants to psychologically immerse in natural environments may be an important practice for facilitating PWB outcomes in adventure tourism. Although considerable evidence illustrates the importance of nature connection for well-being, research has yet to directly and thoroughly unravel *how* and *when* nature connectedness influences specific PWB outcomes for adventure tourists, guides, or communities.

Basic psychological need fulfilment: Key well-being mechanisms in adventure tourism

Houge Mackenzie et al. (2021) argued that identifying key mechanisms underpinning eudaimonic and hedonic well-being can help us better understand the conditions required for adventure to facilitate well-being. Both eudaimonic and hedonic well-being outcomes have been directly linked to the fulfilment of basic psychological needs for autonomy, competence, and relatedness. *Autonomy* means having an authentic sense of self-direction and volition and being the perceived origin of one's own behaviour rather than being externally controlled. *Competence* involves feeling effective in interactions with one's environment and having opportunities to exercise one's capacities. *Relatedness* entails feeling connected to others and having a sense of belonging with individuals and one's community. It should be noted here that these basic psychological need constructs are part of a larger, leading theory of PWB: self-determination theory (Ryan & Deci, 2017).

This theory proposes that people are intrinsically motivated by desires for personal growth, self-direction, and affiliation and that intrinsic, self-determined motivation stems directly from the inherent enjoyment and satisfaction of an activity or behaviour, without external rewards. Achieving psychological growth is predicated on the degree to which we can experience mastery (i.e., competence), connect to others (i.e., relatedness), and feel in control of our behaviours and choices (i.e., autonomy).

A robust body of literature demonstrates that satisfying these basic psychological needs is vital for psychological growth, subjective vitality, and overall well-being. Recently, research has explored whether *beneficence*, that is having a positive influence on others, might be an additional basic psychological need (Martela & Ryan, 2020). Although Martela and Ryan's initial analyses suggested that beneficence enhances PWB but to a lesser degree than autonomy, competence, or relatedness, research with adventure guides highlighted beneficence as a primary PWB determinant in that context. This was linked with an enhanced sense of meaning and purpose (Houge Mackenzie & Raymond, 2020). Thus, the role of beneficence, alongside autonomy, competence, and relatedness, requires further examination in adventure contexts. Despite the potential significance of self-determination theory constructs for understanding and designing adventure tourism experiences, this framework has not been widely applied to understand how adventure tourism impacts the well-being of people and places.

Future focus: Realising the potential of adventure for enhancing psychological well-being

This chapter has highlighted the need for adventure research and practice that (1) explores a broader range of participation motives and outcomes beyond risk and thrill-seeking and (2) seeks to facilitate a wider range of well-being outcomes more inclusively and deliberately by drawing on emerging research linking nature and basic psychological needs with PWB. Reframing adventure tourism in research and practice will require increasing researchers' and practitioners' awareness of, and focus on, how adventure promotes PWB via a range of mechanisms. This approach offers the potential to make adventure tourism more accessible and broaden adventure offerings by developing experiences that go beyond thrill to foster deeper, more meaningful connections to people and places. The following sections highlight just some of the many ways we can reconceptualise and realise the potential of adventure (see also Chapter 1) to enhance PWB via research and practice.

Future research opportunities

Future research should employ longitudinal and mixed methods approaches across a wider range of adventure activities, participants, guides, and diverse destination communities to better understand *how* adventure fosters PWB (i.e., key mechanisms). Adventure research could benefit from drawing upon larger complementary disciplines (e.g., psychology, education, public health) to enhance these understandings and develop refined models of adventure and PWB. Fruitful avenues of study include investigations of how adventure tourism (1) supports basic psychological needs for autonomy, competence, and relatedness; (2) fosters nature connectedness; (3) provides opportunities for beneficence; and (4) how these elements foster both hedonic (e.g. positive emotions, flow) and eudaimonic (e.g. meaning, purpose) outcomes. In addition, research should examine how the social environment impacts inclusiveness in adventure tourism, particularly in relation to groups who have traditionally been underrepresented in adventure research, guiding, and participation (e.g., women; individuals with diverse ability levels). Another relatively unexplored area lies in understanding how climate change may impact guide and client PWB. In an exploratory study, Valdivielso-Martínez and Houge Mackenzie (2022) found that climate change and environmental impacts associated with adventure could negatively impact guide and client well-being. Similarly, investigating how and if "microadventures" (i.e., localised, smaller scale, human-powered pursuits) can foster similar well-being benefits to traditional adventure travel (i.e., far-flung, larger scale, resource-intensive) can spark innovative approaches to fostering adventure benefits more inclusively, with less environmental impact (e.g., Houge Mackenzie & Goodnow, 2020). Exploring how these issues may impact nature connection and other mechanisms linked to PWB outcomes is critical as these tensions will become increasingly central to the entire adventure tourism ecosystem. Overall, research that examines a broader range of adventure benefits, as well as a more nuanced understanding of how and when adventure provides these benefits, is needed to advance theory and, ultimately, improve adventure experiences for larger populations.

Future practice: Strategies to support well-being via adventure tourism

Improving inclusivity in adventure tourism should be a key priority if the potential PWB benefits of adventure are to be realised. While refining models of adventure to incorporate a broader scope of motivations, PWB mechanisms, and outcomes associated with adventure is a necessary step toward this goal, it is not sufficient. Translating this knowledge into clear strategies that increase access and improve experiences for participants, guides, and adventure communities is critical. This means adventure operators need to consider both the physical and psychological needs of diverse client groups and rethink how they can design more inclusive experiences. In Aotearoa New Zealand, Making Trax is an organisation leading this movement by helping businesses to create more inclusive, rather than simply physically accessible, experiences across the adventure tourism industry (for details, visit www.makingtrax.co.nz). There are also inspirational exemplars of organisations dedicated to enhancing gender equity in adventure tourism, such as 3 Sisters Adventure Trekking (www.3sistersadventuretrek.com; see Case study 12.1) and its non-governmental organisation (NGO) Empowering Women of Nepal, GOOD Travel (www.good-travel.org), and Intrepid Group (www.intrepidgroup.travel). While these organisations vary in size and focus, they all showcase successful approaches to improving gender equity in adventure tourism among participants and guides, and ensuring that communities benefit from more inclusive adventure tourism practices (for analysis of how each organisation fosters basic psychological needs, see Houge Mackenzie & Raymond, 2021).

Enhancing PWB outcomes for broader populations can also be achieved by purposefully designing adventure experiences to support basic psychological needs and foster nature connection. This means fostering autonomy and competence by providing "challenge by choice" options and offering varied options for challenge within the same trip (e.g., multiple lines of travel for whitewater rapids), providing rationales for tasks, empowering guides to develop trips according to their own interests and skills, and minimising ego involvement for participants and guides (e.g., emphasising self-referenced goals or group goals). Purposefully designing activities that build connections between

---

Case study 12.1   3 Sisters Adventure Trekking and Empowering Women of Nepal

3 Sisters Adventure Trekking is a trekking organisation created for women, by women. Founded by three Nepalese sisters, 3 Sisters Adventure Trekking and its NGO Empowering Women of Nepal supports women's well-being through adventure tourism and associated training. Since 1994, more than 2,000 Nepalese women have completed the NGO's training and over 100 guides are now employed by 3 Sisters Adventure Trekking. Empowering Women of Nepal uses adventure tourism as a vehicle for empowering, educating and creating employment opportunities for disadvantaged Nepali women and girls. Their programmes exemplify how support for basic psychological needs (autonomy, competence and relatedness) can help foster well-being, independence and self-sufficiency for disadvantaged women. The NGO fosters autonomy by supporting women to gain financial and social independence via training and employment. In addition, during the month-long intensive training programme, they provide opportunities for women to build relatedness with other women on the same journey, and competence across a range of new skills. This is achieved through practical training in skills such as first aid and geography, as well as educating women in diverse topics such as leadership, team building, cross-cultural communication, mountain culture and people, and women's human rights.

---

participants and amongst guides will also enhance PWB outcomes by fostering relatedness and beneficence. In addition, ensuring that participants and guides have dedicated time to immerse in, and connect with, natural environments also supports a sense of nature connection, a key PWB mechanism. While space limitations preclude an expanded discussion of practical strategies for supporting basic psychological needs and nature connection in adventure contexts, fuller discussions are available in the literature (e.g., Houge Mackenzie et al., 2021; Ryan & Deci, 2017).

A third pillar of improving adventure practice, and research for that matter, is better understanding how operators, guides and clients can enhance the well-being of adventure destinations, that is the people and places adventure tourism depends on. There is an abundance of sustainable tourism literature outlining clear strategies for replacing consumptive adventure tourism practices with approaches that better support the well-being of natural places and associated communities. For instance, Beames et al. (2022) analysed a range of guidance operators can use to enhance the well-being of the people and places involved in adventure tourism. However, a mindset shift amongst adventure operators is required to achieve this. Rather than relying on client satisfaction and profit as the sole barometers of success, refocusing on how adventure tourism can enhance the well-being of host communities and natural environments that sustain adventure tourism operations is needed. This means genuinely engaging with local communities, rather than seeing them as merely part of an "adventure product"; actively recruiting and training local guides, particularly women; and understanding the unique environments, cultural contexts and community aspirations of adventure destinations. This approach will ultimately enhance the success of adventure operations by supporting the development of more meaningful, high-value adventure experiences; enabling more sustainable, knowledgeable guiding pools; and supporting long-term community well-being. In summary, de-emphasising risk, diversifying guiding pools and participants, and redesigning adventure tourism experiences in terms of (1) personally referenced or group challenges, (2) opportunities to give back to communities and experience beneficence, and (3) building relationships with others and natural environments, are key strategies for improving the well-being outcomes of adventure and affording those benefits inclusively.

Conclusion

This chapter refutes traditional approaches that frame risk and thrill as the central elements of adventure experiences and notions that adventure is inherently beneficial for participants. Rather, it identifies a wide range of potential benefits adventure tourism can offer participants, guides, and communities and suggests key mechanisms that may explain how these benefits are achieved. Framing adventure activities in terms of underlying psychological mechanisms (i.e., nature connectedness, basic psychological need satisfaction) improves our understanding of *how* and *when* adventure tourism can foster hedonic and eudaimonic PWB. This, in turn, has important implications for the value individuals and societies place on adventure, and how it is practised. Reconceptualising adventure tourism via a wider well-being lens and purposefully designing adventure tourism experiences accordingly will help realise the untapped potential of adventure. Hopefully, this shift will enable more inclusive, meaningful adventure tourism opportunities for participants, guides and communities.

Review questions

1 What are the key differences in traditional versus contemporary conceptualisations of adventure?
2 What are the psychological well-being benefits of adventure tourism for participants, guides, and communities?
3 What are the key psychological mechanisms proposed to underpin adventure benefits?
4 How can adventure tourism become more inclusive?

Note

1 Human well-being is multidimensional and holistic (e.g., Huppert, 2014). Please note that the primary focus of this chapter is on psychological well-being, which forms one key part of overall well-being.

Further reading

Reid, P., & Brymer, E. (Eds.). (2022). *Adventure psychology: Going knowingly into the unknown.* Taylor & Francis.
Ryan, R. M., & Deci, E. L. (2017). *Self-determination theory: Basic psychological needs in motivation, development, and wellness.* The Guilford Press.

References

Barlow, M., Woodman, T., & Hardy, L. (2013). Great expectations: Different high-risk activities satisfy different motives. *Journal of Personality and Social Psychology, 105*(3), 458–475.

Beal, B., & Smith, M. (2010). Maverick's: Big-wave surfing and the dynamic of "nothing" and "something". *Sport in Society: Cultures, Commerce, Media, Politics, 13*, 1102–1116.

Beames, S., Houge Mackenzie, S., & Raymond, E. (2022). How can we adventure sustainably? A systematized review of sustainability guidance for adventure tourism operators. *Journal of Tourism and Hospitality Management, 50*, 223–231.

Beedie, P. (2003). Mountain guiding and adventure tourism: Reflections on the choreography of the experience. *Leisure Studies, 22*(2), 147–167.

Berman, M. G., Jonides, J., & Kaplan, S. (2008). The cognitive benefits of interacting with nature. *Psychological Science, 19*(12), 1207–1212.

Brymer, E., & Houge Mackenzie, S. (2015). The impact of extreme sports on host communities' psychological growth and development. In Y. Reisinger (Ed.), *Transformational tourism: Host perspectives* (pp. 129–140). CAB International.

Brymer, E., & Oades, L. (2009). Extreme sports: A positive transformation in courage and humility. *Journal of Humanistic Psychology, 49*(1), 114–126.

Brymer, E., & Schweitzer, R. (2013). Extreme sports are good for your health: A phenomenological understanding of fear and anxiety in extreme sport. *Journal of Health Psychology, 18*(4), 477–487.

Buckley, R. C., & Brough, P. (2017). Nature, eco, and adventure therapies for mental health and chronic disease. *Frontiers in Public Health, 5*, 220.

Capaldi, C. A., & Dopko, R. L. (2015). A review of the benefits of connecting with nature and its application as a wellbeing intervention. *International Journal of Wellbeing, 5*(4), 1–16.

Castanier, C., Le Scanff, C., & Woodman, T. (2010). Beyond sensation seeking: Affect regulation as a framework for predicting risk-taking behaviors in high-risk sport. *Journal of Sport & Exercise Psychology, 32*, 731–738.

Eigenschenk, B., Thomann, A., McClure, M., Davies, L., Gregory, M., Dettweiler, U., & Inglés, E. (2019). Benefits of outdoor sports for society: A systematic literature review. *International Journal of Environmental Research and Public Health, 16*(6), 937.

Ewert, A., & Hollenhorst, S. (1989). Testing the adventure model: Empirical support for a model of risk recreation participation. *Journal of Leisure Research, 21*, 124–139.

Ewert, A., & Yoshino, A. (2011). The influence of short-term adventure-based experiences on levels of resilience. *Journal of Adventure Education & Outdoor Learning, 11*, 35–50.

Ewert, A., Zwart, R., & Davidson, C. (2020). Underlying motives for selected adventure recreation activities: The case for eudaimonics and hedonics. *Behavioral Sciences, 10*(12), 185.

Filep, S., & Laing, J. (2019). Trends and directions in tourism and positive psychology. *Journal of Travel Research, 58*(3), 343–354.

Frühauf, A., Hardy, W. A., Pfoestl, D., Hoellen, F. G., & Kopp, M. (2017). A qualitative approach on motives and aspects of risks in freeriding. *Frontiers in Psychology, 8*. https://doi.org/10.3389/fpsyg.2017.01998

Garcês, S., Pocinho, M., Jesus, S. N., & Rieber, M. S. (2018). Positive psychology and tourism: A systematic literature review. *Tourism & Management Studies, 14*(3), 41–51.

Hanna, P., & Kimbu, A. (2019). Active engagement with nature: Outdoor adventure tourism, sustainability and wellbeing. *Journal of Sustainable Tourism, 27*(9), 1355–1373.

Houge Mackenzie, S., & Goodnow, J. (2020). Adventure in the age of COVID-19: Embracing microadventures and locavism in a post-pandemic world. *Leisure Sciences, 43*(1–2), 62–69. https://doi.org/10.1080/ 01490400.2020.1773984

Houge Mackenzie, S., Hodge, K., & Filep, S. (2021). How does adventure sport tourism enhance well-being? A conceptual model. *Tourism Recreation Research, 48*(1), 1–14.

Houge Mackenzie, S., & Kerr, J. H. (2013). Stress and emotions at work: Adventure tourism guiding experiences in South America. *Tourism Management, 36*, 3–14.

Houge Mackenzie, S., & Raymond, E. (2020). A conceptual model of adventure tour guide well-being. *Annals of Tourism Research, 84*, 102977.

Houge Mackenzie, S., & Raymond, E. (2021). Women's adventure guiding experiences: Challenges and opportunities. In M. Baker (Ed.) *Leisure activities in the outdoors* (pp. 90–102). CABI.

Huppert, F. A. (2014). The state of well-being science: Concepts, measures, interventions, and policies. In F. A. Huppert & C. L. Cooper (Eds.) *Interventions and policies to enhance well-being*. Wiley-Blackwell, Oxford, p. 1–49.

Huta, V., & Waterman, A. S. (2014). Eudaimonia and its distinction from hedonia: Developing a classification and terminology for understanding conceptual and operational definitions. *Journal of Happiness Studies, 15*, 1425–1456.

Kar, D., & Tarafder, S. (2020). Sensation Seeking and Aggression in Adventure Sportspersons: A Comparative Study. *Journal of Psychosocial Research, 15*(1), 137–149.

Kerr, J. H., & Houge Mackenzie, S. (2012). Multiple motives for participating in adventure sports. *Psychology of Sport and Exercise, 13*, 649–657.

Kerr, J. H., & Houge Mackenzie, S. (2020). 'I don't want to die. That's not why I do it at all': Multifaceted motivation, psychological health, and personal development in BASE jumping. *Annals of Leisure Research, 23*(2), 223–242.

Llewellyn, D. J., & Sanchez, X. (2008). Individual differences and risk taking in rock climbing. *Psychology of Sport and Exercise, 9*(4), 413–426.

Lynch, P., & Dibben, M. (2016). Exploring motivations for adventure recreation events: A New Zealand study. *Annals of Leisure Research, 19*(1), 80–97.

MacGregor, A., Woodman, T., & Hardy, L. (2014). Risk is good for you: An investigation of the processes and outcomes associated with high-risk sport. *Journal of Exercise, Movement, and Sport, 46*(1), 175.

MacIntyre, T. E., Walking, A. M., Beckmann, J., Calogiuri, G., Gritzka, S., Oliver, G., Donnelly, A. A., & Warrington, G. (2019). An exploratory study of extreme sport athletes' nature interactions: From well-being to pro-environmental behaviour. *Frontiers in Psychology, 10*, 1233. https://doi.org/10.3389/fpsyg.2019.01233

Martela, F., & Ryan, R. M. (2020). Distinguishing between basic psychological needs and basic wellness enhancers: The case of beneficence as a candidate psychological need. *Motivation and Emotion, 44*(1), 116–133.

Mathisen, L. (2019). Storytelling: A way for winter adventure guides to manage emotional labour. *Scandinavian Journal of Hospitality and Tourism, 19*(1), 66–81.

Monasterio, E., & Cloninger, C. R. (2019). Self-transcendence in mountaineering and BASE Jumping. *Frontiers in Psychology, 9*, 2686.

Mutz, M., & Müller, J. (2016). Mental health benefits of outdoor adventures. *Journal of Adolescence, 49*, 105–114.

Mykletun, R. J., & Mazza, L. (2016). Psychosocial benefits from participating in an adventure expedition race. *Sport, Business and Management, 6*(5), 542–564.

Parsons, H., Houge Mackenzie, S., & Filep, S. (2019). Facilitating self-development: How tour guides broker spiritual tourist experiences. *Tourism Recreation Research, 44*(2), 141–152. https://doi.org/10.1080/02508281.2019.15821

Pretty, J., Peacock, J., Sellens, M., & Griffin, M. (2005). The mental and physical health outcomes of green exercise. *International Journal of Environmental Health Research, 15*, 319–337.

Ryan, R. M., & Deci, E. L. (2017). *Self-determination theory: Basic psychological needs in motivation, development, and wellness.* Guilford Publications.

Ryan, R. M., & Martela, F. (2016). Eudaimonia as a way of living: Connecting Aristotle with self-determination theory. In J. Vittersø (Ed.), *Handbook of eudaimonic well-being* (pp. 109–122). Springer.

Ryan, R. M., Weinstein, N., Bernstein, J., Brown, K. W., Mistretta, L., & Gagne, M. (2010). Vitalising effects of being outdoors and in nature. *Journal of Environmental Psychology, 30*, 159–168.

Ryff, C. D. (2013). Psychological well-being revisited: Advances in the science and practice of eudaimonia. *Psychotherapy & Psychosomatics, 83*, 10–28.

Seligman, M. E., & Csikszentmihalyi, M. (2000). Positive psychology: An introduction. *American Psychologist, 55*(1), 5–14.

Sharma, P. (2009) Sustainable mountain tourism development in Nepal. In E. Kruk (Ed). *Proceedings of the Sustainable Mountain Development Workshop in Nepal* (pp. 40–47).

Sharpe, E. K. (2005). "Going above and beyond": The emotional labor of adventure guides. *Journal of Leisure Research, 37*, 29–50.

Steiner, C., & Reisinger, Y. (2006) Ringing in the fourfold: A philosophical framework for thinking about wellness tourism. *Tourism Recreation Research, 31*(1), 5–14.

Thompson, C. J., Boddy, K., Stein, K., Whear, R., Barton, J., & Depledge, M. (2011). Does participating in physical activity in outdoor natural environments have a greater effect on physical and mental wellbeing than physical activity indoors? A systematic review. *Environmental Science and Technology, 45*, 1761–1772.

Torland, M. (2011). Emotional labour and job satisfaction of adventure tour leaders: Does gender matter? *Annals of Leisure Research, 14*(4), 369–389.

Tsaur, S. H., Lin, W. R., & Cheng, T. M. (2015). Toward a structural model of challenge experience in adventure recreation. *Journal of Leisure Research, 47*(3), 322–336.

Valdivielso-Martínez, E., & Houge Mackenzie, S. (2022). Climate change and adventure guiding: The role of nature connection in guide well-being. *Frontiers in Public Health, 10*, 946093. https://doi.org/10.3389/fpubh.2022.946093

Varley, P., & Semple, T. (2015). Nordic slow adventure: Explorations in time and nature. *Scandinavian Journal of Hospitality and Tourism, 15*(1–2), 73–90.

Waterman, A. S., Schwartz, S. J., & Conti, R. (2008). The implications of two conceptions of happiness for the understanding of intrinsic motivation. *Journal of Happiness Studies, 9*(1), 41–79.

Weiler, B., & Black, R. (2015). The changing face of the tour guide: One-way communicator to choreographer to co-creator of the tourist experience. *Tourism Recreation Research, 40*(3), 364–378.

Willig, C. (2008). A phenomenological investigation of the experience of taking part in extreme sports. *Journal of Health Psychology, 13*(5), 690–702.

Wong, K., & Wang, C. (2009). Emotional labour of the tour leaders: An exploratory study. *Tourism Management, 30*(2), 249–259.

Zapf, D., & Holz, M. (2006). On the positive and negative effects of emotion work in organisations. *European Journal of Work and Organisational Psychology, 15*(1), 1–28.

Zelenski, J. M., & Nisbet, E. K. (2014). Happiness and feeling connected: The distinct role of nature relatedness. *Environment and Behavior, 46*(1), 3–23.

Zuckerman, M. (Ed.). (2007). *Sensation seeking and risky behavior*. American Psychological Association.

Part III

The supply of adventure tourism

The third section of the book moves to consider adventure tourism from a supply per-spective. Whilst much of the industry growth over preceding decades has been relatively organic, often pioneered by outdoor lifestyle entrepreneurs, the contemporary adventure tourism industry is now noticeably more sophisticated and integrated into the broader tourism sector. With this has come greater complexity in structure, as discussed by Manuel Sand and Sven Gross in the next chapter (Chapter 13). Perhaps as a result of its private sector genesis, the governance of adventure tourism is therefore as likely to be led by businesses as it is by destinations. Indeed, certain industry associations have supported the process of formalisation, most notably the Adventure Travel Trade Association

DOI: 10.4324/9781003393153-16

(ATTA), which is discussed in depth. The ATTA has developed a long way since small grassroots meetings such as that attended by one of the editors in Anchorage, Alaska, in 2000, now hosting the annual Adventure Travel World Summit attended by many delegates in a selected global location (the 2023 meeting was in Hokkaido, Japan). Part of the success of the ATTA has been in an expanded remit (perhaps in line with the diversification of adventure tourism more broadly) into sustainable tourism and products that may have previously come under ecotourism.

Despite the origins of adventure tourism being small scale, today we witness destinations worldwide incorporating these diverse activities into their product offer. In Chapter 14, Jelena Farkić and Steve Taylor examine some of the challenges of achieving collaborative product development in adventure tourism, particularly in aligning values. In common with some of the trends discussed by Sand and Gross, they note the development of slow tourism and forest bathing as examples of more sustainable place-based adventure tourism.

Indeed, earlier chapters have noted the broadening appeal of adventure tourism and the growth of such forms. Likewise, the centrality of risk as a motivation has been questioned, despite its importance to prior adventure tourism scholarship. Nevertheless, from a supply perspective risk management remains a fundamental part of sustainable practice, and indeed legal obligation in many jurisdictions as Arild Røkenes and Sigmund Andersen point out in Chapter 15. They note the conflicting demands that guides may often come under during the activity but note the importance of adequate information and a healthy reporting culture across all domains. Similarly, in his examination of risk management by guides and businesses, Jon Heshka (Chapter 16) echoes the importance of personal experience in risk assessment and the ability to maintain well-informed dynamic evaluations. Both these chapters suggest clear colour coding of activities and risks to provide visual clues to guides and clients alike.

Such a recognition of the importance of personal experience in guiding relates to the growing professionalisation of the sector discussed in the subsequent chapters. Jason King, Chris Webber and Ashley Hardwell underscore the importance of leadership qualities in the adventure tourism professional through the framework of attention to *Safety*, *Enjoyment*, *Learning*, *Legacy* and *Self*. Likewise, Matt Groves, Tamara Griffiths and Peter White assess the career priorities of adventure guides (and thus training programs) in *Becoming*, *Being* and *Inspiring* professionals. Whilst qualifications are important (in this case a study of paddle-sport qualifications), equally important are nurtured experiences and a community of practice.

The final chapters in this section turn to wider trends that have impacted the development of adventure tourism, in particular its marketing. May-Kristin Vespestad (Chapter 19) describes processes of co-creation in adventure tourism through examples of a community-developed bike park and a theme park (illustrating how wide the definition of adventure has become). Immersive experiences for both hosts and visitors can be employed to create both ownership and sustained commitment, as well as experiential marketing, blurring the line between product and promotion. As we saw with consumers in Chapter 11, technology has a key role to play in this, and Cam O'Bierne (Chapter 20) outlines some of the key strategies that adventure businesses need to be aware of moving forward.

13 The structure of the adventure tourism industry

Manuel Sand and Sven Gross

Chapter learning outcomes

1 Learn about adventure tour operators, travel agencies, accommodation providers, and other types of adventure organisations.
2 Acquire knowledge about challenges and solutions of adventure tourism governance.
3 Introduce two case studies on the Adventure Travel Trade Association and Hauser Excursions, an international tour operator.
4 Learn about current issues and their effect on the adventure tourism industry.

Introduction

'Adventure tourism is a complex industry with a wide variety of elements and suppliers' (Swarbrooke et al., 2003, p. 135). This chapter investigates different adventure tourism actors and how they achieve the sustainable management of tourism within and between destinations. In addition, it explores the governance of adventure tourism, which is little researched. While the fundamental aspects of tourism also apply to the field of adventure tourism, there are still a few adventure-specific elements, mostly to do with risk and the outdoors, that make its governance distinctive. Therefore, adventure governance needs to cater specifically to the demands of adventure travellers. According to dos Anjos and Kennel (2019, p. 257):

> the governance of tourism involves an increasingly networked set of inter-relationships between actors in the public, private and third sectors, and should bring together tourists, host communities, businesses and the traditional institutions of the state with an interest in tourism.

Figure 13.1 gives an overview of the three sectors of adventure tourism governance according to dos Anjos and Kennel (2019). The public sector looks at the roles of government and destinations in adventure tourism. Some destinations are better suited to adventure tourism than others and this will influence governmental involvement. Furthermore, networking and cooperation with corresponding tourism organisations are important here. This chapter focuses more on commercial operators, categorised as specialist tour

DOI: 10.4324/9781003393153-17

Figure 13.1 Adventure tourism governance. Adapted from Anjos and Kennel (2019, p. 257).

operators, incoming agencies, adventure accommodation and destination marketing organisations. The services offered by these providers are divided into water-, land-, and air-based activities (Gross, 2017a, 2017b). The third sector comprises non-profit organisations and trade and research associations involved in adventure tourism. A significant organisation is the Adventure Travel Trade Association (ATTA; see Case study 13.2), which is an interest group and political mouthpiece for the adventure travel industry. This association has introduced the Adventure Tourism Development Index (ATDI), which examines countries in terms of their potential for sustainable adventure tourism (ATTA & International Institute of Tourism Studies, 2018).

Given that much adventure tourism was pioneered by the private sector, the following section, first, looks at providers and different adventure tourism settings. Important aspects of the governance in adventure tourism are explained, starting with adventure tour operators.

Specialist tour operators

A tour operator is a provider that bundles together at least two tourism products, often accommodation and transport to the destination, and offers them at a package price. According to Holloway and Humphreys (2022), operators connect suppliers and customers. Tour operators are tourism enterprises or parts of enterprises that predominantly combine services of third parties to satisfy the temporary need for a change of place and other related needs into a new, independent service offer and charge for these services (Dörnberg et al., 2018).

Above all, the range of holiday types or degrees of specialisation provides a clear basis for differentiating tour operators in adventure tourism. The 'generalists' offer various types of travel, for example recreational, cultural or adventure, combined with forms of transport, such as air, rail, bus and car travel to numerous destinations with different types of accommodation. 'Specialists' or 'assortment providers' specialise in some of these aspects (Freyer, 2015, p. 265). Generalist major tour operators include 'active sports,' 'adventure and fun offers' or similar in their product range. The German market leader, TUI AG, for example, offers diving, canyoning, kite surfing, surfing or winter sports for families (TUI AG, 2022). Yet, it is mainly small-scale specialist tour operators that offer different types of soft or hard adventure trips (Beames et al., 2022; Swarbrooke et al., 2003). Established providers, in terms of the number of customers per year, are G Adventures, National Geographic

---

Case study 13.1   Hauser Excursions

Hauser Excursions is a medium-sized company in the German tourism industry, specialising in the organisation and implementation of trekking trips and other forms of adventure tourism. Hauser is one of the leading companies in this field in German-speaking countries. The company offers a wide range of products and services. Under the motto "Experience the world and yourself," Hauser has been appealing to customers who seek international adventure trips since 1973. Company founder, Guenter Hauser, was an adventurer and enthusiastic mountain climber and wanted to make these outdoor activities available to a wider market. Related to this, the company is considered a pioneer of environmentally and socially responsible holidays. Sustainability is a high priority and a variety of projects, such as Climate Trek in Nepal, benefit local people in the destination countries.

Hauser's portfolio includes trekking, hiking, skiing, mountain biking and climbing tours, which are offered worldwide. The alpine programme (climbing, via ferrata, high alpine, skiing) is carried out by the subsidiary Bergspechte. The company makes a distinction between different levels of difficulty. Customers can find information about the tours, the guide, the region, and the culture online via www.hauser-exkursionen.de. Hauser stopped printing catalogues and developed a smaller scale magazine with information and website links to the trips. Detailed content on length, altitude, and the trip's level of difficulty enables customers to plan the tour for themselves. The maximum number of participants is limited to 15 travellers to ensure a better group experience. Thanks to many years of experience, the company has been able to build up a team of reliable partners in various regions of the world who work to high safety standards. The company mainly works with local people who are trained by Hauser Excursions.

---

and World Expeditions, for instance. Increasingly, however, there are other smaller providers which offer specialised products such as canoeing, mountain biking, climbing tours or multi-activity programmes.

In the following Case study (Case study 13.1), Hauser Excursions, a tour operator that can draw on many years of experience and offers a wide range of products, is presented.

Incoming agencies and local providers

Incoming travel agencies are located at the destination and arrange travel services in their region (e.g., package tours, individual services, or tour guides) for tourists and non-local tour operators. In addition, they may also provide on-site services for travellers using international tour operators (Dörnberg et al., 2018). Formally, this is also a package tour, usually excluding travel, but it has been put together in the destination and is therefore – in contrast to a package tour booked in the source country – subject to the legal framework of the destination country (Scherhag, 2021). 'Organisations specializing in handling incoming foreign holidaymakers have a rather different role to that of outbound operators' (Holloway & Humphreys, 2022, p. 520). Some operators are merely ground handling agents who organise accommodation or transfers for foreign tour operators, while others provide a comprehensive range of services including for example adventure tours, activities or restaurant reservations.

In destination countries such as Spain and Greece, the incoming market is often equally as important as the outgoing one, as they are attractive destinations, yet they also have a strong outgoing market (Holloway & Humphreys, 2022). According to the Federal Association of German Incoming Companies (DIU), German incoming companies were responsible for approximately 6 million guests, 14 million overnight stays, and 2.7 billion euros in direct tourism turnover in 2018. In terms of guests, German incoming companies and destination management organisations have a market share of about 16%, and a turnover of about 20% (Bundesverband der Deutschen Incoming Unternehmen - DIVU, 2022).

In the English-speaking world, incoming providers for adventure trips are more widespread, and the destinations within these countries offer a wider range of products (Huddart & Stott, 2020; Swarbrooke et al., 2003). Although adventure incoming agencies have become more important in many destinations around the world, such as South America, Asia, and Africa, there are still countries where adventure is not the focus, especially those in Europe like Germany, Italy, and Croatia (ATTA, 2022b; Gross, Sand & Berger, 2023).

In addition to incoming agencies with a wider range of products, specialist providers offer only single outdoor adventure tourism activities. These can be for example kayak and bicycle rentals, high ropes courses, or paragliding or sailing schools. These smaller operators can be considered as the backbone of the adventure industry, as they provide the initial core product. They can also act as incoming agencies when they combine their own services and/or services from other providers into package programmes. Some operators concentrate on one key product, while others have a wider portfolio of activities. These can be booked at the location, through other agencies, or via the company's website. Most of the time, the company provides specialist equipment on a rental basis, instructors, and guided tours.

Adventure accommodation

The commercial accommodation industry can be divided into serviced and self-catered sectors (see Figure 13.2). The service industry includes hotels, bed-and-breakfasts, guesthouses, and inns. In contrast, the self-catering industry includes all forms of accommodation not classified as classic hotels (Holloway & Humphreys, 2022). In terms of adventure tourism, youth hostels, campsites, caravans, mountain and ski huts, holiday camps, and club hotels are of particular interest.

Accommodation services can be part of an adventure tourism package deal or can be the adventure tourism offer itself, for example mountain huts. In this context, the accommodation services are sold both directly to the adventure consumers, for example via a hotel's own website, and via other companies in the tourism industry, such as tour operators and travel agents. In recent years, reservation and booking options have increasingly been made available via internet platforms, such as booking.com, hotel.com, and trivago.com, sometimes in combination with supplementary travel. Recently, the so-called sharing economy (see also Chapter 20), which is defined as the sharing of goods and services, has become more important (Stors & Kagermeier, 2017). This includes the possibility of sharing overnight accommodation, such as rooms, flats, or houses, with

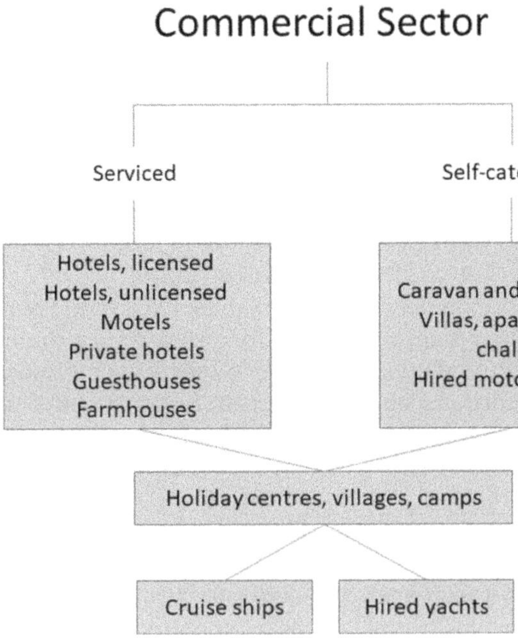

Commercial Sector

Serviced

Self-catering

Hotels, licensed
Hotels, unlicensed
Motels
Private hotels
Guesthouses
Farmhouses

Caravan and campsites
Villas, apartments, chalets
Hired motor-homes

Holiday centres, villages, camps

Cruise ships

Hired yachts

Figure 13.2 The adventure tourism and hospitality industry. (Adapted from Holloway & Humphreys 2022, p. 307).

private owners, otherwise known as "couch surfing." Airbnb (www.airbnb.com) plays a significant role in the sharing-economy market. It offers adventure tourism experiences for different skill levels, based around activity, culture, animals, boating, and culinary themes. These offerings are developed in collaboration with experts from the ATTA. Adventures on Airbnb (see also Chapter 20) are 'a collection of multi-day experiences hosted by local experts. Small groups travel to off-the-beaten-path locations to immerse themselves in unique settings, cultures, and communities' (Airbnb, 2022).

Camping and caravanning

Camping is regarded, on one hand, as an activity and, on the other hand, as a type of self-catered accommodation (Brooker & Joppe, 2013). Camping usually takes place on sites equipped to cater for leisure users and with certain levels of sanitation and supply infrastructure, whereby tents, trailers, or self-propelled caravans serve as accommodation. Leisure camping does not include permanent camping, where people live in a camper van long term (Leser, 2014). Camping tourism is especially popular in North America, Australia, New Zealand, and Europe, due to a long tradition of this activity and well-established infrastructure (Rogerson & Rogerson, 2020).

Caravanning can be considered as camping tourism, according to various definitions in the literature (Gross, 2017b; Widmann, 2006). Regarding the type of accommodation, camping can be divided into tent, motorhome/campervan and caravan, whereby the term *caravanning* is used as an umbrella term for motorhomes (also called campervans) and caravans. A motorhome is a self-propelled motor vehicle and must have minimum equipment in the living area. In addition to permanently installed furniture, the living space must occupy the main part of the vehicle and give the impression of a room suitable and intended for living purposes (Widmann, 2006).

For a long time, camping was considered a cheap way of travelling and the assumption was that it was primarily undertaken for financial reasons. It was not until the mid-1980s that different non-financial motives started to drive the growth in camping. This can be seen in the shift towards people increasingly staying in, or buying, motorhomes and caravans for camping. This involves high investments and, in terms of costs, is quite comparable to a holiday in a hotel (Gross, 2017b; Widmann, 2006).

The worldwide COVID-19 pandemic has led to an increase in the popularity of camping tourism, especially caravanning, as shown, for example, by the high demand for recreational vehicles. The industry is experiencing an influx of newcomers who are interested in a holiday with a motorhome or caravan for the first time, often within their home country and neighbouring ones. In the first half of 2021, approximately 63,000 motorhomes and caravans were newly registered in Germany (a growth of approximately 15% compared to the first half of the previous year). New registrations of motorhomes rose by more than 22% to approximately 48,500 vehicles, reaching an all-time high (Caravaning Industrie Verband e.V., 2021).

Caravans are rented both for private trips and for business purposes, such as trade fairs or exhibitions. In addition to the classic holiday motives, a camping holiday offers familiar surroundings, on one hand, but plenty of space for freedom and spontaneity, on the other. While caravan users want to enjoy their holidays in peace and quiet and staying at a campsite is essential, for motorhome travellers, the focus is often on travelling and flexibility (Figure 13.2), and camping or motorhome sites are only a necessary stopover (Gross, 2017b; Widmann, 2006).

One camping trend that is emerging is the so-called #vanlife, in which more people are converting buses and vans into their own motorhomes. Meanwhile, the rental of high-quality, converted buses and vans has developed into a successful tourism business model for companies like Roadsurfer or Camper Boys (Figure 13.3).

Glamping

For more than a decade, glamping, otherwise known as boutique, luxury, comfort, and posh camping, has been a sub-form of camping. It has become a different style of camping for new generations of campers and a symbol of the reshaping of camping tourism. The term *glamping* is a fusion of the words *glamour* and *camping* and has become synonymous with camping accommodation that is usually owned by campsite operators rather than campers. Glamping can take place in an extraordinary, natural environment or on a "normal" campsite (Milohnić et al., 2019).

The different forms of accommodation in glamping are distinguished as follows (Gross et al., 2022; Petrusa & Vlahv, 2019):

1　Glamping domes and tents – for example, bell tents, yurts, lodge and safari tents, tipis
2　Cottages and cabins – for example, wooden huts, cabins, mobile homes, tiny houses
3　Caravans – for example, caravans for hire, shepherds' caravans, circus caravans
4　Other unusual glamping accommodation – for example, tree house or tent, igloo hut, outdoor bed, pod, wagon

Figure 13.3 Adventure tourism by campervan. Photograph by Adele Doran.

A special form of glamping is 'deep-nature glamping' (Groß & Sand 2022, p. 78). Essentially, this does not take place on a camping or glamping site but, rather, is relatively secluded in a near-natural environment (Gross et al., 2023).

Hut-to-hut hiking

So-called hut-to-hut hiking can be found in several countries and at different spatial levels. A hut, also known as a bothy, is the common term for overnight accommodation related to hiking. Hut facilities are usually simple and often include shared cooking facilities, sometimes cold running water, compost toilets, and bunk beds in shared dormitories. Huts are located in many countries including Australia, Austria, Canada, Finland, France, Germany, Italy, Ireland, Iceland, Japan, Kenya, New Zealand, Norway, Peru, Scotland, South Africa, Sweden, Switzerland, and the US (Gross & Werner, 2018).

Trekking

Trekking is a sub-form of hiking and describes walking from place to place, without returning to the starting point, with an overnight stay outdoors (Hosters, 2019). This often involves long hikes in remote areas offered by specialist operators and agencies. Trekking tourism started in Nepal and spread from there around the globe.

Recently, in line with the microadventure movement, trekking has become more popular close to home. While camping in nature is legal in some countries, such as in Scandinavian countries, other countries have strict regulations, and wild camping in the countryside is prohibited. In Germany, trekking tourism has developed recently in areas such as the Black Forest, Odenwald, Steigerwald, Eiffel, and Sauerland, where there are designated hiking trails and overnight campsites in the forest. This therefore provides a legal framework for spending the night in nature. Most of the sites have an outhouse or compost toilet and a fireplace. Sometimes, they also have running water or a tent platform (Sand, 2019).

Other providers in the commercial sector of adventure tourism

In addition to the previously mentioned operators, other organisations play a role in adventure tourism.

Adventure education providers

Adventure education providers offer training and activities to a range of different groups including schools, youth clubs, companies, and adult groups. These organisations use adventure activities to take their clients out of their comfort zones and confront them with physical, psychological, or social challenges in order to enable them to learn and develop their personalities (Heckmair & Michl, 2008). Adventure education trips can have a positive effect on the host destinations by, for example, participants visiting them again after their respective stay with their family or friends or alone. These providers embed outdoor activities, such as canoeing, climbing, hiking, high ropes, raft building, abseiling, and rafting, into their programmes. Some companies also lend equipment, for example, canoes and bicycles, to their clients and open up their activities for tourists to use too.

Providers of survival training and trekking courses

The offer of, and demand for, survival training and trekking courses have increased in recent years. More and more people want to learn how to survive in nature, set up a bivouac, make a fire (without technical aids), or cook. Others want to learn how to lead a group in an outdoor setting. The training and the shared stays in nature usually take place in groups, in areas and destinations suitable for this type of activity. In some cases, booking at trekking sites or local permits are required.

Experience marketers

Experiences are an important component of adventure tourism, in addition to catering, accommodations, and transportation services. The digital distribution of experience services has been internationally established for some time (see also Chapter 20). Online reservation platforms, which are activity booking engines for experiences, leisure activities, tours, and guides, have existed since at least 2007. Rezgo from Canada and Rezdy from Australia are now the global market leaders in this tourism sales segment but have so far barely been able to gain a foothold in Europe. European countries have seen the rise of local providers for digital experience sales, with the Munich-based company Regiondo in particular, establishing itself as the market leader (Wagner et al., 2019).

Activity booking engines mostly remarket different products from third-party providers. Through joint marketing, the small individual providers achieve a greater reach but pay a commission to the agency for this. The providers take on the role of suppliers and the experience distribution platforms and experience brokers are the large distribution agencies that market them in a similar way to online tour operators. The term *online activity agencies* is used to describe these (Wagner et al., 2019).

Events and adventure parks

The staging of experiences plays a special role not only for experienced marketers but also for events. However, with events, every participant contributes to the overall result (Horch et al., 2004). Events can be highly beneficial in terms of value creation and advertising impact, especially for structurally weak regions (Stroebel et al., 2018). In this way, events can positively influence the level of awareness and the image of destinations. At the same time, people get to know the destination and its special features.

In the outdoor sports sector, in addition to corresponding competitions such as mountain bike races, trail running or climbing competitions, an increasing number of "testivals" are also taking place. Participants can try and test outdoor activities and equipment at these events, mostly for free. In addition, obstacle runs, and mud runs are rising in popularity.

Obstacle and mud runs can be found in adventure parks. One example is the Dorset Adventure Park in southern England (www.dorsetadventurepark.com). This park has an outdoor water park and a mud trail, which is an obstacle course suitable for both adults and children. Other examples offer different activities, such as the DX Adventurepark (www.dxadventurepark.nl) in the Netherlands (adventure water park), Aventura Parks (www.aventuraparks.com) in Dubai (mainly climbing and zipline), and Kolibki Adventure Park (www.adventurepark.pl) in Gdynia/Poland (including off-road driving track, quad biking, climbing, paintball, forest walks) or Area 47 in Austria (www.area47.at). Outdoor film festivals and their videos, which are presented in a positive and emotional way, can also attract adventure tourists to destinations (Buckley, 2010).

Sporting goods and equipment manufacturers

Sporting goods manufacturers primarily produce clothing, shoes and accessories for sporting activities. They also make specific products that are necessary for the practice of certain sports, for example, mountain biking or stand-up paddling (SUP) manufacturers (Swarbrooke et al., 2003). In recent decades, the outdoor sector has experienced growing demand in the sporting goods industry (Fuchs & Hovemann, 2022; Heckmeier, 2011). Special outdoor clothing, matching shoes, and accessories are on offer, alongside equipment such as tents, backpacks, sleeping bags, and camping stoves. One development in the distribution of sporting goods is that the products are no longer sold only through intermediaries, and they are marketed directly in their own shops (e.g., Globetrotter, Jack Wolfskin), and online shops (Riedmueller, 2011).

Sports equipment manufacturers also occupy a prominent position in the supply side of the market because adventure activities could not be practised without, for example, a mountain bike or surfboard. Depending on the activity, the complexity and costs of the equipment vary. They are offered by the suppliers themselves and by corresponding specialist shops (Swarbrooke et al., 2003). In addition, equipment, such as wakeboards, canoes, or mountain bikes can be rented or provided by the operator as part of the offer. Manufacturers are also increasingly organising test events where the products can be tried out. There are even guided experiences lasting several days, during which products can be tested (Heckmeier, 2011). Through cooperation between suppliers, retailers, and destinations, emotional events are created, where the products can be experienced and the brand is enjoyed differently. Among other things, this should encourage consumers to develop a special bond with the products (Riedmueller, 2011).

Adventure destinations

The term *destination* has established itself in academic tourism literature to designate the target area of travel. In everyday language, terms such as *travel* or *holiday destination, tourism* or *tourist destination, holiday country, area, city,* or *destination area* are used. However, the following definition is widely cited:

> Basically, a tourism destination is a geographic area that attracts visitors. ... A geographic area that has an administrative boundary or boundaries. ... States, provinces, territories, regions, counties and cities within individual countries can also be destinations.
>
> (Gowreesunkar et al., 2018, p. 19)

What all definitions have in common is that destinations represent clearly delimitable spaces and are considered from a tourism marketing perspective. Swarbrooke et al. (2003) state that there are several ways to categorise adventure destinations, such as by their geographical features, for example, mountain, coastal, or urban, or the types of activity they offer. Destinations can be perceived differently by individuals from a continental or smaller geographical area perspective.

The majority of providers are located at the destination, which explains why they are referred to as crystallisation points of tourism (Kaspar, 1996). Active service providers and intermediaries represent the destination management organisations. Furthermore, there are individual providers, such as accommodation organisations, mountain or sports schools, incoming agencies, equipment rental companies, tour operators, and transport companies. These providers represent the derived offer, which includes tourism and leisure infrastructure, and the special tourism offers of a destination. In addition, there is the original offer, which comprises the natural resources, for example, landscape and topography; the socio-cultural offer, for example, tradition and customs; and the general infrastructure, for example, communication and transport (Freyer, 2015).

One way of measuring the suitability of destinations for adventure tourism is the Adventure Tourism Development Index. This index examines countries in terms of their potential for sustainable adventure tourism development. In 2018, almost 200 countries – 28 industrialised countries and 163 developing countries – were ranked by experts over 5 years. Existing statistical data from other studies (including the World Health Organization and the United Nations Educational, Scientific and Cultural Organization) is also used to measure a country's adventure tourism potential. The later 2020 study involved 360 experts, 41% women and 59% men. The experts had on average 15 years of experience in tourism and 12 years in adventure tourism. Countries are rated according to 10 pillars (see Table 13.1), anchored in the three key categories

Table 13.1 The 10 pillars of adventure market competitiveness of the Adventure Tourism Development Index (ATTA, 2020, p. 4)

Safe and Welcoming	Adventure	Readiness
Sustainable Development	Entrepreneurship	Humanitarian
Safety	Adventure Resources	Infrastructure
Natural Resources		Cultural Resources
Health		Image

Table 13.2 Adventure Tourism Development Index: Ranking top 10 destinations for adventure tourism (ATTA, 2020, p. 5)

Rank	Developed countries	Developing Countries
1	Iceland	Czech Republic
2	Switzerland	Chile
3	New Zealand	Slovakia
4	Germany	Slovenia
5	Norway	Israel
6	Finland	Estonia
7	Sweden	Poland
8	Canada	Bulgaria
9	Denmark	Romania
10	Australia	Costa Rica

of safe and welcoming, adventure, and readiness (ATTA & International Institute of Tourism Studies, 2020).

The Adventure Tourism Development Index distinguishes between developed and developing countries and ranks the countries into one of the two categories. Table 13.2 shows the leading adventure tourism destinations according to the ATDI index in 2020. Iceland, Switzerland, and New Zealand have the biggest adventure tourism potential among developed countries. The Czech Republic, Chile, and Slovakia lead the list of developing countries.

Political issues

While some countries strategically focus on adventure tourism and provide funding to develop this, for example, New Zealand, Canada, or Scotland, others, including Germany, do not focus on the development of this form of tourism. In particular, countries with natural resources and little tourist infrastructure can benefit from the development and promotion of adventure tourism. Some adventure tourists do not need special infrastructure and prefer remote and natural surroundings. They are often the last tourists to leave and the first to return when there is a crisis in the destination (ATTA, 2013). While many regions around the world promote their adventure potential, a lot of European countries mostly focus on previously established forms of tourism. With increasing popularity and a growing demand for adventure activities, this trend is currently changing.

Case study 13.2   The Adventure Travel Trade Association

The ATTA (www.adventuretravel.biz) is a global organisation that promotes adventure travel and supports providers around the world. 'The ATTA is a privately held, for-profit industry trade group that serves to network, educate, professionalize and promote the adventure travel industry' (ATTA, 2022a). Its mission is 'to empower the global travel community to protect natural and cultural capital while creating economic value that benefits both trade members and destinations' (ATTA, 2022a). The association advocates adventure tourism and supports tour operators around the world. It aims to influence politicians, stakeholders and destinations alike and support the cause of adventure travel, especially in the interest of their members.

Members of the ATTA comprise approximately 30,000 individual guides, tour operators, lodges, travel advisors, tourism boards, destination marketing and management organisations, outdoor educators, gear companies, and travel media, who share a belief and commitment to sustainable tourism. Members come from more than 100 countries representing all continents. Its headquarters are in the US, so the community is historically the largest there. However, in recent years, the association has developed into a global organisation. The ATTA has launched various initiatives in recent years, such as Women in Leadership, Climate Action and Eliminating Plastic. The ATTA supports adventure travel providers in their work, for example, through research results, webinars, networking meetings and Adventure Weeks. At the latter events, destinations introduce themselves on-site and present their offers to specialised tour operators from all over the world.

While the ATTA is an organisation for practitioners, the Adventure Tourism Research Association (ATRA; www.atra.global) is a membership organisation which is assembling a worldwide academic community. ATRA facilitates knowledge exchange and the creation of new relationships in the fields of outdoor leisure and adventure tourism. In doing so, this association helps promote adventure travel and provides support for such holidays. It organises the series of International Adventure Conferences, which are important gatherings for networking and knowledge exchange, both for academics and the industry.

One of the important political topics for the sector is climate crisis and the urge to make travel CO_2-positive. In adventure tourism, the topic of sustainability has played an important role for years (ATTA, 2013). This is not only through the commitment of the providers but also through their nature-based offerings. Related to this, the first standards, which include behavioural guidelines and voluntary restrictions, have already been established. For instance, CO_2 offsetting is already an important aspect for adventure operators.

Furthermore, adventure tourism facilitates empowerment for local guides, cultural exchange between hosts and tourists, as well as economic and social strengthening of the destination region:

> Adventure tourism is not in itself sustainable but by targeting adventure travel that attracts passionate, high-paying tourists to participate in activities specifically suited to local landscapes, communities have the potential to create bespoke tourism products that fit the socio-ecological system and produce clear conservation and sustainable development outcomes.
>
> (Knowles, 2019, p. 1)

Adventure travellers seek remoteness and sometimes do not need luxury or infrastructure to get to their destination. At the same time, they have positive and emotional experiences in natural environments, which create a certain bond with nature conservation and protection taking a higher priority than for other travellers (Gross & Sand, 2019).

A qualitative study (Hanna et al., 2019) concludes that adventure tourism leads to a closer bond with nature, which not only translates into environmentally conscious behaviour but also enhances well-being. See also Chapter 1 for an overview of the inextricable link between adventure tourism and well-being. Participants reported a higher affinity with nature, a greater need to protect it, and environmentally conscious behaviour in everyday life. Environmental awareness and social justice are also becoming increasingly important values for adventure tourists (Buckley, 2018). Not only are these travellers willing to spend more money on their holiday if it has a smaller environmental footprint as a result, but their spending also often benefits the destination itself (Knowles, 2019). Other authors confirm that outdoor sports can lead to a greater appreciation of nature, which, in turn, translates into environmentally friendly behaviour (Brymer et al., 2009; Hanna et al., 2019). Providers and guides in particular play a key role in teaching environmentally friendly behaviour in nature (Rosenberg et al., 2021). However, this presupposes that they are also appropriately trained and sensitised.

Conclusion

Adventure tourism has established itself as a respected form of tourism with growth in specialist operators who are important players in the industry. Due to the distinctive character of adventures, and the interaction with nature and risk, adventure tourism is still a niche market although mainstream operators are starting to include adventure trips in their portfolios. There are diverse types of adventure accommodation such as huts, caravans, tents, and lodges. Often, the type of accommodation is part of the adventure. Some destinations are more suitable for adventure experiences, yet the adventure potential is big around the world. Some destinations are well known for their adventure offers, while others are developing their adventure products. Therefore, other regions are not marketing their full adventure potential as yet. The ATTA is an important advocate of adventure tourism around the world, promoting adventure tourism, connecting players and offering market insights. Future adventure tourism offers will need to be as sustainable as possible.

Review questions

1 Which types of providers make up the adventure tourism industry?
2 What kind of adventure offers can be found on Airbnb?
3 What is the term *glamping* made up of? And what different types of accommodation are included in the glamping category?
4 What is the Adventure Tourism Development Index, and why is it important?
5 What remains to be done to make adventure travel more sustainable?

Further reading

Gross, S., & Sand, M. (2019). Adventure tourism: A perspective paper. *Tourism Review*, 75(1), 153–157. https://doi.org/10.1108/TR-06-2019-0211

Huddart, D., & Stott, T. (2020). *Adventure tourism: Environmental impacts and their management*. Cham: Palgrave Macmillan.
Janowski, I., Gardiner, S., & Kwek, A. (2021). Dimensions of adventure tourism. *Tourism Management Perspectives, 37*, 100776. https://doi.org/10.1016/j.tmp.2020.100776
Sand, M., & Gross, S. (2019). Tourism research on adventure tourism – Current themes and developments, *Journal of Outdoor Recreation and Tourism, 28*, 1–5. https://doi.org/10.1016/j.jort.2019.100261

References

Airbnb (2022). *About Airbnb Adventures*. Retrieved September 29, 2022, from About Airbnb Adventures - Airbnb Help Centre
ATTA – Adventure Travel Trade Association (2013). *Adventure tourism market study*. Seattle.
ATTA – Adventure Travel Trade Association (2022a). *About us*. Retrieved September 29, 2022, from https://about.adventuretravel.biz
ATTA – Adventure Travel Trade Association (2022b). *Adventure travel industry snapshot*. Seattle.
ATTA – Adventure Travel Trade Association, & International Institute of Tourism Studies (2018). ATDI Adventure Tourism Development Index – The 2018 Report, 8th Edition.
ATTA – Adventure Travel Trade Association, & International Institute of Tourism Studies (2020). ATDI Adventure Tourism Development Index, ATDI 2020.
Beames, S., Houge Mackenzie, S., & Raymond, E. (2022). How can we adventure sustainably? A systematized review of sustainability guidance for adventure tourism operators. *Journal of Hospitality and Tourism Management, 50*, 223–231.
Brymer, E., Downey, G., & Gray, T. (2009). Extreme sports as a precursor to environmental sustainability. *Journal of Sport & Tourism, 14*(2–3), 193–204, https://doi.org/10.1080/14775080902965223
Brooker, E., & Joppe, M. (2013). Trends in camping and outdoor hospitality – An international review. *Journal of Outdoor Recreation and Tourism, 3–4*, 1–6.
Buckley, R. (2010). *Adventure tourism management*. London: Routledge.
Buckley, R. (2018). Private conservation funding from wildlife tourism enterprises in sub-Saharan Africa: Conservation marketing beliefs and practices, *Biological Conservation, 218*, 57–63. https://doi.org/10.1016/j.biocon.2017.12.001
BVDIU – Bundesverband der Deutschen Incoming-Unternehmen (2022). *Incoming-Tourismus. Ein Kontakt für viele Leistungen*. Retrieved September 29, 2022, from https://www.bvdiu.org/deutsch/incoming-tourismus/#!
Caravaning Industrie Verband e.V. (2021). Bestes Halbjahr für Caravaningbranche – Neuzulassungen von Freizeitfahrzeugen trotz 'Lockdown' weiter gestiegen, Frankfurt am Main.
Doernberg, A. von, Freyer, W., & Sülberg, W. (2018). *Reiseveranstalter- und Reisevertriebsmanagement – Funktionen, Strukturen, Prozesse* (2nd ed.). Berlin/Boston: De Gruyter/Oldenbourg.
dos Anjos, F. A., & Kennell, J. (2019). Tourism, governance and sustainable development. *Sustainability, 11*(16), 4257. http://doi.org/10.3390/su11164257
Freyer, W. (2015). *Tourismus – Einführung in die Fremdenverkehrsökonomie* (11th ed.). Berlin/München/Boston: De Gruyter/Oldenbourg.
Fuchs, M., & Hovemann, G. (2022). The circular economy concept in the outdoor sporting goods industry: Challenges and enablers of current practices among brands and retailers. *Sustainability, 2022*(14), 7771.
Gross, S. (2017a). Abenteuertourismus in Neuseeland. In C. Pforr & D. Reiser (Eds.), *Tourismus in Australien und Neuseeland* (pp. 279–294). Berlin/Boston: Walter de Gruyter.
Gross, S. (2017b). *Handbuch Tourismus und Verkehr – Verkehrsunternehmen, Strategien und Konzepte* (2nd ed.). Konstanz/München: UVK.
Gross, S., Culemann, J., & Rebbe, J. (2022). *Glamping in Deutschland – Angebot und Nachfrage eines naturtouristischen Konzepts*. Tübingen: UVK.
Groß, S., & Sand, M. (2022). *Draußen erleben! Abenteuer – Outdoor – Tourismus*. München: UVK.
Gross, S., Sand, M., & Berger, T. (2023). Examining the adventure traveller behaviour—Personality, motives and socio-demographic factors as determinants for German adventure travel. *European Journal of Tourism Research, 33*, 3307.

Gross, S., & Werner, K. (2018). Hut-to-hut-hiking trails – a comparative analysis of popular hiking destinations. In C. M. Hall, Y. Ram, & N. Shoval (Eds.), *The Routledge international handbook of walking* (pp. 159–171). New York: Routledge.

Hanna, P., Wijesinghe, S., Paliatsos, I., Walker, C., Matthew, A., & Kimbu, A. (2019). Active engagement with nature: Outdoor0020adventure tourism, sustainability and wellbeing, *Journal of Sustainable Tourism*, 27(9), 1355–1373. https://doi.org/10.1080/09669582.2019.1621883

Heckmair, B., & Michl, W. (2008). *Erleben und lernen, Einführung in die Erlebnispädagogik* (6th ed.). München: Ernst Reinhardt Verlag.

Heckmeier, A. (2011). Mulit Channel-Kommunikation für Bergsport/Spezialisten: Das Beispiel SALEWA. In A. Hermanns & F. Riedmüller (Eds.), *Management-Handbuch sport-marketing* (pp. 447–464). München: Vahlen.

Holloway, J. C., & Humphreys, C. (2022). *The business of tourism* (12th ed.). Harlow: Pearson Education Limited.

Horch, H. D., Heydel, J., & Sierau, A. (2004). *Events im sport – Marketing, management, Finanzierung.* Cologne: Institut für Sportökonomie und Sportmanagement.

Hosters, D. (2019). Eifel-Trekking – zielgruppenorientierte Konfliktvermeidungsstrategie im Naturpark Nordeifel. In R. Forst, M. Porzelt, & V. Scherfose (Eds.), *Konflikte durch Erholungsnutzung in Großschutzgebieten und deren Entschärfung durch innovatives und Besuchermanagement* (vol. 520, pp. 151–159). Bonn: BfN Skripten.

Huddart, D., & Stott, T. (2020). *Adventure tourism: Environmental impacts and their management.* Cham: Palgrave Macmillan.

Kaspar, C. (1996). *Tourismuslehre im Grundriss* (5th ed.). Haupt: Bern/Stuttgart.

Knowles, N. L. B. (2019). Targeting sustainable outcomes with adventure tourism: A political ecology approach, *Annals of Tourism Research*, 79, 102809. https://doi.org/10.1016/j.annals.2019.102809

Leser, H. (Ed.) (2014). *DIERCKE Wörterbuch Geographie. Raum – Wirtschaft und Gesellschaft – Umwelt* (15th ed.). Braunschweig.

Milohnić, I., Cvelić-Bonifačić, J., & Licul, I. (2019). Transformation of Camping into Glamping – Trends and Perspectives, *ToSEE – Tourism in Southern and Eastern Europe*, 5, 457–473.

Gowreesunkar, V. G. B., Séraphin, H., & Morrison, A. (2018). Destination marketing organizations. In D. Dursoy & C. Chi (Eds.), *The Routledge handbook of destination marketing* (pp. 16–34). London: Routledge.

Petrusa, I., & Vlahv, A. (2019). The role of glamping in development of Camping tourism offer – Possibilities and future prospects in the republic of Croatia. In J. Šimurina, I. Načinović Braje, & I. Pavić (Eds.). *Proceedings of FEB Zagreb 10th International Odyssey Conference on Economics and Business*, 12–15 June 2019, Opatija, Croatia. Retrieved March 2, 2022, from https://www.bib.irb.hr/1007232

Riedmueller, F. (2011). Ansätze des kooperativen Marketing für Sportartikelhersteller und-händler. In A. Hermanns & F. Riedmüller (Eds.), *Management-Handbuch sport-marketing* (pp. 413–430). München: Vahlen.

Rogerson, C. M., & Rogerson, J. M. (2020). Camping tourism: A review of recent international scholarships, *GeoJournal of Tourism and Geosites*, 28(1), 349–359. https://doi.org/10.30892/gtg.28127-474

Rosenberg, A., Lynch, P. M., & Radmann, A. (2021). Sustainability comes to life. Nature-based adventure tourism in Norway. *Front. Sports Act. Living*, 3, 686459. https://doi.org/10.3389/fspor.2021.686459

Sand, M. (2019). Die touristische Bedeutung von Micro-Adventures und deren Einfluss auf die Gesundheit. In B. Eisenstein & J. Reif (Eds.), *Tourismus und Gesellschaft. Kontakte – Konflikte – Konzepte* (pp. 519–530). Berlin: ESV-Verlag.

Scherhag, K. (2021). Incomingtourismus. In W. Fuchs (Ed.), *Tourismus, Hotellerie und Gastronomie von A bis Z* (pp. 505–506). Berlin/Boston: De Gruyter/Oldenbourg.

Stors, N., & Kagermeier, A. (2017). The sharing economy and its role in metropolitan tourism. In M. Gravari-Barbas & S. Guinand (Eds.), *Tourism and gentrification in contemporary metropolises*. London: Routledge.

Stroebel, T., Moesch, C., & Buser, S. (2018). Inszenierung von Erlebnissen im touristischen Wettbewerb – Eine Fallstudie am Beispiel des Eiger Ultra Trails. In G. Nowak (Eds.), *(Regional-) Entwicklung des Sports* (pp. 193–206). Schorndorf: Hofmann.

Swarbrooke, J., Beard, C., Leckie, S., & Pomfret, G. (2003). *Adventure tourism: The new frontier.* Oxford: Butterworth Heinemann.

TUI AG (2022). *Aktivurlaub – Urlaub mit sportlichem Kick*, Retrieved September 29, 2022, from https://www.tui.com/pauschalreisen/aktivurlaub

Wagner, W., Schobert, M., & Steckenbauer, G. C. (2019). Erlebnisgenese: Bedeutung und Nutzen? In W. Wagner, M. Schobert, & G. C. Steckenbauer (Eds.), *Experience Design im Tourismus – eine Branche im Wandel Gestaltung von Gäste-Erlebnissen, Erlebnismarketing und Erlebnisvertrieb* (pp. 3–20). Wiesbaden: Springer Gabler.

Widmann, T. (2006). *Wohnmobiltourismus in Deutschland am Beispiel der Destination Mosel.* Trier.

14 Sustainable adventure tourism products and destinations

Jelena Farkić and Steve Taylor

Chapter learning outcomes

1 Understand the concept of sustainable tourism products and destinations.
2 Understand the significance of stakeholder collaboration in developing adventure tourism products.
3 Become aware of the challenges the stakeholders face in developing tourism products.
4 Become familiar with slow adventure and forest bathing as part of the broader adventure tourism product portfolio.

Introduction

In recent times, societies worldwide have been facing various crises and adversities that have prompted discussions on building resilience and developing the capacity to recover (Sharma, Thomas & Paul, 2021). The "transformative turn", particularly in the post-pandemic consumer landscape, has encouraged alternative ways of "doing" tourism. The pandemic has greatly changed the ways in which tourism spaces are utilised, thereby accelerating the need for responsible tourist behaviour at all stages of their journey and across all tourism spaces (Gössling, Scott & Hall, 2020). There has been a greater demand for natural spaces and the repurposing of tourism to address new consumer needs. Tourism scholars and industry professionals in particular have advocated greener forms of travel that empower fragile and less resilient local communities, promote more sustainable business practices, and encourage environmentally and socially responsible tourist behaviour (Brouder et al., 2020; Juvan et al., 2021; Juvan & Dolnicar, 2014). Much greater emphasis has been placed on the enhancement of well-being (see Chapter 12) as paramount in the creation of tourism experiences (Brouder et al., 2020), with safety and security becoming high priorities for destinations and governments globally (Page, Bentley & Walker, 2005).

In order to remain competitive in the marketplace and provide memorable experiences for diverse customer segments, the tourism industry needs to constantly innovate and diversify its product portfolio. In this process, destinations are increasingly directing their focus to sustainable development which must 'deliver clear benefits to residents and sustain the support of all key stakeholders' (Getz, Svensson, Peterssen & Gunnervall, 2012,

DOI: 10.4324/9781003393153-18

p.48). In supporting the regeneration of local environments or communities, they develop "alternative" products that are small scale, are more socially and environmentally sustainable, encourage appreciation of a destination's character, and involve businesses that are locally owned. Such products appeal to more specialist markets, and the lower numbers involved mean that less infrastructure for their operationalisation is required (Novelli, 2022). Importantly, creating adequate conditions for tourism development in natural areas and creating sustainable destinations demands the education of the local population and raising tourist awareness, not only about the fragility of resources but also about how they can benefit from consumption embedded in those spaces. The ensuing chapter discusses sustainable tourism products and destination development more broadly in order to provide the necessary context for the subsequent discussion on stakeholder collaboration in adventure tourism.

Sustainable tourism products and destination development

The notion of the tourism product was initially conceived by travel agencies and tourism organisations as a marketing concept in order to focus supply to specific markets, satisfying their needs and generating profit (UNWTO, 2008; Xu, 2010). The UNWTO (2008) suggested that tourism products are still not sufficiently characterised in a uniform way; hence, there is no international recommendation for using this type of classification. Similarly, McKercher (2016) suggested that due to the fragmentation of the tourism industry, it is difficult to identify and classify tourism products, explaining that a product is defined in its broadest sense as the activities, attractions, and interests consumed by tourists in a destination that satisfies their needs.

Smith (1994) was one of the first researchers to propose the composition of a tourism product and break it down into five elements – physical plant, service, hospitality, freedom of choice, and involvement – while emphasising that the products are essentially experiences. Other definitions suggest that it is a bundle of services, experiences, and benefits marketed and sold to tourists (Leiper, 1990); a package of activities or services that caters to the needs and desires of a tourist (Buhalis & Laws, 2001); a composite of attractions, amenities, and ancillary services that are marketed as a single product or service (Mason, 2003); or an experience that combines a range of elements that appeal to the tourist's senses, emotions, and intellect (Getz, 1997). Following these definitions, a tourism product can be understood as a combination of tangible (transportation, accommodation, food, activities) and intangible (atmosphere, culture, heritage) elements that work together to provide a memorable experience to a tourist and bring benefits for the tourism destination. Also see Chapter 13 for an overview of adventure tourism products.

In attempting to address the sustainable development goals (SDGs), efforts have been made to design and operationalise tourism products with a focus on minimising negative environmental, social, and cultural impacts while maximising positive benefits for everyone involved (Haid & Albrecht, 2021). Commensurate with SDGs 14 and 15 promoting sustainable use of aquatic and terrestrial ecosystems, respectively (UN, 2015), examples of such products are eco-tourism, rural tourism, adventure tourism, and other forms of tourism that prioritise responsible tourism practices. They typically emphasise the preservation and protection of natural resources and cultural heritage while promoting the well-being and prosperity of local communities (Choibamroong, 2011). Equally important is their focus on the provision of deep, rich, and immersive experiences that allow tourists to connect with a destination in meaningful ways; in fostering this attachment, such experiences help to promote participants' mental and physical well-being, the critical importance of which is espoused by SDG 3 (UN, 2015). They benefit both tourists and local communities as they foster greater understanding, mutual respect, and appreciation for the natural and cultural diversity of the world while promoting sustainable livelihoods and economic growth in a way that is economically viable, environmentally responsible, and socially equitable.

Likewise, sustainable destination development is concerned with the management of tourism destinations in a way that maximises its economic, social, and environmental benefits while minimising negative impacts (Liu, 2003). This approach seeks to balance the desires of visitors with the needs of the destination community and environment in order to ensure long-term sustainability. In doing so, it requires a holistic approach and involves a range of strategies, for example, the promotion of responsible tourism practices, the protection and conservation of natural and cultural resources, and the involvement and engagement of local communities in tourism planning and management (Sedarati, Santos & Pintassilgo, 2019). Some of the key principles of sustainable destination development include the following:

- *Environmental conservation*: Minimising negative impacts on the natural environment, including conserving biodiversity, protecting ecosystems, and reducing carbon emissions (see also Chapter 22).
- *Social and cultural authenticity*: Valuing the preservation of cultural heritage and ensuring that tourism activities respect and support local customs, traditions, and ways of life.
- *Economic benefits*: Maximising economic benefits for local communities and businesses, including creating employment opportunities and stimulating local economies.
- *Visitor satisfaction*: Providing visitors with high-quality, authentic, and meaningful tourism experiences that meet, or exceed, their needs and expectations.

In thinking relationally and synergistically in developing tourism sustainably, innovating, customising and diversifying the products based on destinations' principal resources and customers' changing needs is crucial. As a consequence, it can greatly enhance destination image and competitiveness and have many advantages for the profitability of the businesses involved (Benur & Bramwell, 2015). One example of this is (thematic) package tours, which are seen as an important destination planning tool and a way of optimising holiday logistics, thus enabling cost-effectiveness, safety, and convenience of their production and consumption (Mauri & Turci, 2018; Stamboulis & Skayannis, 2003). Package holidays have been represented as innovative ways of designing visitor experiences, enhancing sustainability, competitiveness, employment, and community cohesion (Holloway, 1981; Liao & Chuang, 2020; Theuvsen, 2004; Zillinger, Jonasson, & Adolfsson, 2012). They are popular among tourists who value the convenience of a pre-arranged holiday, as they provide a degree of security for tourists and are regulated in such a way that ensures consumer protection and safety. Furthermore, tourists can select from a range of destinations and itineraries that suit their budget and can also be tailor-made or customised to better match their preferences.

Yet, there are some constraints that destinations experience in the process of product development, such as the following:

- *Lack of resources*: Developing a tourism product can require significant financial and human resources, including funding, expertise, and time.

- *Competition*: The tourism industry can be highly competitive, and developing a new product can be challenging if there are already established products or destinations that offer similar experiences.
- *Infrastructure and logistics*: Developing a tourism product may require the construction or improvement of infrastructure, such as transportation, accommodation, and visitor facilities. This can be a significant challenge in remote or undeveloped areas.
- *Regulations and permits*: Tourism product development may require compliance with a range of regulations and permits, including environmental, health, and safety regulations, which can be time-consuming and costly.
- *Community involvement*: Developing a tourism product may require the involvement and support of local communities, and gaining community support can be a challenge if there are concerns about the impacts of tourism on the local environment or culture (see Chapter 24 on host community involvement in adventure tourism).

Other constraints include seasonality, a lack of expertise, a lack of tourism control, and the potential negative impacts of tourism on the natural environment and host communities.

Stakeholder collaboration in adventure tourism product development

Adventure tourism is often understood in a commercial sense, as guided tours or package holidays, with recreational activities as their main element (Buckley, 2007). When designing an adventure tourism product, tour operators usually focus on its five dimensions: the natural environment, physical activity, the expertise of the guides/instructors, the local cultural experience, and the level of involvement or control, thus catering to different consumer needs, expectations, and levels of preparedness (Janowski et al., 2021). Unskilled clients can access high-volume, low-difficulty products, while those more experienced are more likely to consume low-volume, high-cost products that require prior technical competency and involve significant individual risk. These so-called hard adventure products often operate in more remote and inhospitable areas and involve somewhat more "demanding" activities, such as ice caving, diving, or mountaineering (Buckley, 2007). Between these extremes, there is a vast array of adventure tours with varying designs, durations, locations, group sizes, and prices.

Post-pandemic, the rise in popularity of nature-based tourism is increasingly driven by the desire for self-fulfilment, relaxation, introspection, and mindfulness, performed somewhat farther away from urban commercialism but not necessarily in extreme environments and wild nature (Janowski et al., 2021). Likewise, adventure tourism has shifted away from being solely focused on physical features and adrenaline-induced recreational activities towards paying more attention to the psychological benefits of immersion in nature (Farkić, Filep & Taylor, 2020). To this end, understanding activities in nature as a means of promoting health and well-being and establishing meaningful connections with the environment has particularly influenced the significant role that service providers play in developing adventure tourism products. The adventure tourism sector is well suited to meet new consumer demands by providing services that enable people to pursue these softer and slower experiences while also satisfying their wider interests and need for being in outdoor spaces.

Although natural, wild, or remote places represent an opportunity for sustainable economic growth through the development of region-specific tourism products, some authors argue that adventure tourism is not inherently sustainable (Knowles, 2019). Nevertheless, adventure tourism providers are increasingly striving to implement sustainability principles in their business practices and operate in a more responsible manner, with the goal of benefitting the environment, tourists, and local communities. Product development assumes identifying unique natural or cultural attractions, designing itineraries, creating marketing campaigns, and establishing partnerships with local suppliers and service providers. The aim is to provide tourists with well-rounded yet safe, enjoyable, and meaningful experiences that satisfy their desire for adventure while at the same time contributing to the sustainable development of destinations. Travel operators and activity providers demonstrate their commitment to social responsibility by offering training programmes for their employees or employing guides on a permanent basis, rather than on a contractual one (Wengel, 2021). While this does increase the overall cost of the adventure holiday, there is recognition that a portion of the payment goes directly to local service providers in the destination, of which tourists are aware and to which they consciously contribute (Adventure Travel Trade Association, 2016).

As in any other tourism sector, stakeholders in adventure tourism are represented by individuals, groups, or organisations that have an interest or concern in its activities and outcomes. These stakeholders can include tourists, tour operators, local communities, equipment and clothing manufacturers, sports clubs and associations, government agencies, non-governmental organisations (NGOs), tour guide associations, and environmental groups, among others. Beames, Mackenzie and Raymond (2022, p. 230) explain that these stakeholders:

> have obligations to deliberately and systematically incorporate sustainability practices and principles into their business plans, research agendas, policy documents, and concrete actions. These efforts require the development and adoption of sustainability guidance that takes a truly holistic approach and provides practical strategies across a range of adventure tourism stakeholders.

More specifically, destination management organisations (DMOs), tour operators, and local communities play a critical role as they provide the necessary infrastructure and support for the activities to take place. Government agencies have a responsibility to regulate and manage adventure tourism to ensure safety and environmental protection; NGOs and environmental groups advocate responsible tourism practices that minimise the negative impact on the environment and local communities. Tourists, meanwhile, are seen as the primary stakeholders as they are the ones for whom the products are made and who, by participating in the activities, contribute to the industry's economic success and sustainable development of a destination.

SDG 17 recognises the importance of partnership working as a more effective means of achieving sustainability ambitions (UN, 2015), and collaboration and partnership among stakeholders are key to product and destination development (Erkuş-Öztürk and Eraydın, 2010; Xu, 2010). Good destination governance and the commodification and packaging of distinct cultural landscapes in innovative and resource-efficient ways are crucial to achieving responsible and sustainable tourism development (Nguyen et al., 2019). The appeal of the place is a primary attraction for tourists; however, their satisfaction also depends on the quality and efficiency of related businesses (Porter, 2008). For example, one way of increasing the competitiveness of a regional industry is through clustering and collaborative participation in the creation of tourism products (Delgado et al., 2010; Novelli et al., 2006). Clusters represent spatially concentrated groups of service providers (e.g., accommodation, food and beverage, transportation, or recreational activity providers) and supporting institutions (e.g., public agencies, DMOs, NGOs, and local communities) that are focused on the delivery of adventure tourism services. Clustering has been shown to be especially beneficial for small communities in remote or peripheral areas; the participation of small and medium-sized enterprises (SMEs) collaborating in clusters, along with the involvement of other related stakeholders, such as DMOs, services providers, host communities, academia, and media, as well as tourists themselves, is the key to the successful development of sustainable adventure products (Farkić, Taylor & Bellshaw, 2019).

While stakeholder collaboration can bring many benefits to adventure tourism, it is a sector characterised by a broad range of micro-businesses competing for custom (Garcia & Cater, 2022), and it is important to be aware of the potential disadvantages, working to overcome them through effective communication, trust-building, and inclusive decision-making processes (Minnaert, 2020). Some of the potential disadvantages of stakeholder collaboration include the following:

- *Conflicting interests*: Different stakeholders may have different priorities, interests, and goals, which can lead to conflicts and disagreements. For example, adventure tour operators may prioritise profits, while conservation organisations may prioritise environmental protection.
- *Power imbalances*: Some stakeholders may have more power, influence, and resources than others, which can lead to unequal collaboration and decision-making. For example, large tour operators may dominate the collaboration process, leaving smaller operators or local communities with little say.
- *Communication barriers*: Stakeholders may speak different languages, have different cultural backgrounds, or operate in different industries, which can make communication difficult and lead to misunderstandings.
- *Time and resource constraints*: Collaboration requires time, effort, and resources from all stakeholders, which can be challenging to coordinate, especially if stakeholders have other commitments or conflicting priorities.
- *Lack of trust*: If stakeholders do not trust each other, collaboration may not be effective. For example, local communities may not trust tour operators, leading to tensions and conflicts.
- *External issues*: Adventure tourism may be subject to various laws, regulations, and permits, which can create additional barriers to collaboration. For example, tour operators may need permits from government agencies or permission from landowners to access certain areas.

Despite potential challenges, stakeholders play a critical role in sustainable tourism development, as they have the power to influence decision-making, promote responsible tourism practices, and ensure that tourism benefits are shared fairly among all parties, leading to a more resilient and sustainable tourism industry for both present and future generations. The ensuing two sections provide examples of stakeholder collaboration in developing the tourism product and expanding the adventure tourism product portfolio in a sustainable way.

Slow adventure

The celebration of slowness is viewed as an antidote to the fast-paced lifestyle and has become one of the priorities, not only in improving people's health and well-being (Farkić, Isailović & Lesjak, 2022) but also in enhancing the tourist experience (Varley & Semple, 2015). By slowing down, people can better appreciate the value of time and regain a sense of meaning, authenticity, comfort, and security in a time of technological, environmental, and existential anxieties. Tourists who take the time to fully immerse in a place not only develop a greater appreciation for places and spaces but also have many opportunities to reconnect with the natural world, themselves, and others over an extended time.

As the demand for sustainable tourism grows, slow adventure is becoming an increasingly popular and attractive option for tourists seeking a more authentic and responsible travel experience. In developing slow adventure, the quality of slowness is considered crucial and is credited with providing multiple benefits for those who consume it. It offers an alternative to "traditional" adventure tourism, which often emphasises adrenaline-fuelled and risky activities, such as whitewater kayaking, mountain biking, horseback riding, or mountaineering. Here, emphasis is on experiencing the journey rather than reaching the destination or achieving a goal. As a sustainable product, it focuses on immersive experiences, human-powered travel, making deeper connections with nature and local cultures and potentially living transformative moments. Most importantly, it allows tourists to disconnect from their busy lives and engage with the natural landscape in a more meaningful way. Slow adventure places an emphasis on hyperlocal: services offered by small, locally owned businesses or eco-tourism operators, often incorporating cultural experiences and opportunities for learning and personal growth (see Case study 14.1).

Case study 14.1   Developing, promoting and selling slow adventure experiences

Slow Adventure in Northern Territories (SAINT) was a transnational initiative co-financed by the Northern Periphery and Arctic Programme that aimed to encourage tourism development in European remote and peripheral areas. Based on Varley and Semple's (2015) conceptual work on slow adventure, the project took a step further to focus on collaboration among SMEs at the local level and introduce consumers to an alternative dimension of adventure. In doing so, it aimed to address issues such as remoteness, seasonality, and a lack of development by introducing a less seasonally dependent product that generates year-round employment and income. The project partnered with local communities, businesses, and organisations to identify opportunities for developing slow adventure tourism experiences with a core focus on business clustering for collaborative product development and marketing.

A direct spin-off from the SAINT project, Slow Adventure Ltd, was created to be a commercial arm of the slow adventure movement. The effective commercial implementation of the slow adventure concept in a number of destinations (e.g., Scotland, Sweden, Finland, Iceland, and Italy) can be particularly beneficial for rural destinations that rely on natural resources, as it simultaneously supports local communities and engenders understanding of and interaction with the landscape while minimising negative environmental impacts. This approach goes beyond the principles of sustainable tourism to align more closely with regenerative tourism by encouraging and directly supporting positive change. Through an Impact Fund, customers donate directly to local conservation projects in the area they visit, making a positive impact on the people and landscape in which the experience takes place. Slow Adventure Ltd goes beyond engendering responsible tourist behaviour towards natural or cultural assets to also support the regeneration of local environments or communities and enhance the region's resilience.

Further information:
https://saint.interreg-npa.eu/

The core dimensions which constitute the essence of slow adventure are the recreational activities, food, and accommodation (Farkić, Taylor & Bellshaw, 2019). An emphasis is placed on human-powered, slow journeys through natural places as well as activities that are typical for the region, such as creel fishing, canoeing, northern lights watching, or bike touring, on- or off-road (see Figure 14.1). Furthermore, consuming locally sourced foods, and engaging tourists in the whole process of food preparation is also promoted, from the collection of food by foraging, catching, or hunting to slow or wild cooking and collective eating. These activities take people back to primitive times in essence and exemplify how we have been dependent on nature in order to survive. The accommodation chosen is usually embedded in and reflects the natural setting and may range from tents, yurts, and bunkhouses to five-star hotels.

These three fundamental dimensions are interconnected, and when merged they generate a seamless adventure tourism product that facilitates a more profound connection with the environment, wildlife, people, and their culture. While the slow adventure ethos is a common denominator, the experiences are distinct for every country, showcasing the

Figure 14.1 Making slow progress through Glen Kinglass. Photograph by Steve Taylor.

diverse features of the local landscapes, peoples, and traditions that characterise their own particular slow adventure practices. For example, in Finland, slow adventure focuses on wintertime activities such as snowshoeing, cross-country skiing, dog-sledding, and traditional Finnish sauna, reindeer farm visits, and swimming. In Iceland, unique experiences such as ice cave explorations, glacier trekking, and ice hiking are offered. The primary slow adventure activity in Ireland is road cycling, accompanied by meditation, yoga amidst natural settings, outdoor dining, and culinary tours.

The stakeholders involved in creating, delivering and promoting the slow adventure product have realised the benefits of a product that combines various experiences specific to the region. In particular, local SMEs have recognised the strength of micro-clustering and offering their services as part of a collective, hybrid product. The idea that the overall slow adventure experience should be a collaboration between SMEs and other stakeholders has been further reinforced by the emotional and physical connections with the place, mediated and enabled by certified outdoor adventure guides. However, the success of the slow adventure product is dependent on the level of cooperation, trust, and synergies between the stakeholders involved, which need to share the same values and implement them in a common territory and complementary ethical approach. The SAINT project and Slow Adventure Ltd have demonstrated that sharing the same values is crucial not only to the slow-adventure philosophy and its ethos but also more widely in the sustainable regional development of tourism.

Forest bathing

Forest bathing, also known as *Shinrin-yoku,* is a practice that originated in Japan in the 1980s and has gained popularity worldwide in recent years. It involves immersing oneself in nature and mindfully engaging with the natural environment to promote physical, mental, and emotional well-being (Li, 2018). Forest bathing typically involves a slow, meditative walk through a forest, during which the participant engages all their senses to fully experience their surroundings. Studies have shown that forest bathing is beneficial for people's health and well-being in that it lowers stress levels, reduces blood pressure, boosts the immune system, improves mood and cognitive function, and/or reduces symptoms of anxiety and depression (Li et al., 2006).

Forest bathing is becoming an increasingly popular tourist activity, as people seek diverse ways to disconnect from their busy lives and connect more meaningfully and mindfully with nature (Farkić, Isailovic & Taylor, 2021; Ohe et al., 2017). Many destinations, particularly those with abundant forests and natural landscapes, are incorporating forest bathing into their tourism offer. Forest bathing experiences can range from guided walks with trained forest therapy guides to self-guided walks that allow visitors to immerse in nature at their own pace. In addition to the health and wellness benefits, forest bathing also represents an opportunity for sustainable tourism development of destinations, as it serves as a means of promoting the conservation of forest resources, engaging local communities and strengthening the local economy. More recently, it has been increasingly combined or integrated with other products, such as eco-tourism, adventure tourism, or rural tourism, as a way of diversifying the tourism offer and developing tourism destinations.

Forest bathing is not typically considered an adventure activity. Rather, it is a calming and meditative practice that involves spending time in a forest or natural environment and immersing oneself in the sights, sounds, and smells of the natural world. While forest bathing may involve some physical activity, such as walking, the focus is mainly on relaxation and mindfulness rather than physical recreation, the thrill, or excitement. However, in many respects, it aligns itself with slow adventure and for those who seldom spend time in nature and are connected to a fast-paced and digitalised way of life, forest bathing can be a novel and transformative adventure in its own right. For instance, during the Outdoor by ISPO Adventure conference in Munich, Germany, in 2022, forest bathing was proposed as a valuable addition to the realm of adventure tourism, especially in the context of the post-pandemic era, during which individuals are increasingly seeking slower and more mindful experiences. Notably, forest bathing was ranked as the second most recommended practice for sustainable adventure tourism during the conference (ISPO, 2022). Moreover, forest bathing was experienced as an outdoor activity during the International Adventure Conference in Treuchtlingen, Germany in 2023 (see Figure 14.2). The conference, themed "Being-With and Being-Well in Nature: Our Tentacular Adventures" (ATRA, 2023), featured a group of 12 participants who engaged in forest bathing as part of the conference activities, immersing themselves in the Franconian forest atmosphere. Obviously, there is an interest in demand for practising forest bathing. By incorporating this practice into adventure tourism packages or destinations' tourism offers, tourism operators can offer a holistic experience for tourists. This allows tourists to seamlessly combine outdoor activities with the restorative effects of immersing themselves in forest environments.

Figure 14.2 Expressing gratitude to the forest, International Adventure Conference, Treuchtlingen, Germany, June 2023. Photograph by Jelena Farkić.

Forest bathing can be considered a sustainable adventure tourism practice in several ways:

- *Preservation of natural resources*: Forest bathing encourages people to connect with and appreciate nature, which can lead to a greater understanding and appreciation of the value of natural resources. This, in turn, can promote conservation and preservation efforts that help protect natural habitats and ecosystems.
- *Low-impact activity*: Similar to other soft or slow outdoor activities, forest bathing is a low-impact practice that does not require any special equipment or infrastructure.
- *Non-consumptive activity*: Forest bathing does not involve the extraction or consumption of forest resources, which helps reduce the impact of human activities on the environment.
- *Positive impact on mental health*: Spending time in nature can have a positive impact on mental health and well-being which can promote sustainable living by encouraging people to adopt lifestyles that prioritise mental and emotional health over material consumption and overconsumption of resources.
- *Promotion of tourism*: Forest bathing can be part of a larger adventure tourism industry that promotes sustainable tourism practices, helping create jobs and economic opportunities.

In developing and implementing forest bathing as part of the broader adventure tourism product, there are certain factors to consider. First, it is important to be familiar with the legal and regulatory requirements that may apply; depending on the location, certain permits or other approvals may be needed. Forest bathing providers also need to comply with environmental regulations to ensure that the activities do not cause damage to the environment. Finding suitable locations for forest bathing can be a challenge, as the trails need to be easily accessible, have a good variety of trees and plants, and be safe for visitors. Some locations may be protected or restricted, making it difficult to gain easy access to them. Also, forest bathing guides or forest wellness mediators should not only have experience in leading groups and be knowledgeable about the area but also have specialist skills and knowledge of forest bathing techniques and the benefits of this practice. However, many destinations struggle to manage tourism at the local level, which is normally caused by a lack of understanding of the tourism product and its benefits or the lack of communication between them (Dahlan et al., 2021). To address this, it is crucial to provide training and empowerment to the involved stakeholders. To achieve this, practical guidelines must be developed for planning, designing, and operationalising forest bathing. Incorporating the principles of forest therapy, enhancing the visitor experience through the education of all stakeholders, and promoting forest bathing as a sustainable tourism practice, it can be an impactful tourism product that benefits the local communities, tourists, and the environment.

Stakeholders involved in the development of a forest bathing product can vary depending on the specific context and location, for example, tour operators, local communities, government agencies, forestry institutes, environmental organisations, or natural resource managers (see Case study 14.2). They work closely with local communities who provide support services such as accommodation, transportation, and provisions of local foods and herbs. Locals may also have knowledge and expertise about the forest

---

Case study 14.2   Forest bathing as a sustainable adventure tourism experience

An example of stakeholder collaboration at the destination level and successful implementation of forest wellness into the existing tourism offer, the INTERREG IPA CBC "Natural Selfness – NATESS" project (Natural Selfness, 2020) aimed to promote the health and well-being benefits of being in forest environments in the region between Hungary and Serbia. During the 12 workshops that were organised as part of the project activities, visitors had the chance to engage in guided mindfulness practices, such as forest bathing, art from nature or bio-gardening, as well as utilising a newly established health path with a specific composition of plants that have a positive impact on human health. The reconstruction of an authentic building, the Forest Selfness Centre, further enriched the tourist offer of the region in terms of accommodation capacities. The project also involved training programmes for tour operators and natural resource managers to promote responsible tourism practices and environmental conservation. It raised the stakeholders' capacities for this type of tourism offer by developing specific skills and tools aimed at enabling the successful delivery of tourism and wellbeing services. Equally important, the project brought together the professional and academic community by organising an international conference dedicated to addressing the challenges and opportunities in regional tourism development. By promoting a more conscious and mindful approach to life, the NATESS project encouraged people to make small changes that can have a big impact on their health and the world around them. To this end, the NATESS Forest Selfness Project serves as a good example of how forest bathing can be integrated into sustainable tourism offers of destinations.

---

environment which is invaluable for forest guides. The forest bathing product is also inclusive in that it engages local women or other marginalised groups in the production of handicrafts that are used as forest walk props, such as woollen wellness balls or furoshiki scarves. Furthermore, certain government agencies and environmental organisations may provide regulatory and monitoring services to ensure that undertaking the activities does not have negative environmental impacts. Natural resource managers are responsible for managing the forest areas and ensuring that the forest ecosystem is conserved and protected. Forest owners and managers play a critical role in maintaining and protecting the forest environment and ensuring that forest bathing activities are conducted in a safe and sustainable manner. Other stakeholders, such as scientists and researchers, are also key actors in studying the health benefits of being in nature, as well as the cultural and social significance of forest environments.

Collaboration among stakeholders is essential for delivering both forest bathing and slow adventure experiences. Effective collaboration assumes a shared commitment to enhancing the health and well-being of tourists, local communities, and the environment. Therefore, prioritising such sustainable tourism practices is key at both the SME and destination levels. This can involve promoting responsible tourism practices, supporting local conservation efforts, and working with local communities to develop tourism strategies that seek to benefit all stakeholders and that promote the conservation and regeneration of natural environments (Nguyen et al., 2019). By doing so, adventure tourism can make its contribution to a healthy, resilient, and sustainable future for people and the planet.

Conclusion

The COVID-19 pandemic has had a significant impact on tourism, forcing the industry to rethink and diversify its product offerings. It has accelerated the trend towards sustainable and responsible tourism, as many consumers have become more aware not only of their impact on the environment and local communities but also about the necessity to care for their own health and well-being. When they travel, many tourists are now seeking slower, more meaningful, and authentic experiences that can also have positive impacts on their mental, and physical, well-being, such as slow travel, undertaking outdoor activities, and more mindful practices. Offering products or experiences that address the insecurities and anxieties so prevalent in the postmodern consumer context is becoming of paramount importance in the development of adventure tourism in particular and sustainable tourism development in general. The tourism industry therefore increasingly strives to educate and inspire people to spend more time in natural spaces and adopt a more responsible and sustainable lifestyle, which can have positive impacts on their health, the environment, and society as a whole.

This chapter focused on two such products which promote a more conscious and mindful approach to life. Through consuming slow adventure and forest bathing as part of their holidays, tourists can learn how to take care of themselves and the environment in a more natural and holistic way while also supporting local businesses. The further development of these products can take various forms, including creating new destinations, events, and types of accommodation and expanding the existing experience-scapes, to include, for example, more immersive, even artistic approaches, to being in and experiencing nature. For this process to be most effective, encouraging the collaboration of destination stakeholders and industry partners, such as tour operators, small-scale food producers, farmers, and local communities, to ensure that the product is sustainable, profitable, marketed effectively to key consumer groups and benefits all stakeholders is crucial.

A multidisciplinary approach to adventure tourism product development helps create innovative, sustainable, and competitive tourism offerings that meet the needs and expectations of diverse customer segments. A multidisciplinary approach can help identify and address potential challenges and opportunities, such as environmental impacts, cultural sensitivities, or changing consumer trends, in a holistic and proactive way. By incorporating a range of perspectives and expertise from various disciplines, SMEs and tourism planners can create experiences that are unique, authentic, or transformative and that start to address some of these challenges. Such an approach can lead to the development of meaningful, profitable, and socially responsible tourism products that benefit the adventure tourism industry, its consumers, and the communities and ecosystems within which the adventure tourism product is operationalised.

Review questions

1 What advantages does collaboration bring to the development and marketing of sustainable regional tourism?
2 What are the key challenges faced by sustainable tourism product developers in terms of both structure and agency?
3 How can sustainable tourism make a significant contribution to achieving the UN's SDGs?
4 How can the consumption of forest bathing or slow adventure products help alleviate overtourism issues in popular visitor locations?

Further reading

Slow Adventure: https://www.slow-adventure.com/
Slow Adventure in Northern Territories: https://keep.eu/projects/18571/Slow-Adventure-In-Northern--EN/
Association of Nature & Therapy Guides & Programs: https://www.natureandforesttherapy.earth/

References

Adventure Travel Trade Association. (2016). *International adventure travel guide qualifications & performance standard.* https://www.adventuretravel.biz/education/guide-standard
ATRA (2023). Book of Abstracts "Being-with and being-well in nature: Our tentacular adventures", *10th International Adventure Conference,* Accessed on 2/7/2023 from https://atra.global/proceedings
Beames, S., Mackenzie, S. H., & Raymond, E. (2022). How can we adventure sustainably? A systematized review of sustainability guidance for adventure tourism operators. *Journal of Hospitality and Tourism Management, 50,* 223–231.
Benur, A. M., & Bramwell, B. (2015). Tourism product development and product diversification in destinations. *Tourism Management, 50,* 213–224.
Brouder, P., Teoh, S., Salazar, N. B., Mostafanezhad, M., Pung, J. M., Lapointe, D., … Clausen, H. B. (2020). Reflections and discussions: Tourism matters in the new normal post COVID-19. *Tourism Geographies, 22*(3), 735–746.
Buckley, R. (2007). Adventure tourism products: Price, duration, size, skill, remoteness. *Tourism Management, 28*(6), 1428–1433.
Buhalis, D., & Laws, E. (2001). *Tourism distribution channels: Practices, issues and transformations.* Continuum.
Choibamroong, T. (2011). A stakeholder approach for sustainable community-based rural tourism development in Thailand. In E. Laws, H. Richins, J. Agrusa, & N. Scott (Eds.), *Tourist destination governance: Practice, theory and issues* (pp. 173–185). Wallingford: CABI.
Dahlan, M. Z., Dewi, M. R., & Putri, V. O. (2021, November). The challenges of forest bathing tourism in Indonesia: A case study in Sudaji Village, Bali. In *IOP Conference Series: Earth and Environmental Science* (Vol. 918, No. 1, p. 012012). IOP Publishing.
Delgado, M., Porter, M. E., & Stern, S. (2010). Clusters and entrepreneurship. *Journal of Economic Geography, 10*(4), 495–518.
Erkuş-Öztürk, H., & Eraydın, A. (2010). Environmental governance for sustainable tourism development: Collaborative networks and organisation building in the Antalya tourism region. *Tourism Management, 31*(1), 113–124.
Farkić, J., Isailović, G., & Lesjak, M. (2022). Conceptualising tourist idleness and creating places of otium in nature-based tourism. *Academica Turistica, 15,* 11–23.
Farkić, J., Isailovic, G., & Taylor, S. (2021). Forest bathing as a mindful tourism practice. *Annals of Tourism Research Empirical Insights, 2*(2), 100028.
Farkić, J., Filep, S., & Taylor, S. (2020). Shaping tourists' wellbeing through guided slow adventures. *Journal of Sustainable Tourism, 28*(12), 2064–2080.

Farkić, J., Taylor, S., & Bellshaw, S. M. (2019). Slow adventure in remote and rural areas: Creating and narrating the tourism product. In T. O'Rourke & M. Koscak (Eds.) *Ethical and responsible tourism: Managing Sustainability in Local Tourism Destinations* (pp. 83–95). Routledge.

Garcia, O., & Cater, C. (2022). Life below water; challenges for tourism partnerships in achieving ocean literacy, *Journal of Sustainable Tourism, 30*(10), 2428–2447.

Getz, D. (1997). *Event management & event tourism*. Cognizant Communication Corporation.

Getz, D., Svensson, B., Peterssen, R., & Gunnervall, A. (2012). Hallmark events: Definition and planning process. *International Journal of Event Management Research, 7*(1/2), 47–67.

Gössling, S., Scott, D., & Hall, C. M. (2020). Pandemics, tourism and global change: A rapid assessment of COVID-19. *Journal of Sustainable Tourism, 29*(1), 1–20.

Haid, M., & Albrecht, J. N. (2021). Sustainable tourism product development: An application of product design concepts. *Sustainability, 13*(14), 7957.

Holloway, J. C. (1981). The guided tour a sociological approach. *Annals of Tourism Research, 8*(3), 377–402.

ISPO (2022). 7 tips for sustainable adventure tourism. Retrieved September 2, 2023, from https://www.ispo.com/en/news-markets/7-tips-how-make-your-adventures-more-sustainable

Janowski, I., Gardiner, S., & Kwek, A. (2021). Dimensions of adventure tourism. *Tourism Management Perspectives, 37*, 100776.

Juvan, E., Podovšovnik, E., Lesjak, M., & Jurgec, J. (2021). A destination's social sustainability: Linking tourism development to residents' quality of life. *Academica Turistica-Tourism and Innovation Journal, 14*(1). https://academica.turistica.si/index.php/AT-TIJ/article/view/230

Juvan, E., & Dolnicar, S. (2014). Can tourists easily choose a low carbon footprint vacation? *Journal of Sustainable Tourism, 22*(2), 175–194.

Knowles, N. L. (2019). Targeting sustainable outcomes with adventure tourism: A political ecology approach. *Annals of Tourism Research, 79*, 102809.

Leiper, N. (1990). Tourist attraction systems. *Annals of Tourism Research, 17*(3), 367–384.

Liao, C. S., & Chuang, H. K. (2020). Tourist preferences for package tour attributes in tourism destination design and development. *Journal of Vacation Marketing, 26*(2), 230–246.

Li, Q. (2018). *Shinrin-yoku: The art and science of forest bathing*. Penguin.

Li, Q., Nakadai, A., Matsushima, H., Miyazaki, Y., Krensky, A. M., Kawada, T., & Morimoto, K. (2006). Phytoncides (wood essential oils) induce human natural killer cell activity. *Immunopharmacology and immunotoxicology, 28*(2), 319–333.

Liu, Z. (2003). Sustainable tourism development: A critique. *Journal of Sustainable Tourism, 11*(6), 459–475.

Mason, P. (2003). *Tourism impacts, planning and management*. Butterworth-Heinemann.

Mauri, C., & Turci, L. (2018). From ski to snow: Rethinking package holidays in a winter mountain destination. *Worldwide Hospitality and Tourism Themes, 10*(2), 201–210.

McKercher, B. (2016). Towards a taxonomy of tourism products. *Tourism Management, 54*, 196–208.

Minnaert, L. (2020). Stakeholder stories: Exploring social tourism networks. *Annals of Tourism Research, 83*, 102979.

Natural Selfness. (2020). Retrieved March 19, 2023, from http://www.natess.org/index.php/9-english/52-forest-selfness-centre-en

Nguyen, T. Q. T., Young, T., Johnson, P., & Wearing, S. (2019). Conceptualising networks in sustainable tourism development. *Tourism Management Perspectives, 32*, 100575.

Novelli, M. (2022). Niche tourism. In D. Buhallis (Ed.), *Encyclopaedia of tourism management and marketing* (pp. 344–347). Edward Elgar Publishing.

Novelli, M., Schmitz, B., & Spencer, T. (2006). Networks, clusters and innovation in tourism: A UK experience. *Tourism Management, 27*(6), 1141–1152.

Ohe, Y., Ikei, H., Song, C., & Miyazaki, Y. (2017). Evaluating the relaxation effects of emerging forest-therapy tourism: A multidisciplinary approach. *Tourism Management, 62*, 322–334.

Page, S. J., Bentley, T. A., & Walker, L. (2005). Scoping the nature and extent of adventure tourism operations in Scotland: How safe are they? *Tourism Management, 26*(3), 381–397.

Porter, M. E. (2008). *Competitive strategy: Techniques for analyzing industries and competitors*. Simon and Schuster.

Sedarati, P., Santos, S., & Pintassilgo, P. (2019). System dynamics in tourism planning and development. *Tourism Planning & Development, 16*(3), 256–280.

Sharma, G. D., Thomas, A., & Paul, J. (2021). Reviving tourism industry post-COVID-19: A resilience-based framework. *Tourism Management Perspectives, 37*, 100786.

Smith, S. L. (1994). The tourism product. *Annals of Tourism Research, 21*(3), 582–595.

Stamboulis, Y., & Skayannis, P. (2003). Innovation strategies and technology for experience-based tourism, *Tourism Management, 24*(1), 35–43.

Theuvsen, L. (2004). Vertical integration in the European package tour business. *Annals of Tourism Research, 31*(2), 475–478.

United Nations (2015). Sustainable Development Goals. Retrieved March 19, 2023, from https://www.un.org/sustainabledevelopment/sustainable-development-goals

UNWTO. (2008). *International recommendations for tourism statistics 2008 draft compilation guide.* UNWTO. https://unstats.un.org/unsd/tradeserv/egts/CG/IRTS%20compilation%20guide%207%20march%202011%20-%20final.pdf

Varley, P., & Semple, T. (2015). Nordic slow adventure: Explorations in time and nature. *Scandinavian Journal of Hospitality and Tourism, 15*(1–2), 73–90.

Wengel, Y. (2021). The micro-trends of emerging adventure tourism activities in Nepal. *Journal of Tourism Futures, 7*(2), 209–215.

Xu, J. B. (2010). Perceptions of tourism products. *Tourism Management, 31*(5), 607–610.

Zillinger, M., Jonasson, M., & Adolfsson, P. (2012). Guided tours and tourism. *Scandinavian Journal of Hospitality and Tourism, 12*(1), 1–7.

15 Balancing experience production, safety legislation, and protection of the environment in adventure tourism

Arild Røkenes and Sigmund Andersen

Chapter learning outcomes

1 Be aware of challenges related to safety for the protection of people and nature in adventure tourism activities.
2 Understand legislative safety demands in adventure tourism.
3 Recommend strategies and solutions to manage risk, protect people from accidents, and protect nature.

Introduction

Adventure tourism is growing in many natural scenic regions of the world. If the industry develops within effective frameworks it can contribute to value creation and employment in remote regions with migration problems. For example, in Norway, the government aims to not only contribute to adventure tourism development but also include goals related to the triple bottom line of sustainability. However, there is also a fear of tourism, including adventure tourism, having a negative impact on the environment and local culture. At the same time, increased activity in nature can cause unwanted incidents and sometimes fatalities. This chapter suggests ways to improve competence in producing experiences, avoid accidents, reduce the impact on local nature, and develop well-functioning risk management systems. This knowledge is relevant for researchers, operational management, guides, and governmental institutions working on safety and environmental issues. The chapter discusses two main topics:

- What are the specific and unique challenges related to risk management of both safety for people and the protection of nature in adventure tourism?
- How can adventure tourism companies deal with these challenges within the framework of general Health Safety and Environment (HSE) legislation?

Our reflections are based on documentary analysis, literature, interviews, and experiences from research and development projects aiming to develop a quality assurance system for nature-based tourism companies, including adventure tourism. Our aim in this project was to develop the risk management part of the system. Specifically, we investigated risk management systems and legal regulations in Sweden, Norway, the UK, Iceland, and the US, and these were also combined with workshops with Norwegian adventure tourism companies to identify challenges and solutions (Røkenes & Andersen,

DOI: 10.4324/9781003393153-19

2019). The authors of this chapter have also conducted risk management courses for approximately 100 companies in Norway. On these courses, challenges and practical solutions are discussed, and the companies contributed to further understanding and the examples that follow. In addition, both authors have practical experiences from guiding adventure tourism experiences.

Unique risk management challenges in adventure tourism

Use of nature

Paradoxically, a fundamental challenge for the industry is that tourism poses a threat to nature, yet nature is the main product. In other words, you must balance sales, number of customers, and what you do in nature against its tolerance. This is complicated partly because sales and actual behaviour are individual corporate decisions, while nature's tolerance limit is determined by the total load of all businesses (Bricker, 2015; Mason, 2016).

In adventure tourism in particular, the fact that nature is unpredictable is also considered a fundamental challenge. In other industries, procedures are a key element of safety and are often updated after an unwanted accident. Many guides think that this approach has a limited application in the adventure tourism industry because every situation is unique. Therefore, precise or detailed procedures often become irrelevant due to variations in weather, other external conditions, customer composition, and/or a combination of these (Andersen, 2018). See also Chapter 16.

Inexperienced customers

Both guides and business leaders recognise that there are challenges associated with customers being inexperienced, having limited skills, being in poor physical shape, and lacking risk management expertise. That means customers often make mistakes and represent a safety problem themselves. The problem is most evident on short tours with many people because it is difficult to gauge customer competence. On longer tours, many businesses have arrangements for mapping the skills, fitness, and experience of customers, but the same problem may occur because many customers do not understand the difficulty and overestimate themselves or, in some cases, give false information prior to the tour commencing. The problem is exacerbated if the sales department is separated from those who implement the product in practice, because they are aiming for high sales and they may accept participants with a level below the required standard to make the trip viable. This means that injuries in adventure tourism most often affect customers rather than guides (Clayton et al., 2014). The fact that customers are particularly vulnerable is also because some of them seek out and accept the risk because they want to experience mastery, learning, excitement and fun (Cater, 2006). Several of the guides we have been in contact with also mentioned that customers want the guides to arrange for spectacular pictures to be posted on social media (see Chapter 11) and that this contributes to increased risk. These factors are supported by other research and are related to people's need to build identity, status, and recognition through bold achievements involving risk (Langseth & Salvesen, 2018). Other customers in adventure tourism act with great indifference to risk, either due to personality (Rokenes & Mathisen, 2017) and/or in combination with a lack of competence (Cater, 2006). By comparison, experienced adventure tourists are often better trained to collect information and listen to warnings and consequently are less exposed to accidents (Wang et al., 2019).

Customers buy safety and accept risk

As also noted in Chapter 16, our research indicates that customers buy guide services because they want to detach themselves from responsibility. Previous research confirms such a trend. Customers expect the company or guide to render risky experiences harmless (Cater, 2006; Røkenes et al., 2015). Unfortunately, previous research indicates that many companies and guides do not always meet these expectations. First, the guides are not necessarily safety-oriented. The guiding profession has previously been dominated by young and sometimes immature men with a preference for excitement and adventure, and thus, safety has sometimes been given low priority (Morgan, 2000). Second, McCammon (2003) shows that when a group perceives a person as particularly competent, they will not consider safety for themselves. A guide will often become the person with the "halo", the one people follow without thinking. If the guide does not have enough expertise or the right attitude, the risk level will increase. Both guides and business leaders who we have been in contact with have stressed the importance of good competence and the right attitude towards reducing risk (McCammon, 2003).

Another challenge is the broad acceptance that some risk is integral to adventure activities due to the nature of such pursuits. Whereas other industries have zero vision for accidents, adventure tourists accept that accidents can happen, even those which result in severe damage to health:

> Nobody sincerely believes that all recreational activities can be made free of risk. Indeed, some degree of risk is manifestly one of the attractions of many kinds of recreation, and it is clear that people in general are prepared to accept far higher levels of risk in recreation than they would be at work, say, or as the result of the operation of a nearby industrial facility.
>
> (BMA, 1990, p.146)

This means that the risk management system for adventure tourism may consider a wider acceptance of risk to sell their products. Without risk, some experiences may become boring, with reduced possibility for facilitating mastery and learning (see Chapter 12 for an alternative viewpoint on the role of risk in adventure tourism). All of these are important purchase criteria (Beedie, 2003; Pedersen, 2012). This does not mean that companies are exempt from liability. A company or guide must act responsibly and follow industry standards for safety (Collins, 2005). Logically, the next step means that companies must provide information and training to make sure the customer understands the risk they are exposed to, and they must prepare for and mitigate the consequences of unwanted incidents. The more risk there is at play, the more information and training should be implemented, and the more consequence-reducing measures must be prepared. Based on our work with the adventure industry, we have an indication of companies which demonstrate a clear awareness of the importance of competence in risk understanding and risk management. They see the need to minimise risk, depending on the context; use risk (high or low) as an instrument to facilitate experiences; and handle the consequences if the activities lead to unwanted incidents. Several guides point out that this is a demanding balancing act and that everything they do must be within the legal requirements. However, they are often unsure what the laws demand, and they are afraid customers will sue them if things go wrong.

Guides and companies negotiate risk and safety

In our work with the companies and from interviews with guides, we confirm that a problem in relation to risk-taking is that there are diverse types of role conflicts and situations in which guides undertake complicated internal negotiations with themselves. Guides and other types of tour leaders experience this in several ways. First, negotiation occurs in situations where the company has poor finances or has made large investments. This means that those responsible for trips will be forced to bend the line in relation to cancellations and other changes that result in reduced revenues. In short, objective security assessments can be influenced by the economic situation. Second, negotiations can arise in situations where different companies have different practices and make different decisions. For example, several companies in Svalbard offering comparable products mentioned that it is problematic when one company decides to cancel a trip for safety reasons, while others choose to carry out the same trip. This creates pressure to reconsider if customers observe that other companies are going ahead with their trips as planned. A third type of situation demanding negotiation is connected to customers having limited time at their destination. This means they want to partake in activities even if the conditions are unfavourable and it is not safe to do so. This is intensified when customers have less time and they have invested a lot of money in their adventure experience. For example, this was a prominent issue for American heli-skiing guides where both businesses and customers could lose a lot financially on cancelled trips. The interviews also revealed that guides put internal pressure on themselves. Many guides are motivated by bringing out smiles in their customers, facilitating spontaneous joy. To achieve this, one Norwegian ski guide we spoke to noted: 'I sometimes make decisions that may be at the limit of what I should do.' Finally, negotiations are triggered by cultural influences in some parts of the guide community, where status is built by being tough and bold. This can be a form of competition where safety is challenged. Guides do not want to be the ones who cancelled or did the easy route if other guides made a "bolder" decision. However, we have realised that the culture has changed during our decades-long contact with guide communities. Our impression is that guides are becoming more concerned with safety and aware of their elevated level of responsibility. This might be a consequence of the growing offers of guide education, guide associations, and ongoing professionalisation, as described in Chapter 18.

One challenge, noted by several, is that customers have different prerequisites and different attitudes towards risk. This is a security issue because there can be peer pressure to do things which are geared towards the boldest and fittest (and maybe most stupid!) participants in a group, which are too demanding for those with a lower level of skill. This peer pressure results in two outcomes. First, it may put pressure on people to do something they know they should not do to impress others. Second, they feel obliged to do something they know they should not do because they do not want to be a burden on the group (McCammon, 2002). Challenges related to this are amplified because many customers speak inaccurately about their own skills and form, and the guides may thus have problems setting the right level for the group. Companies have different questionnaires and other measures to predict skill and physical fitness before booking, but they have trouble getting precise and reliable answers.

Organisational, practical, and other challenges

A lack of true feedback on risk decisions is another challenge for those engaged in adventure tourism activities. This has been elucidated in recent research on guided tours in alpine skiing (Hallandvik, Andresen & Aadland, 2017; Johnson et al., 2020; Landrø, 2021). For instance, they could have decided to ski a line with serious avalanche danger without releasing an avalanche. A lack of such feedback (Endsley, 2012) leads inexperienced guides to underestimate risk because they have experienced similar incidents where they have been fortunate that their decisions did not have more grave consequences.

It has also been reported that there are challenges related to organisational and practical considerations where time and convenience are weighed against safety. One example is a guide who when travelling to the adventure activity site remembered that he had forgotten the first-aid kit. To go back for it would mean 4 extra hours of travel. The guide chose to defy the risk and did the wilderness trip without the kit. The negative consequences of such a decision could be significant for both the business and the guide. If an accident happened and the missing kit became critical, the guide would be accused of gross negligence and probably face a strong legal response (Collins, 2005).

The companies in our research also report that most injuries are related to slips, trips, falls, and other everyday accidents, as noted in the literature (Bentley et al., 2010). They are often associated with the customers' lack of experience with nature and the cold climate. It is not easy to develop preventive measures, but companies must have a system for dealing with these accidents. The same applies to adverse incidents caused by lifestyle and mental health illnesses. This is rarely linked to the product being sold, but it is interesting that risk analyses in several companies indicate that these illnesses can increase risk; therefore, they clearly require measures.

Some guides also mention another problem related to dealing with heterogeneous groups. They have learned that specific safety measures do not have the same effect on all people. We think this is related to different motivations or goals for participating in the activity. To understand this phenomenon, we need to consider different theoretical explanations of why people undertake risky activities (Langseth & Salvesen, 2018). We mentioned earlier sociological explanations, such as building personal status and identity and the fear of reduced group affiliation. These concerns may "force" people to do something above their competence level, pushing them to engage in risky behaviour. Other types of motivation related to risky behaviour may have psychological explanations. Thrill, excitement, flow, and mastery are feelings that many individuals cannot reach without pushing their risk limits. An example of this is a tragic helicopter accident in Norway when a young pilot (aged 27) undertook sightseeing with a group of youngsters (19–22). His role was both as a pilot and a guide, but the helicopter crashed, and everyone died. The Norwegian Safety Investigating Authority concluded that, among other factors, the crash was linked to the fact that the pilot flew on and under the legal height for a long part of the flight (Norwegian Safety Investigation Authority, 2022). The reason he did so, was first because he wanted to facilitate a feeling of speed and evoke feelings of thrill and excitement. Second, they imply that the pilot was under the (unconscious) influence of the passengers; in other words, he felt under pressure to fly at the limit. Third, the pilot's and the passengers' ages may have contributed to his behaviour. The last assumption reflects earlier research showing that young male car drivers crash more often than older car drivers, and this increases if the car driver has young passengers (Regan & Mitsopoulos, 2001). They do not say it directly, but maybe this is related to a desire to impress others. This means that the problem of customers establishing status and identity or feeling forced to do things above their competence level may also be valid for guides. Clearly, people may have different motives for taking risks. This is a challenge because different motives will demand different strategies in relation to reducing risk.

Formal legislative demands

To understand how companies at organisational level and guides at an individual level manage the challenges and dilemmas they are facing in relation to safety, it is necessary to examine health, safety, and environment (HSE) legislation. In general, this legislation reduces the number of people getting hurt in accidents, prevents damage to health and reduces the impact on local nature. Businesses have a duty to fulfil these demands. That means that strategies and solutions must be rooted in and fulfil the legal requirements. Our investigation indicates that legal requirements are quite similar in many countries, at least in Europe, North America, Australia, and New Zealand. The differences are mostly about how detailed the overall laws are and how many activity-specific regulations the governments have made. In other words, many principles are similar, only the level of detail and specialisation is different. The common features that companies should do are the following:

- Make a HSE policy or develop HSE goals.
- Conduct risk assessments.
- Design measures to protect the environment, their workers, and others from getting hurt by their activity.
- Engage the employees in the processes.
- Report accidents or unwanted incidents.
- Give information and training to employees and customers who use the business's service or product.
- The system and procedures should be documented in written form.

As noted, the previously mentioned countries have additional laws and regulations, but these often have very broad coverage. In Norway, the most important supplying law is the "Act of Products and Consumer Services" (Produktkontrolloven). This law goes further in relation to defining acceptable safety levels, responsible behaviour, and duties. The law recognises that acceptable risk levels can vary in relation to different activities. It acknowledges that acceptable risk varies in time (because we learn and develop technology), and it recognises that acceptable risk varies in relation to external circumstances (like weather). Hence, it is not obvious when the line between acceptable and not acceptable risks is passed, and it is not evident when the service offered reaches the high level of protection the act demands. However, the law says that services are safe when they satisfy national (business) standards. If national standards are missing, international standards will come into play. If there are no standards or they do not cover the product/activity in question, the safety of the product shall be assessed by considering in particular: other national standards from other business areas, commission recommendations, guidelines for product safety assessment, good practice rules applicable in the relevant area, the current level of technical development, and the level of security that can reasonably be expected of the user. As we understand this in a tourism context, it means that international and national business standards and how other similar businesses or experts act may become important if a court of law challenges safety levels. Implicitly this means that businesses must be proactive in seeking standards and to not do so would be negligent. The next stage would be that the company could be criminally liable for accidents.

Furthermore, the act says that businesses have a duty to behave with care and conduct reasonable measures to achieve the mission of the act. It clearly specifies that the consumer also has a duty to act responsibly when using a product or service. An important requirement within the act is that the company is obliged to obtain the necessary knowledge to evaluate risk and give relevant and sufficient information to customers which allows them to consider safety and ultimately protect themselves. The information should be clear, easily accessible and adapted to the customers' needs. In other words, businesses, their employees, and the customers have a duty to behave responsibly. This opens the possibility of shared guilt if accidents happen. We will assume that the line between responsible and not responsible will follow the same logic as the process of defining what is safe and responsible behaviour. For instance, the US has similar regulations and requirements (Collins, Film 2005).

However, our work in Norway indicated that formalised HSE and risk management have been given low priority in nature-based tourism, and knowledge of the previously discussed regulations is limited among businesses. This does not mean that companies have forgotten safety but that they have worked with it and decoupled it from the legislation. They have worked with this as an important goal, and it is an integral part of product development. The lack of a systematic approach and formalisation can be related to the fact that the industry is new, it is growing rapidly and consists of many small companies, and there are relatively few serious accidents. We would like to emphasise that many of the established companies have high-functioning HSE systems, well within the statutory requirements as we interpret them. An important additional factor contributing to the situation can be connected to limited attention from public supervisory authorities (e.g., health and safety executive) who are often interested in more traditional workplace settings. They have little knowledge about the sector, and thus, few specific requirements and orders. An exception might be the Adventure Activities Licensing Authority in the UK (see Case study 15.1), although it is notable that its inspection functions are outsourced to an external provider with specific expertise.

Case study 15.1   Adventure Activities Licensing Authority (Cater, 2015)

The Adventure Activities Licensing Authority (AALA) is a UK body sponsored by the Health and Safety Executive (HSE) to licence commercial activity providers for young people under the age of 18 years. Concern with the growing commercialisation of adventure activities in the UK led to the introduction of adventure activities licensing in April 1996. Licensing is applied for caving, climbing, trekking and watersports when they are done in remote or isolated places. Whilst licensing only applies to operators working with young people in these specific settings, in practice, many who serve a broader market hold AALA accreditation. There has been a steady growth in the number of licence holders, with 1,161 in the UK in 2019. UK government proposals in 2012 suggested an abolition of the AALA in the review of health and safety legislation, Common Sense, Common Safety. This legislation suggested replacing the current licensing requirement with a code of practice within existing HSE requirements. To some extent, this corresponds to best practice and healthy reporting cultures identified by Bentley et al. (2010). However, following a review of the proposal, and responses from devolved administrations, this proposal was put on hold.

Adventure Activities Licensing Authority www.hse.gov.uk/aala/.

Solutions and handling strategies

In this section, we reflect on and give examples of how guides and companies overcome the previously mentioned challenges, and we link them to the general demands in legislation. First, in some cases, we see that the challenges can be managed within the legal framework, for example better risk assessment. In other cases, we may observe that the solution used lacks anchoring to the legal framework. In other words, the legislation has shortcomings. Third, the legislation may be counterproductive in relation to the aim of better safety.

Safety management in complex and unpredictable situations

Traditional safety management systems assume that experience gained from an unwanted incident should be used to predict future incidents and create or revise rules to avoid the same mistakes in the future. When it comes to adventure tourism and nature-based activities, this reactive approach is complicated because so many variables come into play and make situations unforeseeable. Hence, it is difficult to make rules or suggest actions for all dangerous situations (Faarlund, 2016). Rule-based thinking can, in the worst case, lead to employees making wrong decisions to satisfy their actions. Thus, a key question is whether safety systems in nature-based adventure tourism should be based on rules or should there be greater emphasis on developing competence that reduces the need for these. In other words, guides must have experience and the ability to be aware of all the changing elements of nature. Furthermore, they must be constantly aware of and evaluate the risks for the group and make the right decision with the most positive outcome in the actual situation. This process of decision-making in the field of guiding is related to situational awareness as 'the perception of the elements in the environment within a volume of time and space, the comprehension of their meaning, and the projection of their status in the near future' (Endsley, 2012, p.533). A prerequisite for guides is experience from different situations. This is about continuously evaluating risks and consequences, asking the right questions, and finding the optimal solution (see Case study 15.2).

Case study 15.2   Life and death in Arctic tourism

A story told by a guide illustrates the dynamic decision-making process described earlier:

A group of 25 people and four guides are on a glacier course on Svalbard, the land of polar bears. The polar bear is not only highly protected but also very dangerous. One night a polar bear enters the camp, and it is 25 metres from the tents. The guides are ready, and the set protocols are to shoot the bear, because it does not react as normal and flee after use of flares. There follows negotiation about saving a threatened species and protecting course participants. Polar bears are known to be unpredictable. Time is short. If the guides had followed a rule-based approach, the bear would be shot. But the guides were experienced and used handling competence; they had alternatives. Two guides aimed at the bear while the other two got all the participants out of their tents. They gathered the group on the beach behind the guides. The camp was quickly abandoned and left to the polar bear. The group were thinly clothed, and this could have been a problem, but the guides knew about a cabin close by and headed for that. The bear survived, and the group were all right, although much of the equipment was destroyed.

Figure 15.1 Polar bear in camp on Svalbard. Photograph by Sigmund Andersen.

Our research indicates that this adaptability should be a central approach to risk management. This means that the guide must have the opportunity to deviate from protocols for discovering and adopting novel solutions or that these are designed in such a way that allows for discretion. Action competence is related to transparent guiding (Løvoll & Einang, 2022). Transparent guiding is the guide's ability to make connections to all the customers, understand their needs, predict their capacity, and facilitate interaction. Transparent guides are characterised as inviting their customers to discuss specific safety challenges and explaining the reasons for the tour plan and the decisions. Research on ski guides' trip planning (Løland & Hällgren, 2022) showed that social considerations changed the importance of conditions in nature. Our research indicates that a transparent approach to guiding can enhance the customer's ownership of risk management and facilitate learning and understanding, in contrast to a more instructive method with ready-made solutions. This means that an organisational culture must be created where both formal and informal norms allow guides to be made aware and develop the ability to take responsibility in demanding situations. Adventure companies should have a well-thought-out and defined set of values that lay the foundation for continuously discussing attitudes towards risk, people, and views of nature. In this way, overarching guidelines are laid down for unforeseen issues and new situations rather than fixed rules. The prerequisite for this lies in the fact that the company and its employees have the right competence.

Competence requirements

General legislation is relatively vague when it comes to defining what competence is needed to offer different activities in the adventure tourism industry. On one hand, this makes sense because the laws should be robust in relation to time and different activities and should not be changed too often. However, competence demands change and development, and what was considered enough competence at one point could be quite different at a later stage. For instance, some guide companies now demand that guides should be able to operate an AED (defibrillator), but some years ago, this was not possible because of the high cost of this equipment.

However, many companies and guides operate the same type of guiding without formal competence, or they may have acquired their education from other competing umbrella organisations. This means that there are conflicts in this industry, both in relation to what proper competence is and the balance between formality and lengthy practical experience and in relation to who is competent to educate, certify, or approve guides. In the big picture, competence requirements and certifications are perceived as inadequately defined and practised regardless of activity. Our research indicates that many companies and other stakeholders call for defined clear standards. Such standards are important because a key question in any court case rests on whether the company has expertise and safety systems in line with industry standards (Collins, 2005). This means that court cases can set a precedent for what constitutes justifiable competence. There are international standards that might be relevant in such cases; for instance, the International Organization for Standardization developed a risk management system for adventure operators: https://www.iso.org/standard/54857.html.

Based on this, there are no clear answers in relation to what competence is needed, even if we just focus on one activity. The demands are dynamic and change over time because of technological developments, changed norms and values, or even learning points after serious accidents. To have proper competence will depend on how well companies and guides are updated on the demands of other companies and the demands from tour operators and other customers. Companies and guides need to be updated on relevant changes to regulations and results of court cases. Finally, it is important for businesses to analyse the activities or products they offer; what kind of competence is needed to offer river rafting, for instance? In this process, businesses need to analyse what customers need to know and what kind of competence they need to buy the experiences offered. A supplementary solution to identifying what is "enough" competence is related to benchmarking against other companies or court decisions. For small companies with small administrative resources, this is especially difficult, but it can be improved by more and better inspections from governmental authorities which help companies to understand and implement necessary HSE actions. In Norway, very few companies have experienced inspection that addresses the problematic side of HSE requirements. Unfortunately, this has a connection to low competence in adventure activities in the relevant governmental institutions.

Regardless, there is clearly a need for high competence related to handling risk. Adventure companies must manage and facilitate risk levels to give customers a variety of emotions, offer challenges for both experts and beginners, and never breach the safety measures demanded by legislation. We illustrate this in Figure 15.2.

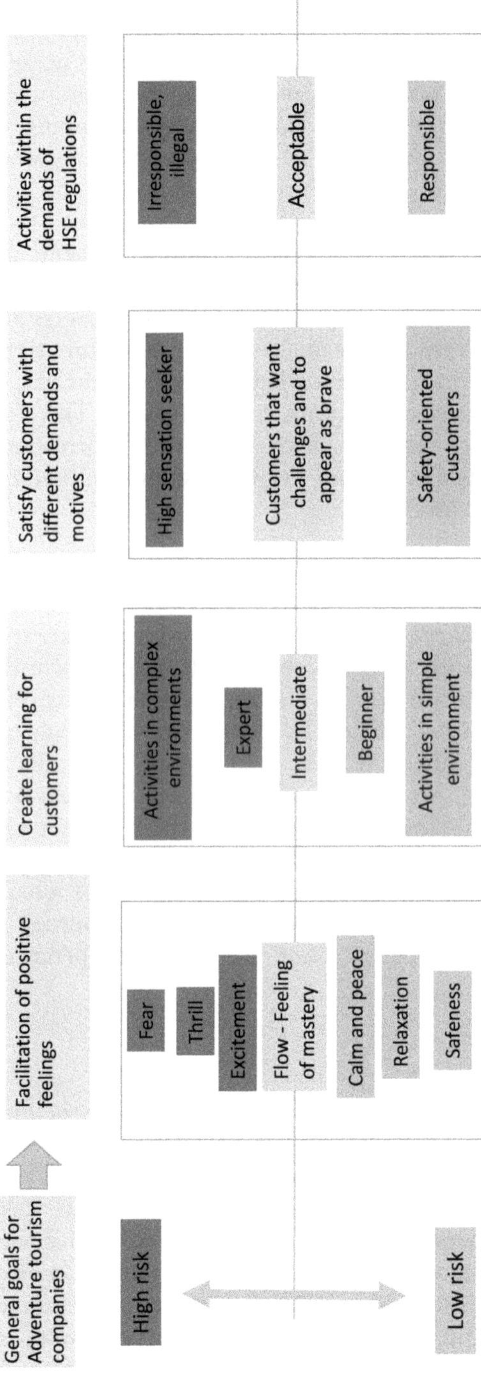

Figure 15.2 Considerations for adventure tourism companies in relation to experience production, learning of customers, general market demands, and legal requirements.

Developing competence

Competence in making the right decisions in challenging safety situations develops through experience with various real-life situations. Klein (1998) explains how experts make good intuitive decisions in complex and uncertain situations. This is on the basis that after years of experience from real-life situations, one has acquired expert intuition. This intuition will most often produce a quick correct solution option. Both Kahneman and Klein (2009) recognise professional intuition decision-making, and that professionals can make exceptionally good decisions. In addition, it is essential to develop the guides' ability to reflect on their own and others' experiences and create an awareness in the guide. The guide must train their ability in real-life situations to evaluate the risks to the group and the consequences of the diverse options for action. This competence cannot be learned only theoretically but requires a well-organised learning community in the field under the guidance of competent and experienced teachers. Such competence is a safety cornerstone for those responsible in a risk-exposed nature setting. It is difficult for an external consultant with in-depth knowledge of schematic risk management to perform an analysis that will cover every conceivable situation. Therefore, a key element in guide development is to learn not to lock oneself into set forms and procedures but to train independently to be aware of, and act in, new and unpredictable situations and ensure the best decisions are taken.

Formulation of goals for HSE compliance

The general HSE regulations require all companies to formulate their own HSE objectives. Through discussions with industry, there is clear agreement that this is an important strategy and a foundation for facilitating a desired organisational culture, in accordance with what was noted earlier. The formulation of goals can provide direction for what kind of company you are. If there is a company that offers high-risk products, it should be stated here. The same is the case if you want to be a company which offers low-risk products. What you choose also indicates which market segments the products are suitable for and hence dictates elements of marketing and sales strategies. Another point is that such goal formulations can provide direction for what the company needs in terms of expertise, equipment, insurance, and other resources. In other words, the more complex products you plan to offer, the more competence is needed in your organisation.

Experience production aspects

However, risk identification and risk analysis must consider the experience production aspects. Risk assessments do cover conduct to minimise the risk of harm to people and the environment but rarely consider risk as part of the experience. The paradox is that large parts of nature-based tourism focus on experiences where risk is an inherent part which is accepted by customers. In our earlier work related to this, we concluded that one solution is to define risk categories that accept the risk, but the more risk is involved, the more risk information and instructions for the customer are required. In addition, high-risk activities will demand more competence and better planning and preparations for handling potential consequences. In practice, this means that the analyses can accept considerable risk if the company compensates for this with training and information about risk to customers and has the resources and expertise to manage risk. Another thing mentioned by companies is that the process of risk analysis raises awareness and contributes to employees and management thinking about consequences daily. This means that risk analyses can play a role in changing the organisational culture in the desired direction.

Colour-coded safety plans for activities

Through our research on adventure tourism companies, safety plans for all products have been suggested as an appropriate way to implement risk management. This can be done in several ways, and each business must consider what is right for it. Nevertheless, good safety plans can solve the challenges associated with dealing with groups with different skills, physical fitness, and lack of competence. These plans incorporate actions that ensure that the company consciously collects information about safety before and during the implementation of a product. Furthermore, safety plans can clarify the guides' autonomy and total responsibility for safety as the priority. This can help curb all the forms of cross-pressure or conflicting demands mentioned earlier in the chapter. Finally, companies believe that these plans should have actions and checklists which help ensure that important aspects of implementation are not forgotten and that employees have clear procedures for crisis situations. For other more complicated issues, there should be guidelines mentioning important aspects to take into consideration. It is also proposed to have an overall description of the product or tour, important risk elements, and a risk category for this specific product. That can also be communicated to customers to ensure that they understand what they are exposing themselves to by participating. We also suggest that such information is colour-coded with a universally accepted green–amber–red scheme to make it easy for participants to recognise. The suggested categories are presented in Table 15.1.

In our suggestion of how such a safety plan can be designed, we propose that safety communication and mapping of customers starts early in the sales process, either as a

Table 15.1 Risk-level communication for adventure tourism activity (Røkenes & Andersen, 2019)

Risk category	Product category
1	Activities with minimum risk beyond normal daily risks. Fits anyone with normal health and physical fitness level.
2	Activities with low risk above everyday risk. Potential for few accidents and minor injuries that will be treated locally. Suitable for people with normal health and fitness levels.
3	Activities with some risk. Suitable for people with good health. Some accidents could occur, and professional help might be needed.
4	Activities with significant risk. Fits people with good health and specialised skills and experience for the activity. Accidents happen and professional help might be needed. If the participant does not follow all orders and instructions serious accidents might happen.
5	High risk activity. Fits people with extreme skills and physical form. Customers must assume full responsibility.
6	Extremely high risk activity which is not normally carried out by tourism companies.

part of the booking system or as the protocol for those who communicate with the customer by email or phone. An essential element in safety planning is that the company should define or clarify what competence is needed to have a lead or subordinate role in producing the experience.

Near-miss reporting

Registration of undesirable incidents and assessment of measures is a requirement in accordance with best practice. This is a valuable tool in terms of both disseminating knowledge and designing better protocols. The companies report that this part of HSE work often fails, especially in relation to "near-miss incidents". Employees do not always write accident or incident reports either because they want to hide their own mistakes or because it is time-consuming and difficult. In the discussions about this, many believed that non-conformity reporting needs to be simplified and have digital solutions designed for mobiles. The discussions also clarified that management has a clear responsibility to help ensure that non-conformity reports are used for follow-ups when they are delivered. Creating a healthy reporting culture is fundamental to risk management and has been shown to be an ongoing challenge for adventure tourism businesses (Bentley et al., 2010).

Conclusion

The chapter describes important challenges in relation to risk management and HSE work in nature-based adventure tourism seen from a research and business perspective. The dynamic nature of risk in adventure tourism does mean that there is no one-size-fits-all approach that we can adopt. However, there are some values behind risk management that are universal, most notably a proactive approach. The risk management challenges are many and may relate to the following:

- The complexity of caring for nature and at the same time using it
- Managing risk in unpredictable and complex environments
- The lack of true feedback on safety considerations in nature
- Building a system matching heterogeneous customers with different skills, competence, and experience
- Risk as an element of experience production, where the balance between safety and thrilling experiences is challenged
- In general, the adventure industry offers products varying from extremely high risk levels to very low risk, and this demands a variety of competence
- Guides and companies must negotiate safety alongside economy, convenience, cultural demands and personal situations
- Companies must deal with injuries that are not always related to the products they offer
- Different motives for risk behaviour demand different safety protocols
- Unclear legislation demands more administrative resources than small businesses may possess

Regarding how these challenges can be handled, the basic step is to gain knowledge of legal demands and implement these in daily operations and the risk management system.

This will require risk analysis and safety planning which aims to both prevent unwanted incidents and manage the consequences if they nevertheless happen. Checklists, protocols, and standard operating procedures are still needed but have significant limitations because of the complexity and unpredictability when conducting activities in nature. Companies should work towards improving organisational culture and facilitating the development of competence in the organisation. By doing so, companies may manage unforeseen risks that are not covered by protocols and procedures. For both companies and researchers, there is a need for an increased understanding of risk behaviour from different theoretical perspectives. Better insight into both sociological and psychological explanations will increase the relevance and accuracy of safety measures. We also propose alternative theoretical contributions when it comes to how adventure products are designed. Both literature and our experiences from different projects indicate that many companies have a hedonistic point of departure, where they aim to satisfy their customers over a short time and fulfil their needs for fun, sensation, and thrill. There are alternatives to these where other motives become more important. One possibility is to draw attention to eudaimonic motives which are more complex and revolve around human flourishing based on meaning, connection to others, self-transcendence, and self-development (Frühauf et al., 2022; Hetland et al., 2018), as discussed in Chapter 12 on well-being. Another possibility is to draw attention to other aspects of "good" experiences. In other words, we should make it less important to achieve pre-announced goals like reaching a specific peak and instead facilitate elements such as connection to nature, enjoying good food, and hospitality (Varley et al., 2018).

It is important to point out that the challenges and solutions presented are not complete. Risk management is a continuous process where new knowledge and experience are being used for improvement. It is quite clear that in this chapter, we have only scratched the surface of the complexity of different contexts. Further research is needed to fill the competence gaps and establish lines to general research within risk management and other disciplines (see Chapter 16), even if the tourism area is unique. We would like to emphasize that there is a need to better understand how risk is used to facilitate emotions, coping, and learning and how this can be adapted to the formal legislation which states quite clearly that risk should be minimised. We see a need to go deeper into the mechanisms that create cross-pressure and role conflicts. We also see a need for an increased understanding of how rule-based systems can be combined with competence management, where in some situations considerable autonomy and clarification of competence requirements are required. Finally, we have not problematised the differences and similarities between real risk, perceived risk and legally defined risk. In one context, all three may be defined differently. Is this a problem? Can it be used in experience production, and how?

Review questions

1 What legal requirements for companies related to health, safety, and the environment can be claimed to be international?
2 What does it mean for companies' management systems that those most vulnerable to accidents and injuries are the company's customers?
3 How can one work to prevent guides from having to negotiate safety and environmental protection against companies' finances?
4 Why is it that a goal of facilitating customer learning in some situations challenges a goal of safe experiences?

Further reading

Cater, C. I. (2006). Playing with risk? participant perceptions of risk and management implications in adventure tourism. *Tourism Management, 27*(2), 317–325. https://doi.org.login.ezproxy.ulapland.fi/10.1016/j.tourman.2004.10.005

Løvoll, H., & Einang, O. (2022). Transparent guiding: Contributions to theory of nature guide practice. *Scandinavian Journal of Hospitality and Tourism, 22*(2), 95–110. https://doi.org/10.10 80/15022250.2021.1955738

References

Andersen, S. (2018). Guiding i polare strøk. In L. I. M. Ed (Ed.), *Friluftsliv og guiding i natur. Teori og praksis* (pp. 189–203). Universitetsforlaget.

Beedie, P. (2003). Mountain guiding and adventure tourism: Reflections on the choreography of the experience. *Leisure Studies, 22*(2), 147–167. https://doi.org/10.1080/0261436032000068991

Bentley, T. A., Cater, C., & Page, S. J. (2010). Adventure and ecotourism safety in Queensland: Operator experiences and practice. *Tourism Management, 31*(5), 563–571. https://doi.org/10.1016/j.tourman.2009.03.006

Bricker, K. (2015). Sustainability in Adventure travel and programming: Maximaizing Benefits for cultural Preservarion, Nature Conservation, and local Residents Well-Being. In R. A. K. B. Black (Ed.), *Adventure Programming for the 21st Centrury* (pp. 459–480). Venture Publishing Inc.

British Medical Association (1990) *The BMA guide to living with risk* (2nd edition). Penguin.

Cater, C.I. (2015) Adventure Activities Licensing Authority. In Cater, C., Garrod, B. and Low, T. (eds.) *The Encyclopaedia of Sustainable Tourism.* CABI.

Cater, C. I. (2006). Playing with risk? participant perceptions of risk and management implications in adventure tourism. *Tourism Management, 27*(2), 317–325. https://doi.org/login.ezproxy.ulapland.fi/10.1016/j.tourman.2004.10.005

Clayton, A., Mustelier, L. C., & Korstanje, M. E. (2014). Understanding perceptions and attitudes to Risk in the Tourism Industry. *International Journal of Religious Tourism and Pilgrimage, 2*(1), 48–57.

Collins, D. (2005). *State of Risk: Risk managment in the outdoor industry.* Knutson and Associate (Film)

Endsley, M. R. (2012). Situation Awareness. In *Handbook of human factors and ergonomics.* (4th edition, p. 1). John Wiley & Sons. https://doi.org/10.1002/9781118131350.ch19

Faarlund, N. (2016). *Samlede Verker, VII: Fjellskiløping og Snø, snøskred og redningstjeneste* (Vol. 7). BoD–Books on Demand.

Frühauf, A., Houge Mackenzie, S., Boudreau, P., Hodge, K., & Kopp, M. (2022). Multiple Motives for Adventure Sport Revisited: A Multi-Activity Investigation. *Leisure Sciences,* 1–23. https://doi.org/10.1080/01490400.2022.2126910

Hallandvik, L., Andresen, M. S., & Aadland, E. (2017). Decision-making in avalanche terrain–How does assessment of terrain, reading of avalanche forecast and environmental observations differ by skiers' skill level? *Journal of Outdoor Recreation and Tourism, 20,* 45–51.

Hetland, A., Vittersø, J., Oscar Bø Wie, S., Kjelstrup, E., Mittner, M., & Dahl, T. I. (2018). Skiing and Thinking About It: Moment-to-Moment and Retrospective Analysis of Emotions in an Extreme Sport. *Frontiers in Psychology, 9.* https://doi.org/10.3389/fpsyg.2018.00971

Johnson, J., Mannberg, A., Hendrikx, J., Hetland, A., & Stephensen, M. (2020). Rethinking the heuristic traps paradigm in avalanche education: Past, present and future. *Cogent Social Sciences, 6*(1), 1807111.

Kahneman, D., & Klein, G. (2009). Conditions for Intuitive Expertise: A Failure to Disagree. *American Psychologist, 64*(6), 515–526. https://doi.org/10.1037/a0016755

Klein, G. (1998). *Sources of power: How people make decisions.* MIT Press.

Landrø, M. (2021). *Why is it safe–enough?* Decision-making in avalanche terrain. https://munin.uit.no/handle/10037/22697?show=full

Langseth, T., & Salvesen, Ø. (2018). Rock Climbing, Risk, and Recognition. *Frontiers in Psychology, 9*(1793). https://doi.org/10.3389/fpsyg.2018.01793

Løland, S., & Hällgren, M. (2022). 'Where to ski?': An ethnography of how guides make sense while planning. *Leisure Studies, 6,* 1–17.

Løvoll, H., & Einang, O. (2022). Transparent guiding: Contributions to theory of nature guide practice. *Scandinavian Journal of Hospitality and Tourism, 22*(2), 95–110.

Mason, P. (2016). *Tourism impacts, planning and management* (3rd ed.). Routledge.

McCammon, I. (2002). Evidence of heuristic traps in recreational avalanche accidents. *Proceedings ISSW*. issw-2002-244-251.pdf

McCammon, I. (2003). Heuristic traps in recreational avalanche accidents. *Work, 406*, 888–926.

Morgan, D. (2000). Adventure tourism activities in New Zealand: Perceptions and management of client risk. *Tourism Recreation Research, 25*(3), 79–89.

Norwegian Safety Investigation Authority. (2022). *Report on air accident in the Skoddevarre mountains, Alta municipality, Troms og Finnmark county, Norway on 31 August 2019 with Airbus Helicopters AS350 B3, LN-OFU, operated by Helitrans AS* (Aviation report 2022/02). https://www.nsia.no/Aviation/Published-reports/2022-02?pid=SHT-Report-ReportFile& attach=1

Pedersen, A.-J. (2012). *Opplevelsesøkonomi: Kunsten å designe opplevelser*. Cappelen Damm Akademisk.

Regan, M. A., & Mitsopoulos, E. (2001). *Understanding passenger influences on driver behaviour: Implications for road safety and recommendations for countermeasure development*. Monash University Accident Research Centre.

Rokenes, A., & Mathisen, L. (2017). Roles of adventure guides in balancing perceptions of risk and safety. In Y.-S. Lee, D. Weaver, & N. Prebensen (Eds.), *Arctic tourism experiences: Production, consumption and sustainability* (p. 199). CABI.

Rokenes, A., Schumann, S., & Rose, J. (2015). The Art of Guiding in Nature-Based Adventure Tourism - How Guides Can Create Client Value and Positive Experiences on Mountain Bike and Backcountry Ski Tours. *Scandinavian Journal of Hospitality and Tourism*, (s1), 62–82. https://doi.org/10.1080/15022250.2015.1061733

Røkenes, A., & Andersen, S. (2019). *Risikostyring og HMS i naturbasert reiseliv – en håndbok* (Vol. 1). UIT- Arctic University of Norway. https://doi.org/10.7557/7.4629

Varley, P., Farkic, J., & Carnicelli, S. (2018). Hospitality in wild places. *Hospitality & Society, 8*(2), 137–157. https://doi.org/10.1386/hosp.8.2.137_1

Wang, J., Liu-Lastres, B., Ritchie, B. W., & Pan, D.-Z. (2019). Risk reduction and adventure tourism safety: An extension of the risk perception attitude framework (RPAF). *Tourism Management, 74*, 247–257. https://doi.org/10.1016/j.tourman.2019.03.012

16 Risk management for guides and adventure businesses

Jon Heshka

Chapter learning outcomes

1 Understand the five steps involved in the risk management process.
2 Determine how to calculate the risk rating of an activity or a trip.
3 Identify how guides make risk management decisions.
4 Detail the factors adventure businesses must account for in managing risk.
5 Understand the paradox of adventure businesses promoting uncertainty to clients but reducing risk in its operations.

Introduction

The International Organization for Standardization (ISO), in its introduction to the ISO 31000 standard on risk management, flatly states: 'Risk is a necessary part of doing business' (p. 1). To riff Mark Twain (Brennan, 2014, p. 37), 'some risk is good, more is better but too much is bad'. Adventure businesses and adventure guides manage risk by monitoring, treating, and transferring it and consciously retaining residual risk at an appropriate level.

Risk as a concept is generally misunderstood and often gets a bad reputation. It can be positive or negative, smart or stupid and is not something to be averse to at all costs but rather carefully managed. This chapter defines risk in the context of adventure tourism and explores its application in the business of adventure and adventure guiding.

The risk management processes discussed in this chapter are not unique to adventure but apply equally to all kinds of organisations and their organisational units. For example, all organisations – whether a factory that makes widgets or a whitewater rafting company that creates experiences – have missions and goals which include a statement of purpose and financial markers like revenue, profit, and market share. They should also have overarching risk management statements concerning their goals (i.e., units sold, profit, growth), risk appetite (i.e., the amount of exposure an organisation is prepared to take on the basis of risk–return trade-offs), its risk philosophy (i.e., its relationship with risk; is it something to be avoided like workplace injuries or is it a pedagogically necessary tool to help participants fulfil their potential?), its processes to assess the risk (i.e., identify, analyse, and evaluate the risks), and its strategies and tactics to treat the risk (i.e., reduce the likelihood of encountering the hazard, reduce the

DOI: 10.4324/9781003393153-20

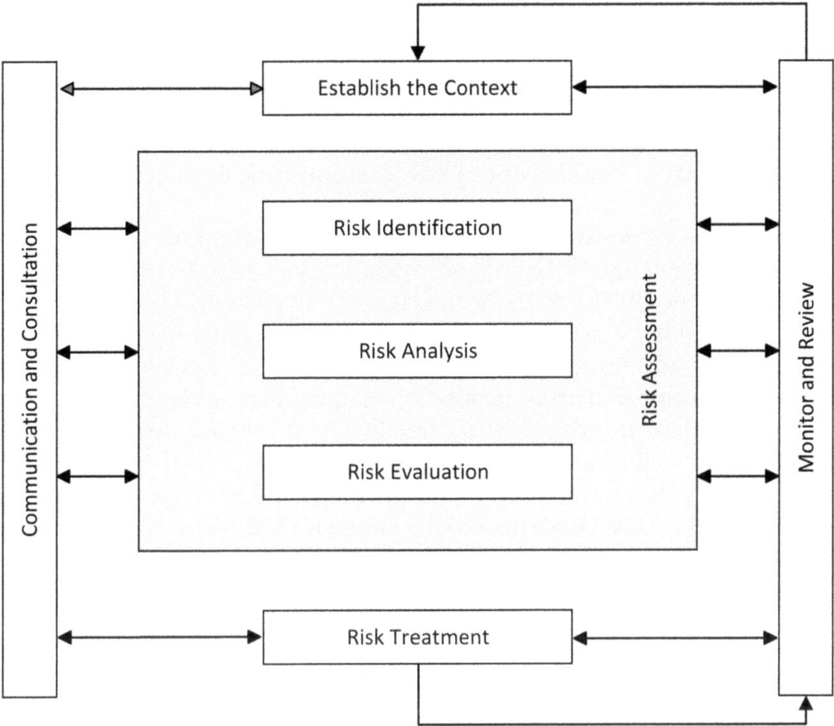

Figure 16.1 ISO 31000 Risk Management Process. (Adapted from ISO, 2018).

exposure to the hazard, and take measures to reduce the consequences – if it happens – of an encounter with the hazard), or transfer the risk (i.e., through insurance or releases of liability).

The processes that an adventure guide uses to manage risk are no different than what an investment broker or insurance agent would do. They each must work in such a way that is consistent with the business's risk philosophy and risk appetite, employ processes to assess the risk, and figure out ways to treat the risk. The obvious differences though are the setting – corporate versus wilderness – and salaries. Figure 16.1 shows the risk management process sanctioned by the Swiss-based International Organization for Standardization (ISO) and this is referred to throughout this chapter.

Risk in adventure

Risk is the effect of uncertainty on objectives (ISO 21101, p. 8). The effect can be positive or negative. Examples of positive risks are buying a winning lottery ticket or buying early into a housing market that increases in value. The negative risks in these examples are the cost of buying the lottery ticket, which is minimal, and the cost of purchasing a home, which is much more substantive. The odds of winning the lottery are extremely remote – although that does not stop millions of people from buying tickets – and while many people have profited from investments in residential real estate, the housing market crash of 2008–2009 is a reminder that that was not a sure thing either.

Risk is always a probability issue (British Standards Institution, 2016). The decision-making takes into account the odds of success versus how much is wagered. It is ultimately a risk–reward or cost–benefit issue, and it boils down to whether the decision-maker determines if the risk is worth it or if it is justifiable due to the benefits obtained. The U.S. Federal Aviation Administration highlights in its risk management training for pilots that 'a key element of risk decision-making is determining if the risk is justified' (2009, p. v).

There is value in adventure and risk. There is worth in climbing a mountain 'because it's there' as George Mallory famously said when asked why he would want to climb Mount Everest (Forbes, 2001). Without risk, there is no growth. The price of choosing to avoid tenuous situations is torpidity, and the result is stagnation (Quinn, 1990). The downsides to risk in adventure include loss of life or limb (i.e., death or critical injury), psychological or emotional distress, property (i.e., gear) loss, and reputational harm. The upsides include feelings of independence, freedom, self-reliance, resilience, community, camaraderie, mental wellness, and physical fitness.

It is these benefits that have spawned burgeoning outdoor recreation and adventure tourism markets. The U.S. Department of Commerce's Bureau of Economic Analysis calculated the economic output of outdoor recreation to be US$862 billion in the United States alone, surpassing industries such as mining, utilities, farming and ranching, and chemical products manufacturing (Outdoor Recreation Roundtable, 2023). Grand View Research, Inc. estimates that the global adventure tourism market was valued at US$282.1 billion in 2021 (Outdoor Industry Compass, 2022). However, the costs or the downsides of risk – death, physical or mental injury, and so on – should be properly considered and truly accepted when participating in adventurous activities. Participants must be mindful that even when proper risk treatment measures are undertaken to reduce the activity's inherent risks that there will remain residual risk.

Distinguishing features of adventure (described in Chapters 1, 2, and 3) are that risk is often integral to the experience (Thomas & Raymond, 1998, p. 278) and is inherently dangerous (Heshka, 1996). Its perils are baked into the activity. The risks are inseparable from the very nature or essence of the activity. Remove the risk and the activity is no longer the same. Drowning and falling are inherent risks of whitewater rafting and climbing. The only real way to prevent any possibility of that ever happening would be to raft on dry land or revoke the law of gravity. The challenge then for recreationists, guides, and adventure businesses is to manage risks in such a manner that the activities can still reward their participants while not unnecessarily exposing them to hazards. A hazard is generally recognised as a source of potential harm or something that increases the consequences and the likelihood of a loss.

How adventure guides and businesses manage risks

It is understandable for ownership and management to avoid programme areas or activities that they consider too risky. It is not disconnected from the idiom 'A fool and his money are soon parted'. Why do things if there is a chance people will get hurt or killed with the consequential blowback including reputational harm, negative publicity, possible lawsuits and payouts of damages, lessened market demand, lower revenue, and lower profits? For example, in 2000, the Swiss-based company Adventure World went out of business after a client died bungee jumping. One year later, 21 clients were killed while canyoning. That story was on the front page of the *New York Times* and other global media in 1999. However, a lack of such opportunities would have the effect of depriving participants of the benefits of the adventure activity. Thus, the need for 'risk management that reduces the deterrent effects of uncertainty about potential future accidental losses by making these losses less frequent, less severe, or more foreseeable' (Elliott, 2018). It has already been established that risk is the effect of uncertainty on objectives and that its outcome can be either positive or negative. The focus of this chapter is on negative risks and, in particular here, how guides manage those risks.

Risk in adventure guiding – and life – is an expression of the combination of the consequences of an event and the associated likelihood of occurrence (ISO 21101, 2014). The word *risk* is fraught with ambiguity because it can mean different things to different people. Its meaning is associated with people's experience and tolerance. Inexperienced people often perceive a risk to be higher than it actually is. Even experts can have different perceptions based on their risk tolerance (Jackson & Heshka, 2011). Because of the challenge of articulating risk – how do we know when something is "too risky" or is "a little risky", and what does that actually mean anyway? – a process has been developed to assess, analyse, and evaluate the risk to determine if the risk is acceptable.

Consequence refers to the outcome of an event, meaning the severity or impact of engaging with the hazard, that is, the degree of injury or harm suffered. This can range from a worst-case scenario of a mass casualty incident with multiple fatalities to critical non-life-threatening injuries like a broken arm or leg to minor injuries like sprains and strains. *Likelihood* refers to the chance of something happening, and it ranges from a near certainty to a risk event actually occurring to it possibly occurring to it almost certainly not happening. This can be scaled and expressed qualitatively or quantitatively. ISO 31000 and ISO 21101 define risk in terms of a combination of the consequences of an event and the associated likelihood of occurrence. The US Federal Aviation Administration similarly defines risk in terms of severity and likelihood.

Likelihood takes into account both the probability of a risk event occurring and exposure to the hazard or the extent to which people are exposed to the hazard whether it is the amount of time, the number of people involved, or the number of times people cycle through possible exposure to the hazard. Some models have probability and exposure subsumed by likelihood so that risk is an expression of only consequences and likelihood. An example of this is the Risk Rating Model shown in Figure 16.2. Other models, however, explicitly include probability and exposure which, when combined with severity (consequence), makes for a SPE acronym (severity, probability, exposure), as used by the US National Park Service's search and rescue programme. The US Occupational Safety & Health Administration (OSHA) also uses SPE. It evaluates 'each hazard by considering the severity of potential outcomes, the likelihood that an event or exposure will occur, and the number of workers who might be exposed' (OSHA, 2023). A further development of SPE is the Risk Analysis Model shown in Figure 16.4. Regardless of the approach undertaken, it is necessary to account for the total amount of time and the number of people possibly exposed to the hazard. It can be scaled and expressed qualitatively or quantitatively.

Risk assessment

Estimating the consequence and likelihood occurs during the risk assessment process (ISO 21101, 2014) which is a three-step process that identifies the risks, analyses the risks, and evaluates the risks. The risk identification step involves establishing and maintaining a systematic and structured process for the continuous identification of risks. It must account for both field activities and business operations. Risk identification in the field is done by the guide/instructor. This is described in detail later on in the chapter. The process of identifying business and programmatic risks should never be done alone but instead by a team of qualified people with diverse backgrounds who can expertly undertake the risk assessment process.

Field activity risk identification includes identifying routine and non-routine field activities, activity sites, the competence and qualifications of the guides/instructors, the background and competence of participants, and equipment and clothing. Business operations risk identification includes the following:

- *Hazard risks* like fires and floods, windstorms and hail, work-related injuries, liability claims, theft, business interruption.
- *Operational risks* like people, processes and systems.
- *Financial risks* like rising interest rates or price increases.
- *Strategic risks* like the broader economic environment (e.g., unemployment rates, inflation), changing demographics, and the political environment (e.g., change in government resulting in changes to rules and regulations affecting the adventure tourism sector)

However, internal processes and systems often do not get the attention they deserve. They include recruitment and retention of staff, staff training, staff professional development, and human relations processes; equipment procurement, inspection, maintenance, and retirement; customer relationship management; and financial management.

Businesses should undertake safety audits as part of their risk assessment. In practice, this could involve, for example, rafting and kayaking guides paddling rivers pre-season looking for and possibly treating hazards like strainers and sweepers or rock guides going pre-season to crags to remove loose rock away from the edge and dislodge loose rock from the cliff face.

Risk rating

The risk rating (Jackson & Heshka, 2011) – alternately called level of risk (ISO 21101) – is a calculation accounting for the consequences of an event and the associated likelihood of occurrence. It can be scaled low–medium–high–extreme or on a numerical scale whereby each of the factors are multiplied. An example of this is shown in Figure 16.2.

LIKELIHOOD

5	LOW	MED	HIGH	EXT	EXT
4	LOW	MED	HIGH	HIGH	EXT
3	LOW	MED	MED	HIGH	HIGH
2	LOW	LOW	MED	MED	MED
1	LOW	LOW	LOW	LOW	LOW
	1	2	3	4	5

CONSEQUENCE

LIKELIHOOD × CONSEQUENCE			
SCORE	0–5	=	LOW
SCORE	6–10	=	MEDIUM
SCORE	12–16	=	HIGH
SCORE	20–25	=	EXTREME

Figure 16.2 Risk Rating Model (Adapted from British Columbia Chief Risk Office, 2022, p. 19).

This rating is for the activity with the existing controls or risk treatments. If the rating is higher than the guide's risk tolerance, then treatments can be undertaken to reduce or eliminate the risk. The risk estimation and rating process is repeated until the score is within the guide's (and the business's) risk tolerance or, if that is not possible, the risk is avoided and the activity is not done.

For example, using the 1–5 scale from the Risk Rating Model in Figure 16.2, foot entrapment is a hazard of kayaking or rafting. Depending on the river, its risk factors (on a scale of 1–5) could be consequence: 3 (medium); likelihood (medium/low): 2; for a multiplied risk rating of 6. The existing risk treatment for foot entrapments are clients being instructed to not stand in moving water if swept away, wearing of proper footwear, and the guide having the skills and equipment to affect a rescue if needed. If the risk rating exceeded the guide's and business's risk tolerance, then other treatment options would have to be explored to reduce the figure, and if that was not possible, then the activity would have to be cancelled.

However, an example of the pitfall of focusing just on likelihood and ignoring exposure in the calculation of the risk rating is the risk of a serac collapsing and causing an avalanche. Climbers on big mountains often cross beneath seracs, which are large and unstable freestanding blocks of glacial ice (Figure 16.3). Serac collapse is a hazard of climbing big mountains. Although unstable, they can remain precariously perched for years without falling. If exposure was not accounted for, its risk factors (on a scale of 1–5) could be consequence: 5 and likelihood 1 for a risk rating of 5, which effectively

Figure 16.3 Climbers crossing glacier face. Photograph by Matt Groves.

means merely pay attention to it. This can have catastrophic consequences if exposure is not accounted for. If exposure is included in the calculation, then its risk factors would be consequence: 5; likelihood: 1; exposure: 1–5, depending on the length of time and the number of people exposed for a multiplied risk rating of 5–25. If the ultimate risk rating was on the high end of the range (i.e., 15–25), it would be materially different than not accounting for the low score adduced by excluding exposure, and it would mean the guide would have to consider reducing or eliminating this risk by waiting for improved conditions, choosing another route, or retreating entirely. This is effectively what happened in 1990 on Lenin Peak in the then Soviet republic of Tajikistan after a serac collapsed and triggered an avalanche that killed 43 climbers (Clines, 1990). They were at Camp 2, a location that had been used by climbers for six decades. A rule of thumb is to minimise the amount of time spent beneath seracs. If exposure was truly accounted for in setting up camp and spending days there, then the risk rating would have been higher and the risks appreciated, and perhaps a decision would have been made to camp elsewhere.

This is the sort of decision making that informed guide Garrett Madison when he rightly called off the climb on Mount Everest in 2019 due to the risk of a serac collapse (Heshka, 2022). This serac, looming about 800 metres above the climbing route between Base Camp and Camp 1, was estimated to weigh 24,500 metric tons and was the size of a 15-storey building. In 2014, a serac collapsed on Mount Everest, which triggered an avalanche that killed 16 Sherpas, three of whom were members of Madison's team at the time, in the same area of the Khumbu Icefall. A client, however, disagreed with the decision and unsuccessfully sued the guide, downplaying the risk by disingenuously saying that there are many seracs on Everest that do not stop people from attempting to climb it (Reimers, 2020). Given the large amount of time climbers are exposed to the threat of the serac collapsing while moving through the Khumbu Icefall, Madison calculated that the risk was unreasonable and cancelled the trip. Madison was not alone in this view. No one summited Everest that autumn.

The Risk Analysis Model in Figure 16.4 uses the same technique of multiplying likelihood, consequences, and exposure but is scaled more discreetly and offers definitions and recommended courses of action. Rather than using the 1–5 scale described earlier, this model uses a 1–10 scale for each of the three variables and prescribes what to do depending on the risk rating.

Figures 16.2 and 16.4 illustrate how if the risk rating exceeded the guide's and business's risk tolerance, then other treatment or control options would have to be considered. The calculations would be redone accounting for the added treatments, and if the risk rating was equal to or less than the risk tolerance, then the activity or trip could proceed. The process would continue until this happened or the activity or trip would have to be cancelled.

How adventure guides manage risk in the field

As noted in the previous chapter (Chapter 15) there is sometimes a false belief that a guide is there to keep clients safe. 'Clients in turn mistakenly believe that nothing wrong could ever happen because they've hired a guide. Isn't that, after all, why a guide is hired? To keep people safe?' (Statham & Heshka, 2021, p. 61). However, safe is a loaded word. What it means is at odds with how it is used. The Oxford English Dictionary (2023) defines *safe* as 'free from hurt or damage; unharmed; unhurt, uninjured, unharmed;

Risk Analysis Model						
Likelihood	×	Exposure	×	Severity =		Risk
Likelihood Scale		Exposure Scale		Severity Scale		
Probable	10	Continuous	10	Catastrophic (many fatalities, critical financial loss, critical public relations problem)	10	**1000**	Very High Risk (consider discontinuing operation)
Might well be expected	9		9		9	**700+**	
Quite possible, could happen	8	Frequent (daily)	8	Disaster (multiple fatalities, critical financial loss, significant public relations problem)	8	**500+**	High Risk (immediate correction required)
	7		7	Very serious (fatality, significant financial loss, significant public relations problem)	7	**300+**	
	6		6		6	**200+**	
Unusual but possible	5	Regular (once a week)	5	Serious (disability results, serious financial loss, serious public relations problem)	5	**100+**	
	4		4		4	**50+**	Substantial Risk (timely correction required)
Remotely possible	3	Occasional (monthly)	3	Important (serious injury, serious financial loss, serious public relation problems)	3	**20+**	
Conceivable but unlikely	2	Minimal (a few times a year)	2		2	**5+**	Possible Risk (non-routine attention required)/ Known Risk (routine attention recommended)
Practically impossible	1	Rare (once a year or less)	1	Noticeable (minor injury, potential financial loss, minimal public relations problem)	1	**1+**	

Figure 16.4 Risk Analysis Model (Heshka, 1996, p. 65).

having escaped or been preserved from some real or apprehended danger; not exposed to danger; not liable to be harmed or lost; secure'.

Risk is a better word to use because it does not imply safety, and risk is inherent in adventure activities. This means clients must be told and they must be in agreement that there is a possibility of injury and realise that, while guides are highly trained to reduce risk, they cannot eliminate it. Clients contend with the residual risk that remains after all reasonable control and treatment measures have been taken (Statham & Heshka, 2021). Guides and businesses should talk about inherent risk, residual risk, and reasonableness with clients. Due to the uncertainty inherent in adventure, clients may not realise that there is always a possibility for an incident even when reasonable decisions are made. It is possible for a guide to be careful and at the same time to make an error which may lead to injury.

Risk communication is an integral element of risk management. Businesses and guides owe it to their clients to be honest about the risks associated with the activities. Rare is the client who will back out and cancel their trip because of this; after all, risk is part of the attraction for doing it. Furthermore, clients may temper their behaviour during the activities, thereby reducing their exposure to unnecessary risk.

The process continues with the matching of a client's goals and objectives with a trip or activity. This matching must account for the client's risk tolerance, experience, and expectations. It makes no sense to take a beginner skier heli-skiing or an expert skier to a beginner slope. It is critical that clients are on the same page as guides in terms of risk exposure and risk tolerance.

Identifying, analysing, and evaluating risks and determining the proper risk controls and treatments is not simply a matter of painting by numbers or connecting the dots. The process appears linear and logical – and to an extent it is – but it is ultimately driven by the way adventure guides deal with uncertainty in their decision-making.

Most hazards encountered by guides have some level of uncertainty of outcome. The uncertainty exists because there are information gaps. The information is often incomplete or inaccurate. Clients over-estimate their ability, weather forecasts are rarely 100% accurate, deep slab avalanche instability in a snowpack is impossible to predict with precision, and the dynamic nature of whitewater paddling means the exposure to hazards is always changing. All are examples of uncertainty in adventure. Guides have to get comfortable with uncertainty and making decisions with incomplete information.

It is a guide's job to take calculated or intelligent risks that are based on that guide's and business's risk appetite. Risk appetite is the total exposed amount that a person wishes to undertake on the basis of risk–return trade-offs for one or more desired and expected outcomes. Risks must be assessed by weighing the potential positive effects or benefits against the potential negative effects or consequences and likelihood of the risk event actually occurring.

As alluded to earlier, decision-making in the face of uncertainty is neither straightforward nor simple. Is a turnaround time on a mountain written in stone if the weather is stable and the clients and guides are feeling strong? Should the guide lead a kayaking course on a shallow river nearby (with greater risk of head injury if a paddler flips) or a river with higher flows (therefore with lower risk of head injury) farther away but with less time on the water?

Business and guides are responsible for identifying and guarding against reasonably foreseeable risks. As mentioned earlier, an example of when things go horribly wrong in identifying, analysing and evaluating the risk is what happened when 21 people died in a canyoning incident in the Saxetenbach Gorge, Switzerland, in 1999. Eighteen tourists and three guides were killed in a flash flood after the company allowed the trip to occur despite the weather forecast of thunderstorms and the guides not stopping the trip in spite of the rising water levels and the water changing colour (Jenkins, 1999). Seven of the eight guides in the canyon on the day of the incident were first-year guides. The only other company offering canyoning trips down the gorge cancelled its late-afternoon trip. Despite claiming that the flash food was unforeseeable, three directors were convicted of negligent manslaughter for not having done a risk analysis and banning canyoning during storms, and three other staff were convicted of not cancelling the trip despite the thunderstorm breaking overhead (Koppel, 2001).

How guides make decisions

Nobel Prize laureate Daniel Kahneman (2011) has written about the two basic ways that people make decisions: unconsciously and consciously. This decision-making framework can be applied to adventure guides managing risk. Conscious decision-making is a deliberate and logical manner of thinking to process information. A key point is that it requires paying attention. Guides must always be paying attention to the weather, their clients, for animals, the environment, and more. Guides making adjustments and fine-tuning their uptrack on a ski tour in response to the terrain features and what they see in front of them is such an example. Thus, meta skills such as foresight and insight are important for the guiding profession (Langmuir, 2013)

Unconscious decision-making is involuntary, intuitive, effortless, efficient, and fast. The more often a task is completed, the more familiar and second nature it becomes; with time and repetition, the decision-making becomes automatic. Unconscious decision-making is a kind of short cut that fast-tracks thinking and decision-making, which is particularly important on busy days in dynamic environments and when time is of the essence. Guides unconsciously make decisions all the time. If a climber above yells "ROCK!" the climber below intuitively and immediately leans into the wall to reduce the likelihood of getting hit. If a climber took the time to consciously hear the shriek and think about what to do, it would be too late (see Case study 16.1).

Case study 16.1   Decision-making in the Alps

An example of both kinds of decision-making in risk assessment took place on the icy north face of the Tour Ronde, a subsidiary peak of Mont Blanc in the French Alps in 1990. The guide was consciously aware that he and his client were behind schedule and moving too slow and that the temperature was hotter than expected (Scott, 1998). He likely unconsciously decided that it was preferrable to move fast to avoid being hit by falling rock, caused by the unexpected heat, rather than spend additional time conforming with "best practice" by placing two ice screws at the anchor. It was not an algorithmic-styled decision but rather one based on his experience in the mountains. In his need for speed, the guide left his client connected to one ice screw and moved quickly out of the fall line of the face beneath a rocky section which offered protection. Unfortunately, he fell, pulling out the anchor to which the client had been connected, and both fell down the face, stopping when the rope snagged on a rock. The client was killed almost instantly. The guide was successfully sued in 1994 for negligence, and the judge found him liable, saying that the guide's concern over the immediate or imminent risk of rockfall was outweighed by the possibility of a foreseeable fall (Liability Issues, 1997). The decision has been controversial with the British Mountaineering Council, saying in response that judgement and experience, not a rulebook, should be used to make the choices that minimise risk (Tour Ronde, 1997). The reality is that decisions should be based on judgement and discretion obtained through training and experience and responding to the immediacy of the circumstances.

Both conscious and unconscious types of decision-making can be influenced and contaminated by cognitive bias. These psychological biases may lead to errors of judgement. Examples of types of bias in adventure guiding include the following:

- Expert illusion: when a guide falsely thinks something complex or complicated in nature is within their skill set simply because they are a guide (Atir, Rosenzweig & Dunning, 2015).
- Expert halo: deferring to the most senior or expert guide merely because of their seniority. Sometimes clients are "guilty" of unquestioningly deferring to guides.
- Dunning-Kruger effect: an inflated sense of one's own competence, for example, teenagers and driving. They think they are great, but statistics show otherwise.
- Continuation bias: when someone is convinced of the efficacy of something because it has worked in the past (even though it might not be the right thing to do). Its effect is to operate closer to the line with a reduced margin of error.
- Complacency or familiarity trap: doing something for so long that they have become desensitised to the need to pay constant attention and consequently let down their guard.
- Confirmation bias: looking for, interpreting and favouring information that is consistent with existing beliefs so that it confirms or rationalises those beliefs.
- Motivational or self-serving bias: decisions are influenced by self-interest, for example, a guide skiing a run that they know will please clients, despite having doubts about snowpack conditions.
- Bandwagon effect (or groupthink): doing something primarily because other people are doing it.
- Sunk-cost fallacy: the tendency to continue with something because of the amount of time, effort, or money into it.

Adventure guides and businesses need to be mindful of the effect of these psychological biases and take steps to minimise them. Strategies to reduce bias include the build-up of a safety culture that incorporates roundtables to discuss ideas to manage risks, encourages honesty to report near-misses, requires debriefs after an incident, and applying a hindsight test to assess if the decisions made were reasonable and the exposed hazards with their resultant consequences foreseeable.

Additional considerations for adventure businesses

Adventure businesses follow the same process to manage risk as guides in the field. The only difference is that their focus is wider to capture other kinds of risks. This chapter focuses on hazard and operational risks. Guides and managers/owners deal with both types of risks. Only managers/owners, however, have to manage financial risks (the effect of market forces on financial assets or liabilities including risks relating to interest rates, credit, and liquidity) and strategic risks (risks that arise from changes and trends in the economy and society such as the mainstreaming of "extreme" or alternative sports like climbing and skateboarding and the effect of baby boomers' purchasing patterns in adventure; Patterson, 2011). Financial and strategic risks are not dealt with here. It is beyond the scope of this chapter to examine them in sufficient detail.

This chapter has focused on the management of risk in the field but also describes adventure business risk management considerations. These businesses ought to adopt a systems-based approach to their management of risk. This accommodates the seven systems of risk management that all intersect with the programme activity at its centre (Jackson & Heshka, 2011). The seven systems are staffing/human resources, business management, organisation planning, programme planning, client information, equipment management, and crisis management. Rather than treating each element as a silo, separate and independent from one another, systems-based thinking recognises that they interact and affect the risk rating of the programme area.

The staffing/human resources system includes recruitment and retention, staff training, guide qualifications, and guide-to-client ratios. Every business wants to hire the best staff. In this case, the best staff are the most qualified, which, in turn, translates to offering clients the best experience while minimising their unnecessary exposure to risk (i.e., being as "safe" as possible). To get the best guides, businesses must offer a highly competitive wage, benefit package, and superlative working conditions. This means that businesses that cannot afford to hire the best guides must hire less experienced and/or younger guides, which affects their ability to manage risks in the field. Astute businesses will recognise this and scale back their trips' exposure to hazards commensurate with their guides' background, and, if possible, offer internal training, shadowing, and mentoring to improve their guides' skills. This will help retain guides and, if it is seasonal work, have them keep coming back. Finally, some sectors of the adventure tourism industry have recommended guide-to-client ratios. However, this is not regulated, so it is not always required. Nevertheless, businesses that do not conform to these "best practices", by operating outside these ratios and hence having more clients, means not only more revenue but also more risk exposure.

The organisation planning and programme planning systems deal with the business's risk philosophy, mission, risk tolerance, management, objectives, planning process, and build-out of trip or activity offerings. In short, philosophy describes the business's relationship with risk in the context of adventure. While seemingly straightforward, it is not hard to imagine instances where an organisation's board of directors and its field staff are not on the same page. The description of a business's mission must use concepts or language that incorporate its risk appetite. For example, like high-risk, high-reward stocks, an adventure business can be aggressive in its programming and push the envelope and rely on, hopefully, robust risk management practices while earning higher revenues. Alternatively, the business can be more conservative in its approach thereby appealing to a broader market of clients, having a wider safety cushion and greater margin for error.

This is connected to the amount of exposure it is prepared to take based on risk–return trade-offs, or its risk appetite. In turn, its objectives and goals (i.e., the types of trips and activities offered) are generated by the business's risk philosophy, mission, and appetite.

Leadership (see Chapter 18) is essential in all systems, but in combination with chain of command, it is critical in the crisis management system. Other key features of a business's crisis management system include its emergency response plan, communications plan, resource requirements, and business continuity and recovery.

It is essential that risk is properly communicated to clients pre-trip and, if necessary, post-incident. The business's marketing must do more than show carefree clients with ear-to-ear smiles but also mention on its website and printed collateral about the activity's risks. In the unfortunate event that a client is injured, risk and adventure must be contextualised to the media using concepts like inherent risk, residual risk, and risk treatments or controls. Transferring risk through insurance or back to the client via a release are techniques of treating risk. The insurance should include commercial general liability coverage and commercial property coverage and possibly search-and-rescue coverage. In many, but not all, jurisdictions, a release of liability is a complete defense to liability. In those countries or provinces/states, it is possible for a defendant business or guide who has caused an incident or has been negligent to not be liable for the injuries sustained if the plaintiff client signed a release.

Content of a risk release

A release should contain the following elements:

1 A prominent heading saying something like "Release of Liability, Waiver of Claims, Assumption of Risk Warning and Indemnification Agreement". The text can be accentuated by writing it in bold and all caps. It can be further highlighted by putting it inside a rectangular box, infilled yellow and bordered in red. The point here is to make all reasonable efforts to bring to the attention of the client that which they are about to sign.
2 A small box in a corner of the large rectangular box for the person to initial, indicating that they have read it.
3 A section that defines all the activities covered by the waiver.
4 An acknowledgement of the risks section that identifies the possible hazards to be encountered and injuries to be sustained. The list of activities, hazards and possible injuries does not have to be exhaustive, but it should be sufficiently broad and descriptive to cover everything reasonable to be encountered.
5 A declaration that the person voluntarily agrees to participate in the activities.
6 A section that spells out the waiver, release of liability, and holds harmless language.
7 A section that says the agreement is binding on the person's heirs, next of kin, and others.
8 A jurisdictional clause that identifies the laws and courts of what province or state (which ordinarily is the location where the operator is based) would govern in the event of a dispute.
9 The name, signature, date of birth, and contact information of the client.
10 The name and signature of the witness. The witness must be an adult. It is recommended that the witness be an employee of the operator.

The release needs to be written in plain language that is not obscure or indecipherable. The release should be one page in length, two pages maximum if the list of assumed hazards is lengthy and deemed important to be very specific, and in a font that is not near microscopic, which makes it hard to read.

Administration and delivery of the release involves the following basic steps:

1 Provide advance notice. Before a client commits with their booking, they must be notified that signing a release is a requirement to participate.
2 Make the release available in advance; it can be posted on the operator's website.
3 Educate clients on the contents of the release. Encourage them to read it carefully.
4 Give clients time to read it. A staff member cannot force a client to read the document, but they cannot rush them or deny them the opportunity to study it if they choose.
5 Explain its terms. It is reasonable for the staff member to explain the hazards, risks and possible injuries but not attempt to answer any questions concerning the law, that is, what negligence means.
6 Provide an atmosphere that is conducive for the person to read, understand, and execute the agreement. Give them the time and space necessary to read and sign the document.
7 The document must be a stand-alone waiver and not combined with other forms like a medical questionnaire, gear rental agreements, or a questionnaire about accommodation and menu preferences.
8 The waiver must be signed when a booking is made (i.e., online) or otherwise made clear that with a booking a release will need to be signed before participation.
9 Ensure the person properly filled out the release, initialled the box at the top (some businesses have boxes alongside other sections), and signed it at the bottom. Online forms can be set up so that the person cannot proceed with the booking until all clauses are agreed to or initialled.
10 The waiver cannot be altered. The person signing it cannot add or cross out anything on it. The waiver must be signed as is.
11 The person should sign the waiver in front of a witness (if in person) or a staff member confirms the online signature upon arrival or sign-in. It is preferable that employees witness the signature.
12 Electronic releases can be valid but need different treatment in terms of user interface, searching, storage, and retrieval of signed releases.
13 Given the ease of scanning and storing documents, it is recommended that releases be retained indefinitely.
14 Have a system in place to ensure that all clients have signed a release and catch and prevent people who have not.

Paradox of adventure

Adventure is premised on people purposefully situating themselves in positions of uncertainty. Whereas people non-afflicted by the adventure "virus" run away from risk, those with adventurous spirits run towards it. It is not unreasonable, after all, for people to avoid activities or doing things that will hurt them. Furthermore, systems are designed to prevent people from getting injured. This is why seatbelts and guardrails exist. In adventure, however, people willingly choose to participate in an activity where not all the variables can be controlled and the outcome has a degree of uncertainty. For example, on an activity as simple as a backpacking trip in the mountains, the hikers cannot control the weather or cannot afford to buy the best gear or predict the animals and bugs possibly encountered on the trail, and the trip's success and enjoyment are, in part, dependent on these. More ominously and obviously, weather and wind are often not 100% accurate on big open-water crossings, nor can avalanches be predicted with 100% certainty. Yet people who seek out adventure are aware that they cannot control the things that can cause them harm and that success is never a guaranteed thing. Some people are not afraid of risk to the extent that they pursue activities that are inherently dangerous and that a market exists to guide those who have the interest but not the skills to do it on their own.

Set against this backdrop of the desirability of risk and uncertainty in adventure guiding is the paradox or contradiction that risk and uncertainty in adventure businesses are to be avoided. Whereas risk in adventure is essential, risk in business is to be shunned. Businesses benefit from certainty. The less uncertainty, the better it is for the bottom line. It is helpful and beneficial for businesses to know that their supply chains are reliable, that their staff are dependable, and that there is minimal variance in the quality of the products being offered. This paradox – risk and uncertainty are good in adventure guiding but bad for adventure business – is unaffected by the principles and processes that drive risk management. Either way, the risk management practices of risk identification, analysis, and evaluation apply.

Another paradox is that adventure guides are both the creators and destroyers of risk. Guides take clients into perilous situations, and guides take them out. Guides lead clients down raging rapids, up rock walls, or across avalanche-prone slopes and – almost all of the time – appropriately manage the risks with the end result that clients have an amazing experience. A guide's job is effectively to create risk simply by being in those environments rife with uncertainty and manage the risks through a series of carefully calibrated risk treatments to reduce the uncertainty to an appropriate or acceptable level. Being a good guide is more than being able to ski, paddle, or climb hard. In addition to client management, it requires selecting an objective for the client that matches their risk tolerance, knowing how to assess the risk in the field on an ongoing basis and how to treat and reduce those risks.

Finally, the role of perceived risk in the adventure experience is an example of yet another paradox. Perceived risk, or what individuals believe to be the risk, can be manipulated (Jackson & Heshka, 2011). Most new or novice participants inflate the real or actual risk due to their inexperience. Beginner climbers, for example, on a top-rope perceive that if they let go, they will fall to the ground and get hurt, not trusting the rope, never mind that the rope and hardware in the climbing system has a tensile breaking strength above 22kN or 5,000 pound-forces. Guides can heighten or play up the perceived risk to get participants to pay more attention or to build anticipation or lower or play down the perceived risk to reduce unnecessary anxiety and improve the adventure experience (Cater, 2006). The paradox here is participants think adventure is absurdly dangerous even though statistically it is not. They think that danger is part of the marketability of what adventure tourism businesses sell. But, if it was truly that dangerous the businesses would not last, and guides can inflate or deflate the perceived risk depending on what the circumstances warrant.

Conclusion

This chapter described risk in the context of adventure, its unique risk management challenges when risk is part of the allure, and the risk assessment process. This process involves identifying, analysing, and evaluating the risks, and this, in turn, accounts for the variables of consequence, likelihood, and exposure. The manner in which guides make risk management decisions and cognitive biases to guard against were also discussed. It also described business considerations and seven systems of risk management and particularised the preparation and presentation of a release of claims.

Review questions

1 What are the five steps involved in the risk management process?
2 What is the difference between likelihood and exposure in the estimation of risk?
3 Define *residual risk* and *risk rating*.
4 What can guides and businesses do to help reduce the effect of bias in decision-making?
5 How should businesses deal with risk and safety in their marketing?

Further reading

Bentley, T. A., Cater, C., & Page, S. J. (2010). Adventure and ecotourism safety in Queensland: Operator experiences and practice. *Tourism Management, 31*(5), 563–571. https://doi.org/10.1016/j.tourman.2009.03.006
Jackson, J., & Heshka, J. (2011). *Managing risk – Systems planning for outdoor adventure programs*. Direct Bearing.

References

Atir, S., Rosenzweig, E., & Dunning, D. (2015) When knowledge knows no bounds: Self-perceived expertise predicts claims of impossible knowledge. *Psychological Science, 26*(8), 1295–1303. https://doi.org/10.1177/0956797615588195
Brennan, S. (Ed.). (2014). *Mark Twain on common sense: Timeless advice and words of wisdom from America's Most-Revered Humorist*. Skyhorse Publishing.
British Columbia Chief Risk Office, Risk Management Branch & Government Security Office. (2022, August). Risk Management Guideline for the B.C. Public Sector. https://www2.gov.bc.ca/assets/gov.british-columbians-our-governments/services-policies-for-government/internal-corporate-services/risk-management/risk-management-guideline.pdf
British Standards Institution. (2016). *Risk Management – Manage Your Risks*. https://www.bsigroup.com/globalassets/localfiles/en-in/certification/iso-31000/iso-31000-risk-management.pdf
Cater, C. I. (2006). Playing with risk? Participant perceptions of risk and management implications in adventure tourism. *Tourism Management, 27*(2), 317–325.
Clines, F. X. (1990, July 18). Avalanche Kills 40 Climbers in Soviet Central Asia. *The New York Times*. https://www.nytimes.com/1990/07/18/world/avalanche-kills-40-climbers-in-soviet-central-asia.html
Elliott, M. W. (Ed.). (2018). *Risk management principles and practices* (3rd ed.). Global Risk Management Institute, Inc
Forbes (2001, October 29) Because it's there. *Forbes*. https://www.forbes.com/global/2001/1029/060.html?sh=39a1be762080
Heshka, J. (1996). *Adventure Education and Liability: Walking the Razor's Edge*. [Unpublished master's thesis]. Simon Fraser University.
Heshka, J. (2022, February 25). *Mount Everest guide cleared of charges of fraud and breach of contract*. Sports Litigation Alert. https://sportslitigationalert.com/mount-everest-guide-cleared-of-charges-of-fraud-and-breach-of-contract
International Standards Organization. (2014). *ISO 21101: Adventure tourism – Safety management systems requirements*.
International Standards Organization. (2018). From "Risk management — Principles and guidelines," *ISO 31000: Risk management*.

Jackson, J., & Heshka, J. (2011). *Managing risk – Systems planning for outdoor adventure programs*. Direct Bearing.

Jenkins, M. (1999, November 1). A Storm in the Distance. Mark Jenkins. *Outside*. https://www.outsideonline.com/outdoor-adventure/exploration-survival/storm-distance

Kahneman, D. (2011). *Thinking, fast and slow*. Farrar, Straus and Giroux.

Koppel, N. (2001, December 11). Six Guilty in Swiss Canyoning Deaths. *AP News*. https://apnews.com/article/ccd7c732604420cb49c114c75e1f1298

Langmuir, E. (2013). *Mountaincraft and leadership: A handbook for mountaineers and hillwalking leaders in the British Isles* (4th ed.). Mountain Training Boards of England and Scotland.

Liability issues in outdoor pursuits – The case of *Woodroffe-Hedley v Cuthbertson* (1994) W 1400 G (unreported). (1997, September/October, Vol. 4, Issue 6). *Sports Law Administration & Practice*.

Mountainclients.org.uk. (1997). *Tour Ronde – Hedley v Cuthbertson*. https://mountainclients.typepad.com/mountain_clients/tour-ronde-hedley-v-cuthb.html

OSHA (2023). Hazard Identification and Assessment. Retrieved May 2, 2023, from https://www.osha.gov/safety-management/hazard-Identification

Outdoor Recreation Roundtable (2023). National Recreation Economic Data. Retrieved May 2, 2023, from https://recreationroundtable.org/economic-impact

Outdoor Industry Compass (2022). Adventure tourism market to reach $1.009 trillion in size by 2030. Retrieved May 2, 2023, https://www.oicompass.com/market-statistics/adventure-tourism-market-to-reach-1009-trillion-in-size-by-2030/97589

Oxford English Dictionary (2023) Safe. https://www.oed.com/viewdictionaryentry/Entry/169673

Patterson, I. (2011). Baby boomers and adventure tourism: The importance of marketing the leisure experience. *World Leisure Journal*, 44(2). https://doi.org/10.1080/04419057.2002.9674265

Quinn, B. (1990). The essence of adventure. In J. C. Miles & S. Priest (Eds.), *Adventure education* (pp. 145–148). Venture Publishing.

Reimers, F. (2020, October 11). A tech CEO suing his guide could change Everest travel. *Outside*. https://www.outsideonline.com/outdoor-adventure/climbing/madison-mountaineering-zac-bookman-everest-lawsuit

Scott, D. (1998). Fame and fortune – Risk, responsibility and the guided climber. *The Alpine Journal*, 103, 189–196.

Statham, G., & Heshka, J. (2021, May 12). *Post-critical incident management plan*. Association of Canadian Mountain Guides. https://www.acmg.ca/05pdf/Post-Critical_Incident_Management_Plan-12.05.2021.pdf

Thomas, S., & Raymond, C. (1998). Risk and provider responsibility in outdoor adventure activities. *Teacher Development*, 2(2), 265–281. https://doi.org/10.1080/13664539800200054

U.S. Department of Transportation, Federal Aviation Administration. (2009). *Risk Management Handbook*. https://www.faa.gov/regulations_policies/handbooks_manuals/aviation/media/faa-h-8083-2.pdf

17 Adventure tourism leadership that SELLS

Jason King, Christopher Webber and Ashley Hardwell

Chapter learning outcomes

1 Evaluate adventure tourism delivery from a leaders' perspective.
2 Provide an overview of different leadership roles in adventure tourism operations.
3 Introduce the SELLS framework as a tool for developing and evaluating leadership in adventure tourism operations.

Introduction

Adventure tourism activities typically involve leadership and expertise in the safe guidance of adventure activity participation. While a plethora of research considers broader organisational leadership (see Achua & Lussier, 2013; Daft, 2011; Northouse, 2018), the specific and nuanced adventure tourism leadership context is seldom considered. Group disparity and disjointedness; participant experience and diversity; limited time frames; and the nature of the activity and its environment may all be key considerations for an adventure tourism leader. Therefore, a fresh approach is required for leadership in this unique and dynamic environment.

This chapter offers a delivery model useful for adventure tourism leaders. *Safety, enjoyment, learning, legacy,* and *self* (the SELLS model) recognises the dynamic environment in which adventure tourism leadership operates and the limited time in which leader–follower relationships form. Established notions of any adventurous activity experience are incorporated by considering the *safety, enjoyment,* and *learning* fundamental to any adventure tourism experience (Ferrero, 2006). However, often neglected, are the final two elements of *legacy* and *self*. These two key features are subsequently considered. Adventure tourism experiences are often extraordinary for participants, leaving lasting memories or a *legacy*, and the model only works through the adventure tourism leader keeping up to date with professional practices and skill proficiency – *self*. The following sections unpack the SELLS model, offering leaders a range of considerations for the successful delivery of adventure tourism experiences.

DOI: 10.4324/9781003393153-21

Another leadership day

Adventure tourism leaders fulfil many roles and responsibilities in a typical day. As the SELLS model is explored, the term *adventure tourism leader* is used to refer to experienced and qualified individuals tasked with providing meaningful adventure tourism experiences that meet client needs. Their role is multifaceted. For example, they plan and adapt sessions commensurate with expected group needs and quickly identify skills and competence, matching challenges with these. They also confidently progress the activity, knowing their technical expertise and facilitation skills can be drawn on to deliver engaging and meaningful experiences. The adventure tourism leader must exude humble confidence, instil this in others, always ensure the safety of the group, and provide appropriate support and guidance for clients. Too much and clients may feel controlled and stifled; too little and they may feel endangered and at risk. Finding the correct balance often requires the juggling of several roles.

Adventure tourism leaders must occupy these multiple, and often simultaneous, roles for the successful delivery of an adventure tourism experience, with these included in this chapter under the umbrella term of *leadership*. They may include, but are not limited to, instructor (Mees et al., 2020), coach (Eastabrook & Collins, 2020), guide (Rokenes et al., 2015), or facilitator depending on the group and environment. The SELLS model aims to stimulate reflection on adventure tourism delivery through five key components. The chapter begins with an adventure tourism scenario (see Case study 17.1) that is referred to throughout to illustrate prominent issues in all five SELLS components.

Case study 17.1   Adventure tourism scenario

I look up at the rain-laden cloudy sky and wonder if the forecast for the day can be relied upon. Heavy rain overnight has left me doubting whether the planned remote canyoning trip with a group of six clients from as many countries who have never met before is the best way forward. I know the canyon well. I have been an adventure tourism leader for five years, yet still, I question my confidence as I think through the myriad difficulties the weather, the group, and the activity might bring.

Canyoning is fantastic. It demands the client's multiple skills in mixed environments and a steady head for heights, and it can provide outstanding adventure tourism experiences. These are my musings as I walk out to the neatly packed minibus already thinking about decision-making. I know there are only 15 minutes of easy river work, scrambling and traversing in which to assess group skills and ability before the first of three 40-metre abseils completely commit the group to our 6-hour day. What if the group is not up to it? What if the water is too high? What if an accident happens? What if. ... What if?

During the short bus ride to the start, I elect to sit in the back with clients, learn names and assess their understanding of English. As we get off the bus, I listen to my own voice while almost autonomously moving through essential briefing information. Despite emphasising the length of the day and the two-hour uphill trek back to our pickup point, clients dismiss this information, preferring to focus on the canyon, what it has to offer, and the three spectacular abseils that will take us into its belly.

I observe everyone as they enter the head of the canyon and glide on their backs down the first short water shoot diligently following safety protocols from the brief. I make eye contact as I closely manage this first section and see joy, worry, and concern. I continually assess individual and whole-group capability, as, in 10 minutes, I will make the decision to commit to the 6-hour trip or ring base for the bus. I lie in a water shoot, look up at the sky, and wonder why the cloud has not cleared as the forecast promised.

I bring the group close to me in a tight circle so they can hear the abseil briefing over the roar of the water. I sound too confident through the brief overcompensating for my worries about high water levels. 'OK – first person', I shout. As expected, the more experienced Arafa volunteers. As she negotiates the tricky start with ease and confidence, the draining away of fear and trepidation from other less experienced clients is almost palpable. I encourage a less experienced client to go next knowing Arafa will be at the bottom to receive. As we all abseil into the depths, excitement and concern are present for us all in differing proportions. These emotions ebb and flow throughout the long canyon journey seemingly at times without end.

Figure 17.1 Canyoning, Imp Grotto Canyon, Haast Pass, New Zealand. Photograph by Jason King.

When we are least expecting it, the gorge abruptly releases us from its grip into an open landscape. A faint trail marks our way back up the escarpment. Spits and spats of rain patter down from a battleship-grey sky. Everyone has been engrossed in the thrill of the canyon journey, but now we are faced with two hours of walking with water-soaked equipment. I bring the group together; I am honest with them about what lies ahead and watch shoulders slump as the passing storm further dampens spirits. Each group member resigns themselves to the slow uphill trudge lasting the best part of two hours.

'There's the bus', Arafa shouts to the group. Immediately, postures change, heads lift, and smiles break from grimacing faces. The pace quickens, and soon doors are flung open, equipment deposited on the back seats, and frenetic chatter and animated stories of the day fill the bus. As we drive towards the centre, I hear the song lyrics from the Bluetooth mobile – 'what have you done today to make you feel proud', and I hear Arafa shotgun first shower before the evening meal.

Leadership that SELLS

The intention of the SELLS leadership model (Figure 17.2) is to provide two avenues for consideration. First, it recognises several established principles within adventure tourism activities: *safety*, *enjoyment*, and *learning* (Ferrero, 2006). It extends these to consider sustainability beyond the experience by recognising legacy and self as important components of the leadership process. Second, SELLS offers practical applications. It is intended to be memorable through its acronym and procedural through its logical representation. The following section explores each component of the model and makes links to various adventure tourism leadership examples.

Safety

The preceding scenario highlights the constant need for leader vigilance in a range of physical and psychological safety concerns. Physical factors include the dynamic operating environment, individual and group ability at any given time, and leader capability.

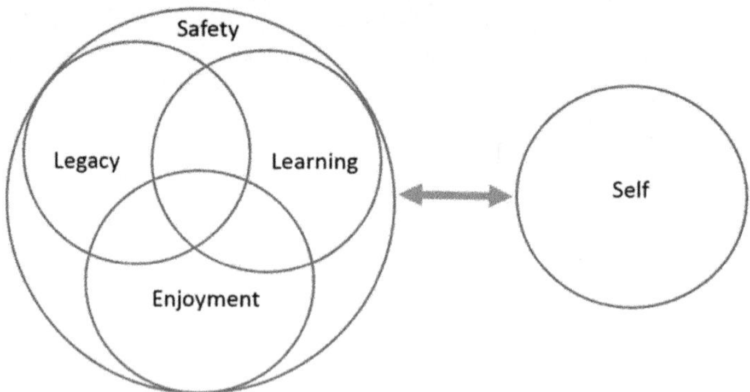

Figure 17.2 A framework for leadership in adventure tourism.

Psychological factors include individual and group confidence, readiness for the activity, and trust in the instructor. A simple definition of risk is 'losing something of value' (Priest & Gass, 2018, p.30). However, as we have seen in Chapters 15 and 16, in adventure tourism, this is too simplistic. As Jackson and Heshka (2011) point out, risk also provides unique opportunities for growth such as developing self-confidence and resilience through these unique experiences. Early recognition by both leader and participant that physical risk in adventure tourism can never be wholly mitigated against, is key to the safety of the group and leader. However, while real risk is always present and can lead to harm, a hallmark of adventure tourism is the participants' attraction to risky situations (Dickson & Dolnicar, 2004). This means the leader needs to be able to identify the optimal balance between "real" and "perceived" risk to which the participant is exposed during an activity. The crux of the canyon trip is its incredible abseils. While perceived risk is high in these sections, reliable equipment provides significantly reduced real risk. The leader will understand this, whilst inexperienced participants may not. Such situations, through careful management, may often provide unforgettable experiences.

Real risk

Real risk is often considered as the physical harm that may result from confronting danger. But it is much more than this. Adventure tourism leaders must be acutely aware of the psychological, emotional, social, and financial risks of poor decision-making. Adventure tourism activities operate in environments where real risk is always present. Typically, situational attributes may include height, water, natural environments, and weather conditions – all of which were constant companions during the canyoning scenario earlier. Real risk also carries the potential of psychological harm from overstretching participants. Anxiety, trepidation, fear, and feelings of inadequacy may all be consequences. These may be addressed by using risk assessment tools considering the activity and its potential outcomes discussed in Chapters 15 and 16. While static risk assessments have a place, these should be coupled with dynamic equivalents to identify emergent risks during activities. Excellent dynamic risk assessing and decision-making come with experience (Collins & Collins, 2013). The canyoning scenario allows insight into an adventure tourism leader's constant appraisal of the situation and possible group consequences.

Perceived risk

Perceived risks are subjective appraisals of the real risks presented and are heavily influenced by previous experience and primary socialisers (Dickson & Dolnicar, 2004). Factors including upbringing, education, lifestyle, technology, and media can all shape perceptions of risk. Leaders also perceive risk differently with the goal of considering the match between challenge and capability to ensure meaningful adventure tourism experiences for participants (Csikszentmihalyi, 1990). The delicate role of the adventure tourism leader is to find the balance between real risk, perceived risk, and the desired risk each participant is seeking (Weber, 2001).

Managing risk forms a substantial part of the role of the adventure tourism leader providing the foundation upon which all other outcomes depend. In Figure 17.2, this is depicted as the safety circle in which other important components operate. Effectively managing risk requires a "whole systems" approach (Curtis, 2008). This encompasses leader, group ability, experience, and readiness for the activity and is combined with static and dynamic risk assessments, alongside an awareness of the unique dynamic operating environment (Jackson & Heshka, 2011).

Enjoyment

The SELLS model depicts safety as the all-encompassing domain. Consequently, adventure tourism leaders must quickly develop rapport, engagement, and trust with their clients. Arguably, without strong client-leader bonds and excellent group cohesion from the outset, enjoyment of the experience will be reduced, as would safety.

Establishing rapport and trust

Enjoyable adventure tourism experiences rely as much on the group as the activity. Using every element of the adventure tourism experience must be considered for this. In the canyoning scenario, building camaraderie, trust, and cohesion occurs while travelling to the venue. On arrival, group rapport is already developing. Music on the bus, group attentiveness on the drive, short team development exercises, and opportunities to learn basic technical competence through group work exercises are all ways of ensuring early group interaction. Briefings should be fun, meaningful, and convey essential information for the safe delivery of the activity (Outdoor Instruction.co.uk, 2011). Early individual and group assessments of competence and confidence are essential for enjoyment through optimal experience.

Providing a rewarding experience

During the adventure, the adventure tourism leader is constantly attending to individual and group needs to create enjoyable experiences. Knowing the venue well and providing 'challenge by choice' is key (Allan et al., 2014, p.37). Choosing the level of challenge in the canyon relies on knowing the environment. Empathy is crucial as an adventure tourism leader. Reflecting on how it feels to abseil into a waterfall for the first time will help manage the client experience. Recalling successful experiences with other groups may help with leader–client confidence. In the canyon, identifying experience within the group helped establish collective confidence on the first abseil.

Regular triggering of positive experience

Canyoning is an excellent example of an adventure tourism activity with multiple elements to consider. The client is stimulated by water, rock, scrambling, climbing, traversing, abseiling, and jumping. It also fosters group encouragement. Bringing all elements together to provide appropriate individual and group challenges is a complex yet rewarding task for the leader. *Cognitive dissonance* refers to the feeling of being challenged and out of one's comfort zone, which often occurs in novel situations (Panicucci, 2007). Contentiously, it has been suggested that appropriate challenges are required to create enjoyable and memorable experiences (Priest & Gass, 2018). But sometimes, the canyoning experience, along with many other adventure tourism activities, require overcoming a crux or difficulty as well as being involved in more mundane tasks. Their careful management to maintain overall enjoyment is important.

This is the very essence of the adventure tourism experience; the feeling of overcoming challenges and arousing feelings of enjoyment and satisfaction. It is important for the adventure tourism leader to recognise the difference between these (Ewert & Sibthorp, 2014). Satisfying experiences may not necessarily be immediately enjoyable. The crux of the long abseil through water to access the canyon may seem a step too far for some clients. On reflection, this may be the most satisfying part of the experience, providing the chance to rise to the occasion, overcome fears, and complete the journey. At the time, it may feel overwhelming and not at all enjoyable. Similarly, the two-hour return trek to the bus will test clients' resolve. Empathetic and understanding adventure tourism leaders manage such situations through constant reassurance and attendance to individual client needs (Cater, 2013).

Threats to enjoyment

Adventure tourism clients often seek novelty through challenging activities (Cater, 2013). It is up to the adventure tourism leader to recognise early opportunities to foster trust, camaraderie, and group cohesion by putting in place the cornerstones for an enjoyable experience. Recognising that, by its very nature, adventure tourism may offer challenge, yet always ensuring this is coupled with enjoyment, provides the foundation for a positive individual and group experience. The adventure tourism leader needs to be acutely aware of the following:

- Group dynamics – recognising group dynamics early and guarding against domination and gung-ho attitudes
- Ebbs and flows – always paying attention to individual and group enjoyment
- Taking breaks – knowing when and where to meaningfully stop or re-energise the group
- Trigger positive emotions – capitalising on moments when emotions are high

Safety will always require constant attention, after which, the second leader's priority is to consider an enjoyable and rewarding experience. In the context of adventure tourism, enjoyment does not have a universal meaning. For some, it could be the importance of cultural experience, for others, it may be more adrenaline-fuelled (Buckley, 2010). The adventure tourism leader must ensure that clients feel comfortable with challenges, allowing positive emotions to be triggered, nurturing feelings of enjoyment, and facilitating learning from the experience.

Learning

Adventure tourism is well placed to generate powerful learning experiences (Blume et al., 2010). As outlined in the canyoning narrative, activities are often novel and occur in unfamiliar places, which creates robust memories (Kesteren & Meeter, 2020).

Information from multiple senses like sight, sound, and touch are combined to further strengthen these (National Academies of Sciences, Engineering, and Medicine [NAoSEM], 2018). Because experiences often involve uncertainty, challenge, and risk, this triggers a range of emotions adding to the richness of learning encounters (Haskell, 2001). Adventure tourism experiences also afford opportunities to draw on existing knowledge and deliberately practise skills. Immediate feedback from the environment and group members can be used to refine this existing knowledge and capability (Darling-Hammond et al., 2019).

Ubiquitous impact

Well-orchestrated outdoor and adventurous experiences routinely elicit a range of personal and social development outcomes (Williams & Wainwright, 2016). Intrapersonal outcomes include generating self-awareness through discovering new preferences, updating aspects of self-concept – like self-efficacy – by achieving success and comparing oneself to others, and enhanced self-regulation and problem-solving skills. Interpersonal equivalents include cultivating teamwork skills, leadership skills, and strengthening relationships. Equally, adventure tourism experiences enable individuals to develop a range of specific procedural capabilities such as tying knots, managing clothing systems, learning how to fit a harness, and abseiling techniques. Interaction with the natural world also stimulates knowledge acquisition about this environment and environmental literacy includes recognising hazards, interpreting weather conditions, and understanding local ecology and history.

Aspiring for residual effects

Learning outcomes are important because these semi-permanent brain changes are carried with individuals into future settings and situations (Ransford et al., 2001). Modifications to cognitive architecture influence how people perceive and interpret new information and experiences (Foley & Kaiser, 2013). Consequently, these changes affect how individuals reason, act, solve problems, make future predictions, and learn new information (NAoSEM, 2018). However, for learning to endure, it must be reactivated through conscious recall or application to genuine tasks.

Pulling the levers

Because learning involves physical changes to long-term memory, this requires an inordinate amount of cognitive energy and effort to achieve (Kesteren & Meeter, 2020). Adventure tourism affords several unique characteristics that support the development of vivid and robust memories. Nonetheless, practitioners can exert additional influence on this complex biological process via several techniques. First, cultivating a psychologically safe climate is conducive to learning because it reduces fear and anxiety, and encourages individuals to reveal vulnerabilities and persist without certainty (Darling-Hammond et al., 2019). Second, exploring participants; existing knowledge and skills allows practitioners to build a competency blueprint, pitch sessions at appropriate levels, and enhance relevance and relatability by referring to this (Allan et al., 2012). Third, framing can be used to highlight how experiential aspects are applicable to life beyond the immediate situation (Schenck & Cruickshank, 2015). From the canyoning example, this could involve regulating one's emotions and trusting others to help. Finally, reflection post-experience can consider pertinent elements and how these can be used in future (Burke & Hutchins, 2007). Although this can occur independently, it is more effective when supported.

Legacy

Outdoor and adventurous experiences routinely stimulate emotional outcomes among participants (Rickinson et al., 2004). However, these affective states are ephemeral and quickly dissipate post-activity (Fredrickson, 2004). Nonetheless, adventure tourism often generates positive experiences that individuals remember for long periods as episodic memories. This knowledge is drawn upon when considering engaging in similar experiences in future. For some, positive emotions and encounters can fuel motivation to develop additional knowledge and skills in pursuit of mastery and independent engagement. This foundation of previous experience, knowledge, and memory is transportable and ready to be built on within successive adventure tourism–related activities (Haskell, 2001).

Transcending contextual boundaries

Individuals with positive experiences of adventure tourism may intend to persist with associated activities, and/or apply elements of learning from engagement in alternate settings and situations. However, there is often a gap between people's intentions and corresponding actions (Ariely, 2009). This application challenge is exacerbated when development contexts are dissimilar to one's everyday environment – like adventure tourism (Ransford et al., 2001). Nonetheless, deliberate planning in a psychologically cold state – affording rational thought about the future – can support the transition of self-awareness, social skills, and other intentions from these discrete events (Behavioral Insight Team, 2020). Example features of this process include contemplating details of when/where something can be applied and how plans can be made public. Equally, factors that may derail intentions and possible strategies to overcome such environmental hurdles all require consideration. Finally, strategies for preventing forgetting, and contemplating how the environment can be manipulated to support desired actions are all important steps towards intentionality.

An example of abstracting meaning from the adventure tourism scenario may include individuals realising that despite the uncertainty of their abseiling ability, and significant negative emotions linked to perceived risks, they successfully completed three descents and achieved more than they anticipated. Similarly, during the water slides, leaning on others for support was an essential skill to ensure favourable experiences and outcomes. This learning can be applied to an upcoming presentation at work and shared with the other contributor. Potential disruptive forces may include being too busy to discuss some self-regulation tips, which could be mitigated by booking a 15-minute coffee break next Tuesday to exchange ideas. This event could be added to a digital calendar, with a daily reminder to nudge engagement. Finally, on the day of the presentation, a message highlighting "you are capable of more than you realise" could be added as a phone background to cue desirable attitudes and temporarily boost self-efficacy.

Strategies to reactivate memories

Memories of past events and learning about oneself and/or the environment typically fade without reactivation (EEF, 2021). Recalling this information strengthens the memory itself and pathways to retrieve it, thereby increasing its applicability (Butler et al., 2017). The exceptions are procedural skills such as specific paddle strokes, using a compass, and reading maps – which are more resistant to decay (Weinbauer-Heidel, 2018). Nevertheless, several techniques can be used to retrieve pertinent knowledge, reinstate previous experiences, and delay knowledge/skill fade. Conveniently, the 21st century affords ample opportunities to capture aspects of adventurous experiences and technological solutions to interrupt and remind one's future self. These include scheduling reminders into calendars to avoid forgetting, using photographs as backgrounds on digital devices to refocus attention, using social media platforms to share experiences (see Chapter 11), and discussing previous events and future plans with others.

Self

The first part of this chapter focused on how adventure tourism leaders consider a range of key elements when delivering sessions. This section explores ways in which adventure tourism leaders can develop and maintain the required knowledge, skills, and experience. As the canyoning scenario illustrates, adventure tourism leaders must respond positively to a range of different challenges within dynamic environments (Priest & Gass, 2018). Working alone with groups and rarely observing other colleagues is common. This professional isolation is compounded by the self-employed nature of the adventure tourism industry. While many adventure tourism leaders operate under the umbrella of larger organisations, individuals are typically self-employed, and responsible for their own development and making sure their practices align with current standards (e.g., following health and safety guidelines for equipment maintenance). As a self-employed worker, this is challenging. Much of their time is allocated to delivering activities, with little available for professional development – an important and necessary factor of the job. Whilst challenging, three broad areas currently provide opportunities for self-development; certification, personal experience, and continued professional development. When accessed, these offer adventure tourism leaders time and space to step back from leadership positions and reflect on their practice with relatable others.

Certification

Certification plays a key role in the training and assessment of leaders who operate in the adventure tourism industry. A wide range of National Governing Body (NGB) awards exist, with these typically employing a similar system that involves components of training, consolidation, and assessment (see Chapter 18). This format is a tested and effective way of providing leaders with the technical skills for delivering sessions for participants in their specific disciplines and is commonly used across the world for certifying leaders (Priest & Gass, 2018). While certification plays a fundamental role in establishing a technical base for leaders to be able to operate in their discipline, this is only one element of the role of the leader. As identified by the SELLS model, leadership in this context requires more than just technical skills when working with a group.

This perspective is echoed by a growing critique of certification, predominantly focused on delivering technical aspects of leading, and less on the more subtle interpersonal or soft skill requirements (Hickman & Stokes, 2016). These are equally important in delivery, providing insight into the thoughts and feelings of individuals and the group. These are vital skills, ensuring that leaders can gauge how the activity is progressing and adjust the activity through active observation. Interpersonal skills are typically developed over time, by working with a range of diverse groups, and observing peers, all of which adds to their rich corpus of knowledge, skills, and practices.

Experience

Experience plays a pivotal role in how an outdoor leader can build their knowledge and skills (Martin et al., 2017). While personal experience is a vital tool for leaders in developing practice and provides a key aspect of development, leaders who rely solely on their own experience are in danger of being lured into the heuristic trap of experience bias. This occurs when an individual processes experiences in isolation, drawing conclusions without considering other influencing factors. For example, the leader in the canyoning scenario (Case study 17.1) avoids this trap by constantly accessing available information whilst also questioning the suitability of the group to progress on their journey. In this context, the leader has adopted a systems approach, by considering previous experience and experience from others, and matching this against the current environment as it is presented. To do this requires the leader to not only adopt a reflective mindset but to also have an open mind. One method for enabling this is through accessing continued professional development activities organised by associations linked to particular disciplines.

Continued professional development

Continued professional development, such as new industry practice workshops, plays a key role in maintaining contemporary practices and adventure tourism leader development. Continued professional development allows access to contemporary industry practices, shared experiences, and the development of a broader knowledge base. For example, in the United Kingdom, several professional bodies are now integrating continued professional development as part of being able to demonstrate current professional practice is maintained. While many adventure tourism leaders have short-lived careers, there are a growing number of professional career adventure tourism leaders who proactively seek out continued professional development as part of their career trajectory (Wagstaff, 2011).

Conclusion

Adventure tourism is a rapidly growing industry that provides unique client experiences. This requires leaders to deploy a range of roles to optimise individual and group experiences. The dynamic operating environment of adventure tourism makes this a complicated process. The SELLS framework distils this complexity by focusing leaders' attention towards five key areas. The first element, *safety*, recognises that risk-free adventure is impossible. Indeed, risk is often synonymous with adventure and the very thing that attracts adventure tourism clients. This requires leaders to manage the precarious balance between inevitable risks and safe practices. The second aspect, *enjoyment*, accentuates rapport and trust as essential ingredients for negotiating adventure. Groups with stronger bonds are better able to manage dynamic environments. Attendance to these sensitivities enables the realisation of personally significant enjoyment and challenges to be overcome. The third area, *learning*, illustrates that novel, multi-sensory, and emotional experiences create robust long-term memories. Adventure tourism offers an effective vehicle for personal and social development. Individuals often develop self-awareness and regulation, procedural skills, environmental awareness, and group work competencies. The fourth component, *legacy*, is inextricably entwined with learning. The extraordinary nature of adventure tourism experiences ensures a legacy by creating lasting memories. But legacy is much more than this and underscores the importance of transferring learning into other contexts and strategies to support this. Finally, the fifth consideration, *self*, acknowledges that ongoing professional development is required to be a safe and effective adventure tourism leader. The independence, autonomy, and dynamic nature of the role make this especially important. Collectively, the SELLS framework offers a simple, robust, and memorable tool for adventure tourism leaders to apply in this rapidly evolving sector to generate positive client experiences and outcomes.

Review questions

1 The chapter views the SELLS model largely from the leader's role; however, how might it be viewed from a participant's point of view?
2 How could an adventure tourism leader better target safety/enjoyment/learning/ legacy/self in their daily work routine?
3 How can adventure tourism leaders ensure they remain knowledgeable, skilled, and current within their chosen field?

Further reading

Hannah Clinch, C., & Filimonau, V. (2017). Instructors' perspectives on risk management within adventure tourism. *Tourism Planning & Development, 14*(2), 220–239.

Jenkins, I. (2019). *Adventure tourism and outdoor activities management: A 21st century toolkit.* CABI.

Martin, B., Bruenig, M. Wagstaff, M., & Goldenberg, M. (2017). *Outdoor leadership: Theory and practice* (2nd ed.). Human Kinetics.

Schenck, J., & Cruickshank, J. (2015). Evolving Kolb: Experiential education in the age of neuroscience. *Journal of Experiential Education, 38*(1), 73–95.

Sibthorp, J., Furman, N., Paisley, K., Gookin, J., & Schumann, S. (2011). Mechanisms of learning transfer in adventure education: Qualitative results from the NOLS transfer survey. *Journal of Experiential Education, 34*(2), 109–126.

Reference

Achua, C., & Lussier, A. (2013). *Effective leadership* (5th ed.). South-Western: Cengage Learning.

Allan, J. McKenna, J., & Hind, K. (2012). Brain resilience: Shedding light into the black box of adventure processes. *Australian Journal of Outdoor Education, 16*(1), 3–14.

Allan, J. F., McKenna, J., Buckland, H., & Bell, R. (2014). Getting the right fit: Tailoring outdoor adventure residential experiences for the transition of schoolchildren. *Physical Education Matters, 8,* 37–42.

Ariely, D. (2009). *Predictably irrational: The hidden forces that shape our decisions.* Harper Collins.

Behavioural Insight Team (2020). *How behavioural insights can help college and university students learning remotely due to COVID-19* [Online]. Retrieved March 27, 2020, from https:// www.bi.team/blogs/how-behavioural-insights-can-help-college-and-university-students-learning-remotely-due-to-covid-19

Blume, B., Ford, K., Baldwin, T., & Huang, J. (2010). Transfer of training: A meta-analytic review. *Journal of Management, 36*(4), 1065–1105.

Buckley, R. (2010). *Adventure tourism management* (1st ed.). Butterworth-Heinemann/Elsevier.

Burke, L., & Hutchins, H. (2007). Training transfer: An integrative literature review. *Human Resource Development Review, 4*(3), 263–296.

Butler, A., Raley, N., Black-Maier, A., & Marsh, E. (2017). Receiving and applying knowledge to different examples promotes transfer of learning. *American Psychological Association, 23*(4), 433–446.

Cater, C. (2013). The meaning of adventure. In S. Taylor, P. Varley, & T. Johnson (Eds.), *Adventure tourism: Meanings, experience and learning* (pp. 7–18).

Collins, L., & Collins, D. (2013). Decision making and risk management in adventure sports coaching. *Quest, 65*(1), 72–82. https://doi.org/10.1080/00336297.2012.727373

Csikszentmihalyi, M. (1990). *Flow: The psychology of optimal experience.* Harper Collins.

Curtis, R. (2008). Risk assessment & safety management (RASM): The complete risk management model for outdoor programs. Retrieved March 17, 2023, from https://www.outdoored.com/articles/ risk-assessment-safety-management-rasm-complete-risk-management-model-outdoor-programs

Daft, R. L. (2011). *Leadership* (5th ed.). South-Western: Cengage Learning.

Darling-Hammond, L., Flook, L., Cook-Harvey, C., Barron, B., & Osher, D. (2019). Implications for educational practice of the science of learning and development. *Applied Developmental Science, 24*(2), 97–140.

Dickson, T., & Dolnicar, S. (2004). No risk, no fun - the role of perceived risk in adventure tourism, CD Proceedings of the *13th International Research Conference of the Council of Australian University Tourism and Hospitality Education (CAUTHE 2004)*.

Eastabrook, C., & Collins, L. (2020). Why do individuals seek out adventure sport coaching? *Journal of Adventure Education and Outdoor Learning*, 20(3), 245–258. https://doi.org/10.108 0/14729679.2019.1660192

Educational Endowment Foundation (2021). Cognitive Science Approaches in the Classroom: A Review of the Evidence [Online]. Retrieved July 22, 2021, from https:// educationendowmentfoundation.org.uk/education-evidence/evidence-reviews/cognitive-science-approaches-in-the-classroom

Ewert, A., & Sibthorp, J. (2014). *Outdoor adventure education: Foundation, theories and research*. Human Kinetics.

Ferrero, F. (2006). *British Canoe Union coaching handbook*. Pesda Press.

Foley, J., & Kaiser, L. (2013). Learning transfer and its intentionality in adult and continuing education. *New Directions for Adult and Continuing Education*, 137, 5–15.

Fredrickson, B. L. (2004). The broaden-and-build theory of positive emotions. *Philosophical Transactions of the Royal Society of London, Series B: Biological Sciences*, 359(1449), 1367–1377.

Haskell, R. (2001). *Transfer of learning: Cognition, instruction, and reasoning*. Academic Press.

Hickman, M., & Stokes, P. (2016). Beyond learning by doing: An exploration of critical incidents in outdoor leadership education. *Journal of Adventure Education and Outdoor Learning*, 16(1), 63–77.

Jackson, J., & Heshka, J. (2011). *Managing risk: Systems planning for outdoor adventure programs*. Direct Bearing.

Kesteren, M., & Meeter, M. (2020). How to optimize knowledge construction in the brain. *NJP Science of Learning*, 5, 1–7.

Martin, B., Breunig, M., Wagstaff, M., & Goldenberg, M. (2017). *Outdoor leadership*. Human Kinetics.

Mees, A., Sinfield, D., Collins, D., & Collins, L. (2020). Adaptive expertise – A characteristic of expertise in outdoor instructors?, *Physical Education and Sport Pedagogy*, 25(4), 423–438. 10.1080/17408989.2020.1727870

National Academies of Sciences, Engineering, and Medicine (2018). *How people learn II: Learners, contexts, and cultures*. The National Academies Press.

Northouse, P. (2018). *Leadership theory and practice*. Sage.

Outdoor Instruction.co.uk (2011). C.L.A.P.S. and S.A.F.E.: Two of the most useful acronyms for any leader in any environment in the outdoors. Retrieved November 1, 2022, from https://www. youtube.com/watch?v=KBPL55QLdiY

Panicucci, J. (2007). Cornerstones of adventure education. In D. Prouty, J. Panicucci, & R. Collinson (Eds.), *Adventure education: Theory and applications* (pp. 33–48). Human Kinetics.

Priest, S., & Gass, M. A. (2018). *Effective leadership in adventure programming*. Human Kinetics.

Ransford, J., Brown, A., Cocking, R., Donovan, S., Bransford, J., & Pellegrino, J. (2001). *How people learn: Brain, mind, experience, and school*. National Academy Press.

Rickinson, M., Dillon, J., Teamey, K., Morris, M., Choi, M. Y., Sanders, D., & Benefield, P. (2004). *A review of research on outdoor learning*. National Foundation for Educational Research.

Rokenes, A., Schumann, S., & Rose, J. (2015). The art of guiding in nature-based adventure tourism – How guides can create client value and positive experiences on mountain bike and Backcountry Ski Tours, *Scandinavian Journal of Hospitality and Tourism*, 15(1), 62–82. 10.1080/15022250.2015.1061733

Schenck, J., & Cruickshank, J. (2015). Evolving Kolb: Experiential education in the age of neuroscience. *Journal of Experiential Education*, 38(1), 73–95.

Wagstaff, M. (2011). Do outdoor leaders have careers? An introduction to the outdoor leader career development model. *Journal of Outdoor Recreation, Education, and Leadership*, 3(2), 116–119.

Weber, K. (2001). Outdoor adventure tourism: A review of research approaches. *Annals of Tourism Research*, 28(2), 360–377.

Weinbauer-Heidel, I. (2018). *What makes training really work: 12 levers of transfer effectiveness*. Tredition.

Williams, A., & Wainwright, N. (2016). A new pedagogical model for adventure in the curriculum: Part one – advocating for the model. *Physical Education and Sport Pedagogy*, 21(5), 481–500.

18 The adventure tourism professional

Becoming and being the guide

Matt Groves, Tamara Griffiths and Peter White

Chapter learning outcomes

1 Recognise the antecedents and emergent culture of risk-based decision making as the core framework of employment and career progression in adventure tourism.
2 Consider the inherent paradox and tensions between the "business as usual" of the adventure tourism industry, and the need for values-based philosophical underpinnings embedded in its professional culture.
3 Understand and reflect on the adventure career transition from focusing on social capital and individual progression towards a community of practice and inspiring environmental justice.

Introduction

Dear Reader,

We imagine you have either arrived here having read this book rigorously from cover to cover, or perhaps your attention was caught momentarily by the title and what it means for you, especially if you are in direct pursuit of that most rewarding of careers – the "adventure tourism professional".

Well, step in and enjoy (cue music).

Visualise that great sunbathed landscape around you as you soak up the feelings of the infinite office and look down on the happy and admiring paying customers you are guiding through this wilderness encounter …

Stop! (Needle *scra-ape!*)

Sorry, that wasn't the right record. To be fair, this picture might have resonated with some true experiences of adventure tourism, and "living the dream" is surely a laudable motivation and a possibility for us all. However, as we explore in this chapter, certain stereotypes around the role and performance of being a "guide" portrayed in this image could be challenged by ethical and philosophical scrutiny as perpetuating an unsustainable culture and practice in our relationship with the land and people around us. We begin with an examination of the expertise of the professional guide and how this is

DOI: 10.4324/9781003393153-22

intrinsically linked to constructs of risk in adventure, as well as being a core element of career progression and the culture of professionalism within the industry. We also interrogate concepts related to professionalism and professional identity, deepening our understanding of what it means to be a professional practitioner and challenging the "business as usual" ethics and values.

The professional roles in adventure tourism resist simple definition, because adventure tourism encompasses a vast array of possible activities and takes place in all possible global habitats, as well as delivering many niche markets, modalities, and purposes, from half a second of freefall on a bungee jump to many weeks on expedition. By regarding this vast diversity of activity cultures under the banner of "adventure tourism", we may glimpse the problem of stereotyping; as with all stereotypes, variation is the norm.

We will often refer to the role of the guide as a *professional*, or *adventure tourism professional*. Such terms are intended to be used interchangeably, along with other terminology commonly found in the sector, such as *leader*, *instructor*, or *facilitator*.

Part 1. Becoming a guide: Entering and operating within the adventure tourism arena

Taking our earlier comment on the diversity of adventure tourism a step further, the field of adventure tourism is a major theme within a broader domain of professional outdoor practice (Collins & Collins, 2012), which has become a global community connected by a range of common practices applied across specialised fields, such as outdoor and adventure education, coaching and instruction, and therapeutic practice. Despite there being a strong base of research into outdoor careers, very little has been focused on careers in adventure tourism. Most studies have revolved around outdoor education (Hall & Jostad, 2020; Thomas, 2001), the specific experiences of women working in the outdoors (Allin & Humberstone, 2006; Wright & Gray, 2013), adventure sports coaches (Collins & Collins, 2012), or outdoor instructors (Mees et al., 2021).

Whilst this work is highly transferable to adventure tourism career roles, they have not specifically addressed the issues and context that are found within adventure tourism careers. For example, it is important to acknowledge that a professional career in adventure tourism is not limited to the role of guiding, because innovative enterprises and shifting demand have led to a greater emergence in the market of "self-guided" and other progressive models of adventure tourism where different niches may intersect, such as family adventures, well-being tourism, and microadventures (Sand & Gross, 2019). This has brought a great divergence and competition to the sector, for which travel, digital, finance, and marketing skills are key, redefining and segmenting the adventure tourism industry (ATTA, 2021a). While these aspects are important for an adventure tourism career, the main perspective for this chapter is on becoming and being a professional adventure tourism guide.

A post-colonial, post-modern inheritance

In the adventure tourism field, it is common to identify as a *guide*, a title that carries global historical significance in travel, trade, and exploration in wildlands, as well as denoting a key factor in the relationship to the client (Carnicelli-Filho, 2013; Laing & Frost, 2014; Rokenes et al., 2015). The roots of adventure tourism reach deeply into predominantly Western narratives of "adventurous" travels, via myths and legends and into the later colonial era and the romantic search for the sublime, with notable examples peppered throughout Western history (Beames *et al.*, 2019). Laing and Frost (2014, p. 9) explore early 20th-century accounts of travellers and tourists seeking authenticity by re-enacting the heroic, relating it to 'a transformation, testing themselves against adversity and encouraging self-reflection, even escape'. See Chapter 2 for detailed insights into the history of adventure tourism.

These historical roots and narratives coalesced with structural social reform through the inter- and post-war decades as an emergent outdoor professional practice (Laing & Frost, 2014; Ogilvie, 2013). During the 1980s and 1990s, the process of commercialisation established this into the adventure tourism industry that remains familiar today. Bell (2017, p. 281) positions this as a post-colonial cultural inheritance, remarking on the misappropriation of the classical values of adventure into a material commodification of romantic notions of risk:

> a sensibility to "adventure" has been turned into a cultural practice which generates the conditions for adventurers to emerge as risk-taking subjects, willing to collude in the calculation of consequences and a context in which risk-taking is consumed for status.

Under the cultural inheritance of Western traditions of heroic endeavour and shaped through the history of colonialism, both society and academic literature have settled on meanings of the word *adventure* that are centred on notions of "drama", "risk", "challenge", and "jeopardy" (Weber, 2001, p. 363). Arguably, this inheritance leads to assumptions that may narrow and over-simplify a concept that offers wider interpretations and meanings through critical examination.

The task of the adventure tourism professional – making meaning, playing safe

Society's complex relationship with adventure and risk has been explored in prior chapters (see, e.g., Chapter 3) and examined in detail by many authors, including Lyng (2004), Beck (1992), Breivik (2007), and Giddens (1991). However, we are more concerned here with the professional "risk management" culture that has naturally dominated outdoor practice through the public preoccupation with seeking the thrills of adventure in commodified experiences and specialised activities where complete trust in the expertise of the guide is demanded (Cater, 2006; Holyfield, 1999) and implicitly in the entire socio-economic system that puts them there.

While perceived risk might be a preoccupation of the public and of the sector, it has a rational basis. Adventure tourism, involving physical activities in dynamic outdoor environments, presents both the client and the practitioner with real, ontological risks through the exposure to potentially harmful factors (e.g., gravity, height, momentum, hard surfaces, water, extreme temperatures, altitude, pathogens). This can lead to real accidents, sometimes with tragic outcomes, so it can be said that the cultural response of the outdoor professional community is a pragmatic one, as explored by Collins and Collins (2013) through approaches to accepting and exploiting risk in the risk–benefit analysis. As described in Chapters 15 and 16, an intriguing difficulty of trading off risk with benefit is the compatibility of the currencies involved – the risk to human life versus the benefits of learning, growth, and health. How much risk is worth the associated benefit in this analysis, and who decides? Also, the embodied pragmatism and management of risk (Gilbertson & Ewert, 2015) involved in practitioners' adventurous encounters has created a unique culture that sets itself apart from the constraints of a more orderly social model and places the locus of control on the individual (Lyng, 1990), adding an allure and attraction for people seeking the "other". This portrays an image of highly skilled practitioners embedded in their professional context and in control of the risks (Beedie, 2003), in contrast to the client, who may well lack sufficient experience and knowledge, but may be attracted by the social capital bound up in the adventure (Beames et al., 2019).

The role of the adventure tourism professional, as explored by Holyfield (1999) through whitewater rafting guides, is, in part, to act as a conduit between the organisation and the customer, negotiating the tension between the *inexperienced* consumer and the playful yet emotionally intense adventure ethic they hope to emulate. Beedie (2003, p. 150) found not only similar tensions but also opportunities emerging through this relationship:

> Mountain guides, therefore, are both guardians of the tradition that has established them as professionals and the medium through which adventure tourists, that is people who buy adventure holidays, experience mountains. Experience of mountaineering offers a frame of reference through which people can make sense of being a mountaineer. Guides will have an established frame whilst clients, who recognize their limitations as independent mountaineers by employing guides, are more likely to bring degrees of framing to their adventure holidays.

The reality of employment in adventure tourism is that the role reflects a complex interplay between the expertise required to keep the clients safe, the management and structuring of an experience that feels "authentic", and the bottom line of business profitability.

Understanding and growing expertise

Tasked by their employer with generating exciting, thrilling, light-hearted adventures with next-to-zero risk of harm ever occurring on their watch, adventure professionals must display a wide range of capabilities, experience, judgement, and knowledge to comply with society's expectations and support the aims of the organisation. This paradoxical position sets up a stereotype image of expertise as a constant and enduring standard – the guide fulfils an expectation of effortless expertise, which is simply accepted and admired. However, experience, knowledge and judgement are properties that develop progressively over time as a response to the person's socialisation and exposure to learning opportunities, suggesting that expertise is far more subjective and individual (Collins et al., 2018). When does the expert emerge? The stereotype does not fit with either the variety of specialisms in the sector or the range of people who work and function in it.

Regardless which activity typology workers might operate in, it is common to see similar patterns in complexity and difficulty of the environment or task, placing varying demands for the professional to have the necessary skills and competencies to not only cope with the degrees of uncertainty and challenge presented but also have spare capacity to extend foresight and care for their clients. This spectrum is illustrated and explored in Case study 18.1, which outlines examples of paddle sports qualification structures from a selection of nations, to illustrate the core roles that personal experience and skill play in the process of development towards fully qualified and competent status. It is worth also briefly noting at this point that the physical environment is only one dimension in which professional working can gain complexity and demand expertise; specialisms (and complexity) also very much exist around the human or social context of the work, where skill is required in observation, knowledge, empathy, and interpretation, such as place-based interpretive working, well-being tourism, and other aspects of emotional labour (e.g., as explored by Carnicelli-Filho (2013) and Farkic (2018)).

Case study 18.1   A comparison of international paddle sports qualifications

Table 18.1 shows clear parity in the broad principles of skill development as a professional coach or leader in canoeing across international systems. The equivalence of the awards is inferred by the authors in terms of scope and remit of the award holder, as described in Table 18.1.

The most immediate similarity that the different frameworks show is the key role that the environment plays in differentiating the level of ability. The primary features that mediate the demands placed on the paddler are the water surface (sheltered, open, moving, or featuring whitewater and obstacles), as well as the strength of the wind and the distance from shore. These environmental factors modify the complexity (or predictability) of the environment and the level of skill needed to successfully navigate through it, where more complex, fast-moving, high-volume, and unpredictable environments contribute to increased likelihood of accidents and worsening potential consequences. This demonstrates very clearly that the fundamental currency of progression is skill and experience as a factor to mitigate against risk. The UK-based system of river "Grades" is broadly equivalent to the "Class" scales used in the US and New Zealand, but they have all evolved by national consensus to respond to their typical geographies, so there is often a different "feel" that paddlers would need to adapt to if translating their skills between countries.

The associations that arrived at these qualification frameworks are national-based organisations that consist of professional guides, leaders, and coaches from the home nations, whose accumulated experience is shaped by a myriad of factors. The degree of similarity between nations positions the grading and standardisation of professional qualifications as built on a true international consensus on the human constraints of canoeing in different environments.

This consensus is embedded in the culture of the sport, both at national and global levels, which has largely evolved via encounters between practitioners and their environments. The fact that a common pattern is visible is also indicative of the global culture that surrounds this adventure sport, where deeper embeddedness and progression through the sport is often reflected in proficiency in the skills and capacity to cope with greater challenges. Early-career qualifications offer an accessible entry point to manage groups in simple environments and begin encountering the additional complication of responsibility for others. At any level, however, the practitioner's personal competence must have developed beyond the ceiling of that remit to facilitate enough spare capacity to be a suitable leader of others in that place. The onus of progression is on constant personal skill development, understanding and managing risks, and looking towards the next level.

Table 18.1 Summary of approximate equivalences between canoe-based qualifications frameworks and progressions in the UK, New Zealand, and the US

UK British Canoeing*	New Zealand New Zealand Outdoor Instructors Association	US American Canoe Association		
Paddle-sport Leader *Mixed fleet of craft, inland open water no more than 200m from shore, in winds no more than Force 3, and simple ungraded moving water*	**Canoe Leader** *Can lead canoe activities on flat or sheltered water with little or no wind present or forecast. Locations include estuaries, river deltas, sheltered beaches, lakes, coastal inlets and swimming pools. This also includes rivers when the flow and the hazards present no greater risk than flat water.*	**Level 1 – Introduction to Canoeing** *Flat water, protected from wind, waves and outside boat traffic, with current less than 0.5 knots, and within swimming distance of shore*		
Paddle-sport Touring Leader *Mixed fleet of craft, inland open water no more than 500m from shore, in winds no more than Force 4, and simple moving water up to Grade 1 (2)*		**Level 2 – Essentials of Canoe Touring** *Protected water near shore with winds up to 10 knots, waves up to 1 foot (.3 metre) or current up to 1 knot* **Level 2 – Essentials of River Canoeing** *Moving water on rivers up to and including class I sections*	**Levels 2–5 – River Canoe Day Trip Leading** *Easy moving water up to and including class I–II rapids*	**Levels 2–5 – Canoe Camping** *Not to exceed Instructors level of certification*
	Canoe Guide *Can guide multi-day canoe trips on rivers with up to class 2 white water. Can work independently without direct supervision.*	**Level 3 – River Canoeing** *Rivers rated up to class I–II, where limited manoeuvring in current may be required to avoid obstacles*		
Canoe White Water Leader *Moving water up to Grade 2 (3)* **Canoe Open Water Leader** *Open water no further than 500m from shore, in winds up to Force 4*	**Canoe 1** *Can organise, guide and instruct multi-day canoe trips on rivers with up to class 2 white water. They can also manage and teach canoe river rescue skills on class 2 white water.*	**Level 4 – Whitewater Canoeing** *Class I and II rapids on sections of rivers rated up to Class II-III*		
Advanced Canoe White Water Leader *Moving water up to Grade 3 (4)* **Advanced Canoe Open Water Leader** *Large exposed open water in winds up to Force 5*		**Level 5 – Advanced Whitewater Canoeing** *Sections of rivers rated class II-III-IV, in rapids where precise manoeuvring in current will be required to avoid obstacles*		

* British Canoeing also have equivalent coaching qualifications for each award described.

Figure 18.1 Kayaking guide instruction. Photograph by Matt Groves.

As this case study indicates, the complexity of the environment needs to be matched by not only the physical skill to perform in it but also the ability to understand it, and adventurous situations with complex or subtle attributes require a combination of effective knowledge structures and intuitive pattern recognition, which can only come from a depth of learning and experience (Dreyfus, 2004; Cannon-Bowers & Bell, 2009). Hogarth et al. (2015) describe such environments as "wicked" learning environments because the cues may be subtle or ambiguous and involve a poor feedback relationship to decisions or actions taken (i.e., the same decisions or actions may have different consequences on subsequent occasions). They require extensive time to develop the tacit knowledge and complex rules or concepts to a sufficient level to function, and indeed, they may continue to be wicked to expert practitioners through the sheer ambiguity of information and feedback, and the added human factors of cognitive habits, biases, and heuristic traps (Johnson et al., 2020; McCammon, 2002). Johnson et al., (2020) provide a critical analysis of these perceptual and decision factors in avalanche terrain, which is an extensively researched domain known for its *wickedness* as a learning environment. This field of study is important to inform our understanding of fallibility in judgement and decision-making, but it is separate from the journey of transition from novice to expertise itself.

From the perspective of developing expert skills, knowledge, and understanding, however, Dreyfus's (2004) work in stages of skill acquisition, and other research, including Groves and Varley (2020) and Mees et al. (2021), supports the view that early-stage practitioners may lack the perceptual skills to recognise the cues in more complex environments, which might appear as mere "noise" (Johnson et al., ibid., p. 14), causing important details to go unnoticed, or misleading through drawing attention to non-relevant information. There is a broad indication from the literature in both judgement and decision-making and developing expertise that, in the early stages of employment, with all the responsibility for monitoring safety of participants, this *noise* needs to be contained and substituted by a predictable structure. This is done to give the employee as clear and unambiguous information to work with as possible, coupled with effective decision-making strategies and a progressive exposure to domain specific knowledge and contextual experiences (Cannon-Bowers & Bell, 2009). In this light, early-stage professionals may bear a stronger resemblance to the more novice client discussed earlier than to the stereotypical "expert". They may lack the requisite perceptual skills and embedded cultural knowledge to function independently or confidently in more challenging environments.

There is a clear skill-based progression route for professional practitioners, which also manifests in a spectrum of industry self-regulation and practice that we see both in employers and through the training and qualification process, from simplified, highly structured, rule-based procedures at one end to unstructured, dynamic decision-making in complex environments at the other. Interestingly, this spectrum of activity involvement mirrors the inherent paradox in managing ("confecting") an authentic adventure "experience" for novice clients (Cater, 2006; Pomfret & Bramwell, 2014; Varley, 2006). Holyfield (1999) describes the culture of whitewater rafting guides as one built on a 'semi-structured' adventure, wrapped up in a codified language of social interactions to manage the high emotional demands of the environment. Holyfield (1999) and Hansen et al. (2020) both draw parallels between the work done in adventure tourism and that of amusement parks, offering a view of adventure tourism far removed from its adventurous ethic roots.

This throws into light another perspective on the lines of separation between "guide" and "client", which is a key point in understanding a fundamental attribute of the role and skill required of you. The journey to become an adventure tourism professional will be marked with many personal experiences and a growing embeddedness in the culture and practice of your chosen domain. This can widen the degrees of separation from the inexperienced client that may have unintended by-products. For example, if, as noted earlier, mountain guides are the medium through which adventure tourists experience mountains (Beedie, 2003), then the product that the tourist is buying is an embodiment of key elements of that mountain culture, as well as a safety frame to substitute for the complexities of personal decision-making, planning, and personal local knowledge. This implies a responsibility on the shoulders of the professional community to consider the values that this culture represents, which we explore later in this chapter.

Part 2. Being: Professionalisation and belonging within the adventure tourism community

In many respects, professions have been a long-standing feature of Western society, often associated with traditional roles such as law and medicine (Beaton, 2010). In today's evolving society, the clearly defined lines between a profession and an occupation (i.e., non-professional) have become challenging to retain. A *profession* could instead be seen as a knowledge-based category of an *occupation*, one which normally follows a period of formal training and experience (Evetts, 2003). Complementing the knowledge-based aspect are the ethical and moral dimensions of a profession, as 'professionals are from the outset involved in the practice of activities and endeavours whose ends and purposes are matters of genuine ethical controversy' (Carr, 2000, p. 29). Considering what is expected of the role as an adventure tourism professional – their knowledge, skills, and experience (ATTA, 2021b) – and the need for them to exercise their own ethical judgements, then it becomes clear that it is indeed a modern-day profession (Cheetham & Chivers, 2005).

Intertwined with this terminology is the concept of *professionalism*. What does professionalism mean? *High standards? Punctuality? Preparedness? A skilled expert in your craft?* Whilst these examples can be evidence of professionalism, they also suggest a deeper level of engagement and quality of what you do *as a* professional, which is often externally governed by a set of ethics and standards (Carr, 2000). Professionalism can be thought of as being enacted through either bottom-up or top-down factors (Evetts, 2003). The former emerges through a common identity from shared education, experiences, and memberships, whereas the latter sees professionalism as a tool for social control for "professional" workers, often dictated by powerful membership bodies.

The emergence of professionalism within the adventure tourism industry can be attributed to a wide range of potential factors, but it is crystallised through the rise of formal qualifications and associate and representative member bodies. One of the most universally recognised qualification standards is the International Federation of Mountain Guides Associations (acronyms associated with this organisation include IVBV/UIAGM/IFMGA), which has existed since 1965, providing regulation and coordination of national professional member associations. This could be argued as the archetype for professionalisation of adventure activity leadership, having established the role of "Mountain Guide" as a globally recognised vocation, and as a protected title and regulated profession in many nations (International Federation of Mountain Guides Associations, 2017), although the remit is strictly defined around technical operating criteria. Within the UK, among many diverse sport governing and regulatory bodies, the National Governing Body for paddle sports (British Canoeing) has developed a "Guide Endorsement" to recognise and enhance the experience and skills required for guiding, particularly within an adventure tourism context (British Canoeing Awarding Body, 2018). This was a shift away from their core approach of managing and delivering a formal qualification framework to one of recognising the needs of a growing industry and working with them to establish a form of national recognition.

We can also see the rise of professionalism through the increased prominence of member bodies that represent distinct groups. Within the UK, the Institute for Outdoor Learning is long established as a general representative body of the outdoor sector. They have developed criteria for recognition as an "outdoor professional" (IOL, 2021), which, whilst they suggest could encompass someone within the field of adventure tourism, still leans more towards the education and learning aspect. A more specific member body may be the Adventure Travel Trade Association, which initially only represented organisations within the industry but has now opened individual professional membership (ATTA, 2022). Alongside their development of the Adventure Travel Guide Standards, it has both allowed recognition of the diversity of what it means to be an adventure travel guide as well as raising the professionalism of both individuals and organisations within the adventure tourism industry.

Standards such as these will have strongly driven your professional behaviours and expectations but are there other dimensions to *being* an adventure tourism professional beyond professional membership? How do *you* define your professional identity and values? Perhaps, in parallel to your progression through qualifications' frameworks and external standards, your professional identity is something that emerges organically through your own career, with the complex array of influences it presents.

At its core, our identity is our response to the question, 'Who are you?' revealing identities that are individual, relational, and collective (Vignoles et al., 2011). The concept of a professional identity is therefore one that we personally hold for ourselves but is developed through our individual core, our relationships with others and the influence of being part of a wider collective group – potentially one that we are choosing to adhere to a certain set of ethics and values. We can see from this that the roles of bottom-up and top-down professionalisation are both required, as the former is willingly enacting who you are (e.g., personal motivation to climb), while the latter is a set of values and behaviours that you must conform to (e.g., code of conduct in a professional association), to enable you to identify as being part of a particular group. This leads us to understanding the importance of these processes of socialisation within developing a professional identity, as they enable a transition from legitimate peripheral participation (personal and shared experience) through to full participation (formally qualified, validated) within a community of practice – all while developing a sense of belonging to that community (Wenger, 1998).

A term appearing more in North American literature is *outdoor leader*, which Wagstaff (2016, p. 76) defines as 'an individual who leads groups into natural settings via a variety of activities and modes of transportation, such as walking, camping, biking, canoeing, caving, ropes courses, kayaking, and mountaineering, for a variety of purposes and outcomes'. To understand what it means to *be* an outdoor leader, Lewis and Kimiecik (2018) suggest individuals build an integrative lifestyle around optimal inner experiences – such as developing their own passion for adventures and places, particularly *before* being inundated with technical skill development. This is closely linked to Wagstaff's (2016) five stages of outdoor leadership career development (Table 18.2) which indicated the importance in the early stages of developing personal passions and desires within the context of outdoor leadership. In the later stages of career development, Wagstaff (2016)

Table 18.2 Summary of Outdoor Leader Career Development Model (Wagstaff, 2016)

Stage 1 Exploration	• Enthusiastic individuals with little experience • Engaging with the outdoors for the first time • Unaware of the larger outdoor industry
Stage 2 Contemplation	• Possibility of a career in the outdoors becomes a realistic option • Focus on developing technical skills • Important stage promotes passion for their work
Stage 3 Engagement	• Firmly grounded in an outdoor career path • Signified by important, formative leadership experiences • Self-efficacy increases as they embrace higher levels of responsibility
Stage 4 Professional Involvement	• Develop a sense of professional establishment within the industry • Involvement in professional communities adds to a sense of professional identity
Stage 5 Career Professional	• Display a strong sense of professional identity with high levels of self-efficacy • Generally able to self-determine their roles • Represents individuals who can establish long-term careers

identified a strong sense of professional identity as being evident within outdoor leader career professionals. This concept becomes critical to understanding what it means to *be* an outdoor professional – or an adventure tourism professional.

Of course, *being* an adventure tourism professional does not occur in isolation and is tightly woven within both society and the adventure tourism industry – an industry that can be seasonal. Traditionally associated with the summers or traditional tourist periods, this created a dominant "story" that those involved in the outdoors enjoy and pursue this transitory and self-fulfilled lifestyle it promotes (Lewis & Kimiecik, 2018). The reality, however, is that this pattern of "peaks and troughs" can create difficulties – within the wider outdoor field long hours, time away from home, and the effect they had on relationships have all pointed to reasons for high turnover (Marchand et al., 2009; Thomas 2001). The challenge is that if our identity as an adventure tourism professional is primarily based upon this "self-fulfilled lifestyle", then our dominant philosophical perspective (and therefore values and ethos) is likely to be inward-focused, putting us at the *centre* of our personal narrative – not necessarily allowing others around us to influence us in ways which may be unanticipated. Therefore, the understanding that adventure tourism is seasonal and attracts transient workers may not be that helpful in either staff retention (Barnes, 1999) or in allowing individuals to grow into truly *being* an adventure tourism professional.

In our own research of a range of mature UK outdoor professionals (unpublished), clear evidence of a strong sense of professional identity appears. The development of their professional identity seems influenced by several factors, such as (1) skills, competencies and understanding, (2) role modelling of others, and (3) being part of a community. Using Vignoles et al.'s (2011) three identity classifications, we can see how their professional identity was developed through doing (individual), watching (relational), and belonging (collective), as developed in Figure 18.2.

Through developing these three dimensions of professional identity, we can develop a more holistic approach to who we are as an adventure tourism professional. For example, rather than the temptation to focus purely on the technical skills required, we may attempt to become more involved in our communities – both professional and local – which would help us in developing this sense of belonging and being part of something bigger (Varley et al., 2018). If we look at ATTA's 'Adventure Travel Guide Standard' (ATTA, 2021b, p. 2), we see an important expression of this shift visible through the '3 core responsibilities' and the '5 core competencies', where *sustainability* is placed number 1 in both parts, above *technical skills* and *safety and risk management* (Figure 18.3). This is a dramatic change of emphasis from its predecessor, the 2015 'International Adventure Travel Guide Qualifications and Performance Standard', which placed *Sustainability* at number 5 (of 5), and *Technical Competence* as number 1 (ATTA, 2015). Whether this marks a deliberate signal and change of emphasis would be speculation, but it clearly suggests evidence of a paradigm shift from the traditional risk society (Beck, 1992; Giddens, 1991) to a culture that values sustainability and ethical standards that underpin that. To meet this shift, to be an adventure tourism *professional* requires more than the established mindset and practice that is successfully passed down through generations of peers. Rather, an increasingly critical understanding of these prominent values and ethical judgements a professional is required to make is needed, both individually and within the professional community – as shown through the increasing importance of our collective responsibility towards inspiring pro-environmental behaviours in others.

Doing

(Individual)

- Gaining qualifications and accreditations that provide levels of validation, recognition, and credibility.

- Developing a depth of understanding of what, why and how you are doing what you do – recognising the deeper ethical values that draw the community together, rather than specific individual competencies.

- Often established through a mixture of reflection, belonging to a community of practice and higher education.

Watching

(Relational)

- What we absorb from our relationships with others in the field, particularly modelling behaviour we see as good and being observant of poor behaviour.

- Relationships (such as with peers or mentors) allow opportunities for developing personal awareness, helping us tune into our own thoughts and values.

Belonging

(Collective)

- Through growing a network of connections and experience, we become more aware of the who and what of the sector, which helps us understand the situation we find ourselves in and enables us to identify what we belong to.

- Linked to growing relationships (ie. Watching), this enables us to know others, improve others knowledge of ourselves and helps to develop a level of security that indicates that we are "allowed" to be in that place.

Figure 18.2 Developing professional identity (adapted from Vignoles et al., 2011).

Part 3. Inspiring: Promoting sustainability and pro-environmental behaviours

Is it fair that the huge challenge of inspiring intrinsic pro-environmental behaviours in clients falls on an individual practitioner, a person who might simply be trying to make a living doing something they enjoy? The ATTA (2021b, p. 2) suggests that this is the adventure tourism professional's key competency. To shed light on this crucial competency one must possess, this section, first, identifies a contextual framing of sustainability and inspiring pro-environmental behaviour. Then within a process of personal and professional identity we challenge the deeper implications of what this may entail for a professional practitioner. In conclusion, Case study 18.2 offers an inspiring example of a personal journey leading to an embodied and quotidian relationship with sustainability.

THE CORE COMPETENCIES OF ADVENTURE TRAVEL GUIDES

To fulfill core responsibilities, an adventure travel guide must possess specific skills and qualifications. Adventure travel is extremely diverse in terms of geography, cultural context and types of activities. The following five competencies have been identified by guides and adventure travel experts around the world as essential for adventure travel guides regardless of geography or activities:

1. Sustainability

2. Technical skills

3. Safety and risk management

4. Customer service and group management

5. Natural and cultural history interpretation

Figure 18.3 Adventure Travel Guide Standard Core Competencies (adapted from ATTA, 2021b, p. 3).

Put simply, tourism has failed to achieve the level of sustainability needed within our 'business as usual' industry, so a 'sustainable future's paradigm' has emerged 'contrasting its features with those of the established mindset,' (Dwyer, 2018, p. 30). However, as Dwyer (2018) highlights, the sustainable alternatives are 'grafted onto the established mind set without basically changing its key elements' (p. 30) largely because we remain underpinned by the dominant model of neo-liberal, anthropocentric growth at any cost (Jackson, 2017). We are most familiar with the UN and the WTO definition of sustainable development, created upon the limited 'three pillar perception' (Cohen, 2018, p. 367). In addition to being lodged within pure anthropocentricity, prominent scholars criticise the UNWTO for being 'a growth advocacy platform' (Gössling et al., 2016; Hall, 2022).

The issue facing the adventure tourism professional is how more than four decades of pro-environmental research have documented a gap between knowledge and action inspiring pro-environmental behaviours (Boyden, 1987). This infamous gap signifies the lack of ability to connect a pro-environmental behaviour belief, with a pro-environmental behaviour action (Fischer et al., 2017, p. 545). Reviewing 44 studies, Carfora et al. (2017) conclude that stimulating deep pro-environmental behaviour remains unrealised yet is one of the 'major concerns of the 21st century' (Carfora et al., 2017, p.92). This "gap" is mirrored in tourism:

The implementation of sustainability of tourism remains difficult despite decades of theoretical and practical discussions. In addition, there is a gap between the attractive conceptual idea of sustainable tourism and its influence on practice, which is clearly related to tourism irresponsibility or irresponsible tourism behaviour.

(Mohamadi et al., 2022, p. 401)

What can one person, as an adventure tourism professional, hope to do if decades of research have not found a solution? With sustainability increasingly questioned, the concept of "Responsible Tourism" is being framed as a fresh viable approach (Mohamadi et al. 2022, p. 402). ATTA (2021b, p. 6), for example, would like to steer you towards responsible tourism: 'Having less of an impact is no longer enough. To be truly sustainable, our travel experiences must contribute to solving environmental and societal problems'. In responsible tourism, the implied *onus* is on individuals, rather than corporate or global players. Without challenging our capitalist growth imperative, responsible tourism is "shifted" onto local actors:

> This shift in emphasis has taken place because not much progress has been made on realizing sustainable tourism since the Earth Summit in Rio. This is partly because everyone has been expecting others to behave in a sustainable way. The emphasis on responsibility in tourism, RT [Responsible Tourism], means that everyone involved in tourism – government, product owners and operators, transport operators, community services, NGO's (Non-Governmental Organization) and CBO's (Community Based Organization), tourists, local communities, industry associations – are responsible for achieving the goals of responsible tourism.
>
> (Nikhildas & Jagadeesan, 2021, p. 1935)

Scotland Outlook 2030 (Visit Scotland, 2020) endorsed by the Scottish First Minister as 'Responsible Tourism for a Sustainable Future' recognises this endogenous approach, stating: 'Residents have now become as important to a destination as visitors' (Visit Scotland, 2020, 1), with a strategy of engaging locals in managing their tourism. Unfortunately, 'the attitude–behaviour gap in responsible tourism is wide (Juvan & Dolnicar, 2014, p. 76). Given how 'adventure travel has been critiqued for contributing to the very environmental, social and economic issues it purports to mitigate' (Mackenzie & Goodnow, 2021, p. 64), the need to stimulate pro-environmental behaviours is pressing. COVID-19 has increased demand for experiences in nature (John Muir Trust, 2021) and stimulated tourism studies calling for 'accelerated' transition to sustainability (Gössling et al., 2021, p. 15), with urgent perspectives regarding empowering local people (Rastegar et al., 2021) and climate change (Prideaux et al., 2020).

Despite the articulation of these needs, responsible tourism has a faltering foundation relying on individuals' altruistic impulses, within capitalism's axiomatic normalisation of crises. This might explain why one can feel disheartened. We have no *training* to induce profound pro-environmental behaviours because tourism studies haven't engaged with ethics within philosophy and this intangibility excludes it from the remit of governing bodies. Our neoliberal industry mentality hinders us from reaching a necessary philosophical foundation to formulate deeper discussion of values and ethics. Without a philosophical ethical base adventure practitioners are left with superficial trends, or simply hope for the best.

Despite a penchant for Heideggerian phenomenology in tourism studies, primarily in terms of existential authenticity (Brown, 2013; Cohen, 1979; Kirilova et al., 2016; Munar et al., 2021; Shepherd, 2015) 'the exploration of more profound ethical problems inherent in contemporary tourism as a social phenomenon is still rarely undertaken' (Cohen, 2018, p. 359). Without any philosophical underpinning, we remain in a void, groping blindly in the "gap", within an industry which is an epistemological barrier that limits engagement in ethics and philosophy. New interest in critical post-humanism in tourism (Cohen 2019) may begin to represent a light at the end of the tunnel.

Given this situation, inducing profound pro-environmental behaviours requires more than purchasing a training package. The work begins at home, inside oneself, in an ontological and practical sense. How can we 'Leave No Trace' (LNT, 2021) in mountains and then return to our abode and habitually embrace our destructive ecological footprint? Some adventure professionals called for leave no trace at home (Beames et al., 2019). Pro-environmental behaviour training begins with ontological questioning: Who am I living on this planet, and what lasting value am I giving/making? These answers induce years of compromise, balancing, sacrifice and further questioning. In transformative tourism it is argued one cannot offer transformative experiences to clients, without having experienced transformation in nature (Sheldon, 2020). Likewise, one needs deep personal proclivity for pro-environmental behaviours before inspiring others. Returning to the start of this section, if one simply wants to make a living, light-heartedly, being an adventure practitioner in nature faces new dissonances and is embedded in 'genuine ethical controversy' (Carr, 2000, p26) as explored in the previous section of this chapter.

Mackenzie and Goodnow (2021) write that the post-pandemic adventure is a journey not to seek new landscapes but to have "new eyes" (p. 67). To find our own "new eyes" each practitioner must make their own path, within a philosophical lacuna, facing self-questioning in the quotidian. It is argued the essence of mankind is *Homo faber* rather than *Homo sapiens*. If humans have a distinctiveness, it may be "making" and not "knowing" (Malafouris, 2013). Making one's way, making one's own *knowing*. The Case study 18.2 offered here is unique, but each person who finds their own way will be.

Case study 18.2   Anna at Wild Roots Highland Guiding

Anna grew up in rural Yorkshire and her deep love for the countryside as a child inspired her to study environmental science at university. The more Anna understood about our environment, the more she wanted to protect it. This manifested in becoming involved in environmental actions, attending marches and protest camps in a fight to speak up against climate change. Although she threw herself into activism, she worried she was alienating people rather than inspiring them to be more environmentally focused. It took another decade of life outdoors and continued study, to evolve into a professional who inspires pro-environmental behaviours in adventure.

While in Edinburgh, studying a master's in outdoor environment and sustainability education, Anna devoted herself to climbing in the Highlands and found a new confidence within.

'A realisation grew that maybe this could be my way of contributing to a more sustainable future, by offering experiences to enable others to benefit from developing close relationships with wild places.'

In 2020, Anna started Wild Roots Highland Guiding, founded on a belief that close, personal relationships with place are vital for our health and for the planet. Anna collaborates with a life coach and counsellor specialising in working outdoors to create unique experiences interweaving moments of learning about geology, botany, and human history with mindfulness activities, allowing emotion and self-reflection. Feedback from clients demonstrates that they often come away with an altered perspective on their place in nature and how they act in relation to it.

A recent conversation with a return client highlighted inspiring pro-environmental behaviours:

'The client disclosed how participating in our journeys gave her a personal connection with the landscape that now changes her choices in daily life. Prioritising the environment has become central to her decision making and her political stance. Green credentials are now her main criteria when choosing what products to buy, she votes Green, recycles more and has become a walk leader with her Rambler's club to inspire others as she has been.

> *Being in the wild for three days and two nights really helps me to lose myself. All the other stuff from my daily life disappears a bit and I am just me. They are moments of growth that I can't un-grow, un-learn if you like, I can't forget because they have become intrinsic to who I am. I also remember sitting with you talking about the landscape, the geology, history and I don't maybe remember all the details, but I have the feel for it you know, and this idea that I can see it differently now. I have also used lots of things that Shona has shown us when I have been on my own to help me ... feel a bit more connected. So, it's very part of me, all of the times I've been out with you, I now carry with me quite consciously.'*

For Anna, mountains teach us how to live more with less.

Figure 18.4 Wild Roots Highland Guiding logo. Courtesy of Wild Roots Guiding.

Conclusion

Reflecting on the opening vignette of this chapter, the reader may find that to truly *be* an adventure tourism professional requires more than technical understanding and competence. This chapter highlights how we must grapple with our own values and challenges traditional notions of training and development within a career progression, where the *ethos* can become chasing the "next qualification".

While deepening embeddedness and growing skill in your chosen adventure pathway are essential attributes for professional identity and progression, one way to sidestep our philosophical vacuum is having the opportunity to assemble a collective set of values and practices which frequently find the space to evolve within an academic context, enabling the individual to study to create a critical analysis of adventure. We are unlikely to grow and develop in isolation; instead, our development is like a web, with multiple connections interlacing theory with lived experience.

Becoming, being, and inspiring are not learned through training packages alone. Now is the time to consider the ontological aspects of a career in adventure tourism and accept that the most necessary and challenging adventure lies *within*.

Review and reflection questions

1 What are your own progression priorities in your field of adventure practice? How have they evolved over time?
2 Consider the most important skills and knowledge that have emerged for you through your active involvement in outdoor and adventure professional development. What foundations did you need before each aspect became evident and familiar to you? What key experiences have generated the most growth and change?
3 What ethical, philosophical, or environmental justice values are of greatest importance in your life? How can you embody those values through the work you do to inspire others and be sensitive to any conflict that might occur with the demands of the activity?
4 Can you think of a time when you have informed the practice of a fellow adventure professional? What influence do the values and actions of your professional peer group have on your own professional priorities?
5 What would be the *best* things you would want others to say about you as a professional and the work that you do?

Further reading

ATTA. (2021). Adventure Travel Guide Standard. https://learn.adventuretravel.biz/guide-standard
Varley, P., Farkic, J., & Carnicelli, S. (2018). Hospitality in wild places. *Hospitality and Society*, *8*(2).

References

Allin, L., & Humberstone, B. (2006). Exploring careership in outdoor education and the lives of women outdoor educators. *Sport, Education and Society*, *11*(2), 135–153. https://doi.org/10.1080/13573320600640678
ATTA. (2015). International Adventure Travel Guide Qualifications and Performance Standard. https://learn.adventuretravel.biz/guide-standard
ATTA. (2021a). Creating, Communicating, and Connecting: Technology in Adventure Travel. https://learn.adventuretravel.biz/research/creating-communicating-and-connecting-technology-in-adventure-travel
ATTA. (2021b). Adventure Travel Guide Standard. https://learn.adventuretravel.biz/guide-standard
ATTA. (2022). Professional Membership. https://membership.adventuretravel.biz/professional
Barnes, P. (1999). *The motivation of staff in the outdoor education industry*. Unpublished PhD thesis, University of Strathclyde.

Bell, M. (2017). The romance of risk: Adventure's incorporation in risk society. *Journal of Adventure Education and Outdoor Learning, 17*(4), 280–293. https://doi.org/10.1080/1472967 9.2016.1263802

Beames, S., Mackie, C., & Atencio, M. (2019). *Adventure and society.* Palgrave Macmillan.

Beaton, G. (2010). *Why professionalism is still relevant.* Australian Council of Professions. https:// www.professions.org.au/wp-content/uploads/Why_Professionalism_is_still_Relevant_Beaton. pdf

Beck, U. (1992). *Risk society: Towards a new modernity.* SAGE Publications Ltd.

Beedie, P. (2003). Mountain guiding and adventure tourism: Reflections on the choreography of the experience. *Leisure Studies, 22*(2), 147–167. https://doi.org/10.1080/0261436032000068991

Boyden, S. (1987). *Western civilization in biological perspective: Patterns in bio history.* Oxford Science Publications.

Breivik, G. (2007). The quest for excitement and the safe society. In M. J. McNamee (Ed.), *Philosophy, risk and adventure sports* (pp. 10–24). Routledge.

British Canoeing Awarding Body. (2018). Guide Endorsement. https://www.britishcanoeingawarding. org.uk/guide-endorsement-2

Brown, L. (2013). Tourism: A catalyst for existential authenticity. *Annals of Tourism Research, 40,* 176–190.

Cannon-Bowers, J. A., & Bell, H. H. (2009). Training decision makers for complex environments: Implications of the naturalistic decision making perspective. In C. Zsambok & G. Klein (Eds.), *Naturalistic decision making* (pp. 99–110). Routledge.

Carfora, V., Caso, D., Sparks, P., & Conner, M. (2017). Moderating effects of pro-environmental self-identity on pro- environmental intentions and behaviour: A multi-behaviour study. *Journal of Environmental Psychology, 53,* 92–99.

Carnicelli-Filho, S. (2013). The emotional life of adventure guides. *Annals of Tourism Research, 43,* 192–209. https://doi.org/10.1016/J.ANNALS.2013.05.003

Carr, D. (2000). *Professionalism and ethics in teaching.* Routledge.

Cater, C. I. (2006). Playing with risk? Participant perceptions of risk and management implications in adventure tourism. *Tourism Management, 27*(2), 317–325. https://doi.org/10.1016/J. TOURMAN.2004.10.005

Cheetham, G., & Chivers, G. (2005). *Professions, competence and informal learning.* Edward Elgar Publishing.

Cohen, E. (1979). A phenomenology of tourist experiences. *Sociology, 13*(2), 179–201.

Cohen, E. (2018). The philosophical, ethical and theological groundings of tourism – An exploratory inquiry. *Journal of Ecotourism, 17*(4), 359–382.

Cohen, E. (2019). Posthumanism and tourism. *Tourism Review, 74*(3), 416–427.

Collins, L., & Collins, D. (2012). Conceptualizing the adventure-sports coach. *Journal of Adventure Education and Outdoor Learning, 12*(1), 81–93. https://doi.org/10.1080/14729679.2011.6112 83

Collins, L., & Collins, D. (2013). Decision making and risk management in adventure sports coaching. *Quest, 65*(1), 72–82. https://doi.org/10.1080/00336297.2012.727373

Collins, L., Carson, H. J., Amos, P., & Collins, D. (2018). Examining the perceived value of professional judgement and decision-making in mountain leaders in the UK: A mixed-methods investigation. *Journal of Adventure Education and Outdoor Learning, 18*(2), 132–147. https://doi.org /10.1080/14729679.2017.1378584

Dreyfus, S. E. (2004). The five-stage model of adult skill acquisition. *Bulletin of Science, Technology & Society, 24*(3), 177–181. https://doi.org/10.1177/0270467604264992

Dwyer, L. (2018). Saluting while the ship sinks: The necessity for tourism paradigm change. *Journal of Sustainable Tourism, 26*(1), 29–48.

Evetts, J. (2003). The sociological analysis of professionalism: Occupational change in the modern world. *International Sociology, 18*(2), 395–415. https://doi.org/10.1177/0268580903018002005

Farkic, J. (2018). Outdoor guiding as hospitality work. *Annals of Tourism Research, 73,* 197–199. https://doi.org/10.1016/j.annals.2018.04.004

Fischer, D., Stanszus, L., Geiger, S., Grossman, P., & Schrader, U. (2017). Mindfulness and sustainable consumption: A systematic literature review of research approaches and findings. *Journal of Cleaner Production, 162,* 544–558.

Giddens, A. (1991). *Modernity and self-identity: Self and society in the late modern age*. Polity Press.

Gilbertson, K., & Ewert, A. (2015). Stability of motivations and risk attractiveness: The adventure recreation experience. *Risk Management, 17*(4), 276–297. https://doi.org/10.1057/rm.2015.16

Gössling, S., Ring, A., Dwyer, L., Andersson, A., & Hall, M. (2016). Optimizing or maximizing growth? A challenge for sustainable tourism. *Journal of Sustainable Tourism, 24*(4), 527–548.

Gössling, S., Scott, D., & Hall, M. (2021). Pandemics, tourism and global change: A rapid assessment of COVID -19. *Journal of Sustainable Tourism, 29*(1), 1–20.

Groves, M. R., & Varley, P. J. (2020). Critical mountaineering decisions: Technology, expertise and subjective risk in adventurous leisure. *Leisure Studies, 39*(5), 706–720. https://doi.org/10.1080/02614367.2020.1754887

Hall, C. M.(2022). Tourism and the Capitalocene: From green growth to ecocide. *Tourism Planning & Development, 19*(1), 61–74.

Hall, J., & Jostad, J. (2020). The role of pay and sense of community in the turnover of outdoor adventure education field staff. *Journal of Outdoor Recreation, Education, and Leadership, 12*(3), 333–349. https://doi.org/10.18666/JOREL-2020-V12-I3-9974

Hansen, M., Fyall, A., & Spyriadis, T. (2020). Adventure or amusement? Image and identity challenges for the aerial adventure industry and implications for positioning and policy. *Anatolia an International Journal of Tourism and Hospitality Research, 31*(3), 423–435. https://doi.org/10.1080/13032917.2020.1741408

Hogarth, R. M., Lejarraga, T., & Soyer, E. (2015). The two settings of kind and wicked learning environments. *Current Directions in Psychological Science, 24*(5), 379–385. https://doi.org/10.1177/0963721415591878

Holyfield, L. (1999). Manufacturing adventure: The buying and selling of emotions. *Journal of Contemporary Ethnography, 28*(1), 3–32. https://doi.org/10.1177/089124199129023352

International Federation of Mountain Guides Associations. (2017). About IFMGA. Retrieved from https://ifmga.info/%3Cnolink%3E/about-ifmga

IOL. (2021). The Outdoor Professional. Retrieved from https://www.outdoor-learning.org/Jobs/The-Outdoor-Professional

Jackson, T. (2017). *Prosperity without growth*. Routledge.

Johnson, J., Mannberg, A., Hendrikx, J., Hetland, A., & Stephensen, M. (2020). Rethinking the heuristic traps paradigm in avalanche education: Past, present and future. *Cogent Social Sciences, 6*(1). https://doi.org/10.1080/23311886.2020.1807111

John Muir Trust. (2021). Management Report. Retrieved from https://www.johnmuirtrust.org/assets/000/001/444/Visitor_Management_Report_09.03.2021_final_lr_original.pdf

Juvan, E., & Dolnicar, S. (2014) The attitude–behaviour gap in sustainable tourism. *Annals of Tourism Research, 48*, 76–95.

Kirillova, K., Lehto, X., & Cai, L. (2016). Existential authenticity and anxiety as outcomes: The tourist in the experience economy. *International Journal of Tourism Research, 19*, 13–26.

Laing, J., & Frost, W. (2014). *Explorer travellers and adventure tourism* (Vol. 40). Channel View Publications.

Lewis, P., & Kimiecik, J. (2018). Embracing the mystery box: How outdoor leaders discover and sustain their way of life. *Journal of Outdoor Recreation, Education, and Leadership, 10*(3), 304–322. https://doi.org/10.18666/JOREL-2018-V10-I4-8745

LNT. (2021). Leave No Trace. Retrieved from https://lnt.org/

Lyng, S. (1990). Edgework: A social psychological analysis of voluntary risk taking. *American Journal of Sociology, 95*(4), 851–886.

Lyng, S. (2004). Edgework and the risk-taking experience. In *Edgework* (pp. 3–15). Routledge.

Marchand, G., Russell, K. C., & Cross, R. (2009). An empirical examination of outdoor behavioural healthcare field instructor job-related stress and retention. *Journal of Experiential Education, 31*(3), 359–375. https://doi.org/10.1177/105382590803100304

Mackenzie, S., & Goodnow, J. (2021). Adventure in the age of COVID-19: Embracing microadventure and locavism in a post-pandemic world. *Leisure Sciences, 43*(1–2), 62–69.

Malafouris, L. (2013). *How things shape the mind: A theory of material engagement*. MIT Press.

McCammon, I. (2002). Evidence of heuristic traps in recreational avalanche accidents. *International Snow Science Workshop, Penticton, British Columbia* (pp. 244–251). http://www.snowpit.com/articles/articles.htm

Mees, A., Toering, T., & Collins, L. (2021). Exploring the development of judgement and decision making in 'competent' outdoor instructors. *Journal of Adventure Education and Outdoor Learning, 22*(1), 77–91. https://doi.org/10.1080/14729679.2021.1884105

Mohamadi, S., Abbas, A., Habib-Allah, R.-K., & Kazem, A. (2022). Conceptualizing sustainable–responsible tourism indicators: An interpretive structural modeling approach. *Journal of Environment, Development and Sustainability, 24*, 399–425.

Munar, A.-M., Meged, J., Bødker, M., & Dam Wiedemann, C. (2021). Existential walking. *Annals of Tourism Research, 87*, 1–14.

Nikhildas, T., & Jagadeesan, P. (2021). Is responsible tourism (RT) a socialistic concept. *Elementary Education Online, 20*(5), 1933–1938.

Ogilvie, K. (2013). *Roots and wings: A history of outdoor education and outdoor learning in the UK.* Russell House Publishing.

Pomfret, G., & Bramwell, B. (2014). The characteristics and motivational decisions of outdoor adventure tourists: A review and analysis. *Current Issues in Tourism, 19*(14), 1447–1478. https://doi.org/10.1080/13683500.2014.925430

Prideaux, B., Thompson, M., & Pabel, A. (2020). Lessons from COVID-19 can prepare global tourism for the economic transformation needed to combat climate change. *Tourism Geographies, 22*(3), 667–678.

Rastegar, R., Higgins-Desbiolles, F., & Ruhanen, L. (2021). COVID-19 and a justice framework to guide tourism recovery. *Annals of Tourism Research, 91*, 103–161.

Rokenes, A., Schumann, S., & Rose, J. (2015). The art of guiding in nature-based adventure tourism – How guides can create client value and positive experiences on mountain bike and back-country ski tours. *Scandinavian Journal of Hospitality and Tourism, 15*(1), 62–82. https://doi.org/10.1080/15022250.2015.1061733

Sand, M., & Gross, S. (2019). Tourism research on adventure tourism – Current themes and developments. *Journal of Outdoor Recreation and Tourism, 28*, 100261. https://doi.org/10.1016/j.jort.2019.100261

Sheldon, P. (2020). Designing tourism experiences for inner transformation. *Annals of Tourism Research, 83*, 102935.

Shepherd, R. J. (2015). Why Heidegger did not travel: Existential angst, authenticity, and tourist experiences. *Annals of Tourism Research, 52*, 60–71.

Thomas, G. (2001). Thriving in the outdoor education profession: Learning from Australian practitioners. *Journal of Outdoor and Environmental Education, 6*, 13–24. https://doi.org/10.1007/BF03400740

Varley, P. (2006). Confecting adventure and playing with meaning: The adventure commodification continuum. *Journal of Sport and Tourism, 11*(2), 173–194. https://doi.org/10.1080/14775080601155217

Varley, P., Farkic, J., & Carnicelli, S. (2018). Hospitality in wild places. *Hospitality and Society, 8*(2), 137–157. https://doi.org/10.1386/hosp.8.2.137_1

Vignoles, V. L., Schwartz, S. J., & Luyckx, K. (2011). Introduction: Toward an integrative view of identity. In S. J. Schwartz, K. Luyckx, & V. L. Vignoles (Eds.), *Handbook of identity theory and research* (pp. 1–27). Springer.

Visit Scotland (2020) Scotland Outlook 2030; Responsible tourism for a sustainable future. https://scottishtourismalliance.co.uk/wp-content/uploads/2020/03/Scotland-Outlook-2030.pdf

Wagstaff, M. (2016). Outdoor leader career development: Exploration of a career path. *Journal of Outdoor Recreation, Education, and Leadership, 8*(1), 75–95. https://doi.org/10.18666/JOREL-2016-V8-I1-7284

Weber, K. (2001). Outdoor adventure tourism a review of research approaches. *Annals of Tourism Research, 28*(2), 360–377. https://doi.org/10.1016/S0160-7383(00)00051-7

Wenger, E. (1998). *Communities of practice: Learning, meaning, and identity.* Cambridge University Press.

Wright, M., & Gray, T. (2013). The hidden turmoil: Females achieving longevity in the outdoor learning profession. *Journal of Outdoor and Environmental Education, 16*, 12–23. https://doi.org/10.1007/BF03400942

19 Experiential marketing and value co-creation in adventure tourism

May Kristin Vespestad

Chapter learning outcomes

1 Comprehend the changing nature of adventure tourism marketing.
2 Know how worldwide adventure tourism events can be marketed and provide value at small-scale adventure destinations.
3 Recognise the importance of experiential marketing in adventure tourism.
4 Understand the relevance of an intracommunity adventure activity image in adventure tourism marketing.

Introduction

The changing nature of adventure tourism marketing is evident in our everyday lives, as we have moved into an experience society, relying on an experience economy with the mantra 'time well spent' rather than 'time well saved' (Pine & Gilmore, 2020, p. xii). We find ourselves at a point where we look for adventure during leisure, especially during our holidays. As explained by Cater in Chapter 2 of this volume, the nature of adventure itself has changed. Activities such as mountain biking, surfing, and skiing are being used to promote a wide variety of things, such as jobs, clothing, food, or accommodation. Seemingly adventurous activities, at least soft adventures, are becoming more mainstream (Fossgard & Fredman, 2019; Janowski et al., 2021).

The overlap between participation in and exposure to adventure-filled activities in leisure, through a variety of marketing channels and tools, also contributes to the blurred boundaries between adventure tourism and everyday life (Vespestad & Lindberg, 2011). It seems that adventure activities are taking a more prominent position, due to increased interest from special interest communities (e.g., surfers, climbers) forming consumer tribes (Cova & Shankar, 2018) and their extension into the larger marketing and tourism sphere (Haukeland et al., 2021).

Image-building in social media has become important (Sand & Gross, 2019), and self-branding, understood as the 'metaphorical expansion of branded goods and services into the realm of individuals' (van Nuenen, 2016, p. 194) is growing. This form of branding of individuals refers to positioning oneself relative to others in the adventure community or in-group, and it enables multidirectional communication that empowers consumers in comparison with traditional marketing (Della Corte & Del Gaudio, 2020; Seeler & Schänzel,

DOI: 10.4324/9781003393153-23

2020), where businesses take the lead of the narrative being told. In adventure tourism literature, however, less attention is given to the *marketing* from the supply side. Instead, adventure tourism destination development and planning, adventure tourism operators, and adventure tourism experiences have received more attention (Cheng et al., 2016).

This chapter investigates adventure tourism marketing and the ways in which it has acquired a new role and meaning in the sense that it has become more mainstream and covers a broader interpretation of adventure. The aim is to shed light on how adventure tourism marketing can attract tourists in the future. Two case studies illustrate how adventure tourism marketing can enable value co-creation, along with the dynamics and scope of such marketing. The first Case study 19.1 is a Red Bull Pump Track World Championships Qualifier event, while the second (Case study 19.2) is the adventure-filled world of Captain Sabretooth in an adventure theme park.

Co-creation of value in consumer communities

This chapter is grounded in axiology, the theory of value, where consumer value is defined as 'an interactive relativistic preference experience, where the relationship of consumers to products operates relativistically to determine preferences that lie at the heart of the consumption experience' (Holbrook, 1999, p. 9), meaning that the design of marketing strategy is shaped by consumer value. Holbrook's (1999) consumer value typology refers to efficiency, excellence, status, esteem, play, aesthetics, ethics, and spirituality value in the consumption experience. Value develops from interaction and hence differs from traditional supply and demand thinking (Galvagno & Dalli, 2014). Instead, it draws on seminal works by Prahalad and Ramaswamy (2004) and Vargo and Lusch (2004, 2008, 2017). This approach is appropriate in adventure tourism marketing, where the search for adventure lies at the heart of consumption, and special interests of consumption predominate and form strong consumer communities, such as in surfing.

Consumer communities are at the core of consumer culture theory (CCT; Arnould et al., 2021; Arnould & Thompson, 2018; Thompson et al., 2018), and therefore provide a useful lens for understanding the marketplace cultures of adventure tourism, through communitas (Cova & Shankar, 2018). Communitas is understood as 'the sense of community that transcends typical social norms and convention' within a group, a sense of camaraderie when people share a common bond of (liminal) experience (Celsi et al., 1993, p. 12; Turner, 1969).

Drawing on experiential marketing (Schmitt, 1999) and the underlying assumption that value resides in the object of purchase, in the surrounding hedonic and experiential elements, and in the experience of consumption itself (Schmitt & Zarantonello, 2013), this chapter describes how adventure tourism marketing can lead up to and be a part of the adventure. It can do this by building on value co-creation frameworks (Campos et al., 2018), such as by emphasising co-creation between professional marketers and consumer communities. Co-creation can be defined as the 'joint, collaborative, concurrent, peer-like process of producing new value both materially and symbolically' (Galvagno & Dalli, 2014, p. 644). In this way, the concept can be useful in an adventure tourism context, as it demonstrates how value can be created through, for example, the increased importance of imagery (Sand & Gross, 2019), both from the marketer side and within consumer communities, to create a sense of communitas.

One example of how value can occur in this context is through the increased publicity of adventure tourism images. Consumers are posting their adventures on social media

(Sand & Gross, 2019), giving prominence to the importance of the "Insta-bility" (the ability to attract attention on Instagram) of adventure tourism destinations, products, and events, particularly within special interest consumer communities, such as climbers (Vespestad & Hansen, 2020). The increased use of social media and technology is, therefore, part of the change in adventure tourism marketing, as tourists have become active contributors and co-creators (Rather et al., 2022) of experiential adventure tourism marketing content and reach by posting, for example, photos from adventures they have participated in or links to adventure events or places of interest (see Chapters 11 and 20). Technology also makes on-site marketing more attainable as the use of apps during adventure tourism experiences allows for further interpretation and involvement and enables immersive experiences where value is co-created (Mandić & McCool, 2022).

There are also examples of negative consequences of co-creation. For example, marketing of adventures in mountains and scenic spots has in some areas been taken to task for its uncritical use of photos of people posing in unlikely postures in spectacular surroundings (e.g., in Norway as shown in Figure 19.1), on the edge of a cliff or on a mountain peak (aftenposten.no, 2016). While this is eye-catching and, from a tourism marketing perspective, has been effective in attracting tourists (e.g., innovationnorway.com), it also normalises and belies the adventure in a way that has created the unexpected effect of hordes of tourists wanting to pose in the exact same way and to be filmed or photographed at that spectacular moment. Tourists' dedication to taking the best selfies has

Figure 19.1 Snowboarder in the Norwegian mountains. Photographer K. Wærnes.

also resulted in deaths. Consequently, national marketers, such as visitnorway.com, now focus more on safety.

Experiential adventure marketing and consumption

Although adventure tourism has been connected to risk and risk seekers, this view has somewhat changed. Adventures are more about challenges (Sand & Gross, 2019), which may be physical (see Case study 19.1) or psychological (see Case study 19.2). This is relevant for adventure tourism marketers, as people tend to avoid risk, particularly when they make decisions on behalf of others. We want marketing activity to ensure that we are making the right choice; the right adventure, an adventure that immerses us, engages us, allows us to feel the flow (Csikszentmihalyi, 2008), and, finally, enhances our well-being. This could be achieved through experiential marketing. As experiential marketing relies on happiness, joy, and well-being, even transformations (Cooper & Buckley, 2021), these are aspects to be involved in adventure tourism marketing.

Adventure activities are a motivation for travel and for choosing destinations (Funk & Bruun, 2007) and are an area of growing importance also for families (Pomfret, 2021). As an example, by highlighting adventure as learning, an adventure quality valued by women, and promoting intrinsic values, such as positive emotions, achievement, and self-development, marketers could be more successful in attracting female adventure travellers (Clarke et al., 2021) and consequently families, as women often make decisions on holidays. Although there is potential for attracting families as a growing adventure niche (Jamal et al., 2019; Pomfret, 2021; Sand & Gross, 2019), family adventures have received less attention in adventure tourism marketing literature. Case study 19.2 on marketing pirate adventures attempts to showcase adventure tourism marketing for the family market segment, from an experiential marketing point of view, associated with feelings, senses, and emotions more than cognition and intentions (Same & Larimo, 2012).

In exploring the experiential aspects of consumption (Holbrook & Hirschman, 1982), an experiential perspective recognises that consumers not only think and do but also feel, emphasising symbolism and a need for pleasure and fun (Holbrook, 1999). The feeling of happiness can be seen as an essential construct in consumption and is associated with broader aspirations in life (Schmitt, 1999), such as well-being, hope, love, optimism, and flow (Seligman & Csikszentmihalyi, 2000). Tourists search for positive feelings, joy and whatever brings them happiness. However, unless consumers see the relevance to issues they personally care about, they are unlikely to value marketers' efforts to educate them (Martin & Schouten, 2014), that is, how the adventure tourism experience can bring them value. Recent studies in tourism have adopted Holbrook's framework to identify dimensions of value associated with tourists' use of guidebooks (Mieli & Zillinger, 2020) and value in climbing experiences (Vespestad et al., 2019). This could indicate that, in adventure tourism marketing, the sense of value is even more crucial, since consumers are so strongly engaged and have a strong sense of commitment to the adventure.

Experiential marketing is based on the idea that experiences arise in marketing, that is, marketing can initiate and add experiential value. By contrast, consumer culture theory focuses on the consumers and the cultures surrounding consumption (Arnould et al., 2021) and therefore forms the backdrop of this chapter. A notable dimension here is the perceived value of marketing one's adventures in social media, where recognition from one's peers provides added experiential value (e.g., status or esteem) through their likes and comments (Vespestad & Hansen, 2020). Furthermore, value is also co-created

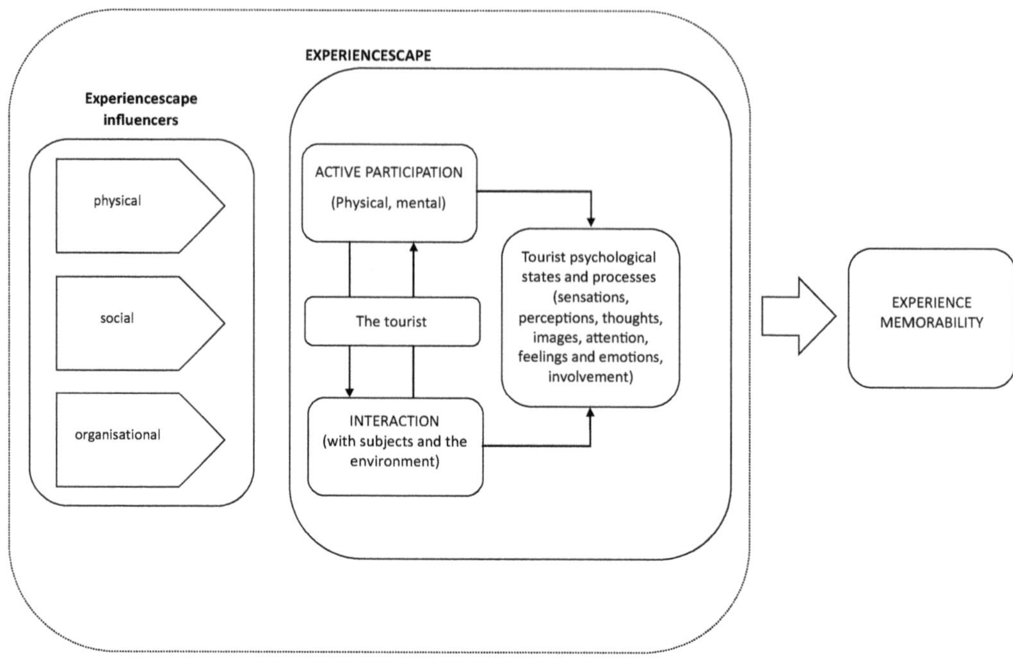

Figure 19.2 The tourist's on-site co-creation experience: A conceptual framework. (Adapted from Campos et al., 2018).

through the involvement of both consumers and other commercial collaborators in the marketing process. Value co-creation frameworks (see Figure 19.2) are, therefore, deemed useful in providing guidance to marketers, managers, and academics in the field of adventure tourism.

Figure 19.2 shows how the experiencescape is co-created by different stakeholders, and how physical, social, and organisational structures influence the co-created tourism experience, which might ultimately result in positive memorability (Campos et al., 2018). Extending on this, drawing on consumer culture theory (Arnould et al., 2021), value that derives not only from adventure tourism experiences but also from the marketing before, during, and after the adventure provides value formation in adventure communities. The sense of communitas is strengthened in all these phases. Although destination marketing organisations can work to create adventure tourism images, image formation within a special interest community with a focus on the adventure activity could be just as strong and necessary, in order to develop and establish an authentic and more sustainable adventure tourism image.

The following sections present two case studies to illustrate how adventure tourism marketing can co-create value through experiential marketing. The case studies are based on data collected autoethnographically by the author in Norway in the summer of 2022. Autoethnography enables knowledge of consumer communities and draws on the researcher's own experiences (Cova & Cova, 2019). The first case study illustrates value co-creation in practice by discussing the marketing of the Red Bull Pump Track World Championships event. The second case study delves into how adventure tourism marketing can be implemented in family adventures.

Case study 19.1   Value co-creation in developing adventure facilities and events: The Red Bull Pump Track World Championships Qualifier event

In Harstad, a small Norwegian town at 68 degrees north, above the Arctic Circle, the worldwide Red Bull Pump Track World Championship Qualifier took place in August 2022, one of 11 qualifiers this year, most held in cities, such as Berlin and Edinburgh. The organiser responsible for marketing the event is the local cycling club, which, over the last few years, has enhanced the town's potential of becoming an adventure tourism destination, with the largest pump track in Scandinavia as one of the main attractions of a new and still developing bike park (Figure 19.3). The marketing of the event takes place in various ways: advertising with posters, newspaper articles, and more importantly, the sharing of information on events and sub-events in social media (Facebook and Instagram). The club has aimed to transform this worldwide event into something attractive yet also attainable to the local community. Anyone 18 and older could participate, and postings on social media especially invited women to join, aiming to market this also as a "local derby," a competition within the competition. The club marketed the event by arranging training sessions in the pump track along with unplanned sessions during the summer, attracting domestic tourists and locals alike who wanted to learn how to best pump the tracks. This created electronic word of mouth and lowered the threshold of entering what to many seems daunting. Particularly important in adventure tourism marketing was promotion on YouTube, organically shared within the consumer community in social media, which proved to attract both domestic and international tourists.

Figure 19.3 Harstad Bike Park. Photograph by Harstad Cycle Club.

> About the event: https://www.redbull/com/int-en/events/red-bull-uci-pump-track-world-championships-Harstad
>
> Post-event marketing: https://www.pinkbike.com/news/68-degrees-north-introducing-harstad-sykkelpark.html

Joint value co-creation is exemplified in this first case study, where Harstad Cycle Club and significant sponsors formed the basis for the establishment of the bike park where the 2022 Red Bull Pump Track World Championships event took place. Excellence (financial) and social value (Holbrook, 1999) were co-created through clever and successful marketing, emphasising adventure, where the marketing campaign activities challenged locals to "walk for the bike park". The aim was to reach 200 walks every week, and 1,800 registrations during nine weeks. Initially, this aim was thought to be ambitious, but the marketing campaign was hugely successful and people from across the local community – young, old, kindergartens, school classes, handball teams – participated regardless of their interest in cycling. The aim was to initiate activity and engagement in the local population and if reached, the sponsored amount of EUR 840,000 from a local bank and sponsor of the campaign would be released as a contribution towards the development of the bike park. The target was reached after only six days. Yet people kept walking during the period, resulting in 4,891 walks being registered.

This adventure marketing campaign is interesting as it portrays how the marketing aimed at fundraising for a bike park can be done to engage and enthuse both the adventure community and the broader local community, even those not interested in biking. Through collaboration and joint efforts, the peer-like process created both material and symbolic value (Galvagno & Dalli, 2014). Value co-creation was also at the heart of the cycling club when they specifically marketed the Red Bull Pump Track World Championships event. For example, the official opening of the bike park initiated involvement from the local community, the biking community, and professional athletes, and served to promote the pump track event, as it was held the day before the championships. With the staggering number of about 450 children participating in a race, it was a family event, including free dinner while watching the pros perform in the jump lines and skills parks. By inviting people beyond the cycling community to join in the pre-event activities and marketing the championship event during the official opening, building up expectations, for example, by marketing the arrival of the reigning world champion, seamlessly led visitors towards the specialised event the next day.

The co-creation of the different stakeholders before, during and after these events contributed to achieving the underlying strategic aim of the cycling club "to create value for all" (not only for its members) through, for example, social value (Holbrook, 1999) for the local community. Therefore, insight into the special interest communities, which form the basis for marketing adventure events, is essential to be able to promote the adventure in a way that is also attractive to a broader, soft adventure tourism segment in order to come across as authentic.

Co-creation amongst the event organisers drives the marketing, along with the cycling community spreading the word, for example, not only on social media and special interest forums online (the websites of Red Bull, cycling magazines, and Facebook groups), serving as important marketing channels (from both the supply and demand sides), but also by posters being hung up around town by volunteers. The value co-creation in the marketing of this event took place through the information and pictures being shared among members of the cycling community. The cycling club in this case relies on volunteers from its members, which therefore is essential in facilitating value co-creation for the wider biking community (e.g., mountain bikers, pump track bikers, or other users of the park). Because special interest communities (e.g., mountain bikers) are global in nature (visiting the same online forums worldwide, e.g., pinkbike.com), the (electronic) word of mouth has great reach.

This case thus demonstrates how large events and global businesses can have an impact and contribute to value co-creation in collaboration with small local stakeholders in the context of adventure tourism. Marketing can create mutual value which can bring emerging adventure tourism destinations onto the international adventure tourism scene and develop their adventure tourism destination image. Furthermore, post-event marketing can also make a significant contribution, as in this case on pinkbike.com and YouTube.

In arranging large events, challenges can emerge as the professional staff of an international business and event organiser meet the volunteers, a special interest organisation, sponsors, and consumers. Collaboration on the adventure tourism event relied heavily on co-creation between all actors but how to balance the different expectations and aims was difficult. The volunteers were led by enthusiasm and passion. Equally so, Red Bull has its international standards to be maintained at all the adventure tourism destinations it visits. To accommodate these at times opposing interests, balancing the perceived value (Holbrook, 1999) of the professional athletes (e.g., excellence and status value, free ticket to the world championships) versus that of the local participants and volunteers (e.g., social and play value), requires creativity and stamina to pull through. The success of the campaigns relied on the community culture taking ownership of the event. Consumer cultures and communitas (Cova & Shankar, 2018), therefore, play an important role in the marketing of adventure events at a destination. Value that is co-created as in the marketing of this case, thus provides space for a successful experiential marketing outcome.

Seamless transitions in marketing family adventure tourism

The following case illustrates the theory of value and experiential marketing in practice related to family adventure tourism in a theme park. It also brings in the dualism of marketing directed at parents versus children. Marketing towards children is debated, with some claiming that children are particularly vulnerable to commercial marketisation, whereas others claim that marketers certainly do view children as both capable and critical consumers (Hjemdahl, 2003). Regardless of which stance one takes, child and family adventure marketing will need to hit the "sweet spot" of both children and adults and allow for co-creation between them, in order to be successful and create a sense of communitas in the family.

Case study 19.2   Marketing of pirate adventures

Kristiansand Zoo and Amusement Park is one of Norway's most frequently visited tourist attractions and markets itself as creating adventures for children and families through a variety of theme worlds. One of them is the pirate village based on the stories of Captain Sabretooth, with themed accommodation, such as the pirate village Abra Harbour. The adventure marketing here paves the way for adventure experiences and living as pirates. Upon arrival one is invited to be woken up by pirates returning from looting on their ship *The Black Lady*, ready to raise their flag in the harbour in the morning. This markets immersion into the adventure. A pamphlet laid out in the pirate housing with information about the day in Abra, sets the scene for the adventure. An invitation to sign the pledge to obey Captain Sabretooth's law to become true pirates and follow the treasure maps and puzzles to encourage people to be involved and engage in the adventures of the pirate bay is also a clever marketing move which serves as experiential marketing. It adds to the adventure instead of interfering with it. Also, an SMS with a recorded personal message from Captain Sabretooth himself, sent both upon arrival and after check-out was an immediate marketing success for preparing the transition into the adventure, as well as a confirmation of accomplishments during the stay, making sure that the guests remember they are true pirates and promoting satisfaction and intention to revisit.

Weeks of studying the theme park's online marketing as a researcher and a mother of two clearly set my expectations. How could we, as a family, build our expectations and somehow start the adventure, and how did on-site marketing (Figure 19.4) interfere with or adjust our plans for the adventure? Before booking, the different possibilities for accommodation, along with the web pages of the park, were the main source of creating expectations and what led us to make decisions. Our children clearly preferred the pirate accommodation of Abra Harbour, probably because this option was marketed with a film promoting the stay and the adventure of sailing with the pirates, living among them, and taking part in their pirate life, that is, becoming a pirate. Such engaging and experiential marketing through videos and a downloaded app enabled continued engagement with our adventure options by 'liking' all our favourites. Our children took great joy in placing hearts on what they wanted to experience, the marketing became an experience for them, and the experiential marketing led the pre-transition into adventure. We experienced added social value and interaction (Galvagno & Dalli, 2014), as well as experiential value from this exercise as we started to discuss the coming adventure and what we would expect and what we would like to do. We experienced a sense of communitas (Celsi et al., 1993).

During our stay in the park, it became evident that the interplay and co-creation between children, staff, and other visitors were a part of the adventure that served as experiential marketing of other shows or attractions in the park. The park's attention to experience design of adventure experiences, along with on-site and online (e.g., app) experiential marketing proved important. For example, the way the actors in plays not only announced what you could participate in next but also physically led the visitors through certain areas of the park, seamlessly promoted and experientially led to new adventures through interaction and co-creation between actors and visitors.

Contrary to my expectations, the on-site marketing was not so much about being pushed in and out of the different adventures of the park, buying products, or adding experiences in the park, it was done in a way that felt subtle, following

Figure 19.4 On-site marketing with treasure map. Photograph by May Kristin Vespestad.

experience design thinking. The walkways and the "transition areas" did contain marketing posters of adventures taking place and made a seamless transition between the different adventure worlds. Moreover, the plays and opportunities to meet different characters served as marketers and led us in and out of adventures in a way that felt natural and added to the perceived value. The experiential marketing (Schmitt & Zarantonello, 2013) was predominant for the family as a whole and was not, to a great extent, disturbed by a sense of being exposed to continued commercialisation. There might be a catch-22 in juggling adventure tourism marketing in a way that enables and promotes the immersion (Pine & Gilmore, 2020) of both children and parents into the adventure as opposed to interrupting being in the moment. Nevertheless, it is necessary to strive for seamless transitions in and out of adventures within the themes and spaces of tourism attractions (Li et al., 2021). Marketing can facilitate this.

Website of case study 19.2: Kristiansand Zoo and Amusement Park

Abra Harbour: Sjørøverhotellet Abra Havn

What is it that creates a sense of adventure in the marketing of adventure tourism? Certainly, marketing should build expectations, set the scene and create excitement by providing a taste of the adventures yet to come. This creates a sense of challenge that leaves one wondering whether one will be able to rise to the challenge once there, or even experience flow (Csikszentmihalyi, 2008). By facilitating a transitional phase for consumers (pre-adventure), which promises an adventure ahead that will provide happiness through play, marketing will contribute to the perceived value for the consumer. If consumers through experiential marketing are promised a sense of achievement, whether to fulfil expectations of becoming a true pirate by going on a treasure hunt or keeping up one's speed and flow through the pump tracks or simply experiencing, these adventures might also provide added value from a sense of well-being (Farkić et al., 2020; Hung & Wu, 2021; Pomfret, 2021), not only in the moment but also after the adventure. Moreover, the perceived value from achievement and well-being will live on through intracommunity marketing, in social media, and by word of mouth and become indirectly part of the adventure tourism marketing.

Value in adventure tourism marketing

In terms of the consumer value theory, one could say that in the two case studies, value is created in the adventure marketing in different phases and for different stakeholders. Nevertheless, there are also similarities in the experiential marketing component of both case studies. For example, using experiential marketing to promote adventure tourism implies that adventure value not only arises during the adventure at a specific adventure tourism destination (which is the foundation for adventure tourism) but could also emerge in the marketing itself, for example, by an experiential YouTube video of mountain biking trails. Therefore, an emphasis on co-creation between different stakeholders of marketing before, during, and after adventure tourism experiences can provide experiential value in adventure tourism marketing, as exemplified in Table 19.1.

Experiential marketing towards families facilitates bonding, expectations, and a preliminary transition into the adventure (see Chapter 7). As one of the commercials says, 'You are giving your kids and family the adventure of a lifetime'. Being exposed to the marketing ahead of experiencing the adventure creates and increases pre-travel value and

Table 19.1 Linking experiential marketing and value co-creation in adventure tourism marketing

Experiential marketing examples facilitating value co-creation

Value Co-creation Before	*Value Co-creation During*	*Value Co-creation After*
Value through expectations, e.g., a promotional film about becoming pirates, and learning and "liking" favourite experiences through an app.	Value through following a treasure map, apps, games, plays, also serving as marketing of activities/shops in the theme park, enabling seamless transtion in and out of adventures as they unfold.	Value through a personal voice message from Captain Sabretooth reminding you that you are a true pirate and therefore can revisit at any time as part of the pirate community.
Value created by the consumer community engaging in promoting an event, within and outside the community, using social media and technology.	Value through participation in "soft adventure events" ahead of the specialised Red Bull event enables full immersion into the main event.	Value through sharing in social media both by the community and professionals.

prepares tourists to immerse into the adventures. Consumers are pre-trained through marketing cues (Pine & Gilmore, 2020), learning how to become adventurers. On-site adventure tourism marketing leads to further immersion into the adventure or possibly interrupts the adventure if it is not done in a way that enhances the adventure theme. An awareness of the possible dilemmas and opposing views and interests of consumers (e.g., adults and children) and other stakeholders should be at the forefront for marketers, particularly those that are on-site and are hands-on with experiential marketing.

As can be seen, value co-creation and marketing rely on close collaboration, knowledge, and competence to promote the transition into the adventure, often relying on the joint efforts of special interest communities in promoting adventure tourism. A key aspect of adventure tourism marketing is communitas, and it can be strengthened in the joint volunteer marketing efforts and the image being promoted (i.e., value co-creation). Consumer communities share their interests and stories, such as discussing equipment, trails, or waves on online forums and in comments on websites or social media. In this way, they also serve as image creators and intracommunity marketers. In this manner, value co-creation spreads amongst special interest communities and creates an intracommunity image, whereby the image of the activity gains importance along with the adventure sought, more so than the destination image (Vespestad & Hansen, 2020). Thus, intracommunity image formation becomes important from an experiential marketing perspective. A potential also lies in reaching out and promoting adventure tourism destinations and events (e.g., the Red Bull Rampage in Utah) to the broader public as spectators of the extraordinary adventures, and as "invited members" of the community.

The road ahead for adventure tourism marketing

People's continued search for adventures raises possibilities and challenges for the adventure tourism industry. What then lies ahead for adventure tourism marketing? There could be several scenarios here. After years of coping under completely changed circumstances with a pandemic threatening the tourism industry, adventure tourism businesses have been forced to be creative to be resilient (Buhalis, 2022). They have innovatively come up with new campaigns, where the focus had to be on experiences for local and domestic markets. In the aftermath of the COVID-19 pandemic, our views of adventure may have changed in that we might see risks differently. Adventures can be, but do not have to be, risky or physically demanding. Experiential marketing may become even more important as a way of taking part in and making a transition into adventures, even before leaving home. This can function as a reassurance that the experience will be adventurous (yet safe). Adventure tourism marketing has taken new forms with an altered focus, relying more on the personal and social rewards of the adventure along with the renewed possibility to co-create experiences. Perhaps adventure tourism marketing in the years to come will therefore look in new directions by promoting adventures that lead to happiness and well-being and enhance quality of life.

What then would be the take-home message for those involved in practical adventure tourism marketing? How can one indeed stand out and attract adventure-oriented tourists in the future? The answer to that might not be straightforward, but in this chapter, we have looked at some possible ways forward. Experiential marketing, with a focus on the co-creation of value for stakeholders in the marketing process itself, could be one way. Moreover, it is crucial to note that reliance on and involvement with consumer communities, creating a sense of communitas, could come across as more authentic and would possibly also be a more sustainable marketing strategy.

Conclusion

As can be seen, adventure tourism marketing has blurred boundaries to the leisure sphere, for example through intracommunity co-creation in social media and online forums. The result is that marketers use experiential marketing to reach consumers and engage them in adventure. An example is the Faroe Islands Tourism "sheep view campaign" (where streaming cameras were attached to several of the island's sheep) which became a hit on social media. Moreover, businesses, marketers, tourists, and locals will have to co-create to determine what adventure might mean to different consumers, and how one can engage in promoting adventures to maximise experiential value for all stakeholders. Through co-creation between different stakeholders, adventure in marketing seems to have gained importance beyond the context of adventure tourism, contributing to our identity projects and community belonging. One example is adventure sports used to market clothing to trigger the modern consumer's emphasis on image. Experiential adventure tourism marketing therefore also serves to bridge the gap between leisure and tourism.

The way ahead might not be obvious. However, given the current trends and developments in adventure tourism, we clearly see that consumers tend to want greater involvement in all aspects of their holidays and in life. Therefore, marketers who aim to continue to attract tourists to adventure tourism must facilitate the co-creation of value with consumer communities. It is important to adopt an experiential marketing focus, merging intracommunity image formation with professional attention to facilitating immersion into the adventure. Furthermore, aiming for seamless transitions in and out of adventures, placing the holistic adventure tourism experience at the core and really emphasising *adventure* in adventure tourism marketing could make marketing part of the adventure.

There seem to be opportunities for focusing on experiential marketing for the full development of the untapped potential that lies in co-creation with consumers of adventure, thus creating joint value through marketing. Studies aimed at identifying where and when value co-creation in adventure tourism marketing emerges could also provide more long-term sustainable solutions for both adventure tourism events and adventure tourism marketing actions, such as when value is co-created between volunteers and professionals. This knowledge could be used to facilitate adventure tourism marketing processes and could also lead to more lasting contributions to local communities in terms of building adventure tourism destinations.

Review questions

1 What are some of the main changes in society and tourism that have influenced the shift in adventure tourism marketing?
2 In what ways has adventure tourism marketing changed?
3 How can experiential marketing be understood in the context of adventure tourism?
4 How can adventure tourism marketing contribute to value co-creation?
5 What might the future of adventure tourism marketing look like?

Further reading

Dolnicar, S., & Ring, A. (2014). Tourism marketing research: Past, present and future. *Annals of Tourism Research*, *47*, 31–47. https://doi.org/10.1016/j.annals.2014.03.008
Pomfret, G. (2021). Family adventure tourism: Towards hedonic and eudaimonic wellbeing. *Tourism Management Perspectives*, *39*, 100852. https://doi.org/10.1016/j.tmp.2021.100852
Schmitt, B., & Zarantonello, L. (2013). Consumer experience and experiential marketing: A critical review. In N. K. Malhotra (Ed.), *Review of marketing research* (Vol. 10, pp. 25–61). Emerald Group Publishing.

Weber, K. (2001). Outdoor adventure tourism - a review of research approaches. *Annals of Tourism Research*, 28(2), 360–377.

References

Aftenposten.no. (2016). Retrieved at https://www.aftenposten.no/reise/i/kJMRXA/turistnaeringen-vil-fjerne-farlige-bilder-fra-trolltunga-reklame

Arnould, E., Crockett, D., & Eckhardt, G. (2021). Informing marketing theory through consumer culture theoretics. *AMS Review*, 11(1), 1–8. https://doi.org/10.1007/s13162-021-00198-9

Arnould, E. J., & Thompson, C. J. (Eds.). (2018). *Consumer culture theory*. Sage Publications.

Buhalis, D. (Ed.). (2022). *Encyclopedia of tourism management and marketing*. Edward Elgar Publishing. Retrieved 6 Jan 2023, from https://www.elgaronline.com/view/book/9781800377486/9781800377486.xml

Campos, A. C., Mendes, J., Valle, P. O. D., & Scott, N. (2018). Co-creation of tourist experiences: A literature review. *Current Issues in Tourism*, 21(4), 369–400. https://doi.org/10.1080/13683500.2015.1081158

Celsi, R. L., Rose, R. L., & Leigh, T. W. (1993). An exploration of high-risk leisure consumption through skydiving. *Journal of Consumer Research*, 20(1), 1–23.

Cheng, M., Edwards, D., Darcy, S., & Redfern, K. (2016). A tri-method approach to a review of adventure tourism literature: Bibliometric analysis, content analysis, and a quantitative systematic literature review. *Journal of Hospitality & Tourism Research*, 42(6), 997–1020. https://doi.org/10.1177/1096348016640588

Clarke, J. F., Previte, J., & Chien, P. M. (2021). Adventurous femininities: The value of adventure for women travelers. *Journal of Vacation Marketing*, 28(2), 171–187. https://doi.org/10.1177/13567667211038952

Cooper, M.-A., & Buckley, R. (2021). Tourist mental health drives destination choice, marketing, and matching. *Journal of Travel Research*, 61(4), 786–799. https://doi.org/10.1177/00472875211011548

Cova, B., & Shankar, A. (2018). Consumption tribes and collective performance. In E. J. Arnould & C. J. Thompson (Eds.), *Consumer culture theory* (pp. 87–106). Sage.

Cova, V., & Cova, B. (2019). Pain, suffering and the consumption of spirituality: A toe story. *Journal of Marketing Management*, 35, 565–585.

Csikszentmihalyi, M. (2008). *Flow. The psychology of optimal experience* (Harper Perennial Modern Classics). HarperCollins Publishers.

Della Corte, V., & Del Gaudio, G. (2020). Traditional versus experiential marketing in tourism. In S. K. Dixit (Ed.), *The Routledge handbook of tourism experience management and marketing*. Routledge. eBook ISBN9780429203916. https://doi.org/10.4324/9780429203916

Farkić, J., Filep, S., & Taylor, S. (2020). Shaping tourists' wellbeing through guided slow adventures. *Journal of Sustainable Tourism*, 28(12), 2064–2080. https://doi.org/10.1080/09669582.2020.1789156

Funk, D. C., & Bruun, T. J. (2007). The role of socio-psychological and culture-education motives in marketing international sport tourism: A cross-cultural perspective. *Tourism Management*, 28(3), 806–819.

Fossgard, K., & Fredman, P. (2019). Dimensions in the nature-based tourism experiencescape: An explorative analysis. *Journal of Outdoor Recreation and Tourism*, 28, 100219. https://doi.org/10.1016/j.jort.2019.04.001

Galvagno, M., & Dalli, D. (2014). Theory of value co-creation: A systematic literature review. *Managing Service Quality*, 24(6), 643–683.

Haukeland, J. V., Fredman, P., Siegrist, D., Tyrväinen, L., Lindberg, K., & Elmahdy, J. M. (2021). Trends in nature-based tourism. In P. Fredman & J. V. Haukeland (Eds.), *Nordic perspectives on nature-based tourism* (pp. 16–31). Edward Elgar.

Hjemdahl, K. M. (2003). *Tur-retur temapark: oppdragelse, opplevelse, kommers* [Return trip to the theme park: Education, experience, commerce]. Høyskoleforlaget.

Holbrook, M. B. (1999). *Consumer value: A framework for analysis and research*. Routledge.

Holbrook, M. B., & Hirschman, E. C. (1982). The experiential aspects of consumption: Consumer fantasies, feelings, and fun. *Journal of Consumer Research*, 9(2), 132–140.

Hung, H.-K., & Wu, C.-C. (2021). Effect of adventure tourism activities on subjective well-being. *Annals of Tourism Research*. https://doi.org/10.1016/j.annals.2021.103147

Jamal, S. A., Aminudin, N., & Kausar, D. R. (2019). Family adventure tourism motives and decision-making: A case of whitewater rafting. *Journal of Outdoor Recreation and Tourism, 25*, 10–15. https://doi.org/10.1016/j.jort.2018.11.005

Janowski, I., Gardiner, S., & Kwek, A. (2021). Dimensions of adventure tourism. *Tourism Management Perspectives, 37*. https://doi.org/10.1016/j.tmp.2020.100776

Li, B., Zhang, T., Hua, N., & Jahromi, M. F. (2021). Developing an overarching framework on theme park research: A critical review method. *Current Issues in Tourism, 24*(20), 2821–2837. https://doi.org/10.1080/13683500.2020.1849047

Mandić, A., & McCool, S. F. (2022). A critical review and assessment of the last 15 years of experience design research in a nature-based tourism context. *Journal of Ecotourism*, 1–28. https://doi.org/10.1080/14724049.2022.2099877

Martin, D. M., & Schouten, J. W. (2014). Consumption-driven market emergence. *Journal of Consumer Research, 40*(5), 855–870.

Mieli, M., & Zillinger, M. (2020). Tourist information channels as consumer choice: The value of tourist guidebooks in the digital age. *Scandinavian Journal of Hospitality and Tourism, 20*(1), 28–48.

Pine, B. J., & Gilmore, J. H. (2020). *The experience economy: Competing for customer time, attention, and money*. Harvard Business Review Press.

Pomfret, G. (2021). Family adventure tourism: Towards hedonic and eudaimonic wellbeing. *Tourism Management Perspectives, 39*, 100852. https://doi.org/10.1016/j.tmp.2021.100852

Prahalad, C. K., & Ramaswamy, V. (2004). Co-creation experiences: The next practice in value creation. *Journal of Interactive Marketing, 18*(3), 5–14.

Rather, R. A., Hollebeek, L. D., & Rasoolimanesh, S. M. (2022). First-time versus repeat tourism customer engagement, experience, and value cocreation: An empirical investigation. *Journal of Travel Research, 61*(3), 549–564.

Same, S. & Larimo, J. (2012). *Marketing theory: Experience marketing and experiential marketing*. 7th International Scientific Conference 'Business and Management 2012', 10–11 May, Vilnius, Lithuania. https://doi.org/10.3846/bm.2012.063

Sand, M., & Gross, S. (2019). Tourism research on adventure tourism – Current themes and developments. *Journal of Outdoor Recreation and Tourism, 28*, 100261. https://doi.org/10.1016/j.jort.2019.100261

Schmitt, B. (1999). Experiential marketing. *Journal of Marketing Management, 15*(1–3), 53–67. https://doi.org/10.1362/026725799784870496

Schmitt, B., & Zarantonello, L. (2013). Consumer experience and experiential marketing: A critical review. In N. K. Malhotra (Ed.), *Review of marketing research* (Vol. 10, pp. 25–61). Emerald Group Publishing.

Seeler, S., & Schänzel, H. (2020). Co-construction of the tourist experience via social networking sites. In S. K. Dixit (Ed.), *The Routledge handbook of tourism experience management and marketing*. Routledge. eBook ISBN9780429203916. https://doi.org/10.4324/9780429203916

Seligman, M. E., & Csikszentmihalyi, M. (2000). *Positive psychology: An introduction* (Vol. 55, No. 1). American Psychological Association.

Thompson, C. J., MacInnis, D., & Arnould, E. J. (2018). Conclusion: Linking CCT and consumer research: Consumers' mobilization of co-created resources. In E. J. Arnould & C. J. Thompson (Eds.), *Consumer culture theory* (pp. 318–346). Sage Publications.

Turner, V. W. (1969). *The ritual process: Structure and antistructure*. Cornell University Press.

van Nuenen, T. (2016). Here I am: Authenticity and self-branding on travel blogs. *Tourist Studies, 16*(2), 192–212. https://doi.org/10.1177/1468797615594748

Vargo, S. L., & Lusch, R. F. (2004). Evolving to a new dominant logic for marketing. *Journal of Marketing, 68*(1), 1–17.

Vargo, S. L., & Lusch, R. F. (2008). Service-dominant logic: Continuing the evolution. *Journal of the Academy of Marketing Science, 36*(1), 1–10. https://doi.org/10.1007/s11747-007-0069-6

Vargo, S. L., & Lusch, R. F. (2017). Service-dominant logic 2025. *International Journal of Research in Marketing, 34*(1), 46–67. https://doi.org/10.1016/j.ijresmar.2016.11.001

Vespestad, M. K., & Hansen, O. B. (2020). Shaping climbers' experiencescapes: The influence of history on the climbing experience. *Journal of Hospitality & Tourism Research, 44*(1), 109–133. https://doi.org/10.1177/1096348019883685

Vespestad, M. K., & Lindberg, F. (2011). Understanding nature-based tourist experiences: An ontological analysis. *Current Issues in Tourism, 14*(6), 563–580.

Vespestad, M. K., Lindberg, F., & Mossberg, L. (2019). Value in tourist experiences: How nature-based experiential styles influence value in climbing. *Tourist Studies, 19*(4), 453–474. https://doi.org/10.1177/1468797619837966

20 Technology and adventure tourism management

Cam O'Beirne

Chapter learning outcomes

1 Describe different types of technology that are used by adventure tourism businesses.
2 Gain insights into a range of management issues concerning the use of information technology by adventure tourism businesses.
3 Outline barriers to access and entry of using technology in adventure tourism businesses.
4 Consider the implications of future technology developments on adventure tourism businesses.

Introduction

Technological advancements have revolutionised adventure tourism businesses, presenting new opportunities and enriching experiences for the industry and tourists alike. The widespread use of the internet as a commercial medium, along with advancements in e-business strategy, e-markets, and e-commerce systems, offers exciting prospects for the adventure tourism industry. For instance, consumer websites play a vital role in public relations and marketing campaigns, enabling businesses to reach a large global consumer base. Social media has become increasingly influential in pre-trip travel preparation, decision-making, and the sharing of tacit information. Embracing these technologies empowers adventure businesses to enhance their operational efficiency, elevate consumer experiences, and remain at the forefront of this dynamic industry. Adventure tourism occupies a space where people challenge themselves, visit far-flung places, and change their expectations of everyday life. The dichotomy of this escape is the use of technology within adventure tourism to drive consumer engagement, relive experiences at home, or virtually access remote places (Figure 20.1).

This chapter delves into the intersection of technology and adventure tourism management, highlighting disruptive technologies, such as the Internet of Things (IoT), quick response (QR) codes, artificial intelligence, virtual reality, and augmented reality and their pivotal role in supporting operations. Such technologies fundamentally change how consumers, industries, and organisations function as they are undoubtedly superior to the systems they are replacing. As technology continues to evolve, adventure tourism

DOI: 10.4324/9781003393153-24

Figure 20.1 Synchronous video editing suite, AJ Hackett bungee jump, Queenstown, New Zealand. Photograph by Carl Cater.

management can leverage its potential to create unparalleled and unforgettable experiences for adventure seekers worldwide. Overall, the integration of technology has significantly influenced the way adventure tourism operates. It has allowed for greater accessibility, communication, and community building, ultimately contributing to the growth and development of this exciting sector within the tourism industry. This chapter focuses on business that is conducted over networks that use non-proprietary protocols (open to everyone) such as the internet. The reason for this is the increasing dependence on the use of the internet to facilitate management information systems and e-business in the adventure tourism industry. It is also driven by people's ability to access the internet's vast global network at a low cost using existing telecommunications infrastructure.

The role of information communications technology in the adventure tourism industry

Chapter 13 examines the diverse range of adventure tourism providers which play a vital role in delivering unique and thrilling experiences to their clients. These providers utilise various technologies and online platforms to establish a strong presence within the adventure tourism industry, and to foster growth of their businesses. Advances in these technologies, particularly the internet, have changed the way that adventure tourism organisations deliver their services to consumers and clients. They include highly efficient networked computers, digital content services, mobile telephones, personal digital assistants, and intelligent appliances in the home and office. These technologies can assist in developing and delivering business efficiency, and they allow a level of connectiveness across the adventure industry unparalleled throughout history.

Although there is a dearth of research on the use of these technologies within the adventure industry, tourism studies offer some valuable insights. For instance, Pierdicca et al. (2019) suggest that not only do information communication technologies empower consumers to identify, customize, and purchase products but that they also support the globalisation of the industry by providing effective tools for suppliers to develop, manage, and distribute their offerings worldwide. Xiang (2017) notes the strong interconnection between tourism and technology and emphasises that our understanding of this relationship cannot be separated from the larger global impact of technology on society. Khatri's (2019) content analysis of academic studies found that different technologies were most used to fulfil information needs, evaluate behaviour and performance, and manage operation and innovation processes within the tourism and hospitality industries.

In this digital era, social media platforms provided by companies such as Meta (Facebook, Instagram, WhatsApp), Snapchat, TikTok, and X (formerly Twitter), have significantly contributed to redefining the internet's role. Rather than solely serving as a publishing platform, the internet has transformed into a platform for participation, connection, and social networking. Social media platforms have become valuable tools for the adventure tourism industry to engage with their target audiences. Businesses can showcase their offerings, share captivating content, and build meaningful connections with potential consumers and clients. Adventure tourists can actively participate with these platforms, share their experiences, and connect with like-minded individuals, creating a vibrant adventure community. Importantly, tourism operators can use these as part of an overall sales channel as many of the platforms allow direct sales to consumers and clients alike.

Aside from the internet, other technologies that adventure tourism businesses use in their day-to-day operations include mobile devices, tablets, and smartphones. These have become ever-present in daily life, with features that include location and geotagging searches, app-based market segmentation programmes, and social media apps that provide aspirational content on adventure experiences. It is, therefore, unsurprising that online holiday bookings and experiences dominate the market. For instance, in the UK, Mintel Academic (2023) established that in the year following 2023, 84% of tourists plan to book their main holiday online rather than in person.

The e-business concept and adventure tourism

E-business is defined as all types of business transactions and interactions which involve preparing or carrying out business electronically (Talha & Subramaniam, 2003), including communicating with consumers and clients. E-commerce involves financial and commercial transactions, for instance electronic funds transfers, credit and debit card activities, and electronic data exchange. Any type of adventure tourism organisation, ranging from businesses with large financial capital to small organisations with in-house websites, can benefit from conducting their business activities electronically. In e-business, activities may be as simple as sending an email to a colleague requesting information about an upcoming adventure festival, purchasing tickets for an adventure tourism event online, or watching live adventure events streamed directly into your home via a broadband connection. For instance, Outside (https://www.outsideonline.com/) live streams various adventure events, including Ironman 70.3 and the ultra-trail of Mont Blanc (UTMB) (https://www.outsideonline.com/) Adventure+ is a subscription video-on-demand platform. Consumers pay a subscription fee to watch live-streamed events and films, and read online adventure magazines: (https://www.outsideonline.com/)

Operating this e-business system for selling involves content production, the digitisation and storage of content, an electronic network with access to an electronic payment system, and links to physical distribution systems. Buying involves people linking to an electronic network, usually the internet, searching and locating content, that is, goods and services, through search engines, directories, or specific websites, and placing orders through a web browser or mobile interface with a link to or an embedded electronic payment system. There are four main types of e-business based on the type of transaction that occurs (see Table 20.1).

Table 20.1 E-business types (adapted from O'Beirne, 2018)

Categories of e-business	Description and examples of e-business type
Business to Consumer (B2C)	Online shopping websites which offer an additional sales channel for existing organisations or for new pure-play internet companies to have direct round-the-clock access to consumers. *Examples* • G Adventures – Booking and purchasing adventure holidays • REI – Purchasing equipment and clothing; booking and buying adventure holidays • The Adventure Travel Show – Information website for adventure tourism events; online banking facilities
Business to Business (B2B)	Virtual supply chains which link together different distributors, suppliers, contractors, and consultants electronically to facilitate information flow. Used to make business processes more streamlined through creating efficiencies in the business, reducing costs and reaching more customers. *Examples* • Booking.com and TripAdvisor - Booking engines and directories listing business products • Adventure Travel Trade Association – Industry networks
Consumer to Consumer (C2C)	Facilitate opportunities for consumers to directly interact with one another and exchange information, services and products. • Classified advertisement sites • eBay – Multinational e-commerce auction site • Instagram - Photo geotagging locations by consumers on mobile phones
Consumer to Business (C2B)	Individuals offering professional services to others. Allows consumers to control transactions through, for example, a reverse auction, whereby the bidder sets the price for services or goods. *Examples* • Tours By Locals - Independent adventure guides • Fliggy-Alibaba Group – A Chinese-based platform owned by marketplace giant Alibaba that allows Chinese companies to monitor e-commerce, social media, and current trends to create and distribute customer-centric products quickly. Successful products or offerings are scaled up, while unsuccessful ones are swiftly withdrawn, allowing for a rapid response to market opportunities, shortening the idea-to-launch timeline to weeks, not months.

Adventure tourism organisations offer a wide range of online experiences for consumers. Search engines are the most frequently used information sources and booking platforms for these experiences, with consumers using various electronic devices to search. However, using a smartphone to book their main holiday is becoming an increasingly important booking channel for UK tourists, particularly for younger travellers in the 16–34 age group (39%). Adventure tourism organisations offer a wide range of online experiences for consumers, mostly using aggregator websites (e.g., TripAdvisor), which are hubs that provide content and services to a wide market with a global reach. These portals act as gateways for consumers or other users to access information and services. Typically, the portal aims to provide a useful resource and a call to action to encourage users to seek out information about a destination or activity, read reviews, or book an experience or accommodation. Examples of these types of websites are presented in Table 20.2.

Table 20.2 Adventure tourism websites

Website	*Features of website*
Adrenaline	Portal site for booking adventure-specific experiences and accommodations with a focus on location, price, and number of travellers. Visitors can search for an adventure by category, location, and price. The website provides aggregation services around a wide range of activities suited to adventure tourism.
TripAdvisor	A well-known customer review site that categorises experiences and accommodation into location and type based upon a star rating and written reviews from actual travellers. More recently, TripAdvisor has expanded into providing booking services for some adventure activities through its shared online platform with Viator, an online booking engine for travel.
G Adventures	Adventure-specific site for booking experiences and accommodations with a focus on destinations, travel styles, and deals on price.
Wilderness Travel	Adventure-specific site for booking experiences and accommodations with a focus on destinations and travel styles. Uses hero imagery of aspirational travel locations.
Abercrombie & Kent	Adventure-specific site. Provides information through segmentation by journey type and/or locations to a particular demographic of adventure-seeking traveller.
Australia	Government-funded country site to promote and engage potential visitors with a range of experiences and travel options. Provides details on the country and benefits of visiting, hero imagery iconic to the country. The site uses artificial intelligence chatbots to connect with potential customers and rich media content with aspirational videos on demand. The site has a focus on experiential activities by promoting "getting outside" to engage in mainly soft adventure tourism activities.
Margaret River Adventure Company	A small business website that provides experiential and booking information on highly unique adventure activities provided in a very remote part of Australia. The site provides the main advertising platform for potential customers who mainly come from Singapore and require detailed information prior to booking a visit to the Margaret River region in regional Australia.

Case study 20.1   How the South Australian Tourism Commission reached millions and doubled leads on YouTube

South Australia offers a range of adventure tourism activities, including swimming with sea lions, shark cage diving, dark skies, Australia's largest Outback mountain range, the Flinders Ranges, and the world's second National Park City, Adelaide (South Australia)

The borders to Australia were closed to the world during the global COVID-19 pandemic. "Fortress Australia", as it was named, barred visitors into the country for over a year to protect the nation from the ravages of the pandemic. In addition, the domestic tourism industry was only open at an inter-state level (travel across state borders) spasmodically as state governments wrestled with lockdowns and pandemic mitigations and tourism became mostly focused on intra-state travel (travel within state borders).

As restrictions started to ease, once public health measures took effect, the South Australia Tourism Commission began to encourage a wider audience of travellers. Responding to a highly competitive market, the commission decided to highlight the country's best sites and activities in an unprecedented, unedited five-day live stream on YouTube. It implemented this through a live stream activation on big screens in Adelaide as well as streaming concurrently on its YouTube platform so that it was widely available to people at home or elsewhere.

The unedited, live content ranged from iconic visions of kangaroos bounding across desert landscapes, to skydiving, gourmet beach barbeques and frolicking with sea lions. Using an array of high-profile South Australians – from Olympic athletes and football stars to award-winning chefs, musicians, and winemakers – 65 tourism operators, 10 regions, 31 local talent, 18 cameras, 11 modes of transport, and more than 50 local musicians participated in bringing the campaign to life.

The results were generally seen as worthwhile, with more than 11 million views of the content on YouTube and a doubling in tourism leads for South Australian tour operators for bookings. Calculated tourism expenditure rose 20% to AU$7.6 billion.

Further information

The South Australian Tourism Commission website entices domestic travellers on YouTube
South Australia's newest tourism ad is launching at 5pm today

---

Adventure tourism and changing technology

The evolution of different information and communications technologies continues to positively impact the adventure tourism industry. However, most tourism research has looked backwards at technology developments rather than focusing on advances in how organisations are using this to enhance experiences for their consumers (Hughes &

Moscardo, 2018). The following sections focus on technological developments and their influence on adventure tourism management and operations.

M-commerce and new payment platforms

M-commerce involves storing your financial details to buy a product or service online using a mobile phone. Your phone may have this as an app called a wallet. It also refers to using a mobile device to access something or complete a transaction by showing a bar code or authenticated redemption icon on the phone. Mobile devices, such as smartphones, tablets, and smartwatches, allow consumers to access information, booking services, apps, and social media sites on the go. Furthermore, these devices have changed the way we perceive and communicate with each other and with our environment. For example, in Australia, domestic travellers have shifted away from using cash to pay for holidays over recent years (Doyle et al., 2017).

Some of these app-based payment systems and Tap & Go technology offer rapid, secure transactions that move payment to the background, rather than interfering with consumer experiences. This reduces so-called friction at the point of sale, removing barriers at the point of sale to customers, and providing consumers with ready access to goods and services. Some examples include PayPal, PayWave, PayPass, Apple Pay, Union Pay, and Alipay. Consumers can also store tickets on their mobile phone in their wallet app for later use at an event, at the point of travel, to gain venue access or check-in. The purchase of the ticket can be completed by mobile phone, or online at an e-commerce website, via a telephone call centre or, alternatively, at a physical ticket outlet or kiosk.

Interactive and reality adventure TV

Adventure tourism organisations can benefit considerably from interactive TV. It can be classified into two categories. Content-driven interactive TV offers the illusion of choice for consumers as they can select ways in which characters or scenes engage with activities as part of the story line. Accordingly, a story can be told in several different ways and comprise engaging and interesting sub-plots to hold the viewer's interest. An example of this is the reality adventure TV series, *You vs Wild*, shown on the streaming giant Netflix. Adventurer Bear Grylls asks viewers to select specific tasks for him to undertake, all of which have been pre-recorded. For instance, does he climb the mountain or abseil down? Will you put him in a plane or a helicopter?

Another use is that of advertising-driven interactive TV, which encourages viewers to pause programming and choose items of content, which they can find out more about. A good example is Amazon's Prime Video which streams sport in the US and allows viewers to purchase team gear from their couch whilst watching the game. In the very near future, adventure tourism apparel companies may "sponsor" a show and provide the gear for that adventure TV episode. Viewers can see garments and equipment being used in "real life" and tap or click to find out more and purchase a piece of gear that an adventurer is wearing or a piece of equipment they are using. It also includes information about the adventure location where the activity is taking place, with a link to further adventure experiences and how to book.

Reality TV can also influence destination image in determining tourists' decision-making (Fu, Ye & Xiang, 2016). Using a popular Chinese adventure tourism reality TV programme *Where Are We Going, Dad?* the authors reported audience involvement is positively related to future behavioural intentions. This chapter author was involved in the Australian production of the adventure tourism show, which is viewed by 75 million people each week. These numbers are mind-boggling from a Western perspective but demonstrate the number of affluent middle-class people in China yearning for adventure. The Australian show featured visits to remote beaches, partaking in surfing and swimming, visiting animal farms, and cooking barbeques on the beach. The show's production was supported by the government body, Tourism Western Australia (TWA), and anecdotal research by TWA has demonstrated a marked increase in Chinese visitation to the state, particularly to the locations that were filmed and activities that were engaged in by the show's participants.

Location-based services

Location-based services are software services which are based on the user's geographical location, for example, global positioning systems and Wi-Fi tracking. Therefore, businesses and individuals can track and communicate with people based on their location. These services allow delivery of information and services that are specific to a time and place and responsive to the conditions and circumstances at that moment (e.g., weather, crowding, bookings, and special offers). Geo-targeting works by using global positioning system technology or local Wi-Fi networks to deliver targeted marketing content and adverts to consumers, based on their physical location. Geo-targeting will be a key marketing focus for the adventure tourism industry going forward. This will allow adventure tourism organisations to be able to target potential clients and consumers based on their locations. These sorts of services can provide more of an immediate or "on demand" offering to clients for adventure tourism activities and accommodation that are available nearby. See Vignette 20.1 for an overview of the global addressing system what3words.

Vignette 20.1   what3words

what3words (https://map.what3words.com/pretty.needed.chill) is a global addressing system. It has divided the world into a grid of 3m × 3m squares and assigned each one a unique and precise three-word address. what3words states that it enables tourism businesses to share their location accurately every time, and it helps travellers explore the world without getting lost. Lewis (2017) notes that three-word addresses are more accurate than postal addresses because they are easier to recall and share compared with global positioning systems coordinates. Adventure tourism businesses can use what3words to design their trip dossiers and 'venture into uncharted territory, safe in the knowledge that they'll always find camp at the end of the day' (Lewis, 2017).

Further information:

https://what3words.com/business/travel

Sharing economy

The term *sharing economy* 'refers to a set of organizational and business models based on "sharing," "collaborative," "gig," or "access" approaches to the use of resources' (Schlagwein et al., 2020, p. 817). Platforms for the sharing economy are available online, including Airbnb, which provides accommodation services to adventure tourists, and "Airbnb Adventures", which offers tourists the opportunity to dive with a shark expert in South Africa or camp on a cliff in the US, for instance (https://news.airbnb.com/introducing-airbnb-adventures/). Such platforms are successful in offering tourists innovative options for accommodation, transportation, and dining. Concurrently, they have significantly disrupted the conventional business models of established tourism and hospitality providers. Some may suggest that these platforms are only changing the role of intermediaries through the tourism supply chain. However, with reviews and ratings by consumers, organisations such as Airbnb have the potential to saturate existing markets and develop new ones through their social reach. Ride-sharing companies are also branching out into other transport modes beyond the usual car transport and offering vehicles that may be used in the provision of adventure tourism experiences. For example, Uber has launched boat-hailing services in Egypt and Croatia and has operated helicopters in the south of France since 2016 (Salinas, 2018). In India and Pakistan, rickshaws and motorcycles are similarly available on Uber.

The Internet of Things (IoT)

The IoT is the application of smart sensors to a variety of devices in settings that autonomously gather and communicate data. These devices can track and gather data, send this to networks for analysis, and then share and/or act on that data to support efficient management and utilisation of resources and areas. With the use of cloud computing, mobile communication, blockchain, big data, and artificial intelligence, the IoT is the primary technology transforming the tourism industry to improve the visitor experience and to create what is termed "smart tourism" (Verma et al., 2021). Examples of use of the IoT may include predictive repair and maintenance of equipment for the tourism operator, as well as enhancing the guest experience through, for example, electronic key cards sent by the hotel to your mobile phone to access your room directly without wasting time in the hotel reception.

QR codes

Who knew what a QR code was before the global pandemic? Despite having been around for many years prior to 2019, the uptake of QR technology was limited to bespoke and small applications, usually in the areas of supply chain management and logistics. However, the use of QR codes as a government-mandated public safety measure forced people to learn to use this technology through their smartphones to adhere to COVID-19 pandemic measures. This consumer familiarisation helped with the commercialisation of QR technology. The use of QR codes has proliferated in tourism, and they are used to educate and supply information to adventure tourists. Such applications include the following:

- Scanning social media QR codes to easily follow or like the social media pages of an adventure tour operator, and to get real-time updates and discounts.
- The use of digital travel guides as a replacement for maps, for example, Lonely Planet and Rough Guides.
- The use of digital menus available by scanning a menu QR code.
- Seamless and contactless mobile check-in to venues or the acceptance of terms and conditions.

Augmented reality and virtual reality

Augmented reality allows for simultaneous viewing of the real world and an enhanced virtual world, generating environments that are like real life (Mine et al., 2012). Most existing applications emphasise visual senses; however, augmented reality applications that focus on other senses are becoming more widespread. An adventure tourism example is the use of headsets and helmets with a mix of computer-generated noises, hand-held gadgets, and/or gloves, which can react to touch and pressure. Also, images of the past which represent, for example, First Nations' cultural tourism resources, can be superimposed over what is currently visible to project an immersive experience.

Virtual reality is a computer-generated immersive interface that enables users to view a three-dimensional (3D) and animated artificial world through, for example, headsets. The experience is immersive, users can virtually visit adventure tourism sites and places, as well as experience tourism products such as horse riding and surfing (see Vignette 20.2). This form of reality provides a virtual representation of possible experiences to encourage users to visit or participate in them physically. It is created using intelligent hardware and software, allowing users to obtain an immersive, direct experience destination or activity (Çeltek, 2020). It is a computer-generated environment that presents total immersion in the 3D digital world.

However, tourism organisations face significant barriers in using virtual reality and augmented reality. One major challenge for the many small adventure tourism businesses that operate within this industry is the considerable cost of investing in such technology.

Vignette 20.2   UNESCO World Heritage Sites

In the realm of adventure tourism and World Heritage Sites, virtual heritage harnesses computer-based interactive technologies in virtual reality to craft visually captivating representations of monuments, artefacts, buildings, and cultural elements. Its purpose is to openly deliver these immersive experiences to global audiences. Many of the key World Heritage Sites have been "digitised" and made available online for audiences to interact with, either as a precursor to visiting the actual place or for those people who may not be able to visit other places in the world.

Further information:

https://www.unesco.org/en/covid-19/culture-response

Relatedly, there are costs associated with training staff in this technology and its ongoing maintenance. Therefore, the benefits of using virtual or augmented reality commercially and the visitor experience derived from these may be negligible. It is fair to suggest that the purveyors of this technology, such as virtual reality skydiving and abseiling and adventure gaming, are enthusiastic about their products. Consumers can experience something that may be beyond their reach for economic or societal reasons. When using augmented reality, there is a risk that the authenticity of travel-related experiences would be compromised. As a result, the most crucial factor in determining whether this technology will be used in adventure tourism experiences is its quality. With virtual reality, it is possible that "real" travel experiences might only be enjoyed by wealthy tourists because augmented reality could replace having an in-person encounter. While it lacks a complete feeling of place, individuals who are less fortunate will likely travel through readily available and reproducible virtual reality.

But is that what adventure tourism is really about? In this overly complicated socially connected digital life, maybe the focus needs to be on enabling technology to get us to the place where we can actually experience something and be in the real moment. Oh, but don't forget your mobile device to take a picture; otherwise, it didn't happen!

Artificial Intelligence

An artificial-based chatbot is an advanced computer programme designed to simulate human conversations and continuously learn from each interaction. Chatbots have undergone significant improvements, enabling them to provide instant, personalised, and valuable responses to user enquiries. This innovative technology has and will continue to accelerate and have a profound impact on the services offered by adventure trip providers, enhancing their capabilities in sales, marketing, and customer service (Demir & Demir, 2023).

Adventure tourism seekers often have numerous questions about destinations, trip packages, and client policies. Chatbots offer a unique advantage by providing immediate assistance without the need for human intervention. These intelligent bots are equipped with a wealth of information about various adventure destinations, itineraries, and relevant policies. They can swiftly address queries, offer recommendations, and provide real-time support to potential customers, enhancing customers' overall experience.

Chatbots contribute to the efficiency of adventure trip providers by managing routine enquiries, thereby freeing up human resources to focus on more complex tasks. By automating responses and providing instant support, chatbots streamline the sales process, leading to faster conversions and improved customer satisfaction. Additionally, chatbots serve as valuable marketing tools, as they can engage with website visitors, initiate conversations, and provide personalised recommendations based on individual preferences and interests. Chatbots continuously learn from user interactions, allowing them to adapt and improve their responses over time. They can gather valuable insights about customer preferences, trends, and pain points, enabling adventure trip providers to refine their services and tailor their offerings to meet the evolving needs of adventure seekers.

Overall, the integration of artificial intelligence–based chatbots is likely to revolutionise the adventure trip industry, offering immediate and comprehensive support to potential customers. By leveraging this technology, adventure trip providers can enhance their sales, marketing, and customer service efforts, delivering a more seamless and engaging experience for adventure seekers worldwide. Examples of how these chatbots will impact adventure tourism providers include the following:

- *Adventure booking platform*: This involves integrating a chatbot into the organisation's website and mobile app. When adventure seekers visit the platform to explore various destinations and trip packages, the chatbot engages with them in real-time conversations. The chatbot can provide instant information about different adventure activities, answer frequently asked questions, and even suggest personalised itineraries based on the user's preferences. By offering immediate support and personalised recommendations, the chatbot enhances the user experience, increases engagement, and facilitates seamless bookings for adventure seekers.
- *National park visitor services*: A national park that offers adventure tourism experiences implements a chatbot as part of its visitor services. The chatbot is accessible through the park's website and social media channels, allowing potential visitors to enquire about trail conditions, wildlife sightings, safety guidelines, and permit requirements. Adventure seekers can receive instant responses from the chatbot, ensuring they have up-to-date information before planning their visit. The chatbot also helps alleviate the workload of park staff by handling routine enquiries, enabling them to focus on more critical tasks while ensuring visitors receive prompt assistance.
- *Adventure gear retailer*: An adventure gear retailer incorporates a chatbot on its e-commerce platform. When consumers browse the website looking for specific adventure gear or equipment, the chatbot engages with them to provide product recommendations, sizing guidance, and answers to product-related enquiries. It leverages its knowledge base to assist consumers in finding the right gear for their adventure activities, ensuring a personalised shopping experience. By offering immediate support and expert advice, the chatbot enhances customer satisfaction, boosts sales, and improves the overall shopping journey for adventure enthusiasts.

These real-world examples demonstrate how artificial intelligence–based chatbots can successfully be integrated into adventure tourism organisations, providing instant support, personalised recommendations, and improving overall consumer experiences in the industry.

Barriers to adventure tourism organisations using technology

Although many organisations will embrace different technologies and the identified benefits they offer, some will not be able to, or be reluctant to change the way of doing things. This is not a dilemma specific to the adventure tourism industry, as any organisation would have to constantly evaluate, change, or manage its use of technology. However, many adventure tourism providers are small or medium-sized businesses managed and owned by lifestyle entrepreneurs. Therefore, although they may be willing to adopt different technologies within their daily operations, they may not have the financial resources or the "know-how" to embed these into their business. Providers may have multiple roles within their organisation, leaving limited time to focus on embedding technology into their business, apart from perhaps dealing with daily activities, such as responding to emails, and posting promotional content on social media. See Table 20.3 for an overview of key barriers that adventure tourism organisations face regarding the use of technologies.

Table 20.3 Constraints and challenges in the use of information and communication technology (ICT) by adventure tourism businesses

Organisation element	Description
Reluctance to use technology	• Financial resources are required to invest in equipment, install it, maintain it, and train people on how to operate and manage it. • Consumers and visitors who are familiar with and want to use new systems and technologies in locations where they are not supported may become frustrated when businesses choose not to implement them.
Customer focus by staff	• As adventure tourism is a service industry, tourist satisfaction frequently relies on how well consumers are served. • Computers and machine learning, using instruments such as chatbots, may appear to speed up some activities, but this is not always the case. By replacing people with computers, adopting self-service for certain tasks, e.g., choosing adventure destinations, selecting specific gear to use in an adventure activity, as well as pre-planning every part of the tourist's itinerary, the interpersonal and natural aspects of travel may be lost. • Adventure seekers searching for a more personalised experience might then decide to go somewhere else. Any implementation of computer-based management solutions must weigh the value of functional benefits against personal and emotional engagement. This is especially the case in high-engagement activities that are implicit in adventure tourism.
Intermediaries in the supply chain change	• The numerous opportunities for adventure tourists to connect with one another may eliminate the requirement for intermediary tourism operators. This would fundamentally alter the key providers as well as the supply chain. • Although companies like Airbnb and Uber are frequently referred to as disruptors (Gretzel et al., 2015), it is argued that they are merely a new kind of intermediary which links consumers to providers. Without these new intermediaries, there is a chance that the link between consumers and providers would be far more disruptive, especially if it changes the power dynamics among providers at a destination. As an example, there may be multiple providers of similar adventure activities at a ski resort, all offering much the same product with the same pricing and benefits to customers. In this instance, customers may be able to leverage this knowledge and negotiate either a better price or ask for added value from the provider to improve the offering. • The value of direct marketing organisations is changing due to technology disrupting the traditional supply chain. These organisations locate markets, promote, offer details on local tours and accommodation, and guarantee service quality. As these tasks are becoming compressed (Buhalis, et al., 2019), such organisations are increasingly adding other value by managing smart infrastructure, conducting data analyses, and disseminating information for clients.

(Continued)

Table 20.3 (Continued)

Organisation element	Description
Privacy and security Personal and financial data	• The use and dissemination of data that has been gathered and kept, with or without people's knowledge or consent, poses concerns. • The Internet of Things (IoT) and cloud technologies have complexified and accelerated the storage of personal and financial data. IoT devices amass vast data about users, which are then sent to the cloud for processing. This convenience poses a risk to privacy and security, as managing and securing these data are challenging. Cloud technology, integral to the IoT, allows efficient data handling but creates a single point of failure. Complicating matters is the need to manage data across devices and cloud platforms. As data continue to flow through the IoT, robust security measures, encryption, and access controls are crucial to mitigate these risks. • Concerns include identity protection, potential network security breaches, and possible infiltration for criminal or terrorist intentions.

Conclusion

This chapter has emphasised the importance of the internet as a vital communication tool for adventure tourism organisations at every level of sophistication. Using the internet and incorporation of different information communication technologies into adventure tourism businesses brings many challenges that include types of e-business models that work for such organisations, the use of technology, and the application of resources to gain efficiencies from the use of the internet. As media and e-commerce sources continue to collide and ultimately converge, the abundance of technology and associated tools that enable business via the internet will become increasingly entrenched. Already traditional media and businesses are scrambling to develop new revenue models and value creation through using the internet. We will continue to see a wide range of platforms that will enable us to engage with customers, members, and adventure tourism consumers to drive the business of adventure tourism.

Further research opportunities

Khatri (2018) suggests that research is needed to assess whether there are opportunities to adopt innovation and technology on the supply side too. The advent of the internet and the emergence of online portals have brought significant changes to the mass tourist scene. However, when it comes to the adventure tourism industry, the influence is more pronounced on the supply side rather than the demand side. For instance, adventure tourism businesses have harnessed the power of the internet and online platforms to showcase their unique offerings, reach a wider audience, and facilitate seamless bookings. Through visually appealing websites, social media campaigns, and interactive content, adventure tourism providers can attract adventurers from around the world and create an immersive online presence. By comparison, the impact on the demand side of adventure tourism is relatively less pronounced. While potential adventure tourists may utilise the internet to research and gather information about different destinations and activities, the actual demand for adventure tourism experiences is primarily driven by personal motivations, interests, and a desire for unique and thrilling experiences.

Therefore, while the mass tourist scene has experienced significant transformations due to the internet and online portals, the adventure tourism industry has primarily witnessed a transformation in the way adventure tourism businesses operate and market their services rather than a substantial shift in the demand for adventure tourism experiences. Access to a variety of resources, such as wide-ranging, often subjective, and constantly expanding quantity of user-generated evaluations, gives consumers leverage (Stoleriu et al., 2019). The application of technologies and innovation could widen the scope of adventure tourism. There could be many new uses and advantages of technologies in adventure tourism.

Review questions

1 What sort of disruptive technologies could adventure tourism organisations provide to benefit their consumers?
2 What barriers are impacting adventure tourism that may affect the use of information communications technology in these businesses?
3 Take an adventure activity and discuss how technology could enhance the consumer experience.

Further reading

Adventure Travel Trade Association Creating, communicating, and connecting: Technology in adventure travel
Digital Trends https://www.digitaltrends.com/?s=adventure%20tourism
Outside Magazine https://www.outsideonline.com/?s=technology
The Conversation https://theconversation.com/au/search?q=adventure+tourism+technology
There is a lack of academic articles specifically related to the use and application of technology in adventure tourism. However, some valuable topical resources are available at the following URL:
TourismTiger How technology could revolutionise your adventure tourism business.

References

Buhalis, D., Harwood, T., Bogicevic, V., Viglia, G., Beldona, S., & Hofacker, C. (2019). Technological disruptions in services: Lessons from tourism and hospitality. *Journal of Service Management, 30*(4), 484–506.
Çeltek, E. (Ed.). (2020). *Handbook of research on smart technology applications in the tourism industry*. IGI Global.
Demir, S. S., & Demir, M. (2023). Professionals' perspectives on ChatGPT in the tourism industry: Does it inspire awe or concern? *Journal of Tourism Theory and Research, 9*(2), 61–77.
Doyle, M. A., Fisher, C., Tellez, E., & Yadav, A. (2017). How Australians pay: Evidence from the 2016 Consumer Payments Survey. *Research Discussion Papers*. Reserve Bank of Australia. Retrieved from Reserve Bank of Australia: http://www.rba.gov.au/publications/rdp/2017/pdf/rdp2017-04.pdf
Fu, H., Ye, B. H., & Xiang, J. (2016). Reality TV, audience travel, intentions, and destination image. *Tourism Management, 55*, 37–48.
Gretzel, U., Sigala, M., Xiang, Z., & Koo, C. (2015). Smart tourism: Foundations and developments. *Electronic Markets, 25*, 179–188.

Hughes, K., & Moscardo, G. (2018). ICT and the future of tourist management. *Journal of Tourism Futures, 5*(3), 228–240.

Khatri, I. (2018). New trends in adventure tourism: A lesson from the 6th International Adventure Conference, 30 January- 2 February 2018, Segovia, Spain. *Journal of Tourism & Adventure 1*(1), 106–114.

Khatri, I. (2019). Information technology in the tourism & hospitality industry: A review of ten years' publications. *Journal of Tourism & Hospitality Education 9*, 74–87.

Lewis, R. (2017). The perfect partner for the travel & tourism industry. Retrieved August 22, 2022, from https://what3words.com/pt-pt/news/general/what3words-perfect-partner-travel-tourism-industry

Mine, M., van Baar, J., Grundhofer, A., Rose, D., & Yang, B. (2012). Projection-based augmented reality in Disney theme parks. *Computer, 45*(7), 32–40.

Mintel Academic (2023). *Holiday planning and booking process, UK, 2023*. Mintel Academic.

O'Beirne, C. (2018). Information and communications technology and its use in sport business. In D. Hassan (Ed.), *Managing sport business* (pp. 502–514, 2nd ed.). Routledge.

Pierdicca, R., Paolanti, M., & Frontoni, E. (2019). eTourism: ICT and its role for tourism management. *Journal of Hospitality and Tourism Technology, 10*(1), 90–106.

Salinas, S. (2018). Uber and Lyft are racing to own every mode of transportation - they're getting close. Retrieved from CNBC: Uber and Lyft are racing to own every mode of transportation (cnbc.com)

Schlagwein, D., Schoder, D., & Spindeldreher, K. (2020). Consolidated, systemic conceptualization, and definition of the sharing economy. *Journal of the Association for Information Science and Technology, 71*, 817–838.

Stoleriu, O., Brochado, A., Rusu, A., & Lupu, C. (2019). Analyses of visitors' experiences in a natural World Heritage Site based on TripAdvisor reviews. *Visitor Studies, 22*(2), 1–21.

Talha, M., & Subramaniam, M. (2003). Role of e-commerce in economic growth. *Journal of Internet Marketing, 4*(1), 1–7.

Verma, A., Shukla, V., & Sharma, R. (2021). Convergence of IOT in the tourism industry: A pragmatic analysis. *Journal of Physics*: Conference Series, Volume 1714, *2nd International Conference on Smart and Intelligent Learning for Information Optimization (CONSILIO) 2020 24–25* October 2020, Goa, India.

Xiang, Z. (2017). From digitization to the age of acceleration: On information technology and tourism. *Tourism Management Perspectives, 25*, 147–150.

Part IV

Sustainability and inclusivity

An international handbook on adventure tourism would be incomplete without consideration of sustainability and inclusivity. In this section, we bring together chapters that explore the complexity of environmental and social sustainability, drawing on global case examples and considering several of the 17 Sustainable Development Goals. It was not our intention to exclude economic sustainability; rather, contemporary scholarly work is responding to the challenges climate change presents; the absence of the social dimension in adventure tourism sustainability research and reconceptualisations of adventure tourism that focus on cultural interactions and understanding; immersion in the natural environment for enhanced wellbeing; slower forms of adventure tourism; and transformative experiences for personal growth (see Chapters 1, 9, 12, 14 & 25).

DOI: 10.4324/9781003393153-25

This part begins with a focus on environmental sustainability. Claire Backhouse and Steven Gillespie (Chapter 21) offer an examination of the impact of human-induced climate change on adventure tourism, while David Huddart (Chapter 22) and Po Yu Wang, Tamara Young and Kevin Lyons (Chapter 23) consider the impact of adventure tourism activities on sensitive natural environments. Tourist flows are highly sensitive to climate, and changes in natural resources, such as snow cover for winter sports, redirect tourist flows to alternative destinations with more favourable conditions. Likewise, negative environmental impacts caused by adventure tourism, such as footpath erosion, human waste, and wildlife disturbance are incongruent to the 'pull' of environmental quality, such as those found in national parks. These chapters explore these issues and offer practical solutions to mitigate environmental impacts.

Historically, sustainability in tourism tends to emphasise the environmental or economic aspects of sustainability and ignores its social dimensions. This absence extends to adventure tourism, raising questions of social sustainability and inclusion, which are the focus of the remaining chapters in Part IV. For example, many local communities own land within or close to protected natural areas where adventure tourism takes place and are, therefore, either directly or indirectly involved or impacted by adventure tourism. Yet, Chiedza Mutanga, Monkgogi Lenao and Joseph Mbaiwa observe that little research has focused on host community involvement in adventure tourism. In Chapter 24, they use stakeholder theory to examine different types of host community involvement in adventure tourism and their importance in fostering long-term sustainable tourism. In Chapter 25, Juha Saunavaara and Takafumi Fukuyama continue the discussion on local communities, giving specific focus to Indigenous communities, emphasising slow cultural immersion as a soft, culturally oriented form of adventure tourism that should include Indigenous past and modernity. Focus here is given to the potential for adventure tourism to restore, protect and promote Indigenous cultures and incite their agency.

Finally, this section turns to inclusive adventure tourism. Chapters 26 and 27 consider how cisheteronormative adventure tourism can be, yet it offers the opportunity to resist and escape stereotypical gender norms imposed in people's everyday lives. Specifically, Maggie Miller and Jenny Cave (Chapter 26) examine the nexus of gender and adventure across three key interrelated topics: participation and representation, embodied experiences and social sustainability and inclusion. Bart Bloem Herraiz and Elía Velo Camacho (Chapter 27) also consider the ways in which gender norms constrain women's participation but move beyond this to focus on LGBTQ+ communities and the unique conditions adventure tourism presents for them to thrive and find belonging. The persistent belief that adventure is exclusively reserved for those who are white, male, wealthy, and non-disabled can also prevent disabled travellers from participating in adventure opportunities and promote exclusionary atmospheres and attitudes. In the final chapter of this book (Chapter 28), Jasmine Goodnow and Kristen Chmielewski discuss the power adventure tourism providers have in designing products, tours, websites and marketing with universal design and accessibility in mind.

21 Adventure tourism and climate change

Claire Backhouse and Steven Gillespie

Chapter learning outcomes

1 Identify and explain the impacts of climate change on adventure destinations and activities.
2 Evaluate the impact of adventure tourism on climate change.
3 Suggest practical management solutions to mitigate and adapt to climate change.

Introduction

Global climate change presents one of the greatest challenges to the sustainability of adventure tourism. With carbon dioxide (CO_2) levels the highest in nearly 2 million years, rapid warming of the oceans, land, and atmosphere has taken place across the globe. The average land surface temperature has risen by 1.1°C relative to the pre-industrial era, with accelerated warming in recent decades (Intergovernmental Panel on Climate Change [IPCC], 2022). Exposure to changes in marine, freshwater, polar, tropical, and mountain environments and the contribution the sector makes to global greenhouse gas (GHG) emissions (UNWTO, 2021) means that adventure tourism is faced with considerable pressures (Huddart & Stott, 2019). It must adapt to the physical and socio-economic impacts of accelerating climate change, such as altered key resources, and changing patterns of demand whilst also mitigating the cost of a carbon-intensive industry.

Climate influences the relationship between tourist motivation and the appeal of adventure tourism destinations. The climate of the source market influences internal push factors, and the climate of the destination influences tourist behaviours, appeal and the viability of adventure tourism experiences, for example, snow cover for winter sports, coral reefs for diving, and predictable river discharge for water activities. From forest fires to extreme weather events and changes to snowfall, the impacts of anthropogenic climate change have been experienced globally (IPCC, 2022) and directly impact the sustainability of adventure tourism businesses.

Impacts of climate change on adventure tourism: Exposure and vulnerability

Adventure tourism is reliant on a wide range of ecosystem services (Pueyo-Ros, 2018). Aside from provisioning services, such as water and food, essential to basic needs, natural environments provide fundamental physical resources for adventure tourism, for

DOI: 10.4324/9781003393153-26

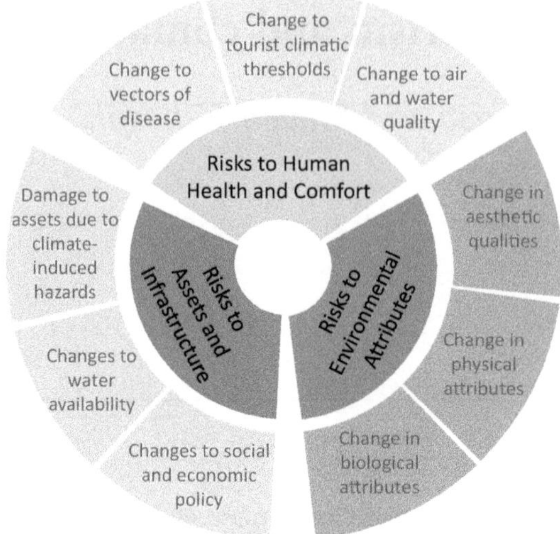

Figure 21.1 Framework of assessing climate risk in adventure tourism.

example, glaciers, rivers, snow cover, reefs, and mountains. These attributes are changing at global, regional, and localised scales (IPCC, 2022). In transitioning towards sustainability, it is imperative that potential consequences are understood, and vulnerability of adventure tourism businesses and destinations is assessed. Alteration of natural assets presents "climate risks", exposing the adventure tourism industry to potentially adverse impacts, representing the greatest challenge the industry will face over the coming decades (Scott et al., 2020). These risks can broadly be grouped into three categories which are explored in this chapter (Figure 21.1): risks to changes in human health and comfort; environmental attributes; and assets and infrastructure (Arabadzhyan et al., 2020).

Changing demands

Tourist flows are highly sensitive to climate (Amelung, Nicholls & Viner, 2007), and change may have knock-on implications for supply and demand. Indeed, by examining the optimum climatic conditions for different adventure activities and applying the IPCC's (2022) climate projection scenarios based on possible future emission pathways, the appeal of destinations and change in demand for activities can be speculated (Askew & Bowker, 2018).

Optimal parameters of climatic comfort exist for different activities. For example, the optimum temperature range for a hiking holiday is thought to be 4–7 °C lower than a beach holiday (Steiger, Abegg & Jänicke, 2016). However, human comfort is based on a greater range of factors than average daily temperature. Maximum and minimum temperatures, sunshine hours, precipitation rates and type, humidity, and wind speed can influence destination choice and the appeal of adventurous activities. Furthermore, there are thresholds beyond which certain activities begin or cease to be attractive, for example, unreliable snow conditions for winter sports (Schneider & Mücke, 2021).

Applying tourism climatic indices, the spatial redistribution of climate resources for tourism becomes clearer. Destinations in higher latitudes and at higher elevations could benefit from improved climatic variables, whereas conditions for general tourist activity may decrease in destinations closer to the equator during peak summer months (Amelung et al., 2007). Whilst heat stress negatively impacts the appeal of destinations such as the southern Iberian Peninsula, coastal Italy, Greece, and others in the peak summer months, these destinations have the potential to benefit from more favourable conditions earlier and later in the year (Grillakis, Koutroulis, Seiradakis, & Tsanis, 2016).

Based uniquely on climate variables, many summer mountain destinations may have improved climatic conditions for summer adventure tourism (Steiger, Knowles, Pöll & Rutty, 2022). Studies focusing on North America and the European Alps found that, in all IPCC GHG scenarios, the number of sunshine days at higher altitudes with thermally comfortable conditions for outdoor recreation activities would increase, coupled with a reduction in the number of days with precipitation (Pröbstl-Haider, Hödl, Ginner & Borgwardt, 2021; Scott et al., 2007). Such climatic changes are likely to increase the appeal of hiking and cycling (Schneider & Mücke, 2021). Hewer, Scott, and Fenech (2016) estimate that visitor numbers could increase by 3.1% (approximately 18,500 additional visitors) for every 1 °C increase in temperature to a national park in Ontario. Urban heat stress may also play an important role in the appeal of mountains driving a rise in overnight stays at higher altitudes (Juschten et al., 2019).

Climate indices are, however, limited in their ability to anticipate changes in tourist flows (Rosselló-Nadal, 2014) and are merely indicators of climate conditions. They fail to consider the broader socio-economic, political, and environmental impacts of climate change which influence tourism supply and demand. Aside from the direct impacts of climate change on optimal weather conditions for adventure tourism, these climatic changes trigger complex impact chains (Arabadzhyan et al., 2020). Changing global and localised climatic conditions pose risks to (1) environmental services, such as water availability and food security; (2) health, such as heatwaves, emerging vectors of infectious diseases, and air and water pollution; (3) climate-induced natural hazards, such as increased flood risk; and (4) social and political stability (IPCC, 2022).

Risks to human health and comfort

Adventure tourism is particularly exposed to changes in water quality, air pollutants, temperature, disease vectors, ultraviolet (UV) radiation, allergens, and extreme weather events, as well as the risk of hazards such as wildfires and landslides. These factors introduce risks to human health and comfort (Figure 21.1) and impact destination appeal (Amelung et al., 2007).

Water quality

Given the value of bodies of water for adventure recreation such as kayaking, sailing and open water swimming, it is important to recognise the increasing risk, incidence, and impact of harmful algal blooms (HABs; Gobler, 2020). An increase in atmospheric CO_2, sunshine hours, and water temperature, along with excessive nutrient pollution, stimulates the growth of cyanobacteria (known as blue-green algae). Globally, the frequency and magnitude of algal bloom events are increasing, creating harmful toxins with implications for adventure recreation and local economies. In 2010, the Great North Swim, a large-scale open-water swimming event in the English Lake District, was cancelled due to cyanobacteria blooms, resulting in an estimated £1.5 million loss to the local economy (Copernicus Climate Change Service, 2019). As a reactive adaptation, the event has since been moved to earlier in the year to avoid the issue recurring (*ibid.*). On a larger scale, in 2011, a HAB on Lake Erie cost the local economy nearly US$71 million in lost tourism revenue (Smith et al., 2019).

Emerging infectious diseases

Further health issues (Figure 21.1) relate to distribution changes of disease vectors (organisms which carry and transmit diseases). Significant shifts are occurring in the population sizes and distribution of vectors due to precipitation changes and rising surface temperature (Müller, Reuss, Kendrovski & Montag, 2019) with important economic consequences (Arabadzhyan et al., 2020). Climate change is influencing the prevalence of several vector-borne diseases such as Zika virus, Dengue fever, malaria, Lyme disease, and leptospirosis (*ibid.*). One such vector is ticks. Milder winters and fewer frosty episodes extend the season in which ticks remain active and reproduce, generating greater densities and range potential (Jore et al., 2014). This increases risk exposure to those participating in outdoor recreation to tick-borne diseases such as Lyme disease (*Borreliosis)* and encephalitis, as well as emerging diseases (Bouchard et al., 2019).

Allergens

With milder conditions earlier in spring and changing ozone levels, the allergenic potential of pollens is likely to increasingly impact health (Damialis, Traidl-Hoffmann & Treudler, 2019). Schneider and Mücke (2021) suggest that by the end of the century, the pollen season in Germany will start 20 days earlier. Climate change is also driving spatial shifts in allergenic plant species, such as ragweed, partly responsible for an increase in the prevalence and exacerbation of respiratory diseases such as bronchial allergic asthma (*ibid.*). Few recent studies have considered the impact of air-borne allergens on tourism. However, the increase in hay fever diagnoses in the late 19th century in New York is thought to have driven tourism demand to the White Mountains in Upstate New York, where sufferers could seek refuge from their symptoms (Mittman, 2003).

Risks to the environmental attributes and infrastructure of adventure destinations

The impacts of climate change on the environmental attributes of a destination are complex, vast, and challenging to forecast with certainty due to inter-dependency on ecological, atmospheric, social, and political systems. Environmental and aesthetic attributes of a destination include resources such as the reliability of snow cover, stable water levels, and the presence of key natural assets such as mountains, beaches, glaciers and coral reefs, and biodiversity. These resources are key assets for the sustainability of adventure tourism. The impact of changes to the quality and quantity of natural resources could negatively influence the appeal and feasibility of adventurous activities (Scott et al., 2007).

Mountain-based adventure tourism

Mountain and upland environments are particularly sensitive to the impacts of a warming climate with higher rates of warming in contrast to lower elevation counterparts. Key climate-induced environmental changes to mountain environments are summarised in Table 21.1. Climate change has had considerable impacts on the cryosphere globally, such as a reduction in the quantity and duration of snow cover – particularly at lower altitudes, and permafrost melting resulting in slope instability and reduced glacial mass in many locations (Hock et al., 2019). The implications for mountain region economies are considerable.

Snow sports

Snow sports are particularly vulnerable to changes in climate and have undergone significant change in the past 30 years. Following several snow-deficient winters in the 1980s and 1990s, coupled with intensifying competition, smaller, lower altitude ski areas struggled to compete with higher elevation resorts and lacked the revenue to invest in artificial snow-making capabilities (Steiger et al., 2019). The trend in the snow sport market has been to consolidate into larger, higher altitude ski areas with far-reaching consequences on businesses and communities previously dependent on winter tourism. In the near term (2031–2050), the IPCC project suggests that snow reliability will likely be maintained in ski resorts in North America, the European Alps, the Pyrenees, and Scandinavia, especially those with extensive snow-making adaptive capabilities (Hock et al., 2019). However, under a high-emissions scenario in longer term projections (2051–2080), aside from a minority of high-elevation resorts, snow reliability is unviable for ski areas under current operating practices (Hock et al., 2019). To demonstrate the significance of the 21 venues that have hosted the Winter Olympics, it is estimated that only 10 will have a

Table 21.1 Summary of climate-induced environmental change in mountain environments in all emission scenarios

	Climate-induced impacts
Cryosphere	• Across all regions, declines in depth and duration of low-elevation snow cover. • In regions with mostly smaller glaciers and limited glacial cover (e.g., tropical Andes), glacier mass will decline, with many glaciers disappearing. • Decreasing permafrost thickness, particularly in European Alps, Scandinavia, Canada, and Asia.
Hydrosphere	• In winter, increased discharge in glacier-fed rivers due to increased rainfall at higher elevations. • In summer, greater local variation with increased discharge due to intensified glacier melt or decreased discharge due to glacier mass reduction. • Increased volume of glacier lakes.
Biosphere	• Extinctions and changes in species composition and abundance.
Natural Hazards	• Glacier retreat and permafrost thaw decreasing stability of mountain slopes, resulting in increased frequency of mass movement events such as landslides and rockfalls. • Higher frequency of wet snow avalanches. • Lower frequency of avalanches at lower elevations. • Higher frequency of wildfires at higher altitudes due to earlier snowmelt. • Higher frequency of intense rainfall events, increasing localised flood risk.

Source: Adapted from IPCC (2022).

sufficiently suitable climate within 30 years under a high-emissions future (Scott, Steiger, Rutty & Johnson, 2014).

Variable snow conditions and a shortening of the season influence customer satisfaction, demand, and competitive advantage. Prettenthaler, Kortschak, and Woess-Gallasch (2022) suggest that there will be a significant loss in overnight stays in lower elevation ski areas in Europe under a stabilised emissions scenario by 2050 as consumer demand shifts to higher elevation snow-reliable ski regions. The winners are high-elevation ski areas, north-facing slopes with lower average humidity, which can capitalise by increasing market share, particularly if they invest in adaptive measures such as efficient artificial snowmaking (Steiger & Mayer, 2008).

The question is whether snowmaking is a viable long-term or environmentally sustainable adaptation strategy. Snowmaking is climatically sensitive; it requires an air temperature of below –3 °C and a substantial volume of water, presenting a challenge to water resource management (Steiger & Mayer, 2008) and wider environmental/ecological impacts. Artificial snowmaking is currently used extensively to lengthen the season and expand skiable areas (Steiger et al., 2022). However, conditions for artificial snowmaking are becoming less reliable. Additionally, operating and environmental costs are significant (Prettenthaler et al., 2022).

Summer mountain tourism

Glaciers are important pull factors of many mountain areas with significant socio-cultural and economic value. Salim, Ravanel, Bourdeau, and Deline (2021a) estimate that glaciers bring in over one million visitors per year to the Haute-Savoie region of France. It is predicted that most glaciers in the European Alps, the Southern Andes, and the Caucasus will disappear by 2100, regardless of the emission scenario (IPCC, 2022). Rapid glacial retreat is of concern to regions with limited economic diversification options. The difficulty in accessing unstable glaciers has resulted in a reduction in guided tours (Welling & Abegg, 2021). Carbon-intensive adaptation approaches, such as helicopter excursions to access glaciers, such as the Franz Josef Glacier in New Zealand (Wilson, Espiner, Stewart & Purdie, 2014), exacerbate the climate crisis.

Retreating glaciers, thawing permafrost and less snow have led to the instability of high mountain environments and some controversial adaptations. In August 2022, following a rockfall on the Goutier route up Mont Blanc, the mayor of Saint-Gervais-les-Bains imposed a €15,000 deposit to anyone wishing to attempt the climb to cover 'rescue and funeral costs' (Giuffrida, 2022). Furthermore, some itineraries may cease entirely as destabilising moraines make routes permanently inaccessible (Salim et al., 2021a). It has also led to difficulty accessing mountain huts and the ability to store food at altitude due to rising temperatures (Mourey, Perrin-Malterre & Ravanel, 2020). Moreover, infrastructure such as footpaths and bridges are increasingly at risk from increasing precipitation intensity and erosion (Pröbstl-Haider et al., 2021). The nature of mountaineering is therefore changing rapidly. It has become a riskier, less predictable activity, and the season has become shorter and shifted to earlier in the year (Mourey et al., 2020). From the perspective of adventure-guiding companies, this changes the characteristics of the profession. Guides increasingly need to be more risk-aware and flexible with route choice and timing to adapt to a changing workspace. This could mean travelling farther, working with clients to find compromises, and diversifying into alternate income streams (*ibid.*).

Case study 21.1   Queyraft Rafting Company: Response and adaptation to climate change

Queyraft, an adventure activity provider in the Queyras region of the French Alps, illustrates the challenges and opportunities of climate change. Over the last 30 years, the company has benefitted from increasingly favourable climatic conditions in the shoulder seasons, thereby extending the length of the season for water-based activities. However, to exploit this opportunity, due to the increasing frequency of low flow conditions, the provider has had to travel further to find suitable conditions to maintain operations and consequently costs have risen. To diversify and further reduce dependency on reliable water levels, the company has invested in training and equipment to increase the portfolio of activities they offer, diversifying into via ferrata and river boarding. Wider societal pressures on energy and water supply are also impacting Queyraft. Plans are currently being considered to dam one of the rivers they use to generate hydroelectricity. This could have potentially positive economic impacts, supporting rafting activities in late summer when river levels would traditionally be lower. However, the suitability of the river for water activities is also likely to be affected by hydropeaking, where the hydroelectricity plant releases large volumes of water (Carolli, Geneletti & Zolezzi, 2017).

For water sports in upland regions, there are greater seasonal variations in the discharge of rivers, particularly glacier-fed systems (Hock et al., 2019). In winter, glacier-fed rivers have higher discharge due to increased rainfall at high altitudes instead of snowfall. In summer, following the initial snowmelt, river discharge decreases as the overall glacier mass decreases. There will also be greater summer variability in discharge due to the increasing frequency of intense rainfall events (Hock et al., 2019). In practice, this means that the fluctuation of water levels in rivers and lakes for water-based activities will impact reliability, access, and egress. This is illustrated in the example of an adventure guiding company based in the French Alps in Case study 21.1. Dedicated paddlers may be prepared to travel greater distances for good conditions, although Bristow and Jenkins (2018) found significant distance decay for recreational paddlers and spectators.

Competition for water resources may also impact adventure tourism providers and consumers. Lac de Serre Ponçon is the largest reservoir in France and the primary water supply to Aix-Marseille. A popular sailing and water sport destination, the lake has more than 100 water sports providers contributing to 12% of the region's tourist revenue (Serre Ponçon Tourist Office, 2021). However, as can be seen in Figure 21.2, following a winter of low precipitation and a summer of record-breaking temperatures and drought, by the beginning of August 2022, the lake level dropped 14m below its optimal level, severely impacting operations and rendering much of the lake-based infrastructure inoperable.

Response to risk: Adaptive capacity

The extent to which businesses and consumers can adapt to climate change is referred to as adaptive capacity. This differs from adaptation in that it relates to having both the

Figure 21.2 Lac de Serre Ponçon August 2022. Photograph by Claire Backhouse.

competencies (skills, knowledge, financial means, technology) and the willingness to employ adaptation strategies. These responses can be either reactive, transitional, transformative, or deferred (Pelling, 2010). Reactive is the response to an immediate risk, such as cancelling a ski race due to poor snow conditions. Transitional is marginal adjustment enabling short-term flexibility, such as education and training. Transformative is where fundamental changes are made to future-proof operations and systems to ensure long-term sustainability. The final response is one of deference, where risk is understood and accepted, but structural, social, and psychological constraints may postpone action. Welling and Abegg's (2021) research in glacier tourism adaption suggests that, despite first-hand experience of the implications of climate change, current strategies are predominantly focused on reactive action. An example of reactive adaptation can be found in Lac de Serre Ponçon, where rolls of carpet are laid on the lakebed to enable access for boat users to the disappearing water resources rather than investing in longer term transformative actions. This adaptation gap is the consequence of issues such as conflicting business priorities, a lack of perceived locus of control, and physical limitations of infrastructure, social norms, economic constraints, and regulatory pressures (Eisenack et al., 2014). Furthermore, from a psychological perspective, it is the decision-making process itself. The adaptive decision-making process essentially has four stages: identification of the issue, evaluation of the risk, decision and response, and, finally, measurement and feedback (Welling & Abegg, 2021).

Knowledge and understanding of environmental issues alone are rarely sufficient to result in attitudinal and behavioural change (Walker & Moscardo, 2014). Climate change is generally well understood and accepted by adventure tourism businesses (Welling & Abegg, 2021; Tapusuwan & Rongrongmuang, 2015). However, agency (the ability to take action) is dependent on behavioural beliefs (attitudes toward the behaviour such as its perceived effectiveness), normative beliefs (the perceived personal and societal pressures to perform in a certain way), and control beliefs (the perceived locus of control; Moser & Ekstrom, 2010). Despite first-hand experience of coral bleaching and the threat to their businesses, Tapusuwan and Rongrongmuang (2015) found that diving operators misunderstood timescales of climate change and lacked locus of control to adapt, resulting in inaction or maladaptation, which unintentionally increase vulnerability (Njoroge & Chemeli, 2022). Agency also needs consensus from stakeholders. Mourey et al. (2020) found that mountaineering guides struggled to respond to changing environmental conditions as many clients were inflexible and unable to perceive the risks climate change posed. Furthermore, changes are generally gradual in human timescales and easily attributed to extreme weather events instead of longer term climatic trends (Lahsen & Ribot, 2022). Daniel Gilbert's acronym PAIN is useful to understand why stakeholders often fail to take action. To implement transformative adaptation the threat must be personal, abrupt, immoral, and happening now (Marshall, 2014). Responses are therefore commonly stimulated by events which pass thresholds of acceptable change and force immediate and abrupt personal adjustment, such as climate-induced wildfires and droughts.

Types of adjustment

Adjustments have three purposes, either to modify vulnerability, potential loss, or modify the hazard itself. The aim of modifying vulnerability is to increase resilience and capacity to cope or potentially benefit from climate change. Modifying potential loss refers to reactive actions and ways in which financial burdens can be shared, such as insurance. Modifying the hazard is mitigating climate change. Hoffman et al. 2009 distinguished three types of adjustments to climate change by ski operators: technical and operational strategies, such as snowmaking; business diversification into areas less exposed to the risk of climate change, such as summer mountain biking; and strategies which aim to share the risks of financial impacts such as insurance and collaboration. Table 21.2 provides an example of adjustments Serre Chevalier, a ski resort in the Haute-Alpes region of France, has taken in response to climate risk and policy. To decrease its sensitivity to unreliable snow conditions, in addition to its snowmaking capacity, the ski area has invested and promoted a range of alternate winter and summer activities to expand its portfolio. As competition from ski areas with glacier skiing and higher altitudes increases, many resorts have rebranded as year-round destinations, taking advantage of existing infrastructure to offer mountain biking and inclusive, easily accessible activities such as zip wires. Mitigation measures include investment in renewable energy technology (Figure 21.3) and a campaign 'Tous Engages' to bring all stakeholders onboard (SCV Domaine Skiable, 2022).

Responses to climate risk: Mitigating the adventure tourism industry

The cost of adventure tourism

Adventure tourism depends on the ecological integrity of natural assets and the stability of social and political systems, and yet paradoxically contributes to tourism's 8–12% of global GHG emissions. Approximately half of tourism's annual 4.5 gigatons CO_2 is

Table 21.2 Summary of adjustments to climate change in Serre Chevalier

Types of Adjustments	*Examples of Responses*
Modify Vulnerability	
• Technological adaptations	• Investment in gondolas to reduce sensitivity to poor snow conditions and dependency on artificial snowmaking at lower altitudes • Improvement in snow-making capability
• Behavioural adaptations	• Tous Engages – Education and collective decision-making to encourage community buy-in and promote sustainability of the destination to an increasingly climate-conscious consumer group, thereby retaining competitiveness
• Strategic managerial adaptations	• Move nursery slopes to a higher altitude • Promotion and development of a broader range of winter and summer activities including a giant zip wire, via ferrata, all-terrain scooter descents, purpose-built mountain biking tracks, nature-focused activities, and branding such as bushcraft and horse-pulled skiing
Modify Potential Loss	
• Reactive	• Closure of ski runs in poor snow conditions
• Financial aid	• Insurance
• Technical aid	• Campaign for regional economic support measures
Modify the Hazard	

• Technological Mitigative measures	• Reduction in greenhouse-gas emissions
	  • Grooming currently uses 90% of the total ski area's energy consumption. Aim to replace 50% of existing grooming vehicles with electric/hydrogen hybrid vehicles. Reduce the extent and frequency ski runs are groomed.
	  • Invest in technology to develop more energy-efficient ski lifts. Run uplift at slower speeds in low demand periods. A 1-minute-longer ski lift = a 20% saving on energy
	  • Reduce dependency on snowmaking where possible by banking snow and use of snow blanket
	• Renewable energy adoption
	  • €3.5 million investment in greener energy production. Aim to produce all the ski area's energy needs from renewable sources on site. 653 m^2 of photovoltaic cells, producing 150,000 kWh/yr, plus 2 × 12m wind turbines on Col Du Prorel, with plans to increase to 30. Hydroelectricity generated via snow-making infrastructure

• Behavioural mitigative measures	• Education campaign to redefine skiers' expectations and encourage consumers to willingly adapt to ungroomed ski runs and unheated seats.

Source: SCV Domaine Skiable (2022).

Figure 21.3 Wind turbines at Col du Prorel, Serre Chevalier. Photograph by Claire Backhouse.

derived from transportation, with shopping, food, and accommodation being other major tourism emission sources (Lenzen et al., 2018). To limit global warming to the 1.5 °C Paris Agreement set by the United Nations Framework Convention on Climate Change (UNFCCC), the United Nations World Tourism Organisation (UNWTO) agreed on at COP26 (Glasgow) to cut tourism emissions by half by 2030 and achieve net zero by 2050 (UNWTO, 2021). The challenge is monumental, especially given that, despite some localised examples of good practice, the travel industry has been slow to deliver the depth and scale of mitigative measures necessary (Scott & Gössling, 2022). A 7.6% decrease per year in GHG emissions is required from 2020–2030 to meet the 1.5 °C target (Scott, 2021). Prior to the COVID-19 pandemic, the annual growth of long-haul flights was 3.3%, with greater frequency and reliance on aviation for tourism purposes (Sun, Lin & Higham, 2020). Atmospheric carbon dioxide has now reached 414.72 parts per million and the planet's natural carbon sinks can only sequester approximately half the current annual CO_2 emissions (National Oceanic and Atmospheric Administration [NOAA], 2022).

Carbon management

To abate climate change, there are essentially three approaches: conventional mitigation strategies which focus on reducing GHG emissions, carbon offsetting which aims to capture and sequester atmospheric carbon, and geoengineering techniques aimed at stabilising global temperatures (Fawzy, Osman, Doran & Rooney, 2020). To achieve a net-zero target, the industry needs to decarbonise; GHG emissions must be reduced to near-zero and residual carbon emissions sequestered by long-term offsetting strategies (UNWTO, 2021; Scott & Gössling, 2022).

Carbon management: GHG reduction

To decouple the adventure industry from a carbon-normative economy requires significant structural, behavioural, and technological change (Gössling, 2011). Baseline data measuring business carbon footprint prior to realistic GHG reduction target setting is essential, then monitoring and evaluation of effectiveness. However, according to the Adventure Travel Trade Association (ATTA; 2020), only 52% of adventure businesses have begun calculating their footprint. This is due, in part, to a research gap in detailed, reliable carbon calculations for the specific elements of operating an adventure tour (Juvan & Dolnicar, 2014). The time and cost of implementation as well as poor cognition of the process are also significant barriers. The considerable number of businesses not calculating their carbon footprint indicates science-based mitigative measures continue to be deprioritised by a large section of the adventure tourism industry.

Carbon management: Offsetting

An alternative and popular strategy for GHG reduction is to buy carbon credits to offset GHG emissions. COP26 and pressure from employees and consumers, as well as legal obligations such as the European Union's Emissions Trading System have pushed many companies to make pledges and seek to publicly redress their carbon burden through offsetting. While this can be viewed as positive climate action, arguably this is the "easy" way out for businesses where you pay to pollute. Further criticism can be levelled at the transparency and effectiveness of many offsetting schemes to deliver meaningful long-term carbon capture.

Environmentally conscious adventure tourists

In contrast to supply-driven mitigative measures, the alternative neoliberal perspective is that consumers have agency to drive sustainable climate policy. Consumers are thought to be increasingly aware of travel emissions and their purchasing power (Han, 2021). Green consumerism, where 'products and services are purchased on the basis of their pro-environmental claims' (Akenji, 2014, p. 13), empowers consumers to self-moralise consumption and take personal responsibility for actions. For adventure tourism enterprises invested in carbon management, it can offer a competitive advantage (Peattie, Peattie & Ponting, 2009). Sustainable practices are becoming of greater importance to the sector and individual travel decision-making (ATTA, 2020). However, it has become clear there is a gap between pro-environmental intentions and consumer purchase behaviour (Odou, Darke & Voisin, 2019). Environmental consciousness and a willingness to pay for carbon mitigation do not necessarily translate into action (Haugom, Malasevska, Alnes & Mydland, 2021). This incongruence is demonstrated in the poor adoption of voluntary carbon offsetting schemes for those taking flights (Gössling, 2011).

Ironically, concern for the natural environment could increase personal carbon emissions (Salim, Ravanel & Deline, 2022). Last-chance tourism, 'a niche tourist market where tourists explicitly seek vanishing landscapes, seascapes, and/or disappearing natural/social heritage' (Lemelin, Dawson, Stewart, Maher & Lueck, 2010, p. 478), has fuelled demand to remote carbon-intensive destinations such as the Great Barrier Reef, glaciers, and polar regions to experience environmental assets before they disappear (Hindley & Font, 2015). Salim, Ravanel, Deline, and Gauchon (2021b) found that a quarter of tourists to glaciers explicitly engaged in last-chance discourse. Concurrently, this poses a chance for adventure tourism to proactively engage tourists through climate education.

Making a positive impact: Fostering pro-climate behaviours

Adventure tourism can play a role in mitigating climate change by fostering pro-environmental behaviours through education. Personal engagement with landscapes threatened by climate change coupled with effective interpretation raises awareness of climate change impacts. Environmental understanding alone, as noted earlier, is insufficient to increase pro-environmental actions which are shaped by emotional connections and responses to an action; conation (perceived behavioural control); subjective social norms; and habitual practices (Han, 2021). In the context of adventure tourism, interactions with guides with effective interpretation and direct encounters with natural environments can yield deeper emotional attachment to place and nature (Sheldon, 2020). This emotional engagement leads to a deeper nature connectedness, which is highly influential in behavioural intention to preserve the natural environment (Yang, Hu, Jing & Nguyen, 2018). Along with role modelling on subjective norms (Prince, 2017), adventure guiding companies can facilitate deeper emotional connectedness to nature and influence values, perceived norms and environmental consciousness.

Moreover, if consumers experience sustainable practices and a change in normative and habitual processes as part of the adventure tourism experience, for example, tours which use public transport instead of air travel and guides who demonstrate pro-environmental behaviours, the perceived descriptive social norm of acceptable behaviours can shift (Giddy, 2016). However, the greatest impact adventure tourism can make is through climate awareness and affective engagement prior to booking (Souza-Neto, Marques, Mayer & Lohmann, 2022). By appealing to the anticipated pride (or contrasting guilt) of choosing pro-environmental products, the adventure industry has a real opportunity to nudge consumer behaviour and motivate tourist behaviours.

Conclusion

Climate change represents the greatest challenge facing contemporary societies including the adventure tourism industry. This chapter sets out some of the key challenges facing the adventure tourism industry from observed changes and threats to the workspace of businesses, possible market reactions, and the adaptation and mitigation responses required. While a comprehensive analysis of all climate change impacts on adventure tourism is beyond the scope of one chapter, it is evident from the number of observed and projected changes presented that the challenge is significant and requires attention. There are research gaps in applying climate science to the sub-sector of adventure tourism; however, there has been substantial growth in the climate change and tourism literature more broadly (Scott & Gössling, 2022).

As illustrated, the adventure tourism sector is facing significant impacts from a changing climate including shifts in consumer demand and destination appeal. Many environmental assets, such as mountains, lakes and rivers, beaches, and forests are experiencing rapid change and new threats from heavy precipitation events, vector-borne diseases, water quality and quantity, increasing "natural" hazards, reducing snowfall and many others. These serve to illustrate the vulnerability and enhanced risk of the adventure tourism industry. If the first step in addressing these challenges is acknowledging adventure tourism is changing, then planning appropriate adaptation responses to ensure the industry has a sustainable future. While a victim of climate change, adventure tourism is also a villain requiring sustained and dedicated responses to address its own contribution to GHG emissions. This will not be easy given the reliance on transport to and from the often-remote spaces in which it operates. Mitigation measures and innovative thinking will be required, as will economic resources, if the industry is to play its role in addressing climate change.

Review questions

1 Name three impacts of climate change on the attributes of adventure tourism and suggest how they impact the appeal and demand of the destination.
2 Define *adaptive capacity* and explain the structural and behavioural constraints to addressing the adaptation gap.
3 State the objectives of the tourism industry by 2050 as agreed at COP26, choose one way in which the adventure industry is mitigating GHG emissions and evaluate the strategy's impact on global emissions.
4 If you were responsible for developing and operating a long-haul adventure tourism product, what would be your priorities and key strategies to future-proof the product from climate and carbon risk, and why?

Further reading

Huddart, D., & Stott, T. (2020). Climate change and adventure tourism. In *Adventure tourism* (pp. 437–469). London: Palgrave Macmillan. http://doi.org/10.1007/978-3-030-18623-4_13
Intergovernmental Panel on Climate Change (IPCC) (2022) *Climate Change 2022: Impacts, Adaptation, and Vulnerability.* Contribution of Working Group II to the Sixth Assessment Report of the Intergovernmental Panel on Climate Change [H.-O. Pörtner, D.C. Roberts, M. Tignor, E.S. Poloczanska, K. Mintenbeck, A. Alegría, M. Craig, S. Langsdorf, S. Löschke, V. Möller, A. Okem, B. Rama (eds.)]. Cambridge University Press. In Press. Available at: https://www.ipcc.ch/report/ar6/wg2/
UNWTO (2021). Transforming Tourism for climate action, available at https://www.unwto.org/sustainable-development/climate-action

References

Akenji, L. (2014). Consumer scapegoatism and limits to green consumerism. *Journal of Cleaner Production, 63,* 13–23. https://doi.org/10.1016/j.jclepro.2013.05.022
Amelung, B., Nicholls, S., & Viner, D. (2007). Implications of global climate change for tourism flows and seasonality. *Journal of Travel Research, 45*(3), 285–296. https://doi.org/10.1177/0047287506295937
Arabadzhyan, A., Figini, P., García, C., González, M. M., Lam-González, Y. E., & León, C. J. (2020). Climate change, coastal tourism, and impact chains – A literature review. *Current Issues in Tourism, 24*(16), 2233–2268. https://doi.org/10.1080/13683500.2020.1825351
Askew, A. E., & Bowker, J. M. (2018). Impacts of climate change on outdoor recreation participation: Outlook to 2060. *Journal of Park and Recreation Administration, 36*(2), 97–120. https://doi.org/10.18666/JPRA-2018-V36-I2-8316

ATTA (2020). The state of climate action in the adventure travel industry: A report from the Adventure Travel Association and Intrepid Travel.

Bouchard, C., Dibernardo, A., Koffi, J., Wood, H., Leighton, P. A., & Lindsay, L. R. (2019). N Increased risk of tick-borne diseases with climate and environmental changes. *Canada Communicable Disease Report, 45*(4), 83–89. https://doi.org/10.14745/ccdr.v45i04a02

Bristow, R. S., & Jenkins, I. (2018). Travel behaviour substitution for a white-water canoe race influenced by climate induced stream flow. *Leisure, 42*(1), 25–46. https://doi.org/10.1080/1492 7713.2017.1403861

Carolli, M., Geneletti, D., & Zolezzi, G. (2017). Assessing the impacts of water abstractions on river ecosystem services: An eco-hydraulic modelling approach. *Environmental Impact Assessment Review, 63*, 136–146. https://doi.org/10.1016/j.eiar.2016.12.005

Copernicus Climate Change Service (2019). Case study Report – Lake District National Park Authority, UK. Retrieved 16/09/2022, from https://climate.copernicus.eu/sites/default/files/2020-01/LNDPA_technical_report_approved.pdf.

Damialis, A., Traidl-Hoffmann, C., & Treudler, R. (2019). Climate change and pollen allergies. In Marselle, M., Stadler, J., Korn, H., Irvine, K., Bonn, A. (Eds.), *Biodiversity and health in the face of climate change* (pp. 47–66). Springer. https://doi.org/10.1007/978-3-030-02318-8_3

Eisenack, K., Moser, S. C., Hoffmann, E., Klein, R. J., Oberlack, C., Pechan, A., & Termeer, C. J. (2014). Explaining and overcoming barriers to climate change adaptation. *Nature Climate Change, 4*(10), 867–872.

Fawzy, S., Osman, A. I., Doran, J., & Rooney, D. (2020). Strategies for mitigation of climate change: A review. *Environmental Chemistry Letters, 18*, 2069–2094. https://doi.org/10.1007/s10311-020-01059-w

Giddy, J. K. (2016). Environmental values and behaviours of adventure tourism operators: The case of the Tsitsikamma, South Africa. *African Journal of Hospitality, Tourism and Leisure, 5*(4), 1–19.

Giuffrida, A. (2022, 04 August). French Mayor to ask Mont Blanc climbers for €15,000 rescue and funeral deposit. *The Guardian*, https://www.theguardian.com/world/2022/aug/04/french-mayor-mont-blanc-climbers-rescue-funeral-deposit

Gobler, C. J. (2020). Climate change and harmful algal blooms: Insights and perspective. *Harmful Algae, 91*, 101731. https://doi.org/10.1016/j.hal.2019.101731

Gössling, S. (2011). *Carbon management in tourism*. Routledge.

Grillakis, M. G., Koutroulis, A., Seiradakis, K. D., & Tsanis, I. K. (2016). Implications of 2°C global warming in European summer tourism. *Climate Services, 1*, 30–38. https://doi.org/10.1016/j.cliser.2016.01.002

Han, H. (2021). Consumer behavior and environmental sustainability in tourism and hospitality: A review of theories, concepts, and latest research. *Journal of Sustainable Tourism, 29*(7), 1021–1042. https://doi.org/10.1080/09669582.2021.1903019

Haugom, E., Malasevska, I., Alnes, P. K., & Mydland, Ø. (2021). Willingness to pay for "green skiing". *Journal of Hospitality and Tourism Management, 47*, 252–255. https://doi.org/10.1016/j.jhtm.2021.03.010

Hewer, M., Scott, D., & Fenech, A. (2016). Seasonal weather sensitivity, temperature thresholds, and climate change impacts for park visitation. *Tourism Geographies, 18*(3), 297–321. https://doi.org/10.1080/14616688.2016.1172662

Hindley, A., & Font, X. (2015). Values and motivations in tourist perceptions of last-chance tourism. *Tourism and Hospitality Research, 18*(1), 3–14. https://doi.org/10.1177/1467358415619674

Hock, R., Rasul, G., Adler, C., Kääb, S. B., Gruber, S., Hirabayashi, Y., Jackson, M., Kääb, A., Kang, S., Kutuzov, S., Milner, A., Molau, U., Morin, S., Orlove, B., Steltzer, H., & Cáceres, B. (2019). High Mountain Areas. *IPCC special report on the ocean and cryosphere in a changing climate*. 131–202.

Hoffmann, V. H., Sprengel, D. C., Ziegler, A. R., Kolb, M., & Abegg, B. (2009). Determinants of corporate adaptation to climate change in winter tourism: An econometric analysis. *Global Environmental Change-Human and Policy Dimensions, 19*, 256–264. https://doi.org/10.1016/j.gloenvcha.2008.12.002

Huddart, D., & Stott, T. (2019). Climate change and adventure tourism. In *Adventure tourism*. Palgrave Macmillan. http://doi.org/10.1007/978-3-030-18623-4_13

IPCC. (2022). Climate Change 2022: Impacts, adaptation and vulnerability. Contribution of working group II to the sixth assessment report of the intergovernmental panel on climate change [H.-O. Pörtner, D.C. Roberts, M. Tignor, E.S. Poloczanska, K. Mintenbeck, A. Alegría, M. Craig, S. Langsdorf, S. Löschke, V. Möller, A. Okem, B. Rama (eds.)]. Cambridge University Press.

Jore, S., Vanwambeke, S. O., Viljugrein, H., Isaksen, K., Kristoffersen, A. B., Woldehiwet, Z., Johansen, B., Brun, E., Brun-Hansen, H., Westermann, S., Larsen, I.-L., Ytrehus, B., & Hofshagen, M. (2014). Climate and environmental change drives *Ixodes ricinus* geographical expansion at the northern range margin. *Parasite Vectors*, 7(11). http://doi.org/10.1186/1756-3305-7-11

Juschten, M., Brandenburg, C., Hössinger, R., Liebl, U., Offenzeller, M., Prutsch, A., Unbehaun, W., Weber, F., & Jiricka-Pürrer, A. (2019) Out of the City Heat—Way to less or more sustainable futures? *Sustainability*, 11(1), 214. https://doi.org/10.3390/su11010214

Juvan, E., & Dolnicar, S. (2014). Can tourists easily choose a low carbon footprint vacation? *Journal of Sustainable Tourism*, 22(2), 175–194. https://doi.org/10.1080/09669582.2013.826230

Lahsen, M., & Ribot, J. (2022). Politics of attributing extreme events and disasters to climate change. *WIREs Climate Change*, 13(1), e750. https://doi.org/10.1002/wcc.750

Lemelin, H., Dawson, J., Stewart, E. J., Maher, P., & Lueck, M. (2010). Last-chance tourism: The boom, doom, and gloom of visiting vanishing destinations. *Current Issues in Tourism*, 13(5), 477–493. https://doi.org/10.1080/13683500903406367

Lenzen, M., Sun, Y. Y., Faturay, F. et al. (2018). The carbon footprint of global tourism. *Nature Climate Change*, 8, 522–528. https://doi.org/10.1038/s41558-018-0141-x

Marshall, G. (2014). *Don't even think about it: Why our brains are wired to ignore climate change*. Bloomsbury.

Mittman, G. (2003). Hay fever holiday: Health, leisure, and place in gilded-age America. *Bulletin of the History of Medicine*, 77(3), 600–635.

Moser, S. C., & Ekstrom, J. A. (2010). A framework to diagnose barriers to climate change adaptation. *Proceedings of the National Academy of Sciences*, 107(51). http://doi.org/10.1073/pnas.1007887107

Mourey, J., Perrin-Malterre, C., & Ravanel, L. (2020). Strategies used by French Alpine guides to adapt to the effects of climate change. *Journal of Outdoor Recreation and Tourism*, 29. http://doi.org/10.1016/j.jort.2020.100278

Müller, R., Reuss, F., Kendrovski, V., & Montag, D. (2019). Vector-borne diseases. In M. R. Marselle, J. Stadler, H. Korn, K. N. Irvine, & A. Bonn (Eds.), *Biodiversity and health in the face of climate change* (pp. 67–90). Springer International Publishing. https://doi.org/10.1007/978-3-030-02318-8_4

Njoroge, J. M., & Chemeli, A. (2022). Maladaptation. *Encyclopedia of tourism management and marketing*. Edward Elgar Publishing.

NOAA (2022). Climate change: Atmospheric carbon dioxide, Retrieved 20/07/2022, from https://www.Climate.Gov/News-Features/Understanding-Climate/Climate-Change-Atmospheric-Carbon-Dioxide

Odou, P., Darke, P., & Voisin, D. (2019). Promoting pro-environmental behaviours through induced hypocrisy. *Recherche Et Applications En Marketing (English Edition)*, 34(1), 74–90. https://doi.org/10.1177/2051570718813848

Peattie, K., Peattie, S., & Ponting, C. (2009) Climate change: A social and commercial marketing communications challenge. *EuroMed Journal of Business*, 4(3), 270–286. https://doi.org/10.1108/14502190910992693

Pelling, M. (2010). *Adaptation to climate change: From resilience to transformation*. Routledge. https://doi.org/10.4324/9780203889046

Prettenthaler, F., Kortschak, D., & Woess-Gallasch, S. (2022). Modelling tourists' responses to climate change and its effects on alpine ski tourism – A comparative approach for European Regions. *Journal of Outdoor Recreation and Tourism*, 39. http://doi.org/10.1016/j.jort.2022.100525

Prince, H. E. (2017). Outdoor experiences and sustainability. *Journal of Adventure Education and Outdoor Learning*, 17(2), 161–171. https://doi.org/10.1080/14729679.2016.1244645

Pröbstl-Haider, U., Hödl, C., Ginner, K., & Borgwardt, F. (2021). Climate change: Impacts on outdoor activities in the summer and shoulder seasons. *Journal of Outdoor Recreation and Tourism, 34*, 100344. https://doi.org/10.1016/j.jort.2020.100344

Pueyo-Ros, J. (2018). The role of tourism in the ecosystem services framework. *Land, 7*(3), 111. https://doi.org/10.3390/land7030111

Rosselló-Nadal, J. (2014). How to evaluate the effects of climate change on tourism. *Tourism Management, 42*, 334–340. https://doi.org/10.1016/j.tourman.2013.11.006

Salim, E., Ravanel, L., Bourdeau, P., & Deline, P. (2021a). Glacier tourism and climate change: Effects, adaptations, and perspectives in the Alps. *Regional Environmental Change, 21*, 120. https://doi.org/10.1007/s10113-021-01849-0

Salim, E., Ravanel, L., & Deline, P. (2022). Does witnessing the effects of climate change on glacial landscapes increase pro-environmental behaviour intentions? An empirical study of a last-chance destination. *Current Issues in Tourism*, 1–19. https://doi.org/10.1080/13683500.2022.2044291

Salim, E., Ravanel, L., Deline, P., & Gauchon, C. (2021b). A review of melting ice adaptation strategies in the glacier tourism context. *Scandinavian Journal of Hospitality and Tourism, 21*(2), 229–246. https://doi.org/10.1080/15022250.2021.1879670

Schneider, S., & Mücke, H. (2021). Sport and climate change—how will climate change affect sport? *German Journal of Exercise and Sport Research*, https://doi.org/10.1007/s12662-021-00786-8

Scott, D. (2021). Sustainable tourism and the grand challenge of climate change. *Sustainability, 13*(4), 1966. https://doi.org/10.3390/su13041966

Scott, D., & Gössling, S. (2022). A review of research into tourism and climate change - Launching the annals of tourism research curated collection on tourism and climate change. *Annals of Tourism Research, 95*(3), 1–13. https://doi.org/10.1016/j.annals.2022.103409

Scott, D., Jones, B., & Konopek, J. (2007). Implications of climate and environmental change for nature-based tourism in the Canadian Rocky Mountains: A case study of Waterton Lakes National Park. *Tourism Management, 28*(2), https://doi.org/10.1016/j.tourman.2006.04.020

Scott, D., Steiger, R., Dannevig, H., & Aall, C. (2020). Climate change and the future of the Norwegian alpine ski industry. *Current Issues in Tourism, 23*(19), https://doi.org/10.1080/1368 3500.2019.1608919

Scott, D., Steiger, R., Rutty, M., & Johnson, P. (2014). The future of the Winter Olympic Games in an era of Climate Change. *Current Issues in Tourism, 18*(10), 913–930. http://doi.org/10.1080/13683500.2014.887664

SCV Domaine Skiable (2022). Tous Engages, Retrieved 20/08/2022 from https://www.tousengages-serrechevalier.com/

Serre Ponçon Tourist Office. (2021). Serre Ponçon Press Kit, Retrieved 30/09/2022, from https://www.serreponcon.com/content/uploads/2021/08/DP-2021_EN-SP-interactif.pdf

Sheldon, P. J. (2020). Designing tourism experiences for inner transformation. *Annals of Tourism Research, 83*, 102935. http://doi.org/10.1016/j.annals.2020.102935

Smith, R. B., Bass, B., Sawyer, D., Depew, D., & Watson, S. B. (2019). Estimating the economic costs of algal blooms in the Canadian Lake Erie Basin. *Harmful Algae, 87*, 101624. http://doi.org/10.1016/j.hal.2019.101624

Souza-Neto, V., Marques, O., Mayer, V. F., & Lohmann, G. (2022). Lowering the harm of tourist activities: A systematic literature review on nudges. *Journal of Sustainable Tourism*, 1–22. http://doi.org/10.1080/09669582.2022.2036170

Steiger, R., Abegg, B., & Jänicke, L. (2016). Rain, rain, go away, come again another day: Weather preferences of summer tourists in mountain environments. *Atmosphere, 7*, 63. http://doi.org/10.3390/atmos7050063

Steiger, R., Knowles, N., Pöll, K., & Rutty, M. (2022). Impacts of climate change on mountain tourism: A review. *Journal of Sustainable Tourism*, 1–34. http://doi.org/10.1080/09669582.202 2.2112204

Steiger, R., & Mayer, M. (2008). Snowmaking and climate change, *Mountain Research and Development, 28*(3), 292–298. https://doi.org/10.1659/mrd.0978

Steiger, R., Scott, D., Abegg, B., Pons, M., & Aall, C. (2019) A critical review of climate change risk for ski tourism. *Current Issues in Tourism, 22*(11), 1343–1379. http://doi.org/10.1080/136 83500.2017.1410110

Sun, Y. Y., Lin, P. C., & Higham, J. (2020). Managing tourism emissions through optimizing the tourism demand mix: Concept and analysis. *Tourism Management, 81*, 104161. http://doi.org/10.1016/j.tourman.2020.104161

Tapusuwan, S., & Rongrongmuang, W. (2015). Climate change perception of the dive tourism industry in Koh Tao Island, Thailand. *Journal of Outdoor Recreation and Tourism, 11*, 58–63. https://doi.org/10.1016/j.jort.2015.06.005

UNWTO. (2021) *Transforming tourism for climate action.* Retrieved 06/07/2022, from https://www.unwto.org/sustainable-development/climate-action

Walker, K., & Moscardo, G. (2014). Encouraging sustainability beyond the tourist experience: Ecotourism interpretation and values, *Journal of Sustainable Tourism, 22*(8), 1175–1196. https://doi.org/10.1080/09669582.2014.918134

Welling, J., & Abegg, B. (2021). Following the ice: Adaptation processes of glacier tour operators in Southeast Iceland. *International Journal of Biometeorology, 65*(5), 703–715. https://doi.org/10.1007/s00484-019-01779-x

Wilson, J. Espiner, S. R., Stewart, E., & Purdie, H. (2014). *Last chance tourism at the Franz Josef and Fox Glaciers, Westland Tai Poutini National Park: A survey of visitor experience.* LEaP.

Yang, Y., Hu, J., Jing, F., & Nguyen, B. (2018). From awe to ecological behaviour: The mediating role of connectedness to nature. *Sustainability, 10*(7), 2477. http://doi.org/10.3390/su10072477

22 Environmental management of adventure tourism

David Huddart

Chapter learning outcomes

1 Understand that adventure tourism impacts can cause environmental damage.
2 Describe some of these impacts on the landscape, animal species, and natural ecosystems.
3 Explain how some management approaches can modify such impacts.
4 Understand that other impacts can be more important or contribute to causing environmental damage.

Introduction

The aim of this chapter is to illustrate that there are important environmental impacts caused by adventure tourism throughout the world's ecosystems. These need to be managed, environmental awareness of the tourists and their guides raised through education, and communities supported for conservation to aid ecosystem monitoring, supporting habitats and restoration.

Environmental impacts caused by adventure tourism

Environmental impacts can damage terrestrial ecosystems like caves and mountain crags, with trail degradation by hill walkers, mountain bikers, and all-terrain vehicles. Skiing, canyoning, and horse riding (see Figure 22.1) can cause environmental damage, and cavers and climbers can destroy environments which take geological time to develop and are impossible to restore. Animals such as caribou and brown bears can be impacted due to hunting, viewing, or habitat disturbance by adventure tourism. Arctic-alpine plants and lichen species can be destroyed by climbers, although the impacts of human-induced climate change are increasingly important (see Chapter 21). In the marine environment, there can be impacts on ecosystems. such as coral reefs by scuba divers and animal species like whales and walruses. Fragile ecosystems (tundra, caves, coral reefs, mountains) are most prone to damage whilst some activities have minimal impacts like aerial adventure sports, kayaking, and surfing.

DOI: 10.4324/9781003393153-27

Figure 22.1 Severely eroded trails caused by trekkers and pack animals, above Ghorepani, Annapurna, Nepal. Photograph by Carl Cater.

Examples of impacts and management related to animal species

Behavioural modification by whale watching in polar regions

Due to the timing of whale migrations, sightings increase gradually throughout the tourist season in Antarctica, and whale watching has become a popular adventure tourism activity. However, research outside the Antarctic has shown that whale-watching boats can alter whale behaviour (Williams et al., 2002). Whales increase swimming speed, make unpredictable direction changes, and take longer, steeper dives from the boats. The conclusion that a single boat can modify whale behaviour is significant because Antarctic tour operators coordinate their itineraries to avoid each other, preserving the impression of remoteness but with only one tour boat in an area there is still a potential for each to affect behaviour.

In Iceland, minke whales also changed their behaviour because of whale watching. Without boats nearby, the whales foraged by performing a series of shallow dives, followed by a longer dive. When boats were present, the inter-breath intervals and dive intervals decreased. This increased erratic movements and metabolic rates, disrupting their ability to feed successfully. However, Williams et al. (2002) and Stamation et al. (2010) have concluded that whales can habituate to boats and ignore them or voluntarily approach boats. While approaching the boats appears to be a positive response, this may disturb the whales, preventing them from engaging in other behaviours, like feeding.

Boat noise might cause problems as light and scent do not easily permeate the sea, but sound moves effectively through saltwater and whales use auditory signals to communicate, navigate, avoid predators and forage. The noise from engines and propellers interferes with communication and evokes varied responses from whales to compensate. Boat noise can mask low-frequency whale calls – especially when a boat is directly in front of a whale – and produce physiological reactions and persistent stress, which can negatively impact whales' health. However, some scientists note that noise pollution is not posing a major threat in Antarctica, although noise produced by vessels may produce reactions like those recorded in whales elsewhere, particularly if ship density and size in the region increases. Erbe (2002) indicated that a two-engine zodiac travelling at 10 km/hr, similar to those that ferry Antarctic tourists to and from landings, could potentially mask whale calls at a distance of 1 km and behavioural responses at a 50 m distance. Even low levels of ship noise may impact whales.

Increased boat traffic raises the probability of whale collisions. Since 1998, there have been one or two collisions each summer season between tour boats and whales in Antarctica, although none is known to be fatal. Whales become more susceptible to collisions when they voluntarily approach ships. Conversely, there is evidence indicating that whales cannot always detect ships or determine their positions and they have been known to turn into the paths of slow-moving ships. This is because noise radiates asymmetrically from ships and varies with depth. The acoustic signature of a vessel is louder from its side and stern than from its front. This produces a bow-null effect acoustic shadow zone and whales that are near the ships' front and close to the surface are most collision-vulnerable.

No research has found evidence that changes in long-term behaviour occur, but several observations indicate that frequent changes in short-term behaviour may be detrimental. Any time whales spend avoiding or approaching vessels prevents them from feeding, which may impact their fitness/health.

The International Association of Antarctic Tour Operators (IAATO, 2007; 2016) has tried to address these issues through guidelines for vessel and zodiac operations. These suggest that operators let the 'animals … dictate all encounters.' Vessels should post a lookout where marine mammals are known to occur so that collisions are avoided. They should not stay near whales for more than an hour and should not circle, separate, scatter, or pursue them and should approach a whale parallel, slightly to its rear and never approach head-on. They list appropriate distances that each type of vessel may approach, not closer than 30 m for small boats or 152 m for ships over 20,000 tons. In practice, approach distances are difficult to achieve or enforce as animals may behave as they wish. Nevertheless, there are potential benefits to whales through the development of ambassadors for Antarctic conservation and whale research. Ambassadors have a connection to, knowledge of, and passion for, the Antarctic as a place, space or idea. They represent and champion Antarctica and support the area through communication and behaviour (Alexander et al., 2020).

Environmental impact of snowmobiles in Svalbard

Snowmobile tours are the most popular adventure tourism winter activity in Svalbard (Figure 22.2) and noise effects on the fauna are likely to be negative. They have been documented for reindeer (Tyler, 1991) and polar bears (Andersen & Aars 2008), although evidence for reindeer is equivocal. Based on the level of snowmobile traffic in 1987, no

Figure 22.2 Snowmobiling on Svalbard. Photograph by Carl Cater.

major negative effect on the reindeer population level was detected. However, individual reindeer show clear avoidance of snowmobile trails, so further research may be needed. A polar bear study found that females with small cubs were especially disturbed, but it was not possible based on individual behaviour to draw conclusions about long-term population-level responses (Andersen & Aars, 2008). To avoid the mountainous terrain, much snowmobiling occurs on land-fast sea ice where polar bears hunt ringed seals. This stable fast ice is particularly important for females with young cubs and is used for movement between hunting habitats and denning areas.

Since disturbance studies are rarely able to assess effects on survival or reproductive success, there is a need to depend on behaviour and physiological responses as indicators. A limitation is that the effect measured on an individual has a short duration. Cumulative population-level effects are difficult to assess in most wild populations, particularly in a long-lived, highly mobile species like the polar bear.

Polar bears can run between 1 and 5 km when disturbed by snowmobiles and can leave the ringed seal breathing holes where they were hunting when vehicles approach. Repeated disturbance could result in increased energetic stress during a time when rebuilding energy stores, which is critical for cub survival. Polar bears running quickly over extended distances, particularly large individuals if pursued over a long time, can be stressed. This could force them to use sub-optimal habitats and spend more time in the water, where they typically take refuge when startled. Nevertheless, most of the denning sites are well away from the main Svalbard adventure tourism areas and in protected areas there is much more concern with chemicals accumulating in these top predators. Brominated flame retardants and polychlorinated biphenyl contamination, which may have an immunotoxic effect and a negative association between organochlorines and retinol and thyroid hormones (Andersen & Aars, 2016), are all potential problems. All accumulate in polar bear fatty tissue whilst radionuclides are contained in muscle and bone tissue. These processes, coupled with major changes in sea ice and the effect of climate change on food sources, are much more worrying for polar bear conservation than the effect of adventure tourism.

Responses of animals to people on foot versus moving in motorised vehicles can be quite different and responses of Svalbard reindeer to snowmobiles and walking humans reveals that they show greater aversion to the latter. Reindeer should not be approached too closely on foot as they need the summer months to graze, building their fat reserves for the winter. If this repeatedly happens, the animal's reserves could be not only far less than they should be, but they also habituate to humans.

An overview of adventure tourism impacts on caribou in British Columbia/Alaska

Snowmobiling

The scientific literature available on the impacts of snowmobile activity and human disturbance on caribou remains limited but suggests that the relative impacts of snowmobile activity on caribou vary with each species (barren ground and woodland), the frequency and speed of snowmobile traffic, noise levels, rate, human scent, visibility, and terrain type (open vs. forested). Relative to other winter backcountry adventure tourism, snowmobiling has the greatest perceived threat to mountain caribou, primarily because high capability, snowmobile terrain can overlap with high capability, caribou winter range, and snowmobile access and affect extensive areas of subalpine winter range. Therefore, the primary concern is related to habitat displacement from preferred late-winter foraging areas, resulting in physical condition decline due to reduced forage intake and increased energy expenditure. Habitat displacement could result in increased mortality by forcing caribou onto avalanche-susceptible, steeper terrain. Snowmobiles create hard-packed trails which allow easier access for predators like wolves and cougars to caribou foraging areas, which are usually not available to them because of the deeper snow conditions at these elevations compared to lower elevation valley bottoms. However, to what extent winter predation contributes to caribou mortality and population dynamics is unclear.

Caribou may also be disturbed while on their early-winter ranges which include mid- and lower-elevation forests as snowmobiling in these forested areas may occur as part of commercial groomed trails. However, the relative magnitude of potential impacts from snowmobiling is partly related to accessibility. Snowmobile areas occupied by caribou can be easily accessed from major highways and/or logging or mine roads, and here the caribou are most vulnerable to disturbance. There is growing demand for fresh powder snowmobiling, resulting in some transportation of snowmobiles by helicopter to alpine areas, which could have potential cumulative effects from both helicopter and snowmobile disturbance as well as from the hard-packed trails.

Heli-skiing

Huebel (2012), Wilson and Wilmshurst (2019), and Wilson (2022) have looked at the effects of heli-skiing on caribou. They have shown that the response varies according to the level of activity, species, season, quality of cover nearby, and the altitude and distance of aircraft from the animal. There is the potential for helicopters to disturb caribou, but the potential for skiers to significantly affect caribou winter habitat use is limited by the steep terrain used by heli-skiers and the spatial area used which is, limited to narrow runs. Caribou may also habituate to benign helicopter activity. Although this suggests impacts are likely localised, there is potential for greater impacts, depending on the location and frequency of use, as most heli-ski operations require between 700–3,000 km^2 of territory to operate a feasible business.

Snowcat skiing

Snowcat skiing is similar to heli-skiing except snowcats with caterpillar tracks are used to transport skiers to the top of the run. Commercial operations target open-bowl areas with deep fresh powder and gladed tree-skiing. Only about half of the caribou sub-populations occur in areas that have high potential for snowcat skiing. The relative impact may be less overall than heli-skiing because it occurs less frequently, involves less vertical skiing during a given time period (usually 5–10 runs per day), and requires less area to operate a commercial snow-cat ski operation (~30–80 km^2). However, there is potential for localised, cumulative impacts due to snowcat trails that provide easy access for snowmobilers and possibly predators.

Ski touring

Ski touring typically involves daily excursions or multi-day trips where participants stay in tents, snow caves, or backcountry cabins. Depending on how accessible the backcountry areas are, ski touring suggests low impacts on mountain caribou, and in general, this activity possesses significantly less threat than motorised activities, but caribou could be disturbed by humans on foot due to their keen sense of smell. Backcountry skiing may also have potentially greater impacts if commercial ski-touring operations (with cabins) access subalpine areas via helicopter.

Overall, the environmental impacts appear slight but potential negative effects are assumed to be greater for motorised compared to non-motorised activities and increase as the size of the affected habitat area becomes larger. It should be noted, however, that the relative importance of each activity will vary among geographic areas, sub-populations and management strategies designed to avoid and/or mitigate caribou-backcountry adventure tourism conflicts.

Environmental management of adventure tourism

The role of different stakeholders such as government agencies, national government organisations, the communities where adventure tourism takes place, and tourist organisations and operators and their employees/guides in environmental management, is crucial. Examples of strategies that are adopted to manage and minimise adverse impacts, and conserve natural ecosystems, allowing sustainable use of these areas, are illustrated. These include education of tourists, company employees/guides, on-site information such as signs, Leave No Trace and the Access Fund's Rock Project Tour information, climbing guidebooks, conservation volunteer schemes and even the provision of specific information on which sunscreens are reef-friendly, or damaging. There can be guidelines for specific activities such as mountain biking and for specific animal viewing.

Management approaches to lessen adventure tourism impacts on bears in Alaska

Public education was listed by Fortin et al. (2016) as the most effective management action to minimise impacts of adventure tourism on bears. Information and education on the impacts of careless, unskilled, and uninformed actions are much more effective than regulations in changing behaviour. Most defensive attacks result from surprise encounters involving humans hiking backcountry off-trail and in areas of natural food abundance for brown bears. Education on response during a bear encounter and the proper use of bear deterrents, such as bear-spray and where bears are likely to occur based on natural food availability, help reduce human-bear conflicts.

Proper food storage and rubbish was the second most effective management action identified. The identification and protection of prime bear habitat for feeding and travel corridors by permanent, seasonal, or daily closures of areas is needed. Campgrounds, trails, and bear-viewing sites should be outside prime bear habitat. Effective bear-viewing programmes are where bears come first in planning facilities, which should not be detrimental to the natural patterns of bear activity. The viewers' movements should be temporally and spatially predictable for the bears. Angling and bear-viewing in the same area do not mix. There should be site-specific guidelines for managing every bear-viewing area, including the behaviour of tourists, their movement, the group size allowed, and the areas that are closed to humans and rules on the minimum human-bear distance. The same approach should be taken with black bears, but education is again stressed as important in management. The wildlife professionals also need educating, and it has been suggested effective education through computerised management simulations, allowing stakeholders and wildlife professionals the opportunity to discuss and learn about management problems and interactions with black bears, is sensible.

Other educational initiatives

There can be memorandums of understanding or agreements and management plans for specific activities such as climbing, caving and canyoning, conferences such as the Adventure Travel World Summit and Adventure Elevate, conservation codes for caving, instructor certification schemes with examples from caving and canyoning, international standards such as ISO standards e.g., ISO 20611, which is good practices for sustainability. There are generally useful codes such as the Global Code of Ethics for Tourism and those in specific areas, such as the Minimal Impact Code for Upper Mustang and the Kathmandu Declaration and Nanda Declaration in the Himalayas. Citizen scientists and early career scientists can aid in ecosystem monitoring, such as the Whales of Guerrero project in Mexico (https://www.whalesinmexico.com).

Other management approaches

Use levels can be managed as in canyons and caves where numbers can be controlled through booking systems, seasonal restrictions in caves and climbing crags, and access limitation through physical barriers. By comparison, clubs can organise restoration and clean-up projects as in canyons and through cave adoption schemes.

Techniques for more resistant footpath surfaces

Resistance can be improved by stonework (see Figure 22.3), aggregates, boardwalks, or adding drainage features like water bars, drains, and treads that dip away from the slope and ditches. Geotextiles, chemical binders, surface glues, and surface moulding have all been used to improve path resistance to trampling pressure. Given time, almost any bare soil surface will be re-vegetated by natural colonisation. Studies of bulldozed Cairngorm hill roads show that even at 1000 m, sites are re-colonised within 5 years. Even these rates are unacceptably slow for most recreation sites, where the aim is to reinstate bare or damaged surfaces rapidly to minimise erosion and visual intrusion. In the past, it has been acceptable to use any kind of vegetation cover, but now management demands higher re-vegetation standards, with native species and landscaping schemes blending eroded areas to the surrounding ground. Transplanting and seeding are widely used.

Overall, site management actions should remain as unnoticeable as possible and should be visually and ecologically less obtrusive than the walkers' impacts that prompted the remedial action.

Coral restoration

Coral restoration can range from simple growing, gardening and planting out to harvesting naturally produced eggs and sperm to create new genetic individuals. The National Oceanic and Atmospheric Administration Restoration Centre in the US works to help corals recover and includes planting nursery-grown coral back onto reefs, making sure habitats are suitable for natural coral growth and building coral resilience to threats not

Figure 22.3 Stone pitching, Yosemite National Park, California. Photograph by David Huddart.

only from scuba divers but climate change too. Examples of their work can be seen in Florida, Puerto Rico, and Hawaii.

Research and education on the possible effects of sunscreens and insect repellents on coral growth

Coral bleaching has increased dramatically, associated with positive temperature anomalies, excess ultraviolet radiation, or altered available photosynthetic radiation and the presence of bacterial pathogens and pollutants, such as insect repellent use and cosmetic sun products. These have potentially important consequences on environmental contamination. The release of these products is linked with the rapid expansion of adventure tourism, particularly in the tropics. Hence, the use of sunscreen products is now banned in a few popular tourist destinations, like marine eco-parks in Mexico.

To evaluate the potential impact of sunscreen ingredients on hard corals and their symbiotic algae, Danovaro et al. (2008) conducted *in situ* studies, adding different concentrations of sunscreens to different coral species. Different sunscreen brands, protective factors, and concentrations were compared, but all treatments caused bleaching of hard corals. Untreated coral branches used as controls showed no change during the experiments. Bleaching was faster in systems subjected to higher temperature, suggesting synergistic effects with this variable.

According to estimates Danovaro et al. (2008) believed that up to 10% of the world's coral reefs would be threatened by sunscreen-induced coral bleaching and this is increasing with time. Downs et al. (2016) investigated the effects of oxybenzone (benzophenone-3), a sunscreen ultraviolet filter, on coral species. This showed that oxybenzone posed a hazard to coral reef conservation and threatened the resilience of coral reefs to climate change. Oxybenzone leaches the coral of its nutrients and bleaches it white, and it is thought to disrupt the development of fish and other wildlife. Now, the US National Park Service recommends using "reef-friendly" sunscreens made with titanium oxide or zinc oxide or not using these products at all when recreationally diving.

Land use zonation

Appropriate land-use zonation may be necessary regionally, as in some American national parks, or locally as in the zonation of recreational walking. Direct physical protection of some areas by banning adventure tourism can occur in some national parks, nature reserves, and areas like Svalbard.

Svalbard adventure tourism zonation

It is clear that there was a need for planning related to adventure tourism in Svalbard. Management agencies needed to limit and control the use of the natural environment for conservation, although the tourist industry had requirements to operate commercially. A compromise was required as tourism was one of the pillars considered by the Norwegian government for development. In the 1982–1991 period, the latter produced documents and legislation that had implications for nature management and adventure tourism in Svalbard. This included protection of the wilderness character, the integration of economic development with environmental protection, and heeding the vulnerability of the Arctic environment. This resulted in a 1995 management plan for tourism and recreation. This plan was revised, and input obtained from tourism stakeholders based on accessibility and regional suitability for different activities and surveys from visitor groups and the local population.

The key points were that the tourism industry must be ecologically sustainable and should ensure continued Norwegian settlement. The main aim was to develop a commercially profitable tourist business through increasing tourism around Longyearbyen and extending its product offerings year-round. Marketing and education based on the high Arctic nature of Svalbard were key elements of this plan. Accordingly, tourism should be controlled and evaluated, and visitors must be informed about and encouraged to respect the special conditions. The plan covered the whole archipelago and the surrounding seas as far as four nautical miles. Its established goals for acceptable levels of use and impacts in different geographical regions structured around a zoning system, with active environmental monitoring.

The plan was based on principles from recreational planning developed particularly in North America, such as the Recreational Opportunity Spectrum and the Limits of Acceptable Change for Wilderness Planning. Five factors were used to identify, characterize, and distinguish the recreation zones in the plan, and it was based on a system of four management zones, divided into nine management areas. Each of the zones has unique goals and management strategies and each of the management areas is defined by a geographical boundary, a management zone and specific management actions. In addition to the nine management areas (see Figure 22.4), two areas close to Longyearbyen were designated, where all motorised travel is prohibited, to provide wilderness-type recreational opportunities close to the major settlement, reducing the conflict between motorised and non-motorised groups.

Two Nature Reserves were established, one in north-east Svalbard, which is a Biosphere Reserve and includes Kvitøya, Nordaustlandet, and Kong Karls Land. The latter is the most important polar bear cubbing area and there is a year-round prohibition of travel, including within 500 m offshore and flying at under 500 m. The second nature reserve is the South-East Svalbard Nature Reserve, including Edgeøya and Barentsøya. Here, there is a limit on tourist vessels, and tourism operates at low levels, with no permanent tented camps for tour operators. Moffen Nature Reserve in North-West Spitsbergen is a gravel island within the national park, established in 1983 after traffic had increased substantially. It is an important walrus hauling-out location and nesting site for birds. The protected area includes 300 m of sea stretching from the island and any of the rocks surrounding it. From 15 May to 15 September, all traffic is prohibited.

The National Parks were established in 1973 when the objective was to preserve the distinctive and largely unspoilt natural environments, securing opportunities for research, teaching, and experiencing nature. Three were established in South Spitsbergen, Northwest Spitsbergen, and the Forlandet National Park. Simple forms of adventure tourism could take place without motorised vehicles or aircraft. The two types of protected areas (reserves and parks) are characterised by size, difficult access, and minimal human impact. All outside visitors must report their travel plans to the governor before visiting and provide insurance in the case of longer expeditions. Three Outdoor Recreation Areas were established to pave the way for adventure tourism without motorised traffic in the vicinity of settlements. The Excursion Area (Management Area 10) is the primary tourism area in central Spitsbergen, including Isfjorden and the settlements of Longyearbyen and Ny-Ålesund, where future tourism development would be developed. Here, there is high accessibility, more economic activity, more traffic, and higher

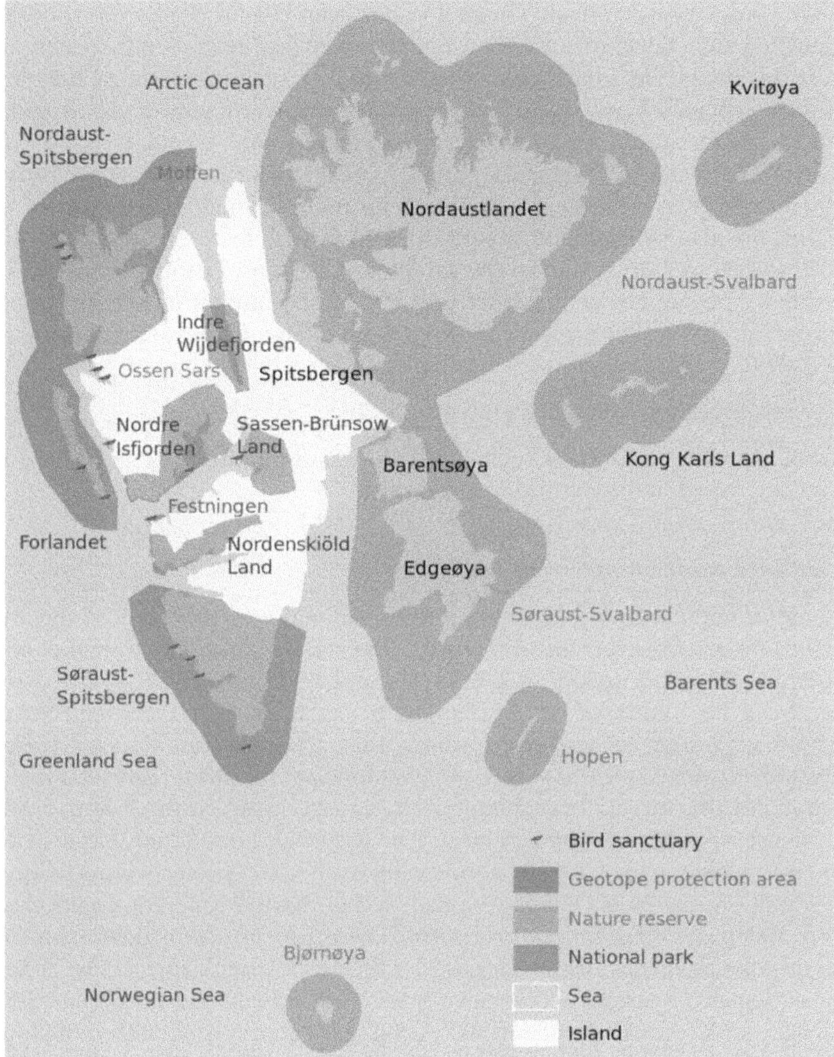

Figure 22.4 Nature conservation zones in Svalbard. Map by Arsenikk, Creative Commons.

tolerance limits. Regulations in both Outdoor Recreation and the Excursion areas are more liberal because they are not protected, and tourists need not report their travel agenda to the governor.

There are 15 special bird reserves, all small islands on the west coast, important for nesting eider, barnacle, and brent geese, and it is forbidden to visit during the nesting period. There are three plant protection areas, where it is forbidden to pick plants: the inner part of Isfjorden, (north and east of Dicksonfjorden and Sassenfjorden), an area between Colesdalen and Adventdalen, and on Ossian Sarsfjellet in the inner part of Kongsfjorden, where it is also forbidden to pitch tents.

There seems to be an acceptance of the regulations set by the government allowing tourism operators to focus on the importance that unspoiled nature plays in their product. This industry is relatively young and has been heavily regulated from the beginning, suggesting that adventure tourism management has been conducted in a holistic and professional way. The tourists also appreciate the area more because they know it is cared for and respected by the local industry and the government, and this makes them feel very exclusive because they are allowed to experience it.

Today, 65% of Svalbard's land area and 87% of its territorial waters are protected by law. All traces of human activity originating prior to 1946, including a zone of 100 m in all directions, are also protected. It is forbidden to disturb historical objects. Camping is prohibited in vegetated areas and in the safety zone around historical remains. Tourists visiting Svalbard pay an environment fee to the Svalbard Environmental Protection Fund that supports initiatives that preserve Svalbard's unique wilderness and cultural heritage.

Alternative resources for adventure tourism

These can be provided artificially such as ski slopes, climbing walls and caves, but they can also be developed in the natural environment.

Artificial reefs and marine sculpture/art trails

These are useful because they take divers away from real reefs and so limit the impacts of scuba diving. An artificial reef is a structure colonised by plant and animal communities which resemble those occurring on a natural coral reef and include scuttled ships and aircraft and can be composed of many types of material. Van Treek and Schumacher (1998; 1999) suggested on aesthetic grounds that artificial reefs do not resemble real reefs, especially during the early stages of development. Artificial reefs (eco reefs), using ceramic modules that mimic branching corals, can be installed in close-packed arrays, creating a dense, spatially complex habitat. The Reef Ball Foundation has also been successful in establishing artificial reefs, conducting over 3,500 projects worldwide, placing over half a million reef balls. These are produced in marine-concrete, engineered to last more than 500 years and cannot overturn. There are pre-formed attachment holes, accepting standardised coral fragment plugs, which have rough surfaces to enhance natural coral settlement. Recreation divers would prefer artificial reef sites relatively close to shore and accessible, which are under 30 m deep and contain sunken ship or shipwrecks. Stolk et al. (2007) found that the artificial reefs when established are as good as, or better than, natural reefs in terms of fish recruitment and overall biodiversity.

Diving trails can be established which are the only sites to be used in an area which restricts recreational diving and concentrates impacts within specific locations. An example comes from Isabel Island in the Mexican Pacific coast, where Riós-Jara et al. (2013) proposed two management strategies: the creation and use of six underwater trails and the estimation of the specific tourism carrying capacity for each trail. There needs to be adequate preventative management if the dive sites are to maintain their aesthetic appeal and biological characteristics in the future. Many other dive locations could establish similar trails. Similarly, marine sculpture and art trails have been created, which are successful.

Human waste disposal

Specific adventure tourism problems need to be managed like human waste disposal in camping, in general, or in high mountain areas, such as the Himalayas, or mountains like Mount McKinley/Denali.

Impact of climbing on Mount McKinley: Human waste

Each year, many climbers attempt Mount McKinley ascents via the West Buttress, located on the Kahiltna Glacier in Denali National Park and Preserve. With this climbing, human waste disposal has become an issue. Due to an increasing number of climbers and based on an average trip length of 18 days and an average daily stool weight of 106 g, climbers generate more than 2 metric tons of human waste annually, most of which is disposed of in crevasses. The management has been difficult, but National Park Service (NPS) Rangers dug a 3- to 4-m-deep snow pit at two heavily used camps (Camp 1 and Camp 5). Each pit was temporarily crowned with an open plywood outhouse and at each climbing season end the outhouses were removed, the waste snow-covered, abandoned, and left to be transported down-glacier. Elsewhere on the mountain, where latrine pits were not provided, climbers were required to dispose of human waste in crevasses, where the waste would be more deeply buried over time. These efforts were effective at keeping the glacier surface clean of most human waste, except above Camps 5 (4,300 m) and 6 (5,200 m), the highest sections of the climbing route, which were notoriously contaminated. Here, deep crevasses were difficult to find, especially by altitude-fatigued climbers, and surface waste disposal was still common.

In 2001, the NPS collaborated with the American Alpine Club to run a pilot programme which tested the use of a small, lightweight portable toilet, the Clean Mountain Can (CMC). By 2004, more than 500 climbers used these cans to remove their waste from the historically polluted Camp 6. The pilot programme was effective in minimising surface snow contamination and evolved into the current policy of requiring all climbers to carry and use these. However, collecting and transporting a climber's waste over the duration of an entire climbing trip was judged impractical, so climbers typically collect their waste in CMCs but are permitted to periodically empty that waste into crevasses at all but two particularly sensitive areas: on the high mountain above Camp 5 and within 0.8 km of landing strips like at Camp 1. This constitutes the current waste management plan (see Case studies 22.1 and 22.2 for further examples of addressing the human waste issue in the Himalyas).

McLaughlin (2005) suggests that faecal pathogens might be a problem. Direct faecal contamination of the water supply (melted snow) was documented by an epidemiological study on Denali when a survey of 132 climbers revealed that 39% saw faecal contamination on the snow in or near camps, 78% reported collecting snow for consumption within 10 m of camp, and 29% suffered from acute gastroenteritis within 1–21 days upon arrival on the mountain. These conditions were attributed to inadequate disposal of human waste and poor hygiene practices.

To assess potential health impacts of the waste management practices, Goodwin et al. (2012) conducted field studies and laboratory experiments to document the faecal bacterial persistence in a variety of glacial microclimates. Low concentrations of faecal bacteria found in water samples collected over two melt seasons from the Kahiltna River support the argument that bacteria can survive in a glacial environment for extended periods. Glacier surface velocities were documented and using flow models to predict the time and place that human waste will emerge in the ablation zone, it was predicted that waste buried in major camps will emerge at the surface in as little as 71 years after 28-km travel downstream. However, waste, according to this model, will reappear 30 km downstream of the burial site in less than 15 years for the case of waste buried at Camp 1 in 1954. The results show faecal microorganisms are persistent and pose a minor human health threat. Buried human waste can be expected to emerge at the glacier surface within decades and to increase with time.

Whether NPS should change its existing waste management policy is debateable. An alternative, pack out policy would have costs, including higher rates of climber non-compliance, increased education and enforcement programmes and the substantial expense, carbon footprint, and safety hazard associated with increased air traffic hauling of collected waste. Any new management policy requires careful cost considerations.

Case study 22.1   Bio-toilets in the Himalaya

The first bio-toilet was set up in Ladakh in 1994, and many more have been constructed since in high-altitude regions. The bio-friendly toilet developed by the Indian Defence Research and Development Organisation is simple: human excrement goes into the special tank below the toilet and then bacteria feed on the faeces. However, no bacteria function in cold, high-mountain conditions. This is why the company sent scientists to Antarctica under India's 13th Antarctic Mission in 1994 to look for microorganisms that break down excreta. The bacteria (*Clostridium* and *Methanosarcina*) that they found can live in cold as well as hot climates, and they feed on waste to survive. When human excreta encounter the bacteria, it is converted into methane and water through a series of steps: anaerobic digestion – hydrolysis, acidogenesis, acetogenesis, and methanogenesis.

However, most scientists, for example, Ells and Lee (2000), and the Leave No Trace Center for Outdoor Ethics (1999), believe that the problem of human faeces should be reflected in sustainable tourism rules: Pack It In–Pack It Out. All collected waste during a trip would then be transported to a disposal point by the tourists. It is difficult to achieve, due to technical transportation problems.

Case study 22.2   The Mount Everest Biogas Project

The Mount Everest Biogas Project was designed as a 100% sustainable solution to address the human waste problem (www.mteverestbiogasproject.org). It is a solar powered, biogas system which is the world's highest altitude, anaerobic digester and unique in that is fuelled by human waste. In 2016, a mini-digester was successfully lab-tested at Kathmandu University, in association with Seattle University, using human waste from Everest Base Camp. This breaks down and produces methane, and the recent final phase of testing determined if pathogens were destroyed. The process will eliminate the annual dumping of 26,000 lb of solid human waste in the vicinity of the Gorak Shep village. Additional environmental benefits are the lessened water contamination risk by faecal coliforms, the reduced reliance on burning wood or yak dung for heating and resultant health risks, and lessened deforestation of the limited wood resources. The methane, a clean fuel, is used for cooking and lighting. The pathogen-reduced or pathogen-free effluent produced is used as fertilizer for crops in the Khumbu Valley. The project won the 2017 Mountain Protection Award for its visionary solution to the decades-long impact of human waste on Everest and other remote, high-altitude locations.

Waste issues on Mount Everest

Increased littering at high altitudes has had a major negative environmental impact, including non-biodegradable rubbish (plastics, glass bottles, tins), batteries discarded along the trails, tents and oxygen bottles at campsites, outside trekking lodges, and at base camps. Inadequately covered toilet pits can be scattered around campsites and trails where toilet paper can take decades to decompose above 4000 m.

The growth in adventure tourism numbers in both mountaineering and trekking has meant that there have been projects to try and address the issue. The 1990 Everest Environmental Expedition invested 709 man-days in cleaning up rubbish on the north Tibetan side of Everest. This included rubbish hauling to a landfill site, establishing stone holding areas, metal bins and paper recycling from garbage at Rongbuk Monastery, recycling beer bottles and aluminium cans, planning follow-up expeditions and the establishment of the Everest Environmental Project.

However, the Nepalese Himalayas have been called the highest junkyard on earth and the Sagarmatha Pollution Control Committee Area was proposed by the Nepalese Mountaineering Association as part of a Clean Himalaya Campaign in 1993. The committee established clear-up campaigns, developed recycling and waste management programmes with sherpa involvement, and is committed to re-afforestation, environmental education and cultural conservation. Since 1993, there have been further waste retrieval schemes, such as the Clean Everest Project (2017), the Big Mountain Cleanup Project (2019) and those established by the Nepalese Army (2022).

In 2014, the Nepalese government passed new laws to regulate the climbing on Everest. However, these policies show that officials do not really grasp what is happening on the Himalayan peaks and more must be done to make the mountains sustainable. The first new rule that climbers must bring back at least 8kg of rubbish with them when descending Everest was puzzling and contradicted the second policy, which was reduced mountaineering fees in 2014 to attract even more climbers, although this has now changed.

The government of Nepal and Everest expedition organisers have launched further clean-up operations after several deadly avalanches. Sherpas and other climbers have been given canvas bags, each holding 80kg of waste to place at different elevations. This includes camp two, a large campsite established in 2014, after an avalanche killed 16 guides. This led to the cancellation of the climbing season, and the following year's climbs were also axed after an earthquake-triggered avalanche swept through south base camp, killing 19 people. The tents and supplies left by rescued climbers needed to be removed. Once full, the bags are helicoptered down the mountain. Removing the sacks by air means sherpa guides do not have to carry heavy waste loads through the Khumbu icefall to the base camp. It is also cheap as the operation uses helicopters that would ordinarily return empty after dumping climbing ropes on the site. Recreational climbers are urged to pick up rubbish along their route, while sherpas who carry equipment up the mountain for their clients will be paid extra to return with rubbish bags.

Community organisation of adventure tourism

This can be important, for example, where the Khanchenzonga Conservation Committee in the Himalayas or the Yukon Ecotourism Service Providers Association provide sustainable livelihoods for Indigenous populations.

Banning of tourism and the creation of Eco Development Committees

Uncontrolled mountaineering activities prior to 1983 on Nanda Devi resulted in the destruction of the biological resources by wild animal poaching, tree-felling by expedition parties, collection of medicinal herbs and rubbish accumulation. To curb this, the Nanda Devi National Park (NDNP) was established, and adventure tourism was banned in 1982. In 1988, the Nanda Devi Biosphere Reserve (NDBR) with an inner core zone (NDNP), surrounded by a buffer zone, was designated. A ban on tourism activities, followed by the reserve designation, helped improve forest cover and density. The status of wild animals, including rare, endangered species like musk deer and blue sheep, was an improvement indicator. From a socio-economic viewpoint, the income loss from adventure tourism forced locals to migrate. In the absence of alternative income sources, marginal agriculture and animal husbandry became the major sources of local income. Nevertheless, the low density of human population kept human pressure under control in NDBR. Promotion of eco-tourism and natural resource-based employment schemes were suggested to compensate for the economic loss and to maintain biodiversity. Improvement in the rural economy – the prime concern of local people – had not received as much attention as legal protection enforcement by the reserve management.

In 2003, the government made major revisions to the park rules. A partial reopening allowed 500 visitors per year to enter a segment of the park's core, although the peak remained off limits. The environmentally sensitive plan called for employment of local guides and porters and was an opportunity for community economic rejuvenation. The Nanda Devi Campaign then focused on capacity building and infrastructure development for community-owned ecotourism.

In 2005, the Nanda Devi Campaign obtained support from the Winterline Foundation to train local youth in mountaineering skills at the Nehru Institute of Mountaineering (NIM). They would become the major stakeholder in Mountain Shepherds, a new community-owned and operated tourism company. Representing the future of their communities and able to take advantage of new opportunities made available by new skills, the NIM training was seen as providing higher paid work opportunities. This allowed them to take greater advantage of adventure tourism that had thus far relegated them to a support service role in the land they knew best, without migration. With the launch of the Mountain Shepherds Initiative, the Nanda Devi Campaign attempted to establish a community-owned operation without human exploitation and environmental degradation.

Eco Development Committees have been established with major aims to minimise the negative impacts of tourism, ensure effective community participation in tourism planning, and manage tourist activities. Community participation is the key to sustainable mountain tourism.

Khangchendzonga Conservation Committee

The Khanchendzonga landscape, comprising the Sikkim and Darjeeling Himalayas, has been a major tourist attraction with mountaineers, hikers, trekkers and nature lovers visiting the Khanchendzonga National Park (KNP) due to its exceptionally high biodiversity and nine ethnic communities. Realising the potential of economic development through ecotourism, several new initiatives were started, addressing the issues of environmental and cultural conservation and emphasising the economic benefits for locals. Opening undiscovered, unspoilt, fragile environments to mass tourism can be environmentally damaging, affecting biodiversity and sustainability. The increased numbers, although economically beneficial for the villagers, caused negative impacts on trekking routes and in the national park, with waste disposal issues and large-scale deforestation through firewood used as fuel.

Unregulated tourism and economic leakage of benefits by outside companies led to income uncertainty and livelihoods and made the locals realise the importance of monitoring and regulation. Hence an active community organisation, the Khanchendzonga Conservation Committee (KCC), was established, focusing on natural resource conservation and education, income generation and rural development. This committee set out to collaborate with all levels, including locals, government agencies, and research institutes to conserve the natural, cultural and historical assets. It is involved in livelihood generation and sustainable ecotourism, such as Village Homestay projects. As of 2017, the committee has helped train approximately 400 homestay operators by selecting lower income family members, who can benefit from additional income, and women for whom financial empowerment would be valuable. The committee handles the marketing and booking requirements and provides a range of hospitality-based training and the establishment of a minimum room tariff for homestays.

It formed a participatory monitoring programme of the trekking trail inside Khanchendzonga National Park and surrounding areas. It monitors the status of waste, the trekking trail condition and campsite facilities. Tourists, trekkers and local tourism operators are involved in the monitoring process. Help in waste collection from the trekking trails and the forests occurs and the committee operates a Waste Segregation Centre and has worked with the Forest Department to create a system by which trekking operators have to declare non-biodegradable waste products and upon return account for these products. Waste is recycled to make products that tourists can buy when visiting the committee office. Yuksam was the first village in Sikkim to ban the use of plastics in 1996, and the Government of Sikkim has since implemented this throughout the state. KCC also implemented and encouraged trekking companies to adopt a prohibition on the use of firewood during treks and trekking groups are required to carry kerosene stoves. It introduced low-impact trekking trails and widely introduced dzos in the 1990s (a hybrid between yaks and domestic cows). Surveys had shown that yaks adversely impacted grazing areas compared to dzos.

The Ecotourism Service Providers Association of Yuksam (ESPAY) is an association that the committee set up in 2004 to provide capacity building to trekking guides, dzo and yak owners, porters and other ecotourism-related professionals. By 2017, it had trained over 100 cooks, 250 naturalist guides, 1000 porters and pack animal operators. ESPAY currently has 250 members and a range of activities, including lobbying at the state level for minimum living porter wages, a community compensation scheme for loss of dzos due to trekking accidents and natural causes, and the provision of vaccinations and veterinary support for livestock.

The KCC office is also an environmental education centre that tourists are encouraged to visit to learn about KNP and low-impact tourism. In 2005–2006, KCC trained around 100 teachers in Environmental Education. With information about the wildlife that trekkers can encounter on trails to recycled trekking waste, the KCC centre remains accessible for all visitors, but given that visiting the Centre is not mandatory for trekkers, bringing in all visitors is difficult.

Enforcing and monitoring stringent trekking regulations has been challenging because of varying support from the Forest Department. Primarily, trekking occurs inside the national park and, therefore, strictly enforcing regulations on collection of wildlife species and forest products and fuelwood use is still largely within the jurisdiction of the Department's core activities. While KCC can push the department to carry out patrolling and monitoring activities, these decisions are beyond the scope of this committee's influence.

Conclusion

Adventure tourism can have a marked impact on the environment in many parts of the world, but it has been shown that sometimes the effects on individual species have been exaggerated. The aim of environmental management has been to mitigate impacts caused by adventure tourism, raise environmental awareness and educate adventure tourists and their guides, create sustainable livelihoods for the areas and people where adventure tourism takes place, increase community support for conservation, aid ecosystem monitoring and support habitat and ecosystem restoration. The biggest problems related to environmental management are a lack of meaningful research and sufficient funding to carry out the aims. There seem to be bigger problems causing environmental impacts, such as climate change and pollution.

Review questions

1 Describe the impacts of adventure tourism on animal species in polar regions.
2 Evaluate the success of various management approaches to mitigating adventure tourism impacts on the environment.

Further reading

Buckley, R. (2014). *Adventure tourism management*. Routledge.
Huddart, D., & Stott, T. (2019). *Outdoor recreation: Environmental impacts and management*, Palgrave Macmillan.
Huddart, D., & Stott, T. (2020). *Adventure tourism: Environmental impacts and management* (p. 475). Palgrave Macmillan.

References

*The last two books should be consulted for detailed references for most of the text.

Alexander, K. A., Liggett, D., Leane, E., Bailey, J. L. Nielsen, H. E. F., Brasier, M. J., & Howard, M. (2020). What and who is and Antarctic ambassador? *Polar Record, 55,* 497–506.
Andersen, M., & Aars, J. (2008). Short-term behavioural response of polar bears to disturbance by snowmobiles. *Polar Biology, 31,* 501–507.
Andersen, M., & Aars, J. (2016). Barents Sea polar bears (*Ursus maritimus*): Population biology and anthropogenic threats. *Polar Biology, 35,* 26029. http://doi.org/1--34002/polar.v.35.26021
Danovaro, R., Bongiornia, L., Corinaldesi, C., Gionvandli, D., Dantiani, E., Astolfi, P., Greci, L., & Pusceddu, A. (2008). Sunscreens cause coral bleaching by promoting viral infections. *Environmental Health Perspectives, 116,* 441–447.
Downs, C. A., Kramarsky-Winter, E., Sehal, R., Fauth, J., Knutson, S., Bronstein, O., Ciner, F. R., Jeger, R., Lichtenfeld, Y., Woodley, C. M., Pennington, P., Cadenas, K., Kushmaro, A., & Loya, Y. (2016). Toxicopathological effects of the sunscreen UV filter, Oxybenzone (Benzophenone-3) on coral planulae and cultured primary cells and its environmental contamination in Hawaii and the US Virgin Islands. *Archives of Environmental Contamination and Toxicology, 70,* 265–288.

Ells, M. D., & Lee, K. J. (2000) The fate of feces and fecal microorganisms in human waste deposited on snow and smeared on rocks in the alpine environment. *National Outdoor Leadership School, National Park Service, Leave No Trace Center for Outdoor Ethics*, Ferris State University, p. 54.

Erbe, C. (2002). Underwater noise of whale-watching boats and potential effects on killer whales (*Orcinus orca*), based on an acoustic impact model. *Marine Mammal Science, 18*, 394–418.

Fortin, J. K., Rode, K. D., Hildebrand, G. V., Wilder, J., Farley, S., Jorgensen, C., & Marcot, B. G. (2016). Impacts of human recreation on brown bears (*Ursus arctos*): A review and new management tool. *PLoS One, 11*(1), e0141983. https://doi.org/10.1371/journal.pone.0141983

Goodwin, K., Loso, M. G., & Braun, M. (2012) Glacial transport of human waste and survival of fecal bacteria on Mt. McKinley's Kuhiltna glacier, Denali National Park, Alaska. *Arctic, Antarctic and Alpine Research, 44*, 432–445.

Huebel, K. J. (2012). Assessing the impacts of heli-skiing on the behaviour and spatial distribution of mountain caribou (*Rangifer tarandus caribou*). MSc thesis, Thompson Rivers University, Kamloops, B.C.

IAATO (2007). IAATO Marine Wildlife Watching Guidelines (Whales, Dolphins, Seal and Seabirds) for Vessel and Zodiac Operators. https://www.env.go.jp/nature/nankyoku/kankyohogo/database/visit/pdf/for_vessel_and_zodiac_en.pdf

IAATO (2016). Cetacean Watching Guidelines https://.scribd.com/document/107134004/Marin-wildlife.

Leave No Trace Center for Outdoor Ethics (1999). Leave No Trace Principles. www.LNT.org.

McLaughlin, J. B. (2005). Gastroenteritis outbreak among mountaineers climbing the west buttress route of Denali - Denali National Park, Alaska. *Wilderness and Environmental Medicine, 16*, 92–96.

Rios-Jara, E., Galvan-Villa, C. M., Rodriguez-Zaragosa, F. A., Lopez-Uriate, E., & Munon-Fernandez, V. T. (2013). The tourism carrying capacity of underwater trails in Isabel Island National Park, Mexico. *Environmental Management, 52*, 335–347.

Stamation, K. A., Croft, D. B., Shaughnessy, D. B., Waples, P. D., & Briggs, S. V. (2010). Behavioral responses of humpback whales, (*Megaptera novaeanglie*) to whale-watching vessels on the southeastern coast of Australia. *Marine Mammal Science, 226*, 98–122.

Stolk, P., Markwell, K., & Jenkins, J. M. (2007). Artificial reefs and recreational scuba diving resources: A critical review of research. *Journal of Sustainable Tourism, 15*, 331–350.

Tyler, N. J. C. (1991). Short-term behavioural responses of Svalbard reindeer *Rangifer tarandus platyrhincus* to direct provocation by a snowmobile. *Biological Conservation, 56*, 179–194.

van Treek, P., & Schumacher, H. (1998). Mass diving tourism – A new dimension calls for new management approaches. *Marine Pollution Bulletin, 37*, 499–504.

van Treek, P., & Schumacher, H. (1999). Artificial reefs created by electrolysis and coral transportation: An approach ensuring the compatibility of environmental protection and diving tourism. *Estuarine, Coastal and Shelf Science, 49*, 75–81.

Williams, R., Trites, A. W., & Bain, D. E. (2002). Behavioural responses of killer whales (*Orcinus orca*) to whale-watching boats: Opportunistic observations and experimental approaches. *Journal of Zoology, 256*, 255–270.

Wilson, S. F. (2022). Effects of helicopter skiing on mountain goats and woodland caribou in British Columbia. https://doi.org/10.32942/osf.io/awqxb

Wilson, S. F., & Wilmshurst, J. F. (2019). Behavioral responses of southern mountain caribou to helicopter and skiing activities. *Rangifer, 39*, 27–42. https://doi.org/10.7557/2.39.1.4586

23 Adventure tour operators in natural areas

Unpacking values in national parks

Po-Yu Wang, Tamara Young and Kevin Lyons

Chapter learning outcomes

1 Analyse the debate surrounding adventure tourism supply in national parks.
2 Evaluate the values of adventure tour operators in national parks.
3 Identify gaps in cooperation between adventure tour operators and national park agencies.
4 Demonstrate the importance of education and communication in promoting sustainable adventure tourism in national parks.

Introduction

The previous chapters have introduced adventure tourism as a rapidly growing industry that is closely tied to natural environments, such as cliffs, rivers, canyons, and diverse fauna and flora. National parks are particularly attractive destinations for adventure tourists due to the abundance of natural features they offer (Wang & Lyons, 2012; Williams & Soutar, 2005). Despite their potential economic benefits, however, the impact of tourism on national parks is a topic of debate among scholars and policymakers (Crouch et al., 2009; Young, 2009). This chapter examines the complex relationship between adventure tourism and national parks and highlights the opportunities and challenges associated with this industry.

The rapid growth of adventure tourism in national parks has created increased pressures on the natural environment. While some argue that adventure tourism puts undue pressure on natural resources (Nepal, 2020), others contend that it can be a sustainable and responsible form of nature-based tourism that supports conservation efforts (Wang & Lyons, 2012). As a result, national park agencies are increasingly working with adventure tour operators to strike a balance between economic development and environmental protection. This chapter explores the values held by adventure tour operators in national parks and how these values can be harnessed to promote sustainable development in these sensitive natural environments.

DOI: 10.4324/9781003393153-28

Adventure tour operators in national parks

Fennell (2008, p. 112) addresses the important relationship between national parks, tourism, environmental education and conservation, noting, 'national parks are broadly mandated with the dual purposes of protecting representative natural areas of significance, and encouraging public understanding, appreciation and enjoyment'. This definition indicates that national parks are established primarily to protect and conserve the natural and cultural environment while also providing opportunities for different uses of the park, such as tourism.

National park management faces the challenge of balancing the protection of ecosystems and the natural environment with providing for a variety of recreational opportunities, particularly tourism (Hall & Frost, 2009a; Moore & Weiler, 2009). Natural areas and outdoor activities are popular tourism attractions, and national parks offer both (Hanna et al., 2019; Reinius & Fredman, 2007). Providing recreational opportunities that are compatible with environmental protection is essential to managing national parks, especially when the conservation of a natural area is the primary objective (Hall & Frost, 2009a; Newsome et al., 2002).

With growing interest in visiting national parks and recognition that environmental quality was an important issue, the idea of marketing unique tourism experiences to protected natural environments in national parks became a popular practice for the tourism industry (Holden, 2008; Wang & Lyons, 2012). Nature-based tourism subsequently became a popular tourism industry development strategy (Sharpley, 2009). National parks are particularly attractive destinations for nature-based adventure tourism because they offer a range of adventure activities, such as rock climbing, four-wheel driving, and bushwalking, in one setting (Reinius & Fredman, 2007; William & Soutar, 2005). Adventure tour operators play a crucial role as intermediaries, connecting tourists with these unique experiences in national parks (Buckley, 2000; William & Soutar, 2005).

Most definitions of adventure tourism imply that it occurs in natural outdoor settings (Giddy & Webb, 2018). Adventure tour operators package adventure activities, the scenic beauty of national parks, adventure technology, and adventure activities into a commodity to meet the demands of adventure tourists (Cloke & Perkins, 2002). However, these authors argue that 'in some ways, these activities will themselves be regarded as spoilers of unspoilt landscapes and environments' (p. 533) for the tension between maximising profits and preserving natural environments is central to any discussion of adventure tour operators in national parks. Despite claims of sustainability, adventure tourism does increasingly bring pressure to the natural environment.

The debate over the extent to which adventure tour operators should prioritise environmental protection alongside providing unique adventure experiences for tourists remains a key issue in tourism management literature. Studies have shown that tourism firms are often constrained in their ability to develop sustainable practices due to their focus on short-term financial gains (Beames et al., 2022; Hall & Gössling, 2009; Hanna et al., 2019). However, scholars also argue that knowledge sharing and long-term economic considerations can assist national park managers to protect the environment, often by developing an ethical code, while maintaining profitability (Beames et al., 2022; Herremans et al., 2005). Therefore, a challenge for adventure tour operators is to strike a balance between conserving their unique working environment and meeting the demands of their clients for unique and satisfying adventure experiences (Appiah-Opoku,

2011). These tensions are not merely technical or economic; they also involve values, norms, and beliefs.

This chapter employs a values approach as a foundation for understanding factors underpinning the behaviours and decision-making practices of adventure tour operators in national parks. The purpose is to offer recommendations that could enhance cooperation and collaboration between adventure tour operators and national park managers.

The role of values in adventure tourism research

A values approach recognises that an individual's fundamental beliefs guide conduct, and direct behaviour, conflict resolution, and decision-making (Fulton, Manfredo & Lipscomb, 1996; Van Riper & Kyle, 2014). This approach recognises that values play a role in determining preferences, choices, and actions. As Fennell (2009, p. 214) contends, 'our values push us – they motivate us to do something'. Values exist in an organised system where each value is ordered in priority with respect to other values (Fulton et al., 1996; Vaske & Donnelly, 1999). Values theory recognises that values are integrated into an individual's value system and used to maintain self-esteem and consistency in conflict situations (Rokeach, 1979; cited in Madrigal, 1995). For instance, an adventure tour operator facing a conflict between profit and conservation, for example, in a decision about taking more clients for extra money or exceeding the carrying capacity of a natural area, will prioritise values based on their hierarchy. An individual's values form a hierarchically ordered system that enables them to manage conflicts associated with different issues.

Debates concerned with the use of nature for tourism often revolve around two primary paradigms grounded in environmental ethics – anthropocentricity and biocentricity (Fennell, 2018). Anthropocentricity is a more dominant world view that regards nature as separate from humans, with humans as the dominant force in relation to the natural environment (Rottman, 2014; Steg & de Groot, 2012). This perspective considers the natural environment as a resource for human needs, including recreation and tourism-related needs (Wearing & Huyskens, 2001). Therefore, any efforts to protect the natural environment are premised on the belief that such efforts ultimately benefit human society (Sharpley, 2009; Wearing & Huyskens, 2001).

Biocentricity, on the other hand, sees nature and humanity as interconnected and of equal value. Biocentricity recognises the intrinsic value of nature to exist for its own sake (Rottman, 2014; Sharpley, 2009; Wearing & Huyskens, 2001). In relation to tourism, biocentric perspectives may question the need for the use of natural resources and challenge the purpose of their use (Gao et al., 2018; Sharpley, 2009; Van Riper & Kyle, 2014).

There can be significant variations in people's attitudes towards nature along this continuum (Gao et al., 2018). Fennell (2018) discusses these core values in relation to different tourist types seeking nature-based experiences. However, those tourists holding biocentric values are 'positively associated with ecotourism, pro-environmental attitudes, and environmental protection' (Fennell, 2018, p. 197). In comparison, those tourists holding anthropocentric (or egotistic) values are 'less interested in nature-based tourism, more interested in hedonistic-type activities, and … less supportive of the conservation and protection of the environment' (Fennell, 2018, p. 197). Biocentric and anthropocentric values are not mutually exclusive. Van Riper and Kyle (2014) assert that these values can be viewed along a continuum with biocentric viewpoints on one end and anthropocentric viewpoints on the other.

Ongoing debates concerned with the impacts of tourism activities on the natural environment are, essentially, values-based (Fennell, 2008; 2018). Understanding these values is, therefore, valuable to comprehend the behaviour of adventure tour operators in national parks. Indeed, as the demand for adventure tourism in national parks grows in popularity, the number of adventure tour operators has increased. The activities of adventure tourism can have adverse effects on the biodiversity and ecosystems of national parks; therefore, contestation regarding the presence of adventure tourism in national parks ensues. These tensions have long been a concern for national park stakeholders (Vaske & Donnelly, 1999). The values that individuals hold towards nature are crucial in shaping attitudes and decisions related to environmental concerns. This chapter now considers the values held by adventure tour operators in national parks.

The values held by adventure tour operators in national parks

Wang (2013) employed the concepts of anthropocentric and biocentric values to study adventure tour operators in national parks in New South Wales (NSW), Australia. These tour operators offered a range of adventure activities from high risk (e.g., rock climbing, whitewater rafting, and skydiving) to lower risk (e.g., bushwalking and whale watching). Wang (2013) conducted interviews with 24 adventure tour operators based in four national parks in NSW and identified several values that underpinned their adventure tourism businesses.

Anthropocentric values

Enjoyment

Enjoyment is often viewed as a construct of an individual's perceived positive emotions such as fun, joy, and pleasure (Kelly, 1983; Podilchak,1991). It is a characteristic of leisure, which is widely recognised as an important element in people's lives (Pigram & Jenkins, 2006). The importance of enjoyment cannot be overstated, particularly when it comes to the values held by adventure tour operators in national parks. Wang's (2013) research showed that adventure tour operators recognised the importance of enjoyment in various aspects, including the natural environment, adventure activities, working environments, and customers. For example, they loved being outdoors because they found the natural environment enjoyable, and this led to their passion to supply outdoor adventure activities for tourists. Furthermore, they recognised that both adventure tourism and the natural environment as ideal tools for people to gain enjoyable experiences – a strongly anthropocentric position. By prioritising enjoyment, these operators were not only able to provide an unforgettable experience for their customers, but they were also able to improve their own attitudes and behaviours towards their businesses. As such, it is clear that enjoyment is a central factor influencing the thoughts, words, and behaviours of these adventure tour operators.

Freedom

Freedom is a significant value for adventure tour operators, as it is for leisure in general (Pigram and Jenkins, 2006). Small-scale tour operators, in particular, are motivated by freedom in their desire for autonomy and control over their operations (Page, Bentley &

Walker, 2005). Wang's (2013) study found that adventure tour operators often held negative attitudes towards working for other people, which led them to pursue greater freedom through running their own businesses in their working life. Operating an adventure tourism business, therefore, became an ideal way for them to fulfil the desire for greater freedom. Through their businesses, operators can also offer activities to give their clients feelings of freedom in natural environments (see Figure 23.1). In addition, several operators even filed complaints with national park management over restricted access to natural areas and wildlife, and the desire for more freedom was evidently the foundation underpinning these complaints. Freedom, therefore, plays a crucial role in the values held by the adventure tour operators in national parks.

Pragmatism

Pragmatism, in a broad sense, is a belief that the practical consequences of a proposition determine its truth or meaning, rather than anything metaphysical (Ulrich, 2007). In entrepreneurship, pragmatism is recognised as a significant value that views entrepreneurship as a means-driven, risk-aversive circular process. This view differs from the mainstream view of entrepreneurship as a rationally planned, risk-taking and linear process of opportunity recognition and exploitation (Kraaijenbrink, 2012). Wang's (2013) study found that adventure tour operators considered being an operator as the most realistic way to fulfil their motivations, such as passion for adventure activities and

Figure 23.1 Freedom as a value expressed by adventure tour operators. Photograph by Po-Yu Wang.

affinity for nature (see also Cloutier, 2003). Moreover, success in running an adventure tourism business was achieved through an accumulation of experiences and working in a pragmatic way. The operators viewed minimum impact as an important rule for their businesses in national parks, but the strategies needed to be pragmatic within their budget. Correspondingly, many operators complained about the inability of the national park managers to realistically consider the situation of an adventure tourism business within the national park structure. These examples highlight pragmatism as an important value underpinning adventure tour operators' behaviours and attitudes.

Biocentric values

Sustainability

Sustainability is commonly defined as meeting the needs of the present generation while not compromising the ability of future generations to meet their own needs (World Commission on Environment and Development, 1987). An ecological focus on sustainability is concerned with 'maintaining the long-term viability of supporting ecosystems' (Cottrell, 2003, p. 121). Adventure tour operators in Wang's (2013) study indicated that their affinity for nature was a significant motivator for engaging in the supply of adventure tourism in national parks (see Figure 23.2). They expressed the importance of maintaining a healthy natural environment for future generations, not only for their enjoyment but also intrinsically for nature itself. Moreover, when the operators expressed their feelings towards the natural environment, the importance of environmental sustainability to their businesses was asserted, together with its significance for future generations. As

Figure 23.2 Expressing biocentric values through interpretation. Photograph by Eric Wang.

such, environmental sustainability is viewed as a core value influencing the operators' supply of adventure tourism in national parks.

Equity

Equity in a biocentric view holds that nature and humans have the same rights to exist and all beings are of equal value (Sharpley, 2009). Wang (2013) found that adventure tour operators identified themselves as stewards of the natural environment, believing that the natural environment has every right to be sustained and protected for its own sake. They also believed that future generations have equal rights to enjoy the natural environment supporting the sustainability argument outlined above. The value of equity was evident in the adventure tour operators' concerns regarding the relationship between humans and the natural environment.

Balancing anthropocentric and biocentric values in adventure tourism supply

A central theme emerges in the preceding discussion: adventure tour operators believe that the sustainability of the natural environment and the operation of their businesses should exist in harmony. While the adventure tour operators interviewed were found to hold biocentrically dominant values of sustainability and equity, they also prioritised anthropocentric values. These values influence their supply of adventure tourism in national parks, with nuanced relationships between them.

A values hierarchy (see Figure 23.3) can aid in balancing and prioritising values. As noted above, values exist in an organised system that reflects an individual's assessment of the importance of different values (Vaske & Donnelly, 1999). In other words, individuals will refer to their own value hierarchy to decide which attitudes and behaviours align with their priorities. For adventure tour operators, the prioritisation of particular values affects how they conduct their businesses and handle conflicts related to adventure tourism supply in national parks.

Wang's (2013) study found that adventure tour operators prioritised anthropocentric values as they viewed adventure tourism and the natural environment as a means of

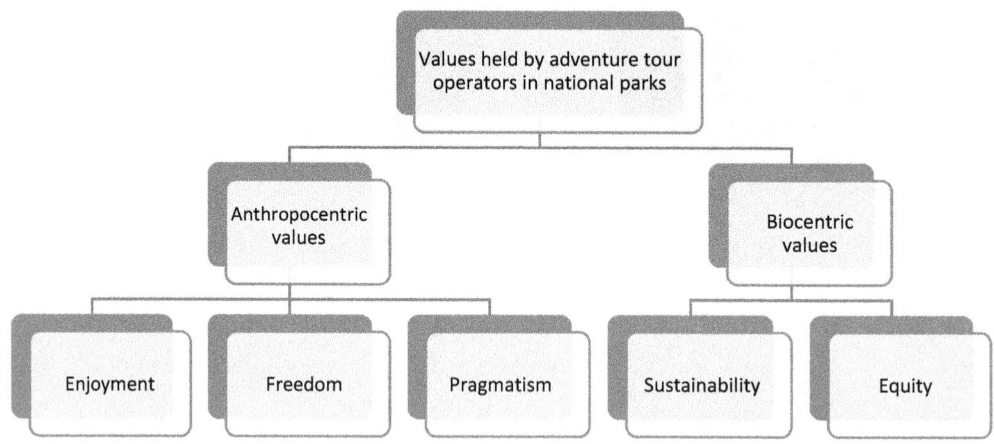

Figure 23.3 Values held by adventure tour operators in national parks.

providing enjoyable experiences for their clients and ensuring their business's success. In the operators' comments regarding access to national park areas and wildlife, the requirement for more freedom was clearly for the consideration of their clients and businesses. Although environmental sustainability was recognised as a key consideration in the supply of adventure tourism, it was secondary to their business interests. Overall, adventure tourism played a more important role than the natural environment in these operators' value systems. These tensions emerge in the coexistence of anthropocentric and biocentric values.

While anthropocentric values dominated the operators' value systems, these values were tempered by biocentric values when dealing with issues regarding adventure tourism supply in national parks. For instance, Wang (2013) argues that the operators' values relating to both the natural environment and commercial profit were not static but dynamic and integrally related. The operators recognised that their businesses and related activities are natural resource dependent, and their desire to protect the natural environment was, in fact, a desire to protect the resource on which their businesses and activities depend (Herremans et al., 2005). Therefore, when they supplied their services, they would consider their impacts on the natural environment. These findings suggest that the operators' biocentric values play an important role in "tempering" the dominance of their anthropocentric values. They demonstrate that adventure tour operators could be potential partners for the conservation of national parks.

Harris and Leiper (1995) and Stern et al. (1999) provide valuable insights into how environmental stewardship and free enterprise can coexist within a sustainable tourism framework. Harris and Leiper (1995), for example, argue that sustainable development requires environmental stewardship and a shift away from an anthropocentric focus, with the goal of preserving natural resources for future generations. As previously discussed, a healthy natural environment is essential for human survival and economic growth. Thus, the benefits of a healthy environment have motivated adventure tour operators to pursue sustainable development practices.

Stern et al. (1999) developed the value–belief–norm theory, which suggests that people's actions towards environmental protection stem from accepting particular values, believing that objects important to those values are under threat, and believing that their actions can help alleviate the threat and restore the values. In Wang's (2013) study, the adventure tour operators held both anthropocentric and biocentric values and viewed the natural environment as threatened by those who do not care about environmental protection. Thus, the operators aimed to use their businesses and their tours to aid in improving environmental protection and protecting the natural environment for personal fulfilment.

Gap between stakeholder goals and values

The protection and conservation of natural areas involves multiple stakeholders (Haukeland, 2011), with some engaging in partnerships while others experience tensions that hinder cooperation and collaboration (Jamal & Stronza, 2009). Different stakeholders may hold diverse goals and values which can lead to potential conflicts of interest (Baxter & Chippindale, 2005). For instance, in Wang's (2013) study reported earlier, adventure tour operators indicated that their values differed from those of national park managers. While national park managers were not interviewed by Wang (2013), it is evident that the adventure tour operators recognised that differences in values can make it difficult for them to collaborate effectively towards conservation goals.

According to Hall and Frost (2009b), biocentric values drive both the establishment and management of national parks. Thus, biocentricity is intrinsic to national park management and environmental codes of conduct governing managers' role in overseeing the protection of natural areas through the development and implementation of management plans (Fennell, 2018; Hall & Frost, 2009b; Wearing & Huyskens, 2001). Consequently, for some adventure tour operators in Wang's (2013) study, national park management was perceived as an obstacle to running their businesses in national parks. According to one adventure tour operator, national park managers hold fundamentally different values on the operation of adventure tourism in natural areas.

It is understandable that national park managers and private tour operators may have distinct goals and needs when addressing the intersection of conservation and tourism in national parks (Wilson, Nielsen & Buultjens, 2009). This divergence in objectives gives rise to tension and conflicts of interest between the two parties (Stevenson, Airey & Miller, 2008). The presence of contrasting fundamental values among stakeholders further complicates collaboration efforts. The primary challenge lies in whether operators perceive conservation as aligned with their own values.

Consequently, it becomes essential for national park managers to find effective strategies to foster cooperation and shared goals with operators. Encouraging operators to see the value of conservation in relation to their interests is pivotal. However, establishing partnerships with adventure tour operators for the purpose of national park conservation can be a difficult task for national park managers. To this end, we recommend education as a way to encourage change in values and practices for the purpose of enhanced cooperation and collaboration for sustainable adventure tourism futures in national parks.

Managing adventure tour operators: An uneasy task for national park management

The presence of adventure tour operators in national parks can be considered as a "double-edged sword" as it requires striking a balance that allows adventure tourism to thrive while safeguarding the ecological, cultural, and recreational values of national parks for future generations. The ongoing debate on this topic is a global concern, crucial to the wider discussion on nature-based tourism. Studies have highlighted the importance of nature-based tour operators, including adventure tour operators, but opinions differ on their suitability for the sustainable development of national parks (e.g., Fennell, 2018; Higham et al., 2016; Wilson et al., 2009). Adventure tour operators inherently generate both positive and negative impacts on national parks, making it challenging to categorise them as either facilitators or barriers to national park management. Tourism operators can offer practical ideas and activities beneficial to national park agencies, particularly when national park managers have limited knowledge of tourism management (Wang & Lyons, 2012; Wilson et al., 2009).

However, concerns persist that adventure tour operators may have limited perspectives on what is truly beneficial for the natural environment. While they may exhibit greater environmental awareness compared to mass tourism operators, their definition of conservation sometimes overlooks deeper interpretations and higher biocentric values. For example, in Wang's (2013) study, the operators frequently reflected a misunderstanding that minimal impact equated to sustainable tourism. There was also a tendency to neglect their own negative influences on the park environment, despite adventure tourism being one of the least environmentally friendly forms of nature-based tourism

(Buckley, 2000; Newsome et al., 2002). These issues can hinder the operators' potential contribution to the conservation of national parks.

Furthermore, although adventure tourism may require less infrastructure development than mass tourism, it necessitates greater interaction with the natural environment for success. When forming stronger partnerships with adventure tour operators and other nature-based tour operators in national parks, there is a risk of tour operators prioritising business interests over environmental concerns, particularly if operators are encouraged to develop tourism in those areas (Higham et al., 2016). Therefore, before including adventure tour operators as partners in national park management, it is crucial to fully understand the operators' conservation education and skills and the underlying values towards their services in national parks.

National park managers must be cautious of the operators' anthropocentric values when expanding their involvement in national park management. It is erroneous to assume that tourism operators will protect the natural environment simply because it is the foundation of their businesses (Eagles, 2009; Fennell, 2018; Wang, 2013; Wang & Lyons, 2012). These scholars assert that tourism operators are often unaware of the environmental impact of their services or possess a limited understanding of the broader implications of their small-scale and profit-driven operations on the environment.

However, this argument does not negate the potential for adventure tour operators to contribute to the conservation efforts of national parks management. Rather, we suggest the importance of bridging the gap between tour operators and park managers and establishing a stronger partnership for the management of tourism and conservation in national parks. The facilitation of such partnerships necessitates (1) developing strategies to regulate the supply of adventure tourism in national parks, (2) recognising the value of adventure tour operators in conservation efforts, (3) improving communication channels between park management and adventure tour operators, and (4) promoting attitudinal and behavioural change through education.

The roles of regulation, recognition, communication, and education in successful partnerships

Fennell (2008) argues that defining the fine line between acceptable and unacceptable human–wildlife interactions has been a difficult task for some time, particularly on issues regarding tourism in protected areas. We contend that government control of nature-based tour operators in national parks is still necessary to temper their anthropocentric values or, in case the biocentric values of operators are not able to moderate their anthropocentric values. We outline four strategies to facilitate successful partnerships between tour operators and national park managers in the context of commercial tourism and nature conservation.

First, national park management should develop specific regulations to govern the operational behaviours of nature-based adventure tour operators, ensuring their activities align with national park goals and do not exceed carrying capacities. For instance, national park agencies could provide a checklist for operators to assess if their tourism activities meet sustainability requirements, such as client numbers, littering prevention, and "leave no trace" practices. These regulations serve as a minimum requirement for commercial tourism management in national parks and promote a clearer understanding of appropriate behaviour in protected areas.

Second, raising awareness of the importance of adventure tour operators in national park management is another crucial step. Increasing the representation of nature-based tour operators in planning meetings of national park management allows their perspectives to be heard and valued, fostering a better partnership between stakeholders. Furthermore, governments should recognise the significance of these nature-based tour operators (Wilson et al., 2009). Failure to acknowledge their importance may create a widening gap between operators and national park agencies, impeding collaboration.

Third, improved communication between national park managers and tour operators is paramount. Wang (2013) found that adventure tour operators may possess misunderstandings about national park management due to a lack of communication. National park managers should actively engage operators in discussions regarding conservation and tourism in national parks. Additionally, national park managers can participate in familiarisation tours ("famil-tour") organised by operators for booking agents. Such an experience enables national park managers to gain a deeper understanding of adventure tourism products and facilitates more effective discussions on adventure tourism in national parks. Communication and recognition should be prioritised in partnerships between tour operators and national park agencies. As noted by Laing et al. (2008), while it is not necessary for partners to share the same visions all the time, they at least need to be amenable to valuing each other's perspectives and keeping communication open. Through communication, mutual respect and understanding can be facilitated and enhanced.

Finally, central to the partnership is the development of adventure tour operators' biocentric values. Scarles and Liburd (2010, p. 153) assert that 'values influence how sustainability is interpreted and implemented in different destinations'. Adventure tour operators' values significantly impact the conservation of national parks, particularly when governments seek their input in improving tourism and conservation in these areas. Although some operators may possess biocentric values and express intentions to protect the natural environment (as found by Wang, 2013), concerns remain regarding the dominance of their anthropocentric values and their level of conservation knowledge. Raising operators' biocentric values would result in greater alignment with national park agencies, a stronger commitment to environmental protection, and the preservation of ecosystems.

Education plays a pivotal role in changing the behaviours and attitudes of adventure tour operators, potentially leading to an evaluation of their biocentric values. As Fennell (2018, p. 199) points out, 'At the heart of the issue are the values we hold concerning the natural world'. National park agencies must take responsibility for educating operators, providing information on environmental conservation and sustainable tourism. Through such education, operators' behaviours and attitudes towards conservation can be transformed, fostering biocentric values within their operations. National park agencies should develop plans to educate tour operators through, for example, sustainable tourism workshops, to enhance their understanding of conservation principles and priorities and promote joint efforts in preserving national parks. It is crucial for national park agencies to consider operators' needs in providing services to customers, the environment, and their businesses. Educational programmes should be tailored to adventure tour operators' specific circumstances, taking into account their biocentric values and business operations. By providing operators with a solid understanding of conservation principles, they can make informed decisions and select strategies that align with the preservation of national park environments.

Collectively, the way national park agencies engage with adventure tour operators can significantly influence the actions of operators in national parks (see Case study 23.1). The responses of tour operators are influenced by their perceptions of environmental regulations, the regulatory agency, and the relationship between operators and regulators. National park agencies play a vital role in fostering cooperation between commercial tour operators and national park managers, ultimately fostering compatibility between tourism and conservation in national parks. By implementing strategies such as

---

Case study 23.1   Whale-watching operators in Port Stephens, New South Wales, Australia

Port Stephens-Great Lakes Marine Park is about 2.5 hours' drive northeast of Sydney. This marine park is well known for its diverse opportunities for adventure tourism and the diverse and unique marine habitat it offers. Commercial whale watching attracts thousands of tourists to the region and has placed Port Stephens on the list of "must-see" destinations for visitors to the region.

Port Stephens whale watching is considered a success story in balancing anthropocentric and biocentric values. This success has been attributed to a Code of Conduct for local operators that is designed to protect the environment that local operators depend on. The ethical code includes, for example, distance rules, time spent with the whales, how many trips per day, and how many boats in one group, among others.

The Whale Watching Code of Conduct was developed through a collaborative effort involving government agencies, local tourism operators, environmental organisations, and the broader community. The primary goal of the code of conduct is to promote responsible and sustainable whale-watching practices to protect marine life and ensure a positive and safe experience for both tourists and the whales. Central to this code was recognition that multiple values held by different stakeholders needed to be recognised through a process of shared communication and engagement.

The Code of Conduct was developed in consultations with stakeholders, including whale-watching tour operators, environmental organisations, local government agencies, and indigenous groups. These discussions helped identify concerns and objectives related to whale watching in the region while enabling shared and competing values to be clearly communicated.

Based on the thorough benchmarking, research and stakeholder values input, a code of conduct was formulated. It included guidelines on issues like approach distances, speed limits, noise, and waste management, all designed to minimize the impact on whales and their habitat. Tourism operators and their staff have been educated on the Code of Conduct to ensure its proper implementation. While authorities and environmental organisations continue to play a role in monitoring whale-watching activities to ensure compliance with the code, it is the operators themselves who are central to ensuring the code is upheld.

The Code of Conduct is a collaborative values-driven approach that balances the anthropocentric values associated with the economic benefits of whale watching with the anthropocentric value of marine life conservation.

regulation, awareness, improved communication and education, successful partnerships can be cultivated, benefitting both the adventure tour operators and the long-term conservation goals of national parks.

Conclusion

In conclusion, the management of adventure tourism in national parks presents a complex challenge that requires striking a balance between promoting tourism opportunities and conserving the ecological, cultural, and recreational values of these natural environments. This chapter has highlighted the need for partnerships between national park managers and commercial tour operators to address these challenges effectively. To facilitate such partnerships, it is crucial to develop strategies that regulate the supply of adventure tourism, recognise the value of tour operators in conservation efforts, improve communication channels, and promote attitudinal and behavioural changes through education. Additionally, fostering operators' biocentric values and raising awareness of their importance in national park management can contribute to a mutually beneficial relationship. However, it is important to acknowledge that further research and practical implementation are needed to refine these strategies and ensure their effectiveness in different national park contexts.

Review questions

1 Why can adventure tour operators be a double-edged sword in national park management?
2 What values may dominate adventure tourism businesses in national parks?
3 What approaches can be employed by national park agencies to have better cooperation with adventure tour operators?
4 How can partnerships between national park managers and adventure tour operators contribute to sustainable management?
5 What strategies can be implemented to regulate the supply of adventure tourism and ensure its compatibility with conservation goals in national parks?
6 What role does education play in promoting responsible behaviour and fostering biocentric values among adventure tour operators in national parks?

Further reading

United Nations World Tourism Organisations (UNWTO) *Ecotourism and Protected Areas* (https://www.unwto.org/sustainable-development/ecotourism-and-protected-areas). The UNWTO website offers insights into sustainable tourism practices in protected areas, including national parks.

Parks Australia *National Parks and Marine Parks* (https://parksaustralia.gov.au). This website offers information and guidelines for adventure tourism operators in Australian national parks.

Ecotourism Australia (https://www.ecotourism.org.au/). This website provides resources, certification programs, and industry news related to ecotourism and sustainable tourism practices in Australia, including national parks.

References

Appiah-Opoku, S. (2011). Using protected areas as a tool for biodiversity conservation and ecotourism: A case study of Kakum National Park in Ghana. *Society and Natural Resources*, 24(5), 500–510.

Baxter, I., & Chippindale, C. (2005). Managing Stonehenge: The tourism impact and the impact on tourism. In D. Leslie & M. Sigala (Eds.), *International cultural tourism: Management, implications and cases* (pp. 157–170). Routledge.

Beames, S., Houge Mackenzie, S., & Raymond, E. (2022). How can we adventure sustainably? A systematized review of sustainability guidance for adventure tourism operators. *Journal of Hospitality and Tourism Management, 50,* 223–231.

Buckley, R. (2000). Neat trends: Current issues in nature, eco- and adventure tourism. *International Journal of Tourism Research, 2*(6), 437–444.

Cloke, P., & Perkins, H. C. (2002). Commodification and adventure in New Zealand tourism. *Current Issues in Tourism, 5*(6), 521–549.

Cloutier, R. (2003). The business of adventure tourism. In S. Hudson (Ed.), *Sport and adventure tourism* (pp. 241–272). Haworth Hospitality Press.

Cottrell, S. P. (2003). Ecologically sustainable development. In J. M. Jenkins, & J. J. Pigram (Eds.), *Encyclopedia of leisure and outdoor recreation* (pp. 121–124). Routledge.

Crouch, D., Marson, D., Shirt, G., Tresidder, R., & Wiltshier, P. (2009). The peak district national park UK: Contemporary complexities and challenges. In W. Frost, & C.M. Hall (Eds), *Tourism and national parks: International perspectives on development, histories and change* (pp. 167–183). Routledge.

Eagles, P. F. J. (2009). Governance of recreation and tourism partnerships in parks and protected areas. *Journal of Sustainable Tourism, 17*(2), 231–248.

Fennell, D. A. (2008). *Ecotourism* (3rd ed.). Routledge.

Fennell, D. A. (2009). Ethics and tourism. In J. Tribe (ed.), *Philosophical issues in tourism* (pp. 211–226). Channel View Publications.

Fennell, D. A. (2018). *Tourism ethics* (2nd ed.). Channel View Publications.

Fulton, D. C., Manfredo, M. J., & Lipscomb, J. (1996). Wildlife value orientations: A conceptual and measurement approach. *Human Dimensions of Wildlife, 1*(2), 24–47.

Gao, J., Zhang, C., & Huang, Z. (2018). Chinese tourists' views of nature and natural landscape interpretation: a generational perspective. *Journal of Sustainable Tourism, 26*(4), 668–684.

Giddy, J. K., & Webb, N. L. (2018). The influence of the environment on adventure tourism: From motivations to experiences. *Current Issues in Tourism, 21*(18), 2124–2138.

Hall, C. M., & Frost, W. (2009a). Introduction: The making of the national parks concept. In W. Frost, & C. M. Hall (Eds.), *Tourism and national parks* (pp. 3–15). Routledge.

Hall, C. M., & Frost, W. (2009b). National parks and the 'worthless lands hypothesis' revisited. In W. Frost, & C. M. Hall (Eds.), *Tourism and national parks* (pp. 45–62). Routledge.

Hall, C. M., & Gössling, S. (2009). Global environmental change and tourism enterprise. In D. Leslie (Ed.), *Tourism enterprises and sustainable development: International perspectives on responses to the sustainability agenda* (pp. 17–35). Routledge.

Hanna, P., Wijesinghe, S., Paliatsos, I., Walker, C., Adams, A., & Kimbu, A. (2019). Active engagement with nature: Outdoor adventure tourism, sustainability and wellbeing. *Journal of Sustainable Tourism, 27*(9), 1355–1373.

Harris, R., & Leiper, N. (1995). Sustainable development and tourism: An overview. In R. Harris, & N. Leiper (Eds.), *Sustainable tourism: An Australian perspective* (pp. xvii–xxxiii). Butterworth-Heinemann.

Haukeland, J. V. (2011). Tourism stakeholders' perceptions of national park management in Norway. *Journal of Sustainable Tourism, 19*(2), 133–153.

Herremans, I. M., Reid, R. E., & Wilson, L. K. (2005). Environmental management systems (EMS) of tour operators: learning from each other. *Journal of Sustainable Tourism, 13*(4), 311–338.

Higham, J., Haukeland, J. V., Hopkins, D., Vistad, O. I., Lindberg, K., & Daugstad, K. (2016). National parks policy and planning: A comparative analysis of friluftsliv (Norway) and the dual mandate (New Zealand). *Journal of Policy Research in Tourism, Leisure and Events, 8*(2), 146–175.

Holden, A. (2008). *Environment and tourism* (2nd ed.). Routledge.

Jamal, T., & Stronza, A. (2009). Collaboration theory and tourism practice in protected areas: stakeholders, structuring and sustainability. *Journal of Sustainable Tourism, 17*(2), 169–189.

Kelly, J. (1983). *Leisure identities and interaction.* George Allen & Unwin.

Kraaijenbrink, J. (2012). The nature of the entrepreneurial process: Causation, effectuation, and pragmatism. In A. Groen, R. Oakey, P. Van Der Sijde, & G. Cook (Eds.), *New technology-based firms in the new millennium* (Vol. 9, pp. 187–199). Emerald Group Publishing Limited.

Laing, J., Wegner, A., Moore, S., Weiler, B., Pfueller, S., & Lee, D. (2008). *Understanding partnerships for protected area tourism: Learning from the literature*. CRC for Sustainable Tourism.

Madrigal, R. (1995). Personal values, traveller personality type, and leisure travel style. *Journal of Leisure Research*, 27(2), 125–137.

Moore, S. A., & Weiler, B. (2009). Tourism-protected area partnerships: Stoking the fires of innovation. *Journal of Sustainable Tourism*, 17(2), 129–132.

Nepal, S. K. (2020). Adventure travel and tourism after COVID-19 – business as usual or opportunity to reset? *Tourism Geographies*, 22(3), 646–650.

Newsome, D., Moore, S. A., & Dowling, R. K. (2002). *Natural area tourism - ecology, impacts and management*. Channel View.

Page, S. J., Bentley, T. A., & Walker, L. (2005). Scoping the nature and extent of adventure tourism operators in Scotland: How safe are they? *Tourism Management*, 26(3), 381–397.

Pigram, J. J., & Jenkins, J. M. (2006). *Outdoor recreation management* (2nd ed.). London: Routledge.

Podilchak, W. (1991). Distinctions of fun, enjoyment and leisure. *Leisure Studies*, 10(2), 133–138.

Reinius, S. W., & Fredman, P. (2007). Protected areas as attractions. *Annals of Tourism Research*, 34(4), 839–854.

Rottman, J. (2014). Breaking down biocentrism: Two distinct forms of moral concern for nature. *Frontiers in Psychology*, 20(5), 905.

Scarles, C., & Liburd, J. J. (2010). Best education network think tank IX: The importance of values in sustainable tourism. *Tourism & Hospitality Research*, 10(2), 152–155.

Sharpley, R. (2009). *Tourism development and the environment: Beyond sustainability?* Earthscan.

Steg, L., & de Groot, J. (2012). Environmental values. In S. D. Clayton (ed.), *The Oxford handbook of environmental and conservation psychology* (pp. 81–91). Oxford University Press.

Stern, P. C., Dietz, T., Abel, T., Guagnano, G. A., & Kalof, L. (1999). A value-belief-norm theory of support for social movements: The case of environmentalism. *Human Ecology Review*, 6(2), 81–97.

Stevenson, N., Airey, D., & Miller, G. (2008). Tourism policy making: the policy makers' perspectives. *Annals of Tourism Research*, 35(3), 732–750.

Ulrich, W. (2007). Philosophy for professionals: Towards critical pragmatism. *Journal of the Operational Research Society*, 58(8), 1109–1113.

Van Riper, C. J., & Kyle, G. T. (2014). Capturing multiple values of ecosystem services shaped by environmental worldviews: A spatial analysis. *Journal of Environmental Management*, 145, 374–384.

Vaske, J. J., & Donnelly, M. P. (1999). A value-attitude-behaviour model predicting wildland preservation voting intentions. *Society & Natural Resources*, 12(6), 523–537.

Wang, P. (2013). *A case study investigation of adventure tourism operators in national parks* [Unpublished PhD thesis]. University of Newcastle, Australia.

Wang, P., & Lyons, K. D. (2012). Values congruity in tourism and protected areas policy: Evidence from adventure tourism supply in New South Wales Australia. *Annals of Leisure Research*, 15(2), 188–192.

Wearing, S., & Huyskens, M. (2001). Moving on from joint management policy regimes in Australian national parks. *Current Issues in Tourism*, 4(2), 182–209.

Williams, P., & Soutar, G. (2005). Close to the "Edge": Critical issues for adventure tourism operators. *Asia Pacific Journal of Tourism Research*, 10(3), 247–261.

Wilson, E., Nielsen, N., & Buultjens, J. (2009). From lessees to partners: Exploring tourism public-private partnerships within the New South Wales National Parks and Wildlife Service. *Journal of Sustainable Tourism*, 17(2), 269–285.

World Commission on Environment and Development. (1987). *Our common future*. University Press.

Young, T. (2009). 'Welcome to aboriginal land': The Uluru-Kata Tjuta National Park. In W. Frost and C.M. Hall (eds), *Tourism and national parks: International perspectives on development, histories and change* (pp. 126–140). Routledge.

24 Host community involvement in adventure tourism

Chiedza N. Mutanga, Monkgogi Lenao and Joseph E. Mbaiwa

Chapter learning outcomes

1 Explain why host communities should and how they can be involved in adventure tourism.
2 Interrogate the different factors affecting community participation in adventure tourism.
3 Outline the importance of host community partnerships, collaborations, and networks.
4 Explain the possible challenges involved in host community participation in adventure tourism.
5 Identify sustainable and innovative ways of overcoming these challenges.

Introduction

Adventure tourism is engaging in activities in uncommon, exotic, secluded, or untamed locations. As such, adventure tourism makes use of the resources found in the natural world, which can be used for a range of activities that offer risk, challenge, sensory stimulation, novelty, and so forth. Many local communities own land that is a part of these natural areas, or they are located close to protected areas where the majority of adventure tourism activities take place, or at their borders (Dumitras et al., 2021). A community, according to Barrow and Murphree (2001), is a group of people who live in a particular location and have a common economic interest in the resources found there. They are socially connected by a shared cultural identity. Host communities in this context are communities that live close to adventure tourism destinations and are either directly or indirectly involved with or impacted by the activities connected with adventure tourism. Host communities actively participate in adventure tourism through a variety of means, such as by providing amenities like lodging and food. By partaking in ancient customs and rituals still prevalent in many communities around natural places, tourists can also interact with the local cultural values (Manea et al., 2019; Szabo et al., 2008).

Host community support is essential to the tourism sector. Therefore, the success of any plan for adventure tourism development depends on the community's successful participation (Inbakaran & Jackson, 2006). This is because the growth of tourism in the

DOI: 10.4324/9781003393153-29

host community frequently has both positive and negative consequences on it. Host community involvement refers to the development of a framework that enables the complete participation of all stakeholder groups in cooperative decision-making and joint ownership of obligations and benefits (Chiang & Huang, 2012; Li & Hunter, 2014). When the local community participates in adventure tourism, it suggests that they have the capacity to influence its direction and take part in its decision-making. Community involvement in the decision-making process is demonstrated by members of the community expressing their concerns and using their collective wisdom and experiences during the early planning stages of tourism development (Timothy & Boyd, 2003). Therefore, in order to achieve sustainable development, comprehending the host community's viewpoint on the growth is essential before any adventure tourism development gets started (Sosa, 2023).

Although the literature is awash with discussion on adventure tourism, host community involvement in this is rarely discussed. This can be attributed to the 'blurring' boundaries between adventurous activities and tourism which are 'evidenced in the diffuse usage of concepts, such as wilderness, safari, nature guiding, adventurous and adventure with regard to tourism' (Rantala, Rokenes & Valkonen, 2016, p. 539). Based on Rantala et al.'s (2018) argument of adventure tourism being more of a category than a concept, this chapter examines host community involvement drawing from related concepts like ecotourism, wildlife tourism, nature-based tourism, and tourism in general.

Stakeholder theory

Stakeholder theory was first conceived by Freeman (1984). In his theory, Freeman emphasises the importance of involving everyone who has an impact on and is impacted by an organisation's activities by including them in any subject related to the organisation's well-being (Nguyen, 2016). These are the organisations or people who decide how well a company accomplishes its goals (Agüera, 2013). Stakeholder theory emphasises the connections between a company's clients, workers, investors, communities, and other parties with an interest in the organisation. According to the theory, stakeholders could and should have a direct influence on any management decision that is made (Jones, 1995). The stakeholders involved in community-based tourism projects include national, provincial, and local governments; accommodation, catering, and transportation companies; the local community; ecological organisations; clients; suppliers; investors; employees; universities; other businesses; tourists; and non-governmental organisations (Agüera, 2013).

To effectively involve all stakeholders in adventure tourism planning and development, planners must first thoroughly understand their perceptions and attitudes in order to avoid creating social, environmental, or political issues that could have a detrimental impact on the area (Khazaei et al., 2015). It follows that an adventure tourism project will only have the support of stakeholders if it also raises their standard of living and does not conflict with their political and economic objectives. Stakeholder theory thus makes it possible to take into account and include a wide range of factors in the tourist system, which has significant benefits for the growth of sustainable adventure tourism.

Tied neatly to the central thesis of an effective stakeholder management approach is the idea of Sustainable Development Goals (SDGs) whose core agenda is the promotion of sustainability in all forms of development initiatives. Adopted by United Nations member states in 2015 to replace the Millennium Development Goals, SDGs were fashioned to act as a constant reminder that no development initiative should receive support from any stakeholder if its proponent(s) are unable to demonstrate how its rollout would benefit the economy, society, and the environment. These SDGs are, therefore, a firm place of advocacy for the protection of all stakeholders, including the mute and silent.

Types and factors affecting community involvement

There are many types of community involvement approaches in tourism projects which can be summarised into three broader categories: spontaneous, induced, and coercive (Table 24.1). From an administrative perspective, Arnstein (1969) divided community involvement approaches into three categories based on the allocation of power: non-participation (manipulating residents), degrees of tokenism (consulting residents), and maximum participation (citizen control). Pretty (1995) divided them into different categories based on the spectrum of resident spontaneity in participation, which includes manipulative participation, passive participation, and self-mobilisation. Using these approaches as a foundation, Tosun (2006) created a typology of community involvement in the context of tourism that includes three all-inclusive involvement approaches: spontaneous participation, induced participation, and coercive participation. These approaches consider the perspectives of both administrative bodies (consultation, manipulation, citizen control) and local communities (spontaneous or passive; Table 24.1).

These three categories of community participation by Tosun (2006) focus on the 'depth' of involvement. Zhang, Cole, and Chancellor (2013) add another dimension that reveals the 'breadth or scope' of involvement. The breadth of involvement, with four scenarios, shows that community members' spontaneity in participating in adventure tourism is heterogeneous: (1) including all community members in decision-making (empowering everyone), (2) including some community members in decision-making (empowering some), (3) excluding community members from decision-making (community members are consulted by authorities rather than making decisions), and (4) excluding community members from any tourism planning effort (non-participation in tourism planning at all).

Factors that affect host community involvement in adventure tourism include depth and breadth of involvement, community commitment and attachment, costs and benefits, and level of education, among others. When they are given the authority to make decisions rather than merely being consulted, members of the community are more likely to recognise the necessity of integrating adventure tourism into the local economy. This raises the chance of sustainable tourism and long-term destination development success (Lee, 2013). Through spontaneous engagement, the community has full authority to decide on the planning processes for adventure tourism (as opposed to induced or coerced participation). This approach not only raises the income of community members, but also boosts their sense of independence, community involvement, and self-assurance in their ability to influence events that have an impact on their lives (Cole, 2006; Zhang et al., 2013). Spontaneous community engagement in adventure tourism thus contributes to the attainment of SDG 16 (peace, justice, and strong institution), by ensuring inclusive and representative decision-making.

Table 24.1 Types of host community involvement

Type of community involvement	Decision-making approach	Description of participation	Power assignment (administrative perspective)	Community spontaneity in participation	Participation in implementation	Participation in benefit sharing	External support
Spontaneous	Bottom-up	Active, voluntary, direct, informal, authentic	citizen control (maximum participation)	Self-mobilisation	✓	✓	X
Induced	Top-down	Passive, indirect, formal, pseudo-participation	Consultation by residents (degrees of tokenism)	Passive participation	✓	✓	✓
Coersive	Top-down	Passive, indirect, compulsory, manipulated, contrived	Manipulation of residents (non-participation)	Manipulative participation	✓	X	✓

Sources: Arnstein (1969); Pretty (1995); Tosun (1999; 2006).

Members of a community are said to be committed to it when they absorb it as their own, are devoted to it, participate in choices that impact it, support its goals and ideals, and try to make it stronger (Moghavvemi et al., 2021; Setiyorini, Andari & Masunah, 2019). In this instance, members who have a strong sense of community; are concerned about identifying and putting into practice programmes, policies, and practices that could foster the growth of adventure tourism in their community; or support programmes that could aid the community in enhancing its present conditions, resolving problems, or generating new chances for adventure tourism. Shaker (2013) defined community attachment as the degree to which people of a community are cognitively or practically attached to one another and their surroundings. Community attachment may thereby promote local adventure tourism in that it combines a person's sense of rootedness and belonging with their involvement in social activities and integration into the community.

Communities' opinions, attitudes, and participation in adventure tourism are influenced by how they weigh the costs and advantages that emanate from it. As a result, the degree to which the host community benefits from the sector greatly influences their support for its expansion. For instance, when adventure tourism improves the local economy, locals are more likely to support and participate in it. The expansion of sustainable tourism and support for projects related to adventure tourism are negatively impacted by perceived costs. Communities will support and participate in tourism efforts more enthusiastically when the perceived advantages are larger and less enthusiastically when the perceived costs are higher (Moghavvemi et al., 2021 Mutanga, 2022).

Education level affects community involvement in all development initiatives. Residents with higher degrees of education view tourism favourably and participate in it more frequently than residents with lower levels of education (Mugizi, Ayorekire & Obua, 2017). Education enhances one's comprehension of the issues related to the conception, development, and sustainability of adventure tourism ventures. Additionally, education broadens one's capacity to access, investigate, and pursue business opportunities in adventure tourism by enhancing one's knowledge, viewpoint, vision, abilities, and self-confidence. A person's employment and business opportunities are typically affected by their educational background.

Importance of community involvement

Community participation can give people more power and inspire them to help the area. By taking part in decision-making, the community will have a better awareness of the benefits and downsides of tourism development, increasing the chances of reducing potential costs earlier in the planning and administration of tourism (Lalicic & Önder, 2018). Participation of the host community in the planning and management of tourism development offers the opportunity to increase benefits and reduce costs of tourist development within their community (Gannon, Rasoolimanesh & Taheri, 2021). Potential advantages include employment at all levels, simple access to tourist attractions and amenities, and equity ownership of facilities and services (see Case study 24.1).

Case study 24.1 Makuleke Ecotourism Project – Pafuri Camp, South Africa

Pafuri Camp is a community-run ecotourism project initiated in 1998. It involves a collaboration between the state (South Africa National Parks – Kruger National Park), a private company (Wilderness Safaris), and local landowners (the Makuleke community). Since the project's beginning, it has also collaborated on community development initiatives with a variety of non-governmental organisations. Pafuri Camp is located in the 24,000-hectare Makuleke concession in the far north of Kruger National Park. The primary goals of the Pafuri Camp are to protect the distinctive ecosystems and wildlife of the area and maintain a profitable ecotourism business that offers the locals alternate means of subsistence and sustainable sources of income. In 2012, the project had 15,000 beneficiaries in three villages. The community-run ecotourism project offers visitors game drives, walks, cultural excursions, bird watching, and hospitality services. Visitors can also go on excursions to the Makuleke village where they can visit Thulamela's archaeological site, partake in a local meal, dance, and consult with a traditional healer. In addition, Pafuri Camp runs a Makuleke Hydroponics Tunnel Farming Project, which grows fresh vegetables that are either used in Pafuri Camp's food preparation or sold to other nearby ecotourism businesses. The community organised an anti-poaching group to protect wildlife in the area. Both herbivore and predator populations have increased as a result of the anti-poaching units' monitoring and surveillance efforts.

Pafuri Camp is the largest single employer of residents of the Makuleke community. The ecotourism business hires at least 90% of its staff from the community and about 74% of these employees are in their first permanent position of employment. The project offers job security and a steady source of income that has stimulated investments in education, housing, and community infrastructure. A premium has also been placed on gender equality, with about 58% of staff members being local women. By 2012, Pafuri Camp had paid out US$538,732 in salary to its permanent staff members and US$461,782 to community members hired for anti-poaching work. Also, by 2012, the Makuleke Communal Property Association had invested US$179,651 in community projects which was accrued in accordance with the revenue-sharing agreement with Wilderness Safaris, in which 8% of lodge revenues is paid into the Makuleke Communal Property Association.

Source: United Nations Development Programme (2012).

Community involvement in adventure tourism can thus be an effective and critical method for achieving some SDGs and facilitating the transformative promise to 'leave no one behind'. For example in the case of Makuleke Ecotourism Project (Case study 24.1), community involvement in adventure tourism also impacts the realisation of SDGs 5 (gender equity) and 8 (decent work and economic growth). While it is crucial to recognise that adventure tourism needs to benefit the local community, decisions about tourist development at the local level also require broad community involvement. For instance, taking into cognisance that adventure travellers of today are more focused on expanding their horizons and discovering new cultures, it is important that industry professionals be aware of and abide by common community practices when incorporating community tourism projects into their adventure travel offerings (e.g., customary law, spiritual beliefs). Furthermore, while the financial motivation to protect the environment and uphold local customs is a potent incentive for the growth of adventure tourism, such activities should be natural and community-driven rather than turning into a show of exploitation. To do this properly, community members must be involved at every stage of the process, from the initial generation of ideas for incorporating adventure tourism into the natural rhythm of a destination and infrastructure planning to risk management and establishing guidelines for what is and is not appropriate for visitors (Haugen, 2018). This ensures responsive, inclusive, and representative decision-making, thus aiding the realisation of SDG 16 (peace, justice, and strong institutions).

Community-based adventure tourism

Community-based adventure tourism is one of the many facets of community-based tourism. *Community-based tourism* refers to tourism that involves the local community and aims to benefit local populations in developing nations by encouraging social and cultural exchanges with tourists (Zielinski et al., 2020). Thus, community-based adventure tourism refers to community involvement in the development, delivery, and administration of tourist activities that involve physical activity, cultural interaction, or outdoor activities. Community-based adventure tourism often consists of softer adventure activities with little risk, little commitment, and basic skill requirements. These include safaris, birdwatching, ecotourism, fishing, horseback riding, and hunting.

Community-based adventure tourism often takes place on a small scale, involving interactions between tourists and host communities, and works best in rural and regional contexts (Figure 24.1). The host communities are the main beneficiaries through the growth of their communities because community-based adventure tourism is governed and owned by the community (Asker et al., 2010). The goal of community-based adventure tourism is to understand and promote the local environment and culture while also boosting regional suppliers and service providers. By endeavouring to protect the natural environment, community-based adventure tourism impacts SDGs 14 (life below water) and 15 (life on land). Community-based adventure tourism has been pursued and supported by communities, local government bodies, and non-governmental organisations alike (Engström & Leffler, 2012). It, therefore, requires community empowerment, engagement, and benefits from adventure tourism in order to be sustainable. Local communities may feel prouder about their natural resources and more in control of their own development as a result of community-based tourism.

Community empowerment, setting environmental and community goals to ensure outcomes are in line with community values, support from enablers (government, funding institutions, private sector), participatory planning and capacity building to strengthen community tourism management skills, collaboration and partnerships, and focus on generating supplemental income, are key elements for community-based tourism success (Dodds, Ali & Galaski, 2018). Community-based adventure tourism promotes "bottom-up" development which provides communities with a voice in decisions affecting their welfare and fosters a sense of ownership and accountability in them (Cooksey & Kikula, 2005; He et al., 2020). Community-based adventure tourism projects also involve a number of stakeholders. When there is reciprocity between them, stakeholder

Figure 24.1 Community based adventure tourism, Dhorpatan valley, Nepal. Photograph by Teryl Brouillette.

involvement in community development projects increases (Mutanga, 2022). Stakeholders also grow to appreciate one another's perspectives as legitimate. Involving a wide range of stakeholders in community-based adventure tourism can result in better decisions that are more in accordance with local social, cultural, and environmental contexts; the establishment of agreement and trust; the decrease of stakeholder conflict; and lower implementation costs (Sterling et al., 2017). Therefore, in addition to maximising returns for investors, community-based adventure tourism also maximises advantages for the community.

Host community collaboration

Partnerships, collaboration, and networks, both formal and informal, are more pervasive in the tourism industry than in any other economic sector (Scott, Cooper & Baggio, 2008). Partnerships, cooperation, and engagement between stakeholders are essential components of sustainable tourism development (Caffyn, 2000). The success of adventure tourism will ultimately depend on the level of support from the various stakeholders. Adventure tourism has a variety of stakeholders, including travellers, businesses, the host community, governments, non-profit organisations, and educational institutions. These stakeholder groups have the ability to affect the development of adventure tourism in a variety of ways, including through regulation, supply and demand, research, management of tourist impacts, and human resources. To accomplish sustainable tourism development, Hao et al. (2018) underlined the significance of stakeholder participation.

Collaboration among stakeholders is thought to involve a larger community and build networks that promote resource sharing and provide experiences and products with greater value and quality, that can increase yield (Fathimath, 2015; see Case study 24.1). Collaboration is the key to minimising the potential long-term consequences that can result from hostility and divisions between distinct groups. Inefficiencies are likely to result when parties harbour mistrust and anger toward one another and toward each problem (Bramwell & Lane, 2000). Where stakeholders are aware of the potential benefits of working together, collaboration can be utilised successfully to settle conflict and advance shared goals (Healey 1998).

The survival of natural and cultural resources, especially in places with a lot of wildlife where adventure tourism is most popular, depends on cooperation with the host community (Jamal & Stronza, 2009). Collaboration is essential for maintaining resource management, responsible tourism marketing, and socio-economic and environmental sustainability. Since government legislation cannot resolve environmental issues on its own, stakeholders from the public, private, and non-profit sectors must collaborate (Erkus-Ozturk & Eraydýn, 2010; Matarrita-Cascante et al., 2010). Case study 24.2 presents a case study of community involvement in soft adventure tourism in Malaysia.

Case study 24.2　Mukim Batu Puteh community's involvement in soft adventure tourism – Miso Walai homestay ecotourism project

The Miso Walai homestay ecotourism project was officially promoted and launched in 2000, with the main objective of involving the local community in community-based ecotourism. The community's location is a part of Malaysia's Lower Kinabatangan Wildlife Tourism Corridor, which is home to a diverse range of natural and cultural resources. One of their sources of revenue is ecotourism. Key stakeholders include four villages of Batu Puteh, Kinabatangan: Kampung Batu Puteh, Kampung Mengaris, Kampung Perpaduan and Kampung Singgah Matahad; the World Wildlife Fund (WWF; for funding), the Sabah State Ministry of Tourism, the Sabah Forestry Department, the Boat Services Association, the Model Ecologically Sustainable Community Conservation and Tourism Cultural Group, and the Tulon Tokou Handicraft Association.

Wildlife observation river cruises, traditional music and dance performances, culinary classes, interpretive excursions through nearby farms and orchards, and tropical fruit consumption, particularly during fruiting season, are some of the soft adventure tourism activities offered by the community. Approximately 208 Mukim Batu Puteh residents are involved in the initiative directly or indirectly. Over the years, the number of adventure tourists visiting the area has increased, which also denotes an increase in revenue from adventure tourism. In 2000, there were about 176 tourists who participated in the program; by 2009, there were 2,943 tourists; and prior to the novel coronavirus (COVID-19) pandemic, tourist arrivals reached 5,000 per year, with approximately 8,000 nights spent. The ecotourism project helped the community change from one that was known for deforestation, illicit fishing, and hunting to one that was environmentally conscientious.

The community's participation in tourism development was the project's primary success factor. The community capacity-building initiative of the non-governmental organisation led to community empowerment. As a result, the development of the

Miso Walai homestay ecotourism project was viewed as a real-world example of a Malaysian community-driven initiative or a project that was established from the bottom up. Building community capacity was the first step in the project's formation, which aimed to increase the abilities and competencies of the people. This is the key element that has ensured the sustainability of the Miso Walai homestay development.

Sources: Strydom, Mangope and Henama (2018) and WTC (2022).

Collaboration thus plays a critical role in the sustainable management and marketing of adventure tourism, particularly when diverse stakeholder interests are involved. As demonstrated in the case study (24.2), community members should be given the opportunity to start plans and have a say in major decisions affecting the development of adventure tourism. However, for them to be successful, they frequently need assistance from other stakeholders, such as political, governmental, and non-governmental organisations, or other strong supporters (Leksakundilok, 2006).

Thinking globally and acting locally in adventure tourism

Benefits to the local community from tourism initiatives are typically hampered by a lack of financial sustainability, inadequate market access, and a restricted ability to satisfy visitor needs (Mutanga, 2022). Suppliers of tourism services have unavoidably been forced to embrace a global perspective because the majority of visitors to a place often originate from distant source markets. Understanding the cultural and linguistic variations of diverse markets as well as researching and utilising the numerous channels of distribution operating in each source country receives a great deal of attention. When it comes to the provision of adventure tourism services, acting locally entails creating time-, place-, and people-specific content and activities in order to engage visitors and enhance their experiences. Communities must create connections between local and global knowledge as a means of "thinking globally" if they are to develop novel strategies for adventure tourism that they can put into practice locally. This is crucial because the markets for adventure tourism vary by nation and culture. For instance, German, British, and American adventure travellers have different wants and needs. Therefore, to think globally, one needs foresight, which is defined as the capacity to establish and maintain a high-quality, coherent, and useful future vision. Foresight also aids in uncovering unconfirmed circumstances, the development of strategies, and an exploration of markets and new opportunities for products and services (Slaughter, 2000). It takes foresight to develop innovations that are driven by tourists, foresee changes in the serviced market, and foresee the rise of new markets (McCardle, 2005). In addition, foresight provides a thorough analysis of how the SDGs are being implemented by identifying potential futures and providing fresh approaches to lowering risks and boosting resilience (Chatkaewnapanon & Lee, 2022).

Communities require investment and capacity building, particularly in entrepreneurship training to enable them to start, market, and manage small adventure tourism companies, in order to deliver and market tourist services. Developing and strengthening the community's skills impacts SDG 17 (partnerships for the goals) in the capacity-building category. Responsible marketing has become crucial for attracting today's adventure tourism market. Responsible tourism marketing is similar to green marketing and environmental marketing (Coddington & Florian, 1993). Thinking globally implies that community-based adventure tourism must adopt the practice of responsible marketing, which entails creating adventure tourism products that match consumers' demands for convenience, affordability, and quality with those for a minimal impact on the environment.

Challenges and implications for sustainability

As alluded to earlier, host community participation is a critical success factor for any type of tourism development including adventure. Stone and Nyaupane (2014) argue that in projects that involve communities, the concept of community as a stakeholder has mostly been poorly conceptualized, and its practical use has become a source of tension and conflict. They decry the tendency to homogenise the community and disregard the demographic differences of constituent members. Such factors as ethnicity, age, gender education, and economic endowments are known to confer varying levels of influence on community members with the result that decision-making processes that should ordinarily be collective and inclusive become discriminatory and exclusive (Lenao, 2014).

Furthermore, it has been noted that, despite the good intentions of arrangements facilitating community participation in tourism development, envisaged benefits have not always trickled down to the intended beneficiaries (Blaikie, 2006). Where some benefits have been realised, they have not always been equal (Mbaiwa, 2004). The unwanted consequence of this is a diminishing allure of these community participation arrangements in the face of community members. For adventure tourism development, therefore, it becomes an even more critical sustainability issue given the nature of the activities involved.

On another note, van der Duim (2011) argues that the various community participation arrangements, especially in developing countries, are characterised by different degrees of centralisation. Lenao and Saarinen (2016) note that in community-based projects, decentralisation is mostly used as a catchword to mask the reality that communities are only being enticed to cooperate in the conservation of resources with almost no control in the decision-making process. The tendency of governments to hold fast to the ownership and control of these resources could be interpreted as a sign of a lack of trust and confidence in the communities' ability to practise sustainable utilisation on their own.

Understanding the priorities and motivations of community members to successfully incorporate them into adventure tourism is critical. The representation of every member of the community must be ensured. In addition, it is crucial to note that no group is entirely homogeneous and that its members are probably from a variety of backgrounds and have a variety of personality traits. For instance, what could motivate one section of the community may not motivate another.

As most community involvement in activities occurs in group settings, there is a need for facilitation, particularly by someone without decision-making authority, to make the planning, implementing, monitoring, and evaluating of the adventure tourism projects more efficient and effective for the group. This is crucial if the community is to be given true decision-making authority and responsibility. In addition, community capacity building is important to help members of the community enhance their abilities and gain confidence. Capacity building through the development of skills and self-confidence is a crucial component in motivating the community.

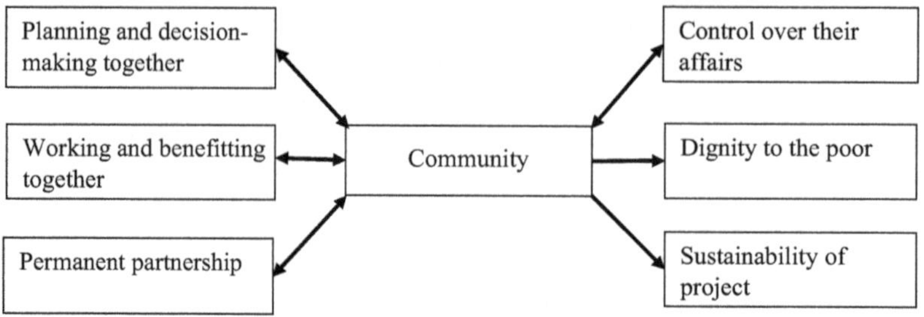

Figure 24.2 Community partnership model.

Source: Reddy (2002).

Sustainable community involvement requires empowering communities through participatory methods where the government, organisation, or agency collaborates with the community to plan and make decisions with long-lasting results (Figure 24.2; Reddy, 2002). Community involvement in adventure tourism initiatives should be viewed in this context as an "end" rather than a "means". When participation is seen as a means rather than an end, it typically takes the shape of mobilisation to accomplish goals rather than a set, measurable development goal. Instead, when participation is seen as an end, the process of enhancing participation's 'meaningfulness' in the development process is the goal. Therefore, it is necessary to pinpoint the mechanism by which engagement as a means might transform into participation as an end (Moser, 1983).

While communities may be open to adventure tourism, a number of factors could make individuals or a community hesitant to participate, including an unfair division of labour or benefits among community members; a highly individualistic society with little or no sense of community; the belief that the government, organisation, or agency should provide the resources; and the agency's or organisation's treatment of community members. The attitude and conduct of the agency or organisation involved are the strongest deterrents to community involvement. As important as rewards are, treating people with respect, hearing what they have to say, and learning from them will go a long way toward fostering long-term community participation in adventure tourism. Thus, on the one hand, the authorities must change their attitude toward the communities, and on the other, it must provide them with guidance so that they can participate and make informed decisions. Authorities typically avoid changing their paternalistic approach to decision-making because of their concern about the uncontrolled empowerment of people and lack of faith in their capacity to make meaningful decisions. Such difficulties in community involvement may be resolved by considering participation from a wider angle and comparing its advantages against constraints.

Conclusion

This chapter has demonstrated that community involvement is a critical success factor in the development of adventure tourism. It asserts that central to the idea of community involvement is the process of stakeholder management, where stakeholders connote all those individuals and/or groups whose decisions and lack thereof have a bearing on the development of adventure tourism and the other way around. The chapter has also outlined some of the challenges which may potentially hamper the successful implementation of community involvement in the development of adventure tourism. These include such factors as the distribution of power and benefits among the various stakeholders involved, a lack of financial viability, and poor market access. The chapter concludes that, if properly thought out and implemented, community involvement in adventure tourism presents a great opportunity for achieving success on the twin objectives of resource conservation and improving human livelihoods. In the same vein, if poorly planned and implemented, community participation has the potential to defeat the same objectives upon which the co-management principle is based.

From the scant existing literature on host community involvement in adventure tourism, there is a perceptible lack of the host community's perceptions of what constitutes involvement in adventure tourism and their subjective experiences in it. A qualitative investigation of the host community's perceptions and experiences in planning, marketing, and management of community-based adventure tourism projects should therefore be afforded greater prominence. Similarly, investigations should also focus on community involvement and benefits from "non-community-owned" adventure tourism activities. This stems from the fact that many tour operators take their clients for adventure activities in natural areas where communities reside or are closer to. As such, these communities are the ones that suffer the negative impacts of tourism in their areas, hence the need to investigate their involvement and benefits from all adventure tourism in their areas – whether community or non-community-owned.

Review questions

1 What is the importance of involving host communities in adventure tourism?
2 Identify any two adventure tourism businesses. How is the host community involved in each of the examples?
3 What is the role that different factors play in fostering community participation in adventure tourism?
4 What is the importance of host community partnerships, collaborations, and networks?
5 What are the major challenges that are often encountered when involving host communities in adventure tourism? How do you suggest these challenges are overcome?

Further reading

Garcia, O., & Cater, C. (2020) Life Below Water; challenges for tourism partnerships in achieving ocean literacy. *Journal of Sustainable Tourism.* https://doi.org/10.1080/09669582.2020.1850747
Mensah, I., & Afenyo-Agbe, E. (Eds.). (2021). *Prospects and challenges of community-based tourism and changing demographics.* Hershey, Pennsylvania: IGI Global.
Singh, S., Timothy, D. J., & Dowling, R. K. (Eds.). (2003). *Tourism in destination communities.* Wallingford: CABI.

References

Agüera, O. F. (2013). Stakeholder theory as a model for sustainable development in ecotourism. *Turydes,* 6(15), 1–17.
Arnstein, S. R. (1969). A ladder of citisen participation. *Journal of the American Institute of Planners,* 35(4), 216–224.
Asker, S. A., Boronyak, L. J., Carrard, N. R., & Paddon, M. (2010). *Effective community based tourism: A best practice manual. Asia Pacific economic cooperation (APEC) tourism working group.* Gold Coast, Australia: Sustainable Tourism Cooperative Research Centre.

Barrow, E., & Murphree, M. (2001). Community conservation: From concept to practice. In E. Barrow & M. Murphree (Eds.), *African wildlife and livelihoods: The promise and performance of community conservation* (pp. 1–30). Oxford: James Currey.

Blaikie, P. (2006). Is small really beautiful? Community-based natural resources management in Malawi and Botswana. *World Development, 34*(11), 1942–1956.

Bramwell, B., & Lane, B. (2000). Collaboration and partnerships in tourism planning. In B. Bramwell & B. Lane (Eds.), *Tourism collaboration and partnerships: Politics, practice and sustainability* (pp. 1–19). Clevedon: Channel View.

Caffyn, A. (2000). Is there a tourism partnership life cycle? In B. Bramwell & B. Lane (Eds.), *Tourism collaboration and partnerships: Politics, practice and sustainability* (2nd ed., pp. 200–246). Clevedon: Channel View.

Chatkaewnapanon, Y., & Lee, T. J. (2022). Planning sustainable community-based tourism in the context of Thailand: Community, development, and the foresight tools. *Sustainability, 14*(12), 7413.

Chiang, C. F., & Huang, K. C. (2012). An examination of antecedent factors in residents' perceptions of tourism impacts on a recreational fishing port. *Asia Pacific Journal of Tourism Research, 17*(1), 81–99.

Coddington, W., & Florian, P. (1993). *Environmental marketing: Positive strategies for reaching the green consumer.* New York: McGraw-Hill Companies.

Cole, S. (2006). Information and empowerment: The keys to achieving sustainable tourism. *Journal of Sustainable Tourism, 14*(6), 629–44.

Cooksey, B., & Kikula, I. (2005). When Bottom-Up Meets Top-Down: The Limits Of Local Participation. In *Local Government Planning In Tanzania.* Special Paper No: 17. Research on Poverty Alleviation. Mkuki Na Nyota Publishers. Tanzania.

Dodds, R., Ali, A. & Galaski, K. (2018). Mobilising knowledge: Determining key elements for success and pitfalls in developing community-based tourism. *Current Issues in Tourism, 21*(13), 1547–1568.

Dumitras, D. E., Mihai, V. C., Jitea, I. M., Donici, D., & Muresan, I. C. (2021). Adventure Tourism: Insight from experienced visitors of romanian national and natural parks. *Societies, 11*(2), 41.

Engström, D., & Leffler, F. (2012). *Perceptions of climate change at ski resorts in midsouth of Sweden* (BTech Thesis, Dalarna University, Sweden).

Erkus-Ozturk, H., & Eraydýn, A. (2010). Environmental governance for sustainable tourism development: Collaborative networks and organisation building in the Antalya tourism region. *Tourism Management, 31*(1), 113–124.

Fathimath, A. (2015). *The role of stakeholder collaboration in sustainable tourism competitiveness: the case of Auckland, New Zealand* (Doctoral dissertation, Auckland University of Technology).

Freeman, R. E. (1984). *Strategic management. A stakeholder approach.* Boston, MA: Pitman.

Gannon, M., Rasoolimanesh, S. M., & Taheri, B. (2021). Assessing the mediating role of residents' perceptions toward tourism development. *Journal of Travel Research, 60*(1), 149–171.

Hao, H., Kline, C., Long, P., & Rassel, G. (2018). Property owners' attitudes toward sustainable tourism: Comparing coastal and mountain counties. *Tourism and Hospitality Research, 18*(4), 429–441.

Haugen, J. (2018, January 30). The Intersection of Adventure Travel and Community Tourism. https://www.adventuretravelnews.com/the-intersection-of-adventure-travel-and-community-tourism

He, S., Yang, L., & Min, Q. (2020). Community participation in nature conservation: The Chinese experience and its implication to National Park Management. *Sustainability, 12*(11), 4760.

Healey, P. (1998). Collaborative planning in a stakeholder society. *Town Planning Review, 69*(1), 1.

Inbakaran, R., & Jackson, M. (2006). Resident attitudes inside Victoria's tourism product regions: A cluster analysis. *Journal of Hospitality and Tourism Management, 13*(1), 59–74.

Jamal, T., & Stronza, A. (2009). Collaboration theory and tourism practice in protected areas: Stakeholders, structuring and sustainability. *Journal of Sustainable Tourism, 17*(2), 169–189.

Jones, T. M. (1995). Instrumental stakeholder theory: A synthesis of ethics and economics. *Academy of Management Review, 20*, 404–437.

Khazaei, A., Elliot, S., & Joppe, M. (2015). An application of stakeholder theory to advance community participation in tourism planning: The case for engaging immigrants as fringe stakeholders. *Journal of Sustainable Tourism, 23*(7), 1049–1062.

Lalicic, L., & Önder, I. (2018). Residents' involvement in urban tourism planning: Opportunities from a smart city perspective. *Sustainability, 10*(6), 1852.

Lee, T. H. (2013). Influence analysis of community resident support for sustainable tourism development. *Tourism Management, 34*, 37–46.

Leksakundilok, A. (2006). *Community participation in ecotourism development in Thailand.* University of Sydney, Geosciences.

Lenao, M. (2014). Rural tourism development and economic diversification for local communities in Botswana: The case of Lekhubu Island. *Nordia Geographical Publications, 43*(2), 1–53.

Lenao, M. & Saarinen, J. (2016). Political ecology of community-based natural resources management: Principles and practices of power sharing in Botswana. In S. Nepal & J. Saaninen (Eds.), *Political ecology and tourism* (pp. 115–129). London: Routledge.

Li, Y., & Hunter, C. (2014). Community involvement for sustainable heritage tourism: A conceptual model. *Journal of Cultural Heritage Management and Sustainable Development, 5*(3), 248–262.

Manea, G., Orbu, L. M., Matei, E., Preda, M., Vijulie, I., Cuculici, R., ... Zaharia, A. (2019). Heritage advantages vs. economic constraints in the sustainable development framing: A study on social perception in Apuseni Mountains Natural Park. *Human Geographies, 13*(2), 173–185.

Matarrita-Cascante, D., Brennan, M. A., & Luloff, A. E. (2010). Community agency and sustainable tourism development: The case of La Fortuna, Costa Rica. *Journal of Sustainable Tourism, 18*(6), 735–756.

Mbaiwa, J. E. (2004). The success and sustainability of community-based natural resource management in the Okavango Delta, Botswana. *South African Geographical Journal, 86* (1) 44–53

McCardle, M. (2005). *Market foresight capability: Determinants and new product outcomes.* University of Central Florida.

Moghavvemi, S., Woosnam, K. M., Hamzah, A., & Hassani, A. (2021). Considering residents' personality and community factors in explaining satisfaction with tourism and support for tourism development. *Tourism Planning & Development, 18*(3), 267–293.

Moser, C. (1983). The problem of evaluating community participation in urban development projects. Evaluating Community Participation in Urban Development Projects, Development Planning Unit Working Paper No. 14.

Mugizi, F., Ayorekire, J., & Obua, J. (2017). Factors that influence local community participation in tourism in Murchison falls conservation area. *Journal of Environmental Science and Engineering A, 6*(4), 209–223.

Mutanga, C. N. (2022). Tolerance for wildlife resources through community wildlife-based tourism. In L. S. Stone, M. T. Stone, P. K. Mogomotsi, & G. E. J. Mogotmotsi (Eds.), *Protected areas and tourism in Southern Africa: Conservation goals and community livelihoods* (pp. 56–70). Milton Park: Routledge.

Nguyen, T. (2016). Stakeholder model application in tourism development in Cat Tien, Lam Dong. *Journal of Advanced Research in Social Sciences and Humanities, 1*(1), 73–95.

Pretty, J. (1995). The many interpretations of participation. *Focus, 16*(4), 4–5.

Rantala, O., Rokenes, A., & Valkonen, J. (2016). Is adventure tourism a coherent concept? A review of research approaches on adventure tourism. *Annals of Leisure Research, 21*(5), 539–552. https://doi.org/10.1080/11745398.2016.1250647

Reddy, G. N. (2002). *Empowering communities through participatory methods.* New Delhi: Manak Publications.

Scott, N., Cooper, C., & Baggio, R. (2008). Destination networks: Four Australian cases. *Annals of Tourism Research, 35*(1), 169–188.

Setiyorini, H. P. D., Andari, R., & Masunah, J. (2019). Analysing factors for community participation in tourism development. *THE Journal: Tourism and Hospitality Essentials Journal, 9*(1), 39–44.

Shaker, C. (2013). Cue-based co-regulated feeding in the neonatal intensive care unit: Supporting parents in learning to feed their preterm infant. *Newborn and Infant Nursing Reviews, 13*(1), 51–55.

Slaughter, R. A. (2000). Futures for the Third Millennium–Enabling the Forward View. *Sydney, Prospect.*

Sosa, M. C. (2023). Tourism development planning as a community industry. *Visión de futuro, 27*(1), 59–72.

Sterling, E. J., Betley, E., Sigouin, A., Gomez, A., Toomey, A., Cullman, G., ... Porzecanski, A. L. (2017). Assessing the evidence for stakeholder engagement in biodiversity conservation. *Biological Conservation, 209,* 159–171.

Stone, M. T. and Nyaupane, G. (2014). Rethinking community in community-based natural resource management. *Community Development, 45*(1), 17–31.

Strydom, A. J., Mangope, D., & Henama, U. S. (2018). Lessons learned from successful community-based tourism case studies from the global south. *African Journal of Hospitality, Tourism and Leisure, 7*(5), 1–13.

Szabo, E. A., Lawrence, A., Iusan, C., & Canney, S. (2008). Participatory protected area management–A case study from Rodna Mountains National Park, Romania. *The International Journal of Biodiversity Science and Management, 4*(4), 187–199.

Timothy, D. J., & Boyd, S. W. (2003). *Heritage tourism.* London: Pearson Education Limited.

Tosun, C. (1999). Towards a typology of community participation in the tourism development process. *Anatolia, 10*(2), 113–134.

Tosun, C. (2006). Expected nature of community participation in tourism development. *Tourism management, 27*(3), 493–504.

United Nations Development Programme. (2012). Makuleke Ecotourism Project - Pafuri Camp, South Africa. Equator Initiative Case Study Series. New York, NY.

van der Duim, R. (2011). New institutional arrangements for tourism: Conservation and development in sub-Saharan Africa. In R. van der Duim, D. Meyer, J. Saarinen, & K. Zellmer (Eds.), *New alliances for tourism, conservation and development in Eastern and Southern Africa* (pp. 83–105). Delft: Eburon Academic Publishers.

WTC. (2022). World Tourism Conference Report. *Tourism Futures Reimagined.* https://www.motac.gov.my/en/download/send/64-ringkasan-eksekutif-laporan-kajian-slaid-pembentangan/686-laporan-world-tourism-conference-wtc-2022&cd=19&hl=en&ct=clnk&gl=uk#35

Zhang, Y., Cole, S. T., & Chancellor, C. H. (2013). Residents' preferences for involvement in tourism development and influences from individual profiles. *Tourism Planning & Development, 10*(3), 267–284.

Zielinski, S., Jeong, Y., Kim, S. I., & Milanés, C. (2020). Why community-based tourism and rural tourism in developing and developed nations are treated differently? A review. *Sustainability, 12*(15), 5938.

25 Indigenous adventure tourism and authenticity

Juha Saunavaara and Takafumi Fukuyama

Chapter learning outcomes

1 Explain typical features of Indigenous adventure tourism.
2 Recognise the different opportunities that adventure tourism can bring to Indigenous peoples and communities.
3 Understand the possible challenges brought about by the introduction of adventure tourism practices within different types of Indigenous communities.

Introduction

Physical activity, cultural exchange or interaction, and engagement with nature are often described as the basic components of adventure tourism experiences (ATTA, 2022a). Other features associated with adventure tourism include, among others, small group sizes, respect for the local culture and natural environment, and the search for and provision of unique experiences and authenticity. Although there are different views concerning risk (real or perceived) as a key element of adventure tourism (Farkić & Gebbels, 2022; Martin & Ren, 2020; Rantala et al., 2018; Rantala et al., 2016), it has been argued that adventure tourism customers are increasingly motivated by and search for cultural understanding, broadened perspectives, expanded horizons and transformative experiences (Beckmann, 2017). However Gilmore and Pine (2007) differentiated between experiences (memorable events that engage individuals in an inherently personal way) and transformations (effectual outcomes that guide customers to change some dimension of their self). The transformative experiences that adventure tourists seek refer to the desire and active endeavour for personal growth and change (Beckmann, 2017).

Like adventure tourism, Indigenous tourism has many definitions. While Smith's (1996) four Hs (habitat, heritage, history, handicrafts) are well-known segments of tourism (re)presenting Indigenous cultures, *Indigenous tourism*, in general, refers to tourism owned, operated, and/or controlled by Indigenous peoples and connected to the local Indigenous community and area (de Bernardi et al., 2017; Huddart & Stott, 2020). The compatibility of adventure tourism with Indigenous tourism seems to depend on the way one understands the former. Indigenous tourism goes hand in hand with definitions of adventure tourism emphasising slow cultural immersion, but it may be incompatible with the gradually diminishing views emphasising risk-taking and physically demanding

DOI: 10.4324/9781003393153-30

activities. Meanwhile, the development of Indigenous adventure tourism may redirect discussion concerning the definition of adventure tourism from the content and components of the experience to the questions of agency and actorhood.

The famous Larrakia Declaration (Adventure Travel News, 2012), originating from the first Pacific Asia Indigenous Tourism Conference, recognised tourism as both a possible driver to restore, protect, and promote Indigenous cultures and a potential force that can diminish and destroy those cultures when improperly developed. Furthermore, the sustainability of tourism was tightly connected to the need to contribute to the well-being of Indigenous communities and the environment. The opportunities, risks and responsibilities concerning the respect and fair treatment of Indigenous communities are recognised by the World Indigenous Tourism Alliance (WINTA) and the Adventure Travel Trade Association (ATTA), which has also joined and established strategic partnerships with WINTA and regional Indigenous tourism organisations (Adventure Travel News, 2022; ATTA, 2022b; WINTA, 2022).

To elaborate on the relationship between adventure tourism and Indigenous tourism, this chapter focuses on the features of authentic Indigenous adventure tourism experiences and adventure tourism's capability to support the sustainable development of Indigenous communities. Therefore, the interaction between hosts and visitors is also emphasised, alongside the authenticity of adventure tourism experiences and products. When Indigenous cultures appear in tourism promotion, services, and souvenirs, references to Indigenousness often fall into the exoticisation of land and people and emphasise tradition and even colonial images instead of contemporary (living) culture. However, Indigenous cultural practices and everyday lives are not static but dynamic and constantly evolving (de Bernardi, 2019; Niskala & Ridanpää, 2016; Olsen, 2008). Therefore, some features of "real" or "authentic" Indigenous adventure tourism experiences can differ from the tradition-oriented expectations that visitors may have. At the same time, there are significant differences among Indigenous communities and their related enterprises in their general attitudes towards tourism and in their attempts to balance the past and the present.

This chapter consists of six parts. The first part introduces the relationships among Indigenous cultures, tourism and issues related to authenticity. The second elaborates on the characteristics of Indigenous adventure tourism experiences and is followed by the third and fourth, focusing on the opportunities for and limitations of Indigenous adventure tourism within two different kinds of settings. The fifth examines the attractiveness of adventure tourism from the perspective of Indigenous communities and is followed by a short conclusion.

Indigenous cultures, tourism, and questions concerning authenticity

The positive effects of Indigenous tourism are often described as a potential remedy for Indigenous communities facing various challenges. Attention is thus paid to the nurturing of arts, cultural creativity, and traditions, as well as language revitalisation. Tourism is also considered a means of empowerment and a tool for challenging stereotypes and representations of otherness as well as for increasing awareness, understanding and people-to-people exchanges (Carr et al., 2016; Zhang & Müller, 2018). Other narratives emphasise employment creation (often secondary or seasonal income) and economic benefits, community development, and the maintenance of a stable population and services in remote areas (Leu et al., 2018; Viken & Müller, 2017). When tourism has been considered a threat to Indigenous peoples, it has been described as a new form of capitalist exploitation of natural and cultural resources. Other negative effects on local communities include, among others, cultural commodification, the disturbance of local land ownership patterns and Indigenous livelihoods, the de-characterisation of local Indigenous populations and the exclusion of locals from decision-making processes (de Bernardi et al., 2017; Leleto, 2019; Pereiro, 2013).

The relationship between tourism and the concept of authenticity is complex (Beer, 2013; de Bernardi, 2019b; de Bernardi, 2020). Authenticity has for a long time been recognised as an appealing factor that motivates people to travel and as a feature that increases tourists' appreciation and attachment to a destination, people, and experience. Early research focused, for example, on questions concerning staged authenticity and pseudo-events satisfying tourists' desire to find something that is pristine, primitive, exotic, natural, and untouched by modernity (Cohen, 1988; de Bernardi, 2019; MacCannell, 1973), but more nuances and different approaches have since been added. While some scholars have tied the authenticity of an object with specific materials and production methods, others have paid attention to the process of 'authentication' through which a role, product, object or event becomes confirmed as 'genuine' or 'real' (de Bernardi, 2019b). Finally, postmodern approaches have questioned whether 'original' or 'authentic' (cf. 'inauthentic') exists as an objective and distinguishable category or whether authenticity is a matter connected with subjective experiences (Cohen, 1988; de Bernardi, 2019b). Related to these discussions are debates on whether authenticity should be understood through the views of hosts or visitors, whether something previously interpreted as inauthentic can, in the course of time, become authentic and whether tourism can stimulate the revitalisation and creation/emergence of new meanings for traditional rituals and objects (de Bernardi, 2019b; Hiwasaki, 2000).

In the context of Indigenous tourism, authenticity is strongly tied with questions on how and on whose terms Indigenous peoples and their culture are portrayed and incorporated into tourism. In the tourism industry, Indigenousness is often objectified, transformed, promoted and consumed through performances, goods/souvenirs, architecture, and experiences whose attractiveness is based on their presumed traditionality and authenticity. Although tourism itself has been described as a 'modernising phenomenon', Indigenous peoples are repeatedly associated with timeless primitive features, and the modern characteristics of the communities in which they live are hidden in touristic contexts (Niskala & Ridanpää, 2016; Olsen, 2008; Pereiro, 2013). Meanwhile, Indigenous peoples' close relationship with nature is usually an inseparable part of the promotional message (Niskala & Ridanpää, 2016). While the tourism industry has been criticised for being a force strengthening distorted stereotypes (de Bernardi, 2019; Viken & Müller, 2017), attention has also been paid to Indigenous communities' tendency to emphasise tradition, and even otherness and primitiveness, to match the (assumed) expectations of tourists (Niskala & Ridanpää, 2016).

According to ATTA (2022c), 'cultural encounters' and 'cultural' are among the top trending motivations for adventure travel and adventure activities. As growing interest in Indigenous communities and cultures seems to exist among both adventure tourism tour operators and their customers, the need for discussion concerning the acceptable and preferable features of Indigenous adventure tourism experiences is greater than ever. The past challenges introduced earlier should be kept in mind when Indigenous adventure tourism is developed.

Indigenous adventure tourism experiences

To demonstrate some of the challenges related to Indigenous adventure tourism from the visitor's side, attention should be paid to Bresner's (2010) account describing the feelings of tourists joining an Indigenous tourism itinerary in Australia's Northern Territories. Describing an encounter with Aboriginal women who had been brought in to sell their handicrafts to a group of visitors guided by a non-Aboriginal guide, Bresner (2010, p. 19) wrote:

> I felt guilt for participating in what I felt like an unequal relationship of power that seemed physically exaggerated as the women sat on the ground painting while white tourists walked around them surveying their work. I was worried I was participating in their exploitation. Or was this a mutual exchange? A positive experience for both these women and myself?

The findings of Ween and Riseth (2017) concerning Indigenous involvement in tourism in northern Norway highlight problems from the perspective of the Indigenous community. Their study showed how nature-based tourism (such as hunting and fishing, tracking and skiing) does not necessarily provide much income for local Indigenous communities or lead to meaningful contacts and knowledge exchanges between the host and visitors. Instead, it increases pressure on limited natural resources and disturbs cultural practices. This example demonstrates that the principles of Indigenous adventure tourism are not necessarily met even if nature, physical activity, and Indigenous culture are part of the same tourism offering.

Adventure tourism in general, and culturally oriented adventure tourism in particular, should maximise the potential of Indigenous engagement in tourism, make the power relations more equal, prevent the materialisation of the risks related to the exploitation of Indigenous cultural resources, and provide further understanding to form the basis for respectful relationships. The other cases from Norway introduced by Ween and Riseth (2017), referring to innovative, interactive, and community-led initiatives, where Indigenous tourism entrepreneurs sell cultural experiences rather than natural resources, can be considered a promising example.

In the case of the tangible elements of Indigenous tourism, attempts to protect and promote authenticity have often taken form in discussions concerning labels. Labelling systems, the sets of criteria to which companies must adhere, have been considered to be ways to give visibility to Indigenous products and services, empower Indigenous communities, ensure greater economic benefits to local Indigenous actors (including entrepreneurs, guides, performers, artisans and other members of the community involved in tourism) and contribute to the overall awareness of Indigenous cultures (de Bernardi et al., 2017; ITAC, 2023). While the aims of these kinds of initiatives are widely supported, they have also revealed some parameters related to authentic Indigenous experiences and products that may be difficult to define. Most importantly, there is a question concerning the extent to which authenticity is tied to historical and cultural accuracy and how much room there is for (modern) artistic freedom and/or individual expression for the Indigenous entrepreneur (de Bernardi, 2019; Viken & Müller, 2017). As Indigenous adventure tourism emphasises both the Indigenous past and modernity, these kinds of freedoms seem quite natural. However, the focus should not only be on the products but also encounters with Indigenous creators sharing the stories and processes behind the artefacts.

Besides the physical items, questions concerning the authenticity and characteristics of Indigenous adventure experiences relate to the intangible elements of Indigenous culture. While the traditional and contemporary features of culture and daily life, shared through oral communication, observation, and participatory activities, are often at the core of experiences, all Indigenous activities are not primarily built around Indigenous culture. According to Butler (2021), involvement has long been a feature of Indigenous tourism, but Indigenous people are mainly considered to be an element that is observed rather than agents who own, control, and develop businesses. This has led to a situation in which Indigenous tourism is almost automatically affiliated with traditional activities and skills. However, Thompson-Carr (2013) argued that tourists (who are usually international) interested in Māori culture do not always want experiences to be tightly linked to tradition and that there is increasing interest in activities that enable personal encounters and the incorporation of cultural values within contemporary and/or nature-based settings. She went on to state that Māori tourism does not have to be based on Indigenous cultural experiences; rather, even the tourism activities that are owned by Māori are special because they give the visitor a unique Māori perspective and interpretation. Although Thompson-Carr did not directly refer to adventure tourism, the open-minded incorporation of Indigenous culture and knowledge with a wide variety of outdoor activities is clearly at the core of Indigenous adventure tourism. Examples of success, such as the development of mountain bike trails by Carcross/Tagish First Nation youths, already exist (Huddart & Stott, 2020; Figure 25.1).

Figure 25.1 Interaction between hosts and guests is at the core of Indigenous adventure tourism. Indigenous sage smoke blessing in sweat lodge, British Columbia, Canada. Photograph by Carl Cater.

Besides recognising the diversity among the different types of Indigenous adventure tourism experiences, the relationship between communities hosting adventure tourists, and the surrounding environment should not be simplified or distorted. The worry over and attention paid to global warming and environmental change have increased interest in Indigenous peoples' ways of living as sustainable alternatives to the destructive way of living in non-Indigenous, developed countries (Hiwasaki, 2000; Mazzocchi, 2020). At the same time, there are often collisions between the non-Indigenous discourses describing the traditional Indigenous homelands as being untouched (and untouchable) wilderness and the fact that many traditional and modern Indigenous livelihoods are leaving (and have left) their imprint on the environment (de Bernardi, 2019; Sámediggi, 2022; Ween & Riseth, 2017). In general, the search for genuine experiences under the brand of adventure tourism necessitates the differences between Indigenous and non-Indigenous cultures to be identified and respected but not artificially created or over-emphasised for the sake of the presumed attractiveness of the exotic Indigenous element.

Developing Indigenous adventure tourism based on traditional livelihoods

Indigenous people often form minorities within societies dominated by non-Indigenous cultures. Besides legal frameworks defining the rights and duties of different parties, the distribution of resources and past and present attitudes towards diversity and multiculturalism, there are also great differences in historical assimilation processes. Therefore, the role that traditional livelihoods play in the everyday lives of Indigenous groups varies. Although links between Indigenous communities and adventure tourism have been established in different parts of the world, this chapter pays particular attention to northern Europe when studying the opportunities for Indigenous adventure tourism based on traditional livelihoods (see Case study 25.1).

Case study 25.1   Developing Indigenous adventure tourism based on traditional Sámi livelihoods

The Sámi are the only Indigenous population of Europe, and their traditional homeland (Sápmi) stretches across northern Norway, Sweden, Finland and Russia. However, a great part of the Sámi population nowadays lives in cities outside Sápmi. Although reindeer herding is still one of their main sources of livelihood, many Sámi have conventional jobs (de Bernardi, 2020; Zhang & Müller, 2018). Tourism has a long history in Finnish, Swedish and Norwegian Lapland, and it has developed into one of the key drivers of the regional economy. The rapid growth in the number of international visitors before the COVID-19 pandemic was mainly due to the attractions built around Christmas/Santa Claus and the aurora borealis, for example. However, there were also significant nature-based tourism service sector offerings and industrialised guided programmes, such as snowmobile excursions, reindeer or sledge dog safaris, fishing and hunting. Rather than being at the centre of these mainstream services, Sámi culture has provided an exotic background to outdoor activities (Niskala & Ridanpää, 2016; Rantala et al., 2018).

The Sámi have established and managed tourism companies, but a great majority of tourism-related activities in Sápmi have been developed and are carried out by non-Indigenous destination marketing/management organisations and companies (Viken & Müller, 2017). The non-Sámi tourism industry has used images related to Sámi culture, and Sámi dolls, shaman drums, and reindeer skin artefacts, for example, are sold at souvenir shops all over Finland. While traditional Sámi costumes, tents (*lavvu*), and a form of song (*joik*) are common motifs, reindeer and reindeer husbandry are the best-known symbols of Sámi culture. The lack of control and power, the misuse of Sámi culture, cultural appropriation, and exoticisation and primitivisation have caused anger among Sámi, who have protested and opposed these kinds of practices. While tourism has offered new possibilities for revenue and markets for traditional handicrafts, for example, it has divided opinions among the Sámi (de Bernardi et al., 2017; Niskala and Ridanpää, 2016; Niskala and Ridanpää, 2016; Sámediggi, 2022).

Traditionally, adventure tourism has not been the framework for Sámi entrepreneurs involved in tourism. Furthermore, both internet-based searches and interviews with local entrepreneurs and destination marketing/management organisations have indicated that adventure tourism is hardly present in tourism conducted in Finnish Lapland. However, both the aspects of responsibility and sustainability attached to adventure tourism and the emphasis on immersion in nature and culture through different outdoor activities are a great match for existing Sámi tourism that is often built around reindeer and reindeer herding as a centuries-old but modernised livelihood. In addition, national tourism promotion organisations (e.g., Visit Finland, Visit Sweden and Innovation Norway) and some companies specialised in nature and/or activity-based itineraries are already actively involved in the international adventure tourism community. Furthermore, UiT – The Arctic University of Norway, one of the leading research and education institutes in the region, has established a bachelor's degree programme focusing on Arctic adventure tourism (UiT, 2023).

The municipality of Inari hosts both Sámi administrative and cultural centers in Finland and a lively tourism industry. The Sámi languages (there are nine of these, of which three are spoken in Finland) and culture are present in everyday life in Inari, and Sámi tourism is carried out side by side with reindeer herding utilising modern technology. Both Sámi entrepreneurs and the local museum dedicated to Sámi culture have introduced indigenous modernity in tandem with tradition as well as new and constantly evolving ways of cultural expression. There are many common features between adventure tourism and the services (such as home visits and mid-winter safaris with a local reindeer herder taking additional food to reindeer grazing freely in the forest) offered in Inari as a part of local Sámi tourism. However, the owners of small family businesses integrating tourism into reindeer herding, which often serves as the primary source of livelihood, and sometimes also into traditional handicrafts did not market their products as adventure tourism. Rather, they associated it with risk-related and physically demanding outdoor activities. Some informants also expressed hesitation towards the concept of 'adventure', as it directs thoughts towards activities such as husky safaris that are strange or even harmful to the local Indigenous culture. While the local indigenous entrepreneurs largely agreed on the desired types of activities and ways of presenting Sámi culture to tourists, there were also differences. For example, some entrepreneurs categorically refuse to wear traditional Sámi clothing (*gákti*) for the sake of tourists, but others utilise traditional clothing in marketing and wear at least parts of the dress when hosting guests.

A foundation for further development of Sámi adventure tourism seems to exist, and there are already some itineraries, mainly in Norway (see Authentic Scandinavia, 2023; Visit Norway, 2023). This foundation is supported by the awareness of the need, and the steps already taken, to educate visitors about authentic Sámi products and services. In practice, this movement has taken form in labels such as Sámi Duodji (the Sámi handicraft trademark, created to show buyers that the producer is Sámi either with an education or strong experience in handicrafts) and Sápmi Experience (created to support the sustainable development of Sámi tourism, facilitate cooperation, and help tourists identify destinations where Sámi culture is being ethically presented (de Bernardi et al., 2017; Keskitalo et al.: 2021).

However, as the earlier Inari example shows, the concept of adventure tourism is not yet familiar to all actors involved in Sámi tourism. Therefore, the meanings attached to adventure vary greatly. The final decision of whether to follow the example of many other Indigenous groups and develop and market their activities under the umbrella of adventure tourism remains to be made by the Sámi entrepreneurs themselves and their possible non-Indigenous collaborating partners.

Adventure tourism in highly assimilated Indigenous communities

Due to long-standing cultural assimilation, the process whereby minority groups of differing ethnic heritage are absorbed into the dominant culture of a society, some Indigenous communities have lost or are on the verge of losing important elements of their culture and/or connection with their traditional livelihoods and lands. This may cause trouble defining what the content of adventure-based indigenous cultural tourism could be. The Indigenous Ainu people, who have historically lived in northern parts of Japan and on islands that are now part of Russia, are an example of an Indigenous group involved in adventure tourism (see Case study 25.2). Adventure tourism offerings are developed despite the long history of oppression, during which most Ainu lost their language, customs, traditional subsistence strategies/means of livelihood (e.g., hunting and fishing), social structures, and living environments (Bellow et al., 2019).

Case study 25.2   Adventure tourism in the assimilated Indigenous Akan Ainu community, Japan

The roots of Ainu-related tourism can be traced back at least to the late 1950s, but the relationship between Ainu culture and tourism has always been complex. While tourism has arguably given economic benefits and livelihoods to many Ainu, helped educate and increase awareness of Ainu culture both within Japanese society and abroad and provided a venue to maintain and reinforce culturally relevant skills, arts, expression, and language, the effects of tourism have often been perceived as negative. Ainu tourism is carried out in different parts of Hokkaido and takes form, for example, in dances, songs, rituals and Ainu products. Ainu tourist centres have been criticised because they allegedly produce and reinforce stereotypes, prejudices and fabricated images and offer yet another tool for ethnic Japanese (*Wajin*) to control the Ainu. While tourism has offered a channel for cultural self-expression and heritage preservation, it has often been limited to forms that consumers find appealing. The Ainu involved in tourism have been blamed for selling their culture and making a show of themselves (Bellow et al., 2019; Cheung 2005; Hiwasaki, 2000).

During the past ten years, adventure tourism has emerged as a new framework for Ainu tourism, especially in the Lake Akan area in eastern Hokkaido introduced below. While adventure tourism in the Akan region leans on nature-based experiences, Ainu culture-based products, such as a forest walk with an Ainu guide and short-term wood carving workshops, have recently been developed. Some courses also involve lunch to introduce traditional Ainu food. Hokkaido also hosted the virtual 2019 Adventure Travel World Summit and will do so again in 2023. While the informants involved in culturally orientated adventure tourism described homestay-like activities as an example of a perfect Ainu adventure tourism experience, they also underlined adventure tourism's role as a business that needs to make profits and satisfy customer demand. The development of adventure tourism has contributed to the critical discussion of authenticity in the context of Ainu tourism, but this debate has been ongoing for decades. Besides the observations made by outsiders, self-reflective voices have also arisen inside the Indigenous community. These discussions have taken the form of debates on wood carvings of bears, often with salmon in their mouths, for example. These items are nowadays the best-known examples of Ainu wood carving, but it is difficult to describe them as the symbols of traditional culture. The idea of a carved wooden bear was, in fact, 'imported' from Switzerland in 1922. The wood carving tradition has also been influenced by the work of *Wajin* craftsmen working side by side with the Ainu and the development of other characters that do not reflect Ainu tradition or cultural roots but the aesthetic tastes of visiting customers. These products have often been sold to tourists alongside products made abroad. The debate concerning the Marimo festival, established in 1950 to symbolise harmony between humans and nature and created around the endangered green Marimo algae that grow into balls, is another example of the critical discussion concerning authenticity. Although considered a representation of Ainu cosmology and a traditional idea emphasising the importance of returning and receiving, it has also been criticised as an 'invented' tradition (Akibe, 2010; Cheung, 2005).

Lake Akan is the home of one of the largest Ainu communities in Japan. The present Ainu living in the area do not share a common origin and ancestors or a long socio-historical relations with the landscape. Instead, a great majority of them or their parents came to Lake Akan from different parts of Hokkaido after the locally powerful Maeda family entrusted the free use of land to the Ainu people. The Akan Ainu community has been tightly involved in tourism since its establishment in the late 1950s (Akibe, 2010). Besides the traditional handicrafts sold in small shops in Ainu *kotan* and dances and religious rituals executed either during special occasions or for touristic purposes, the Ainu culture is also introduced to visitors through experiences involving unprecedented features such as outdoor multimedia entertainment experiences ('Kamuy Lumina' night walk inside the Akan Mashu National Park) and modern dance and digital art ('Lost Kamuy' performance in the Akan Ainu Ikor Theater). While the Akan Ainus's readiness to engage in tourism brought them the nickname 'kankō Ainu' (*kankō* means 'tourism') and exposed them to contempt and criticism among both other Ainu communities and non-Ainu observers, it has also brought income and revitalised, or at least secured, the continuity of the community (Cheung, 2005; Hiwasaki, 2000). The attempts to promote and preserve both the tangible and intangible genuine elements of culture have taken form in the establishment of institutions involved in consulting for, producing and certifying products and services utilising and referring to the Ainu culture (Akan Ainu Konsarun, 2022). The recent emphasis on adventure tourism has contributed to the need to educate Ainu guides to possess the necessary skills to take visitors into a forest environment.

The actors involved in Ainu adventure tourism face the challenge of creating authentic experiences in a situation in which traditional livelihoods have ceased to exist. Some adventure tourists may find an invitation to join a spiritual ceremony or wear traditional Ainu clothes, for example, to be somehow inappropriate. However, adventure tourists desiring to see the 'real' must accept the fact that modern-day Ainu usually live lives that are very similar to those of other Japanese citizens. Meanwhile, the Ainu who, for example, sell Ainu products at their shops during the day and participate in dance shows at night are keeping their traditional culture alive while simultaneously creating new forms of cultural expression. Luckily, as pointed out by Hiwasaki (2000) two decades ago, Ainu have gradually managed to take greater ownership of their cultural expression, and the new initiatives often originate from their own aspiration, rather than from the assumed or proven desires of the visitors.

Akan Adventure Tourism: http://akanat.co.jp/?en

Is adventure tourism an attractive option for Indigenous communities?

Indigenous tourism has already been incorporated within other types of tourism labels and analytical concepts. In the field of ecotourism, for example, attention has been paid to the growing number of ecotourists considering culture and local ways of living as components of their desired experience. As a part of this discussion, concepts such as community-based ecotourism and Indigenous ecotourism, emphasising local/Indigenous knowledge of the environment and learning experiences both to hosts and visitors, have been developed (Walter, 2013). While some Indigenous communities may still be better aware of concepts such as 'ecotourism' or 'community-based tourism', Ainu tourism in Lake Akan is an example of development whereby previous interest in ecotourism (see Akibe, 2010) has paved the way for the more recent development of Indigenous adventure tourism emphasising outdoor activities and visitors' interactions with hosts.

The relationship between Indigenous tourism and adventure tourism is tightly connected with the way one defines the latter. Indigenous tourism seems to be a great match with so-called soft adventure tourism. However, the future development of Indigenous adventure tourism looks difficult if the definitions highlighting the risks and physical challenges form the main narrative for adventure tourism–related discussion. Furthermore, as the Sámi tourism case from Inari demonstrates, the evolving ideas and definitions of adventure tourism shared by academics and the practitioners who are already involved do not automatically reach all the Indigenous communities. Therefore, the opportunities related to adventure tourists, who often stay longer, spend more money locally, and share an interest in remote destinations that may lack developed infrastructure, may not be identified without active interaction between the adventure tourism industry and different Indigenous communities.

Adventure tourism should not be simplified as a trouble-free pathway to success. Adventure tourists' desire to see and experience the 'real' may also make them more demanding. In other words, they may not be satisfied with traditions being externalised into polished and controlled performances or well-known symbols but expect greater dialogue with Indigenous community members (Figure 25.2). Meanwhile, Martin and Ren (2020) offered an opposite example in their study of Chinese adventure tourism in Greenland. They showed that while Visit Greenland as a host had adopted the ATTA's definition of adventure tourism and promoted a dynamic view of local culture consisting

Figure 25.2 Mummified ancestor in indigenous highland community, Wamena, West Papua. Photograph by Carl Cater.

of elements of both tradition and progress, this was not reflected in the actions and expectations of visitors. Martin and Ren (2020) showed how Greenland, where all offerings excluding cruise tourism are marketed as adventure tourism, was seen by Chinese tourists primarily as a destination for natural sights, and the interaction with the local, predominantly Indigenous, population and culture was limited and even culturally insensitive.

The COVID-19 pandemic may also have affected how Indigenous communities see adventure tourism. The pandemic had a huge impact on the tourism industry, including Indigenous tourism. The Indigenous Tourism Association of Canada (ITAC, 2022), for example, reported an estimated 70 per cent reduction in the Indigenous tourism industry's contribution to the GDP and a loss of 21,000 jobs. As discussed by the representatives of Indigenous tourism businesses from Australia, New Zealand, and Papua New Guinea at an ATTA (2022d) event in April 2022, the new situation forced different actors to adapt. Yet, the travel restrictions and sharp decline in tourist flows also challenged many to re-evaluate the sustainability and resilience of the practices carried out during the pre-pandemic rapid growth era. Therefore, the recovery strategies may guide the companies involved in Indigenous tourism to critically consider whether the growth-orientated approaches taken previously are serving the long-term future needs of the community and whether they show respect for the environment and past generations. Here, adventure tourism may offer a sustainable and responsible framework for Indigenous communities to re-engage with the tourism industry.

Conclusion

One can discuss Indigenous tourism without reference to adventure, and it is possible to write articles and books on adventure tourism without making any reference to Indigenous people. However, these two forms of tourism are conceptually and theoretically interlinked. In addition to this, the benefits at the practical level can be seen for various actors in the field of Indigenous adventure tourism, which is developed to facilitate respectful encounters between the hosts and visitors in which both sides are willing and ready to share and exchange. Adventure tourism could be part of the process whereby Indigenous people claim greater rights to represent their own culture, history, and modernity. Yet, all the actors involved need to accept the fact that there are different Indigenous identities, that Indigenous communities are not homogeneous entities, and that there are different views concerning tourism within these communities. Therefore, there are various answers to the question of what components of Indigenous cultures, customs, arts, and ways of living can be shared with tourists and how this sharing process should take place.

The importance of discussing authenticity in the context of Indigenous adventure tourism needs to be recognised. However, the focus should not only be on the search for authentic (or inauthentic) features or characteristics but also power relations. Indigenous actors should have a free hand to decide whether to be involved in adventure tourism, and they should also be in charge of the process of defining what an authentic adventure tourism experience is that they would want to be involved in. It is also necessary to recognise that different Indigenous groups are facing very different realities, and this may reflect both different interpretations of authenticity and desired forms of cooperation with non-Indigenous actors.

Review questions

1 Why has tourism been seen both as a threat and an opportunity in various Indigenous communities?
2 What opportunities can adventure tourism bring to Indigenous communities and entrepreneurs (especially when compared with mass tourism)?
3 Why is it difficult to define what an authentic Indigenous adventure tourism experience is?

Further reading

Brattland, C., Ren, C., Bemborm, E., & Bruin, R. (2022). The domestic turn in post-pandemic indigenous Arctic tourism: Emerging stories of self and other. *Tourism Culture & Communication.* https://doi.org/10.3727/109830422X16600594683337

Butler, R. (2021). Research on tourism, indigenous peoples and economic development: A missing component. *Land, 10,* 1329. https://doi.org/10.3390/ land10121329

Harbor, L. C., & Hunt, C. A. (2021). Indigenous tourism and cultural justice in a Tz'utujil Maya community, Guatemala. *Journal of Sustainable Tourism, 29*(2–3), 214–233. https://doi.org/10. 1080/09669582.2020.1770771

Viken, A. (2022). Tourism appropriation of Sámi land and culture. *Acta Borealia: Nordic Journal of Circumpolar Societies.* https://doi.org/10.1080/08003831.2022.2079276

References

Adventure Travel News. (2012, May 11). *The Larrakia Declaration on the development of indigenous tourism.* https://www.adventuretravelnews.com/the-larrakia-declarationon-the-development-of-indigenous-tourism

Adventure Travel News. (2022, October 12). ATTA joins indigenous tourism collaborative of the Americas as a Founding Member. *Adventure Travel News.* https://www.adventuretravelnews. com/atta-joins-indigenous-tourism-collaborative-of-the-americas-as-a-founding-member

Adventure Travel Trade Association. (2022a). *Values statement.* https://cdn.adventuretravel.biz/ wp-content/uploads/2018/01/Values-Statement-Trade-English.pdf

Adventure Travel Trade Association. (2022b). *Our initiatives.* https://about.adventuretravel.biz/ our-initiatives

Adventure Travel Trade Association. (2022c). *Adventure travel industry snapshot 2022.* https:// cdn-research.adventuretravel.biz/research/3085902385903285901421/Industry-Snapshot-Trends-2022-Report.pdf

Adventure Travel Trade Association. (2022d). *Meet the Experts: Indigenous & Community-based Tourism: Challenges & Opportunities.* https://learn.adventuretravel.biz/webinar/meet-the-experts-indigenous-community-based-tourism-challenges-opportunities.

Akan Ainu Konsarun. (2022). *Akan Ainu Konsarun.* https://a-ainucon.com/consultation/

Akibe, H. (2010). Kankō to ainu minzoku [Tourism and the Ainu people]. Gendai shakai to sen-shūmin bunka – kankō, geijutsu kara kangaeru 1 [Contemporary society and indigenous culture: Consideration through the tourism and art 1). *Dai 24 kai hoppō minzoku bunka shinpojiumu hōkokusho [Proceedings of the 24*[th] *International Abashiri Symposium],* 19–23.

Authentic Scandinavia. (2023). *Tromsø Sami & northern lights adventure.* https://www.authentic-scandinavia.com/tours/tromso-sami-and-northern-lights-adventure

Beckmann, C. (2017, January 31). *Research reveals adventure travelers primarily motivated by transformation.* Adventure Travel News. https://www.adventuretravelnews.com/research-reveals-adventure-travelers-primarily-motivated-by-transformation

Beer, S. (2013). Philosophy and the nature of the authentic. In M. Smith & G. Richards (Eds.), *The Routledge handbook of cultural tourism* (pp. 47–52). Routledge.

Bellow, E., Majd, T., & Casalegno, C. (2019). Creative and sustainable tourism: The case of Ainu in Japan. *Symphonya. Emerging Issues in Management*, 2, 119–132. https://doi.org/10.4468/2019.2.11bellow.majd.casalegno

Bresner, K. (2010). Othering, power relations, and indigenous tourism: Experiences in Australia's Northern Territory. *PlatForum*, *11*, 10–26.

Butler, R. (2021). Research on tourism, indigenous peoples and economic development: A missing component. *Land*, *10*(12), 1329. https://doi.org/10.3390/land10121329

Carr, A., Ruhanen, L., & Whitford, M. (2016). Indigenous peoples and tourism: The challenges and opportunities for sustainable tourism. *Journal of Sustainable Tourism*, *24*(8–9), 1067–1079, https://doi.org/10.1080/09669582.2016.1206112

Cheung, S. C. H. (2005). Rethinking Ainu heritage: A case study of an Ainu settlement in Hokkaido, Japan. *International Journal of Heritage Studies*, *11*(3), 197–210.

Cohen, E. (1988). Authenticity and commoditization in tourism. *Annals of Tourism Research*, *15*(3), 371–386.

de Bernardi, C. (2019). Authenticity as a compromise: A critical discourse analysis of Sámi tourism websites. *Journal of Heritage Tourism*, *14*(3), 249–262. https://doi.org/10.1080/1743873X.2018.1527844

de Bernardi, C. (2019b). A critical realist appraisal of authenticity in tourism: The case of the Sámi. *Journal of Critical Realism*, *18*(4), 437–452. https://doi.org/10.1080/14767430.2019.1654702

de Bernardi, C. (2020). *Authenticity as a compromise: A critical realist perspective on Sámi tourism labels.* Acta Electronica Universitatis Lapponiensis. https://lauda.ulapland.fi/handle/10024/64414

de Bernardi, C., Kugapi, O., & Lüthje, M. (2017). Sámi indigenous tourism empowerment in the Nordic countries through labelling systems: Strengthening ethnic enterprises and activities. In I. B. de Lima & V. T. King (Eds.), *Tourism and ethnodevelopment: Inclusion, empowerment and self-determination* (pp. 200–212). Routledge.

Farkić, J., & Gebbels, M. (2022). *The adventure tourist: Being, knowing, becoming.* Emerald Publishing.

Gilmore, J. H., & Pine II, B. J. (2007). *Authenticity: What consumers really want.* Harvard Business School Press.

Hiwasaki, L. (2000). Ethnic tourism in Hokkaido and the shaping of Ainu identity. *Pacific Affairs*, *73*(3), 393–412.

Huddart, D., & Stott, T. (2020). *Adventure tourism: Environmental impacts and management.* Palgrave MacMillan.

Indigenous Tourism Association of Canada. (2022). *Building back better. Strategic recovery of indigenous tourism in Canada 2022–2025.* https://indigenoustourism.ca/wp-content/uploads/2022/01/ITAC-Building-Back-Better-2022-2025.pdf

Indigenous Tourism Association of Canada. (2023) *The original original.* https://originaloriginal.ca

Keskitalo, E. C. H., Schilar, H., Heldt Cassel, S., & Pashkevich, A. (2021). Deconstructing the indigenous in tourism. The production of indigeneity in tourism-oriented labelling and handicraft/souvenir development in Northern Europe. *Current Issues in Tourism*, *24*(1), 16–32. https://doi.org/10.1080/13683500.2019.1696285

Leleto, N. L. (2019). Maasai resistance to cultural appropriation in tourism. *The Indigenous Peoples' Journal of Law, Culture & Resistance*, *5*(1), 21–34. http://doi.org/10.5070/P651043046

Leu, T. C., Eriksson, M., & Müller, D. K. (2018). More than just a job: Exploring the meanings of tourism work among indigenous Sámi tourist entrepreneurs. *Journal of Sustainable Tourism*, *26*(8), 1468–1482. https://doi.org/10.1080/09669582.2018.1466894

MacCannell, D. (1973). Staged authenticity: Arrangements of social space in tourist settings. *American Journal of Sociology*, *79*(3), 589–603.

Martin, N., & Ren, C. (2020). Adventurous Arctic encounters? Exploring Chinese adventure tourism. *Scandinavian Journal of Hospitality and Tourism*, *20*(2), 126–143. https://doi.org/10.1080/15022250.2020.1743752

Mazzocchi, F. (2020). A deeper meaning of sustainability: Insights from indigenous knowledge. *The Anthropocene Review*, *7*(1), 77–93. https://doi.org/10.1177/2053019619898888

Niskala, M., & Ridanpää, J. (2016). Ethnic representations and social exclusion: Sáminess in Finnish Lapland tourism promotion. *Scandinavian Journal of Hospitality and Tourism, 16*(4), 375–394. https://doi.org/10.1080/15022250.2015.1108862

Olsen, K. (2008). The Maori of tourist brochures representing indigenousness. *Journal of Tourism and Cultural Change, 6*(3), 161–184. https://doi.org/10.1080/14766820802553152

Pereiro, X. (2013). Understanding indigenous tourism. In M. Smith & G. Richards (Eds.), *The Routledge handbook of cultural tourism* (pp. 214–219). Routledge.

Rantala, O., Hallikainen, V., Ilola, H., & Tuulentie, S. (2018). The softening of adventure tourism. *Scandinavian Journal of Hospitality and Tourism, 18*(4), 343–361. https://doi.org/10.1080/150 22250.2018.1522725

Rantala, O., Rokenes, A., & Valkonen, J. (2016). Is adventure tourism a coherent concept? A review of research approaches on adventure tourism. *Annals of Leisure Research, 21*(5), 1–14. https://doi.org/10.1080/11745398.2016.1250647

Sámediggi. (2022). *Ethical guidelines for Sámi tourism.* https://www.samediggi.fi/ethical-guidelines-for-sami-tourism/?lang=en

Smith, V. (1996). Indigenous tourism: The four Hs. In R. Butler & T. Hinch (Eds.), *Tourism and indigenous peoples* (pp. 283–307). Thomson International.

Thompson-Carr, A. (2013). Māori tourism: A case study of managing indigenous cultural values. In M. Smith & G. Richards (Eds.), *The Routledge handbook of cultural tourism* (pp. 227–235). Routledge.

UiT Norges arktiske universitet. (2023). *Arctic adventure tourism – bachelor.* https://uit.no/utdanning/program/360168/arctic_adventure_tourism_-_bachelor

Viken, A., & Müller, D. K. (2017). Indigenity and indigenous tourism. In A. Viken & D. K. Müller (eds.), *Tourism and identity in the Arctic* (pp. 16–32). Channel View Publications.

Visit Norway. (2023). *Sami adventure with Turgleder.* https://www.visitnorway.com/listings/sami-adventure-with-turgleder/130966/

Walter, P. G. (2013). Theorising visitor learning in ecotourism. *Journal of Ecotourism, 12*(1), 15–32. https://doi.org/10.1080/14724049.2012.74209

Ween, G. B., & Riseth, J. Å. (2017). Indigenous hospitality and tourism: Past trajectories and new beginnings. In A. Viken & D. K. Müller (Eds.), *Tourism and identity in the Arctic* (pp. 205–221). Channel View Publications.

World Indigenous Tourism Alliance. (2022). *World Indigenous Tourism Alliance.* https://www.winta.org

Zhang, J., & Müller, D. (2018). Tourism and the Sámi in transition: A discourse analysis of Swedish newspapers, 1982–2015. *Scandinavian Journal of Hospitality and Tourism, 18*(2), 163–182. https://doi.org/10.1080/15022250.2017.1329663

26 Gender and adventure tourism

Maggie C. Miller and Jenny Cave

Chapter learning outcomes

1 Broaden understandings of gender by recognising a multiplicity of femininities and masculinities, particularly in the context of adventure tourism.
2 Critically evaluate the impact of gendered norms on adventure representations and women's participation in outdoor and adventure activities.
3 Identify the implications of gender as it relates to bodily ideals and embodied adventure experiences.
4 Highlight the impacts of inclusion on long-term sustainability, in particular social sustainability, in adventurous experiences.

Introduction

Access to adventure recreation has long been critiqued for a lack of equality and equity, influenced by socio-demographic factors, including gender, race, and socio-economic status (Doran, et al., 2020; Houge Mackenzie & Hodge, 2019; Zink & Zane, 2015). Although many of these are intersecting factors, this chapter focuses its interest on gender as a significant construct with profound implications for adventure tourism experiences and their sustainability. Gender in many tourism studies has been typically studied as a psychological variable for "sex/gender" differences, often viewed from a realist perspective resulting in a compare-contrast discourse (Clarke et al., 2022). Here *gender* is sometimes conflated with *sex* and reproduced as a binary and fixed construct rather than a fluid identity associated with a multiplicity of femininities and masculinities. Findings from these studies provide insights into women's experiences in comparison to men; however, the depth of inquiry can be limited by underlying assumptions of biological sex, which overlooks the complexity of gender and restricts the progress of gender inquiry within tourism. Although this chapter acknowledges these limitations and echoes the need for a more fluid and permeable conceptualisation to think with, and through, gender issues (Clarke et al., 2022; Knijnik et al., 2010) (explored more in Chapter 27), we centre on women and their experiences since this continues to be the primary focus for much of the current adventure tourism scholarship.

In doing so, this chapter examines the nexus of gender and adventure tourism by considering three key, interrelated topic areas: participation and representation, embodied

DOI: 10.4324/9781003393153-31

experiences, and inclusion and sustainability. Each section provides a brief overview of key scholarly developments and debates in the area, beginning with an analysis of women's participation and representation in adventure and outdoor activities. We then explore research that focuses on embodiment in adventure tourism, highlighting Isis Arlene Díaz-Carrión's case study on pregnant Mexican bodies in mountaineering. Finally, our last topic area focuses on gendered inclusion, and the role it plays in the social sustainability of adventure enterprises. Here we feature tourism practitioner Chan Yuen-Li (personal communication, August 1, 2019) and her case story about building a successful adventure tourism business which embodies the intersectionality of gender, culture, and ethnicity. The chapter concludes with insights for future investigation.

Participation and gendered representation

Consideration of gender in relation to adventure and outdoor pursuits has gained significant traction since Henderson's (1990; 1992) foundational work around the meaning of leisure and the outdoors for women. Early work like this concluded that many of the positive aspects and benefits of adventure experiences filtered into other areas of women's lives (Hornibrook et al., 1997). For example, the women in Boniface's (2006) study described the strength and confidence they gained from high-risk adventure experiences, which positively influenced their well-being and self-esteem. One of the research participants explained that through personal challenges found in the outdoors, 'you become a whole person' and that the outdoors is what drives her to do everything (Boniface, 2006, p. 18). In recognising the importance of such experiences, a popular and recurring theme for analysis has been the nature of factors that inhibit women's participation in leisure and the outdoors and, more recently, constraints to participating in adventure tourism (Doran, et al., 2018; 2020; Fendt & Wilson, 2012).

Low, Miller, Doran, and Hardwick (2022) point out that powerful (and alarming!) normatives operate within the outdoors, particularly in mainstream media. These norms and related narratives marshal what we expect a "hiker" or "climber", or more broadly an "adventurer", to be and look like, ultimately impacting who might participate in outdoor and adventurous experiences. Indeed, the adventure traveller has been traditionally associated with masculine attributes of strength and bravery (Myers, 2010). This rests on a historical foundation of colonial exploration defined by male adventurers in which adventure and risk were intertwined in a quest for progress, as described in Chapter 2. Such tracings of colonialism have contributed to the hyper-masculinisation of adventure and constructions of masculinity and femininity, with the former ascribing to aggressive, active, and mobile characteristics and the latter to nurturing, passive, and immobile attributes (Elsrud, 2001). Doran et al. (2018) argue that adventure tourism tends to 'reinforce traditional views of the heroic white male adventurer, conveying masculine features, such as bravery, risk-taking, competitiveness, physical strength, and ruggedness' (p. 403). Consequently, gender stereotypes and the naturalisation of gender roles remain in adventure spaces imposing constraints on women (Díaz-Carrión et al., 2020). For example, if women are included in media representations of adventure, research indicates they are seldom shown alone in the outdoors actively participating in an activity or portrayed as dirty and unkempt. Instead, women are depicted rather passively: walking with children, looking at a view, or holding a piece of equipment rather than participating in the activity that uses this equipment (Kling, et al., 2018; Low et al., 2022).

Although many studies suggest that such gendered representations may limit women's participation in adventure experiences, recent research explores ways that women turn to social media to challenge and resist gendered and hegemonic discourses through "self-presentation" (Gray et al., 2018; Low et al., 2022; Stanley, 2020; Weatherby & Vidon, 2018). Self-presentation is the process of packaging and editing or representing the self to convey particular impressions to others (Goffman, 1959). For some, this curation of a public self occurs online via social media sites. From their examination of user-generated content on Instagram associated with the hashtags #womenoutdoors, #womeninadventure, and #shewentwild, Low et al. (2022) conclude that the ubiquity of social media helps reshape the nature of women's participation in adventure activities. For instance, while a number of scholars recognise the continued use of passive, heteronormative and "soft" portrayals of women in mainstream media (Frazer & Anderson, 2018; Kling et al., 2018), Low et al.'s (2022) findings demonstrate that for many women, self-representation centres on active depictions of "doing" (e.g., see Figure 26.1), highlighting exploration, strength and courage, and "growing" (e.g., see Figure 26.2) through achievement and overcoming challenges. These digital depictions – and the narratives associated with them – can contribute to activism and amplification aimed at diversifying the outdoors (see also Stanley, 2020) in which women might relate to adventure differently, potentially holding implications for increased participation.

Beyond representational debates, however, many women face additional barriers when trying to take part in various adventurous activities. Crawford et al. (1991) have identified and categorised many of these barriers as intra-personal, interpersonal, and structural constraints. Some researchers have adapted these traditional constraint categorisations to conceptualise women's barriers as personal (e.g., sense of guilt; fear), socio-cultural (e.g., social expectations; lack of companions), and practical (e.g., lack of money; need for certain equipment; Doran, 2016; Fendt & Wilson, 2012). Dilley and Scraton's (2010) data suggest that consideration of women's wider lives (e.g., work, attitudes towards childbearing, gender identities, social relationships, and motherhood) is

Figure 26.1 Example of women "doing" – exploration of the outdoors.

Source: Low et al. (2022).

Figure 26.2 Example of "growing" – skills development.

Source: Low et al. (2022).

integral to understanding their participation in, and commitment to, an activity (in their case rock climbing). Furthermore, constraints to participation may vary depending on where the activity falls on the soft–hard adventure tourism continuum (Doran et al., 2020). In their work around mountaineering tourism, Doran et al. (2020) indicate that women's participation can be greatly impacted by their feelings of anxiety and self-doubt due to physical or technical skill deficiencies. Certainly, specific skills and acumen are required for many adventurous activities but perceptions (and self-perceptions) of women being less competent and capable than men are fuelled by gender stereotypes (Brescoll, et al., 2010; Houge Mackenzie et al., 2020). This is evidenced by the continued practices of typecasting and discrimination which not only constrain adventure participation from a demand perspective but have greatly curbed female growth in adventure guiding and leadership positions as well (Hall, 2019; Houge Mackenzie et al., 2020).

Nevertheless, constraint negotiation researchers contend that women are not 'passive victims of constraints', rather active agents who are motivated to pursue their adventurous aspirations despite having identified barriers to participation (Doran et al., 2020, p. 722; see also Doran, 2016; Evans & Anderson, 2018; Fendt & Wilson, 2012). This is perhaps why we have seen a significant growth in female participation in the adventure tourism sector. Many adventure tour operators have reported an increase in women-targeted trips ranging from kayaking to mountain biking, while female-centric adventure travel products expanded by 230% between 2013 and 2018 (The Adventure Travel Trade Association [ATTA], 2018). More recently, the March 2022 ATTA report, surveying adventure businesses within the industry, has indicated that women comprise 57% of the respondents' clients – illustrating a clear demand for adventure travel. Such findings and trends reveal a greater need to support this growing demand by listening to what women want and encouraging equity within and across adventure experiences and enterprises. To do this, it is important to continue to situate participation in the broader social, cultural, and political contexts that frame adventure and the individuals that embody these spaces.

Adventurous bodies and gendered experiences

Adventure tourism can also be understood as embodied sites of human experience and meaning-making. Recent definitions incorporate these elements by framing adventure as 'self-initiated, nature-based physical activities that generate heightened bodily sensations and require skill development to manage unique perceived and objective risks' (Boudreau et al., 2020, p. 2). For some women participating in adventure activities, placing one's body in physically demanding situations provides an opportunity to feel connected to their body in a way that would not normally occur (Dilley & Scraton, 2010). To explore these ideas further, this section focuses on the adventurous body and how these embodied experiences are gendered.

Embedded within the "embodied turn" of social sciences (Shilling, 2017), theorisations of embodiment have increasingly gained traction in tourism studies (Everingham et al., 2021). However, there remain limited understandings of embodiment within, and across adventure tourism (some notable exceptions: Cater, 2006; Cater & Cloke, 2007; Farkić, 2021; Humberstone, 2015), particularly where it is assumed women's bodies do not fit hypothesised norms. As demonstrated in the previous section, adventure tourism is, in effect, a model of heteronormative hypermasculinity, which can silence bodily differences. Accordingly, non-white, female, queer/gay, fat, disabled, and other deviant bodies are rarely represented in mainstream adventure media (Ahmad & Thorpe 2020; Stanley, 2020; Weatherby & Vidon, 2018). Thus, bodily ideals are propagated and reproduced in adventure spaces.

Early work by McDermott (2004) on body perception suggests that one of the primary ways women experience and understand their body is in terms of appearance. This is no surprise when taking into consideration possible cultural discourses at play. For instance, Small's gender analysis of "beach" (2017) and "holiday" (2016) bodies attends to the experiences of women becoming "beach body ready". Here, Small (2017) applies critical discourse analysis to demonstrate how Australian magazines reproduce a normative female body that women should aspire to achieve in particular tourism spaces. This resonates with the sexualised and feminine body posture that is presumed for the female surfers of Knijnik et al.'s (2010) study on the Brazilian beach. Accordingly, the young, slim, white, tanned, able-bodied woman is constructed as an "ideal tourist" that governs how women organise their bodies and tourism practices. In other adventure contexts, women's body practices are seemingly attuned to more traditional, "masculinised" ways of being in order to legitimise their adventurous pursuits and achievements (Hall, 2019; Houge Mackenzie et al., 2020). Influenced by feminist sport sociologist Prikko Markula (2006) and her understanding of the rhizomatic body, Knijnik et al. (2010) consider how 'sport and physical activity have been used to oppress and limit the way in which women expressed themselves corporally' (p. 1172). Their research findings encourage reflection on the possible ways that femininities can be re-modelled, resisting dominant values and beliefs of masculine hegemony, sporting (adventure in our case) subcultures, and the cultural contexts in which these activities occur in the twenty-first century.

Some feminist scholarship emphasises that "ability", functionality, and how the body moves as it relates to physicality as increasingly significant to women participating in adventure tourism (Dilley & Scraton, 2010; Hall, 2019; McDermott, 2004). For example, one participant from Dilley and Scraton's (2010) study explained that she was happiest when her body was strong and fit for rock climbing. This climber, along with other women, did not see themselves as "normal" women because they identified with both traditional masculine and feminine qualities to varying degrees in different areas of their lives. At times this was only realised through engagement with single-gender spaces. For the women of McDermott's (2004) phenomenological research, single-gender spaces enabled women to be more conscious and aware of their bodies in ways that transcend simple scrutiny of aesthetics and appearance. As a result of being challenged through the various components constituting canoeing (e.g., technical skills of paddling, portaging, etc.), women in this study gained a sense of physical confidence, competence, and feeling strong, which contributed to their own sense of identity as 'framed by both the embodiment of a "feminine bodily existence" as well as existing within a male-dominated society' (McDermott, 2004, p. 298).

Although there seems to be an emergence of more "gender-aware" (Clarke et al., 2022, p. 185) research, recognising a multiplicity of femininities and masculinities, risk continues to be a factor that reinforces the gendering of adventure tourism. In many adventure settings, perceptions of corporeal risks contribute to the sense of excitement and adventurousness (Cater, 2006); however, research demonstrates how risk-taking behaviour is received differently by society as it relates to gendered bodies. According to Yang et al. (2017), women's risk-taking behaviours are more likely to be negatively evaluated compared to those of men. This certainly resonates with the featured case study in Case study 26.1, on pregnant Mexican mountaineers. Here socio-cultural constructions of gender and femininity, along with normative antenatal expectations, impose restrictions on the pregnant body (Díaz-Carrión et al., 2020, p. 117–118). Relatedly, Myers's (2010) research raises questions about the socialisation of ageing bodies. Findings suggest that age is influential in women's engagement with certain types of adventure activities and thereby acceptable levels of risk:

The "extreme" adrenaline-rush experiences of skydiving, bungee jumping, river rafting, river sledding, and heli-hiking were sought by a larger percentage of younger women, while the older women had a tendency towards the "softer" adventure activities including tramping (walking in the New Zealand bush), rafting in glow-worm caves, swimming with dolphins, and jet boating.

Indeed, risk is mediated by the body and interrelated emotions. When risk is perceived as being higher than competence and skill, fear and anxiety ensue (Cater, 2006). Fear feels dangerous and, if not properly managed, can itself add increased risk and danger. As women overcome such fear, they accumulate "adventure capital" whereby the bodily risks they engage form a 'strong story about the "self" an "adventure narrative" that can only be highly valued in opposition to something different: the non-adventurer' (Elsrud, 2001, p. 603). As such, risk-taking during adventure activities provides opportunities for personal challenge, achievement, increased confidence, and potential empowerment for women (Myers, 2010). Moreover, adventure tourism can be positioned as a 'risky behaviour providing a way for young women to negotiate and contest dominant discourses around feminine, cultural identities' (Green & Singleton, 2006, p. 853). For instance, in

Case study 26.1 Are Mexican pregnant bodies' adventure bodies?

This case illustrates the ways that gender stereotypes contribute to the masculinization of adventure spaces, in which notions of risk (re)produce gendered dynamics on mountains, rivers, canyons and other adventurous landscapes (Doran et al., 2020; Knijnik et al., 2010). Mexican adventure spaces are no exception: the mountain is for men (el cerro es para los hombres) continues to be a societal norm for economic activities – such as tourism – in Mexico, and participation in outdoor and adventure recreation is restricted.

In this regard, Mexican pregnant women face barriers when engaging and participating in adventure tourism. Without minimising the impacts of physical exercise on women and the foetus, constraints appear closely related to social pressures (Dale & Van Gompel, 2005; Melidosian et al., 2019). Narratives of risk grounded by societal norms based on assumptions or taboos attempt to discourage women from participating in adventure, specifically those considered as high-risk activities during pregnancy. Some Mexican adventure practitioners find that rooted in these assumptions are a myriad of criticisms about women's care abilities (e.g., ¿serás capaz de cuidar a tu bebé? Will you be able to take care of your baby?). Moreover, a *mamá mexicana* (Mexican mother) is expected to make symbolic sacrifices, and the socio-cultural expectation is that pregnant women should stop any activity or engagement in outdoor and adventure activities.

Pregnant adventure tourists not only recognize the importance of avoiding the risks involved but also consider it healthy to maintaining their engagement with physical activity (Díaz-Carrión et al., 2020). As a consequence, in this liminal stage (from a woman to a future mother), Mexican women have to negotiate their pregnancy. To them, experiencing physiological and emotional changes during pregnancy is materialised through complex entanglements. For instance, the need for apparel and equipment adapted to their bodily changes intersects with adventure enterprises' constraints, traditional medical advice and societal questioning. Emotions to manage risks can also be amplified by popular beliefs generating the internalization of fear and worry of always being under surveillance.

Nevertheless, according to Díaz-Carrión et al.'s (2020) research study, some Mexican women are able to negotiate these constraints, especially when having a close relation with their obstetric provider and when family and friends support them and their decisions. In this study, pregnant women found this support fundamental to deconstructing pregnancy and motherhood: "I just had fantastic mates they not only took care of me but also encouraged me all the time" (Evelyn; p. 384). Furthermore, they suggest that some women also recall to experience a less restrictive second pregnancy. Considering this, adventure landscapes can be spaces for transgression to challenge dominant assumptions, and provide pregnancy and subsequent parenting with supportive environments, enabling women to find new ways to maintain adventure practices. Indeed, more research like this is critical to explore womanhood and adventurous bodies and promote adventure and outdoor activities among Mexican women.

Source: Díaz-Carrión & Vizcaino, 2023).

talking about her experiences of soloing, or climbing alone, one of Hall's (2019) research participants called on the 'natural intelligence of the body' to help her climb and navigate risks (p. 163). In doing so, she created a feminised space free from social norms to practice her own form of risk-taking by re-visioning her body as 'strong, capable, active, enduring and sexually powerful' (Hall, 2019, p. 164), resisting conventional notions of femininity.

Inclusion and social sustainability

Sustainability in tourism tends to emphasise environmental or economic aspects of sustainability and ignores its social dimensions (Baum et al., 2016). Moreover, social justice conversations are largely absent in scholarship related to the outdoors (Warren et al., 2014), where issues of underserved populations and investigations of the intersections of gender, race, ability, and sexual orientation remain limited. This absence extends to adventure tourism, raising questions of social sustainability and inclusion. In the final section of our chapter, it becomes evident that social justice, ethical practices, and long-term attitudinal changes need to be in place to empower women as 'working bodies', promote inclusion, and overcome the commonly fragmented notions of sustainability within adventure tourism (Marshall, 2011). Additionally, the re-conceptualisation of self and capabilities beyond masculinist framings is critical for women's continued involvement as participants and entrepreneurs in the adventure tourism domain (Figueroa-Domecq, et al., 2022).

Women's livelihood strategies can be seen as more sustainable and gender diverse than men's, where women work within the available human, natural, and cultural resources that surround an enterprise (Chigamba et al., 2014). For example, women living on outback farms in tropical north Queensland, Australia have initiated a repurposing of farm resources for adventure tourism. Through developing and operating whitewater rafting, horse treks, climbing, and speleology (among other activities), they supplement on-farm agricultural production, thereby ensuring the survival of the farms themselves (Cave, 2010). Here, women undertake care roles within farm operations, such as locating workers, record keeping, and reallocating family and business resources across different enterprise lines while also welcoming and hosting visitors (Savage, Barbieri, & Jakes, 2022). Relatedly, Māori entrepreneurs in New Zealand's adventure tourism industry apply culturally gendered, traditional knowledge to tramping, cycling, fishing, rafting, canoeing, and kayaking activities (Zapalska & Brozik, 2018). Embedding cultural understandings into their design of adventure experiences is an approach that enables women to not only work in the tourism sector but also contribute to social, environmental, and economic sustainability, as well as gender equality through social and educational change.

Despite women's abilities to diversify their livelihoods, the short outdoor tourism seasons, low wages, and external familial responsibilities continue to challenge women's long-term participation as actors in these sectors (Løseth, 2018). Furthermore, contradictory experiences and gendered perceptions of female work, result in ambivalence amongst women, who simultaneously accept and resist their tourism roles (Tucker 2022). For instance, in Houge Mackenzie, Boudreau, and Raymond's (2020) research, they discuss how male workers and clients habitually project negative views of female adventure guides' physical capability, competence, and assumed lack of experience, undermining the effectiveness of their work. Additionally, male managers frequently ascribe passive roles or less challenging work to women, therefore female guides believe they must "outperform" male co-workers and clients to be viewed as equally competent, even in routine tasks (Houge Mackenzie et al., 2020; see also Hall, 2019). Similarly, inaccurate gendered assumptions about the effects of menstrual cycles lead managers to assign tasks requiring less psychological or physical strength (Houge Mackenzie et al., 2020). This places a burden of inauthenticity on their work and affects the social sustainability of women as workers. Unsurprisingly, frustrations with social perceptions of their competence and autonomy manifest amongst female guides as negative emotions towards co-guides, clients, and managers, reflecting their feelings of disrespect and exclusion. For redress, Baum et al. (2016) make a proactive call that occupational stereotyping, low pay, precarity, and discriminatory promotion need to be rewritten to enhance social sustainability and greater inclusion in tourism, and in our case the adventure tourism industry.

Inclusion tends to be emphasised as a "good thing" in workplaces and society, but the notion of social inclusion is also mirrored by actions of social exclusion. Notionally, segregated participation may seem inappropriate to the idea of inclusion. Yet separately conceived and delivered experiences can provide equitable access to new opportunities and engagement with people of similar capabilities and challenges. For example, the by-women for-women, *Three Sisters Adventure and Trekking Company* in Nepal navigates the complexities of gender and entrepreneurial norms in which Nepali women take part in the male-dominated adventure guiding industry (Hillman & Radel, 2021). This connects back to the section on adventurous bodies and how single-gendered spaces can support and empower women. Exclusion, in this case of men, may be an empowered act of resistance to socio-economic systems that continue to replicate and heighten material hierarchies of inequality (Edwards & Miller, 2000). Furthermore, Mitten and D'Amore (2017) indicate how experience design can finely balance capacities and skills with mastery to raise the bar for inclusion. Here, separately conceived and delivered experiences can provide equitable access to new opportunities and engagement with people of similar capabilities and challenges. In this regard, "adaptive" (and adapted) adventure could usefully be framed by the "Inclusion Spectrum" (Aylward & Mitten, 2022), which models the ways that ableist activity design might be altered for inclusion; ideas that link to Chapter 28 of this book.

As this chapter draws towards a close, it becomes apparent that interactions of oppression and privilege transect our societies and render others conceptually invisible on many dimensions: gender, race, age, physical and cognitive strength or ableist grounds. Thus, we need to be alert to the possibilities and opportunities of a multiplicity of femininities and masculinities in terms of inclusion while also incorporating perspectives that interrogate intersectionality (Aylward & Mitten, 2022; Chambers, 2022). Chan Yuen-Li's case story, shared in Case study 26.2 raises these issues and illuminates how the gendered experience is culturally situated and hence, inclusion defined much more broadly, influences the sustainability of adventure. Thus, the social turn in sustainable development argues for inclusion, activation of confidence and capabilities, as well as the ability to overturn unjust structures, mobilise resources, and exert social justice (Scheyvens & Van der Watt, 2021). To achieve this, we argue that openness, diversity, and embeddedness in nature as well as a shift towards "functional bodies" has the potential to create more sustainable and broader enterprises, embedded in communities of practice and participation that remove gendered and culture-bound prejudices.

Finally, we believe that the COVID-19 pandemic played an active role in forcing a reassessment of inclusion and sustainability in adventure tourism philosophies and practices, gendered and otherwise (Houge Mackenzie & Goodnow, 2021). Stonehouse (2022)

<hr>

Case study 26.2   "… because it's an adventure": Crafting an adventure enterprise

I founded Nomad Adventure in Malaysia in 1994, before the internet or mobile phones found traction in my little corner of the world. I felt I was a cutting-edge business because I had a fax machine, a word processor and a dot matrix printer. Adventure sports were so little known back then that once, when I was establishing a new rock-climbing route, concerned passers-by called the fire department to rescue me.

In 1999, I built the first indoor climbing gym in Malaysia, then the first adventure racing series in Southeast Asia. The success of my whitewater rafting company in Gopeng, Malaysia, spawned scores of other businesses, leading to an economic revival of the former tin mining town. Many events now replicate our adventure racing template. Today, my company is one of the more successful adventure tourism businesses in Southeast Asia.

Did my gender influence my professional life? It absolutely coloured my experience, but I cannot say that gender bent the arc of circumstances. I am classic Generation X. A disaffected latchkey kid. Adventure was my way of negotiating a place in the world – rebellion against the status quo. I feel empowered throwing the crux of a 5:10 rock climb or carving my kayak into an eddy in whitewater. Yet I see myself as being somewhat of an accidental adventure businesswoman. I simply wanted to have adventures myself, and the best way was to be in the business. No shops sold rock climbing equipment back then. So, I convinced manufacturers to let me import shipments of gear. This progressed into an outdoor retail shop. When I invited others to paddle rivers with me, I found myself in the role of whitewater guide and outfitter.

The business world was where I learned to negotiate the identity boxes one gets put in. In Malaysia's particular social context, race is a far more prominent box than gender. Today's gender inclusion lexicon means the gender barrier is made permeable. I experience race as inescapable. Social stratification within a clear hierarchy, codified by the country's constitution and enforced by political and social norms. At the top, an ascendant race holds power and opportunity. Below are nuanced layers of racial, religious, wealth, and class combinations in a very unnuanced pecking order.

So what does it matter how our identity box is cast? Which are the boldest lines that define our life's story? Identity matters for a Southeast Asian woman, crafting an adventure business, precisely because of access to opportunity.

When the opportunity arrives here in the global South, it must percolate through layers of social stratification before it reaches you. I have experienced unfairness due to qualities that have nothing to do with virtue but the right to equal access. One can however discover opportunity beyond the borders of where we come from, who our ancestors are, our gender and other accidents of birth. Opportunity is found where you define yourself. And isn't that where we find adventure?

Source: Chan Yuen-Li (personal communication, August 1, 2019) from www.nomadadventure.com

suggests that the unprecedented mobility restrictions post-COVID-19 provided an opportune time to commit to a sustainability worldview for adventure. Perhaps, this would shift the focus of adventure tourism towards becoming 'life affirming, future oriented, and solutions focused … a holistic phenomenon that combines values, knowledge, dispositions, and agency' (Nolet, 2016, p. 63). While many outdoor and adventure experiences traditionally settle around the notion of "trip" or "expedition", a post-COVID-19 awareness of diversity and inclusion draws attention towards localised, everyday microadventures (Stonehouse, 2022) or "hyperlocal" adventure and an ethic of care for local lives (Houge Mackenzie & Goodnow, 2021). We argue that a deeper adoption of diversity, inclusion, and sustainability principles will not only generate closer connections with the places where people live (Stonehouse, 2022) but will also work towards inter-generational and inter-gender equality more broadly.

Conclusion

Adventure tourism is in a state of becoming – shifting in its focus from tacit masculinity and gendered exclusion to intersectional strategies that embrace age, abilities, gender, and social (re)groupings. Early constructions of wilderness and danger in nature of adventure implied its accessibility only to those who could match mastery with risk and physical and mental prowess but today, societal change and shifting gender norms are shaping the nature of adventure.

Skills acquisition and drive are empowering a wider range of embodied experiences, and inclusion is shaping the future by embedding "adventure" activities in communities, placemaking strategies, and the social turn in enterprise which supports sustainability through participation. These factors combined with awareness of the urgency to mitigate climate crises, COVID-19-prompted foci on localism, and the growth of placemaking communities who generate, plan, and activate tourism on their own terms, have widened the range of participants and experiences in the sector.

Review questions

1 How do traditional gendered assumptions and stereotypes shape adventure tourism experiences and women's participation in outdoor activities?
2 In what ways have contemporary societal trends and technological advancements contributed to adventure representations of women in the outdoors?
3 How has risk (real, perceived, actual) contributed to bodily ideals and gendered embodied experiences?
4 In what ways does inclusion impact the sustainability of adventure activities?

Further reading

Aylward, T., & Mitten, D. (2022). Celebrating diversity and inclusion in the outdoors. *Journal of Outdoor and Environmental Education*, *25*, 1–10. https://doi.org/10.1007/s42322-022-00104-2

Díaz-Carrión, I. A., Vizcaino-Suárez, P., & Gaggiotti, H. (2020). Change within the change: Pregnancy, liminality and adventure tourism in Mexico. *Tourism Geographies*, *22*(2), 370–391. https://doi.org/10.1080/14616688.2020.1713876

Houge Mackenzie, S., Boudreau, P., & Raymond, E. (2020). Women's adventure tour guiding experiences: Implications for well-being. *Journal of Hospitality and Tourism Management*, *45*, 410–418.

Knijnik, J., Horton, P., & Cruz, L. (2010). Rhizomatic bodies, gendered waves: Transitional femininities in Brazilian Surf. *Sport in Society*, *13*(7), 1170–1185. https://doi.org//10.1080/17430431003780138

Stonehouse, P. (2022) Sustainable adventure? The necessary "transitioning" of outdoor adventure education. *Journal of Sustainability Education*, *26*, 1–20.

References

Adventure Travel Trade Association. (2018). 20 Adventure Travel Trends to Watch in 2018. Available at: https://cdn.adventuretravel.biz/research/2018-Travel-Trends.pdf.

Adventure Travel Trade Association. (2022). The Influence and Impact of Women in Adventure Travel. Available at: https://learn.adventuretravel.biz/research/the-influence-impact-of-women-in-adventure-travel

Ahmad, N., & Thorpe, H. (2020). Muslim sportswomen as digital space invaders: Hashtag politics and everyday visibilities. *Communication & Sport*, *8*(4–5), 668–691.

Aylward, T., & Mitten, D. (2022). Celebrating diversity and inclusion in the outdoors. *Journal of Outdoor and Environmental Education*, *25*, 1–10.

Baum, T., Cheung, C., Kong, H., Krali, A., Mooney, S., Thi Than, H., Ramachandran, S., Dropulić Ružić, M., & Ling Sio, M. (2016). Sustainability and the tourism and hospitality workforce: A thematic analysis. *Sustainability*, *8*, 809. https://doi.org/10.3390/su8080809

Boniface, M. (2006). The meaning of adventurous activities for 'women in the outdoors'. *Journal of Adventure Education & Outdoor Learning*, *6*(1), 9–24.

Boudreau, P., Houge Mackenzie, S., & Hodge, K. (2020). Flow states in adventure recreation: A systematic review and thematic synthesis. *Psychology of Sport & Exercise*, *46*, 101611.

Brescoll, V. L., Dawson, E., & Uhlmann, E. L. (2010). Hard won and easily lost: The fragile status of leaders in gender-stereotype-incongruent occupations. *Psychological Science*, *21*(11), 1640–1642.

Cater, C., & Cloke, P. (2007). Bodies in action: The performativity of adventure tourism. *Anthropology Today*, *23*(6), 13–16.

Cater, C. I. (2006). Playing with risk? Participant perceptions of risk and management implications in adventure tourism. *Tourism Management*, *27*(2), 317–325.

Cave, J. (2010). *Tourism on the Farm: Opening the Farm Gate to Visitors for 'Agri-profit Not Agro'*. NZ National Fieldays, Hamilton, New Zealand. June 16–19, 2010.

Chambers, D. (2022). Are we all in this together? Gender intersectionality and sustainable tourism, *Journal of Sustainable Tourism*, *30*(7), 1586–1601.

Chigamba, C., Rungani, E., & Mudenda, C. (2014). The determinants of corporate entrepreneurship for firms in adventure tourism sector in South Africa. *Mediterranean Journal of Social Sciences*, *5*(9), 713–723.

Clarke, J. F., Previte, J., & Chien, P. M. (2022). Adventurous femininities: The value of adventure for women travelers. *Journal of Vacation Marketing*, *28*(2), 171–187.

Crawford, D. W., Jackson, E. L., & Godbey, G. (1991). A hierarchical model of leisure constraints. *Leisure Sciences*, *13*(4), 309–320.

Dale, C., & Van Gompel, A. (2005). The pregnant wilderness traveler. *Travel Medicine and Infectious Disease*, *3*(4), 225–238.

Díaz-Carrión, I. A., & Vizcaino, P. (2023). Mexican women's emotions to resist gender stereotypes in rural tourism work. In D. M. Buda & J. G. Molz (Eds.), *Affect and emotion in tourism* (pp. 61–79), Routledge.

Díaz-Carrión, I. A., Vizcaino-Suárez, P., & Gaggiotti, H. (2020). Change within the change: Pregnancy, liminality and adventure tourism in Mexico. *Tourism Geographies*, *22*(2), 370–391.

Dilley, R. E., & Scraton, S. J. (2010). Women, climbing and serious leisure. *Leisure Studies*, *29*(2), 125–141.

Doran, A. (2016). Empowerment and women in adventure tourism: A negotiated journey. *Journal of Sport & Tourism*, *20*(1), 57–80.

Doran, A., Schofield, P., & Low, T. (2018). Women's mountaineering tourism: An empirical investigation of its theoretical constraint dimensions. *Leisure Studies*, *34*(4), 396–410.

Doran, A., Schofield, P., & Low, T. (2020). Women's mountaineering: Accessing participation benefits through constraint negotiation strategies. *Leisure Studies, 39*(5), 721–735.

Edwards, R., & Miller, N. (2000). Inclusion and the denial of difference in the education of adults. Paper presented at *SCUTREA, 30th Annual Conference*, 3–5 July 2000, University of Nottingham.

Elsrud, T. (2001). Risk creation in travelling: Backpacker adventure narration. *Annals of Tourism Research, 28*(3), 597–617.

Evans, K., & Anderson, D. M. (2018). 'It's never turned me back': Female mountain guides' constraint negotiation. *Annals of Leisure Research, 21*(1), 9–31.

Everingham, P., Obrador, P., & Tucker, H. (2021). Trajectories of embodiment in Tourist Studies. *Tourist Studies, 21*(1), 70–83.

Farkic, J. (2021). Challenges in outdoor tourism explorations: An embodied approach. *Tourism Geographies, 23*(1–2), 228–248.

Fendt, L. S. & Wilson, E. (2012). I just push through the barriers because I live for surfing: How women negotiate their constraints in surf tourism. *Annals of Leisure Research, 15*(1), 4–18.

Figueroa-Domecq, C., Kimbu, A., de Jong, A., & Williams, A. M. (2022). Sustainability through the tourism entrepreneurship journey: A gender perspective, *Journal of Sustainable Tourism, 30*(7), 1562–1585.

Frazer, R., & Anderson, K. (2018). Media representations of race, ability, and gender in three outdoor magazines: A content analysis of photographic images. *Journal of Outdoor Recreation, Education, and Leadership 10*(3), 270–273.

Goffman, E. (1959). *The presentation of self in everyday life.* Doubleday.

Gray, T., Norton, C., Breault-Hood, J., Christie, B., & Taylor, N. (2018). Curating a public self: Exploring social mediai of women in the outdoors. *Journal of Outdoor Recreation, Education and Leadership, 10*(2), 153–170.

Green, E., & Singleton, C. (2006). Risky bodies at leisure: Young women negotiating space and place. *Sociology, 40*(5), 853e871.

Hall, J. (2019). *Women mountaineers: A study of affect, sensoria and emotion.* Doctoral dissertation York, England: York St John University. Retrieved from https://ray.yorksj.ac.uk/id/eprint/3793/

Henderson, K. A. (1990). The meaning of leisure for women: An integrative review of the research. *Journal of Leisure Research, 22*(3), 228–243.

Henderson, K. A. (1992). Breaking with tradition: Women and outdoor pursuits. *Journal of Physical Education, Recreation and Dance, 63*(2), 49–51.

Hillman, W., & Radel, K. (2021). The social, cultural, economic and political strategies extending women's territory by encroaching on patriarchal embeddedness in tourism in Nepal. *Journal of Sustainable Tourism, 30*(7), 1754–1775.

Hornibrook, T., Brinkert, E., Parry, D., Seimens, R., Mitten, D., & Priest, S. (1997). The benefits and motivations of all women outdoor programs. *Journal of Experiential Education, 20*(3), 152–158.

Houge Mackenzie, S., Boudreau, P., & Raymond, E. (2020). Women's adventure tour guiding experiences: Implications for well-being. *Journal of Hospitality and Tourism Management, 45*, 410–418.

Houge Mackenzie, S., & Goodnow, J. (2021). Adventure in the age of COVID-19: Embracing microadventures and localism in a post-pandemic world. *Leisure Sciences, 43*(1–2), 62–69.

Houge Mackenzie, S. & Hodge, K. (2019). Adventure recreation and subjective well-being: A conceptual framework. *Leisure Studies, 39*(1), 26–40.

Humberstone, B. (2015). Embodiment, nature and wellbeing: More than the senses? In *Experiencing the outdoors* (pp. 61–72). Brill Sense.

Kling, K. G., Margaryan, L., & Fuchs, M. (2018). (In) equality in the outdoors: Gender perspective on recreation and tourism media in the Swedish mountains. *Current Issues in Tourism, 23*(2), 1–15.

Knijnik, J., Horton, P., & Cruz, L. (2010). Rhizomatic bodies, gendered waves: Transitional femininities in Brazilian Surf. *Sport in Society, 13*(7), 1170–1185. https://doi.org/10.1080/17430431003780138

Løseth, K. (2018). Knowledge development in adventure tourism businesses – the influence of serious leisure, *Annals of Leisure Research, 21*(5), 575–591. https://doi.org/10.1080/11745398.2017.1406808

Low, T., Miller, M. C., Doran, A., & Hardwick, L. (2022). Women's outdoor adventure experiences on Instagram: Exploring user-generated content. *Annals of Leisure Research, 25*(3), 374–398.

Markula, P. (2006). Beyond the perfect body: Women's body image distortion in fitness magazine discourse. *Journal of Sport and Social Issues, 25*, 124–126.

Marshall, J. (2011). En-gendering notions of leadership for sustainability. *Gender Work and Organization, 18*(3), 263–281.

McDermott, L. (2004). Exploring intersections of physicality and female-only canoeing experiences. *Leisure Studies, 23*(3), 283–301.

Melidosian, L., Evans, E., Stewart, K., & Antony, K. M. (2019). Travel during pregnancy: A study of postpartum women in Madison, Wisconsin. *World Medical Journal (WMJ), 118*(3), 114–119.

Mitten, D., & D'Amore, C. (2017). The relationship of women's body image and experience in nature. In D. Vakoch & S. Mickey (Eds.), *Women and nature: Beyond dualism in gender, body, and the environment* (pp. 96–116). Routledge.

Myers, L. (2010). Women travelers' adventure tourism experiences in New Zealand. *Annals of Leisure Research, 13*(1–2), 116–142.

Nolet, V. (2016). *Educating for sustainability: Principles and practices for teachers.* Routledge, Taylor & Francis Group.

Savage, A. Barbieria, C., & Jakes, S. (2022). Cultivating success: Personal, family and societal attributes affecting women in agritourism. *Journal of Sustainable Tourism, 30*(7), 1699–1719.

Scheyvens, R., & van der Watt, S. (2021). Tourism, empowerment and sustainable development: A new framework for analysis. *Sustainability, 13*(22), 12606–12625.

Shilling, C. (2017). Body pedagogics: Embodiment, cognition and cultural transmission. *Sociology, 51*(6), 1205–1221.

Small, J. (2016). Holiday bodies: Young women and their appearance. *Annals of Tourism Research, 58*, 18–32.

Small, J. (2017). Women's "beach body" in Australian women's magazines. *Annals of Tourism Research, 63*, 23–33.

Stanley, P. (2020). Unlikely hikers? Activism, Instagram, and the queer mobilities of fat hikers, women hiking alone, and hikers of colour. *Journal of Mobilities 15*(2), 241–256.

Stonehouse, P. (2022). Sustainable adventure? The necessary "transitioning" of outdoor adventure education. *Journal of Sustainability Education 26*, 1–20.

Tucker, H. (2022). Gendering sustainability's contradictions: Between change and continuity, *Journal of Sustainable Tourism, 30*(7), 1500–1517.

Warren, K., Roberts, N. S., Breunig, M., & Alvarez, M. A. (2014). Social justice in outdoor experiential education: A state of knowledge review. *Journal of Experiential Education, 37*(1), 89–103

Weatherby, T. G., Vidon, E. S. (2018). Delegitmizing wilderness as the man cave: The role of social media in female wilderness empowerment. *Tourist Studies, 18*(3), 332–352.

Yang, E. C. L., Khoo-Lattimore, C., & Arcodia, C. (2017). A systematic literature review of risk and gender research in tourism. *Tourism Management, 58*, 89–100.

Zapalska, A., & Brozik, D. (2018) Indigenous entrepreneurship: Māori female entrepreneurs in the tourism industry and constraints to their success. In S. Yousafzai, A. Fayolle, A. Lindgreen, C. Henry, S. Saeed, & S. Sheikh (Eds.), *Women entrepreneurs and the myth of 'underperformance'* (pp. 72–89). Edward Elgar. Retrieved from https://www.elgaronline.com/view/edcoll/97817864 34494/9781786434494.00015.xml

Zink, R., & Zane, M. (2015). Not in the picture: Images of participation in New Zealand's outdoor recreation media. *Annals of Leisure Research, 18*(1), 65–82.

27 Outdoor and adventure activities as a space of refuge for LGBTQ+ people

Bart Bloem Herraiz and Elía Velo Camacho

Chapter learning outcomes

1 Identify the distinctive traits of adventure recreation that provide opportunities to contest gender normativities.
2 Highlight the benefits of adventure recreation for the LGBTQ+ community.
3 Suggest ways to improve the quality of mountain leadership training regarding LGBTQ+ inclusivity.
4 Propose strategies to create safe spaces for LGBTQ+ people in adventure tourism.

Introduction

LGBTQ+ communities grow and find refuge in the cities, and many researchers have paid attention to these experiences (Guzmán, 1997; Rodó-de-Zárate, 2016; Saz, 2020). Nevertheless, this metronormativity (Halberstam, 2005) is beginning to be questioned, and our[1] presence in other spaces, such as outdoor recreation and adventure tourism, is becoming more visible.

There is currently a dearth of research on LGBTQ+ experiences in outdoor recreation and adventure tourism (Bloem Herraiz, 2019; Meyer, 2010; Meyer & Borrie, 2013; Warren, 2016). Therefore, we have drawn from the broader adventure tourism and outdoor recreation literature. We include outdoor recreation literature in this review, as people participating in both adventure tourism and outdoor recreation use the same spaces and these terms are often used interchangeably. Additionally, terms such as *outdoor activities* and *adventure sports* are also used to refer to these and, thus, are adopted in this chapter.

Authors like Pomfret and Doran (2015) have pointed out that it is possible to find resistance to stereotypical gender norms within adventure activities, where the normative ideas of masculinity and femininity associated with mountaineering are becoming blurred. Other authors have reclaimed the potential of 'wilderness experiences' to allow space for different ways of being and escaping from gendered social constrictions and conventional gender roles (Meyer, 2010; Meyer & Borrie, 2013). These – the possibility to escape gendered social constructions and the opportunity to resist conventional gender roles – are some of the key characteristics of outdoor recreation and adventure tourism that, we argue, create unique conditions for LGBTQ+ communities to thrive and find belonging.

DOI: 10.4324/9781003393153-32

Departing from the queer and feminist methodology of the dialogue (Freire, 1970; Mcphie & Clarke, 2020; Platero & Drager, 2015), we envision a methodology in movement that recognises the ways in which lived experiences, perception, and knowledge production are constructed through place and the spatial practices of sociality and position (Lynch, 2020; Springgay & Truman, 2019). Hence, we depart from our embodied and situated positions (Haraway, 1988) as queer, trans, and non-binary mountain guides and scholars, and we go outdoors to produce 'different knowledge and produce knowledge differently' (Lather, 2013, p. 653).

The chapter is structured in four parts: First, we include a brief state of the art of published literature regarding LGBTQ+ experiences in outdoor recreation and adventure tourism. Following this, we situate ourselves: Where do we stand regarding the LGBTQ+ community, and how do we understand adventure activities and the outdoors? Afterwards, we introduce a dialogue that took place while hiking in the Black Forest, Germany, about the experiences that we had within the two years of training – between 2019 and 2021 in Spain – to become official mountain guides. This conversation leads us to a reflection on LGBTQ+ people's experiences in outdoor recreation and adventure tourism and helps us answer two of our research questions: What singularities do these activities have in relation to gender performativities and expectations?; and what benefits do we find in adventure tourism for queer and trans communities?

Finally, we interconnect our experiences with the results that have been shown in research to make proposals of what is needed in adventure tourism to create welcoming and safe spaces for LGBTQ+ people. Thus, we address the other two of our research questions: How can we enhance the quality of the mountain guide training course?; and what is needed in adventure recreation to create safe spaces for LGBTQ+ people?

State of the art

Here we want to present some of the main key points in adventure tourism research regarding gender and LGBTQ+ communities. This is not a systematic review, but it rather serves as an introduction to the field of LGBTQ+ adventure tourism research that has been done until now.

Many researchers have highlighted the historically male-dominated and male-defined aspects of outdoor recreation and adventure tourism (e.g., Argus, 2018; Díaz Carrión, 2012; Doran, 2019; Doran et al., 2018; Humberstone, 2000; Kling et al., 2020; Martín Talavera & Mediavilla Saldaña, 2020; Meyer, 2010; Stanley, 2020; Warren, 2016). Moreover, outdoor recreation has created a universal subjectivity that is male (Pedersen, 2003) cis, white, and fit; and wilderness has even been described as 'a threat to femininity' (Weatherby & Vidon, 2018, p. 332). In addition, the masculinities represented in outdoor recreation fall within what is called hegemonic masculinities and operate 'within a strict heteronormative gender binary' (Argus, 2018, p. 530). The idea of the outdoors as a place for cis men to penetrate, compete, and conquer (Little & Wilson, 2005) has been perpetrated because of these hegemonic masculinities reinforcing the traditional view of the heroic white cis male adventurer (Doran et al., 2018). As a result of the exclusion of all bodies that don't fit this hegemonic category, the outdoors has been masculinised in such a harsh way that gendered bodies even disappear (Frohlick, 1999; Pedersen, 2003).

This perceived superiority of the cis, white, able-bodied, male subjects in adventure tourism is perpetuated through written and moving narratives. It is known that media, literature, and advertising play an important role in creating representation and an imaginary of the people who can take part in adventure sports, thus creating spaces in which some people may feel included while others will be excluded. 'Just as the outdoors itself is not "natural", there is nothing "natural" about the ways in which outdoor magazines, advertising, and other media portray legitimate outdoors people in very specific ways: White, male, straight, muscular, and able-bodied' (Stanley, 2020, p. 244). The unmarked adventure subject creates an imaginary of a male outdoor person, or even worse, as 'if gender didn't exist in the world of high-altitude mountaineering' (Frohlick, 1999, p. 87). This displaces everyone who is not male, cis, white, straight, and able-bodied to the periphery of the adventure sports imaginary (Frohlick, 2005, p. 175).

Researchers have noted that outdoor leadership programmes are also still predominantly white, male, cis, and heterosexual (Allen-Craig et al., 2020; Jordan, 2018; Lundin & Bombaci, 2022; Warren et al., 2018). 'Outdoor leadership that values the experiences of women and nonbinary gendered leaders, and the contributions of feminist outdoor leadership to the outdoor adventure field need to be considered' (Warren, 2016; Warren et al., 2018, p. 253). Language was identified as a practice that actively affects inclusivity in outdoor leadership (Jordan, 2018; Lundin & Bombaci, 2022; Warren et al., 2018). Jordan (2018) proposes strategies for how outdoor leaders can combat sexism in language, while Warren et al. (2018) propose practices for a more inclusive outdoor leadership environment. Moreover, Allen-Craig et al. (2020) note that a 'collective action for gender equity must therefore be promoted and framed as a common cause, positioning all people, including those who identify outside the binary, as agents of change' (p. 123). Therefore, it is important to identify and amplify the perspectives of a diverse range of outdoor leadership professionals (Allen-Craig et al., 2020). Finally, Lundin and Bombaci (2022) point out that, although there have been efforts to make outdoor leadership programmes more inclusive, 'cultural and procedural changes are still needed to support the Lesbian, Gay, Bisexual, Transgender, Queer or Questioning (LGBTQ+) community when participating in outdoor field experiences'. In their research, they highlight key actions that outdoor leaders can incorporate into their practices to increase LGBTQ+ awareness.

The potential of adventure activities for LGBTQ+ people

What are LGBTQ+ people's motivations, constraints, negotiations, and benefits of going into the outdoors? What does "outdoor" and "adventure" mean for LGBTQ+ individuals? As some studies have observed, 'gender identity determines the opportunities people have to partake in mountaineering' (Moscoso-Sánchez, 2008, p. 183), and a re-conceptualisation of what "outdoor" and "adventure" mean in terms of gender, sexuality, race, and class in contemporary society is starting to be questioned by feminist researchers (Meyer, 2010).

It is known that being in contact with nature has many positive physical, cognitive, and psychological effects (Mitten & D'Amore, 2017). Remarkably, a sense of ecological and social belonging was also reported by Meyer (2010) and Meyer and Borrie (2013) regarding queer women and by Bloem Herraiz (2019) regarding trans people. Meyer (2010) and Meyer and Borrie (2013) have also shown that LGBTQ+ people describe wilderness experiences in terms of connection, integration, and refuge, and a place where it is possible to escape from gendered social constrictions and conventional gender roles. Some studies have also shown that for people who do not identify as cis men, being outdoors provides opportunities to find a place to be and express themselves (Argus, 2018;

Boniface, 2006; Meyer, 2010; Meyer & Borrie, 2013; Overholt & Ewert, 2015). Adventure activities were described as spaces with less societal surveillance and judgement, allowing a space to feel freer in their own bodies, moving and interacting with nature 'away from prescribed and idealised ways we use and present our bodies; and away from the predatory eyes of oppression' (Meyer & Borrie, 2013, p. 314). Furthermore, trans people described wilderness as a refuge from normative gender (Meyer, 2010, p. 186), though the author also pointed out that more research needs to be done in this direction. Finally, a study by Wilson and Lewis (2012) about single-gender outdoor programmes showed that this type of program allows trans people to 'express themselves without constantly needing to react to the presence of the dominant gender binary' (Wilson & Lewis, 2012, p. 233).

Nevertheless, as Graham wrote in 1993 in a newspaper article, being queer and doing adventure activities hasn't been in the imaginary of what an 'outdoor person' is nor what a 'queer person' would do. Almost 30 years later, this situation hasn't changed much, and the presence of queer and trans people in adventure activities is still not taken as a possibility in the outdoor community. Adventure programmes are still cisheteronormative spaces, where queer visibility and participants remain erased or invisible (Argus, 2018; Dignan, 2002; Meyer, 2010; Meyer & Borrie, 2013; Stanley, 2020; Warren, 2016). Moreover, the approach taken by the studies done on gender and adventure activities and the insights they provide into gendered expectations is still very binary (Doran, 2019).

All these articles indicate a dearth of research regarding sexuality and gender identity in the outdoors. Outdoor spaces in which adventure sports take place can be a cisheteronormative and heterosexist environment; however, it has also been shown that they can also be a counter-culture place for alternative femininities and masculinities (Humberstone, 2000), and offer opportunities for new ways of building our bodies (Díaz Carrión, 2012). Promoting 'gender-sensitive' (Warren, 2016) adventure activities are essential to avoid discriminatory situations. Adventure tourism providers should adopt a more feminist approach to adventure recreation leadership that seeks to empower and enhance integrity for all participants. As a final remark, we would like to point out that all the authors were cis-identified people.[2]

The researchers in the research

We approach this chapter from a feminist and queer methodology, from which our position and gaze influence and determine the way we see this world and the knowledge we create. Our knowledge does not pretend to be objective or valid for all realities: instead, we seek situated knowledge (Haraway, 1988), aiming to share our reflections without having a generalising purpose that pretends to unify or equalise the diversity of experiences.

Bart is a trans and genderqueer person with male passing, white, European, with no functional diversity and from the working class, who grew up in Málaga, in the south of Spain. Since school, they didn't fit into the cisheteronormative gender roles which are expected from a girl. They used sports and physical activity – especially outdoor recreation contexts, in which they felt more freedom with gender expression and no need to choose between a girls' or boys' team – as a way of escaping and finding refuge from certain gender normativities. Thus, sports and physical activity became a space in which to evade prevailing gender binarism. From an interdisciplinary place between gender studies and outdoor studies, they venture to queerize adventure tourism, placing dissident corporealities, curiosity, reflection, and affections at the centre.

Elía is a non-binary person with female passing, white, European, with no functional diversity, and from the working class, who grew up in Bonares, a village in the south of Spain. They have always been a very active person, and from an early age, they have felt limited in practising certain sports or joining sport clubs that 'were not meant for girls'. For example, they wanted to join the Scouts Group because it was the only opportunity for them to go to the mountains and practice outdoor activities, but there was always a 'no' as an answer from their parents. Something similar occurred with soccer, because they tried to launch a female soccer team in their village, and when they achieved it, their parents said no 'because soccer is a tough sport, and they may receive lots of kicks'. The team kept going and they were not part of it. Nevertheless, they kept on trying what made them happy despite their parents' impediments: *At least I can play soccer on the streets* – they thought. When playing on the streets they started receiving derogatory adjectives such as 'butch' or 'dyke' without even knowing what they meant or what part of it could be true in their own identity. It wasn't until after their sports sciences bachelor's that they started feeling free to show their active side without fear of being judged as too masculine or not feminine enough. For several years, they have been doing research on outdoor pedagogies, and it is from there, along with their experiences in adventure sports, from where they venture to queerize outdoor recreation and adventure tourism (See Case study 27.1).

Case study 27.1   Walking methodologies

In this section, a case study is presented based on our experiences between 2019 and 2021, amid a global pandemic, participating in a mountain guide certificate course – led by the Andalusian Sport Institute, Spain – to become a Union of International Mountain Leader Association mountain guide. The course aims to prepare the participants to be able to work in the adventure tourism sector as mountain leaders. The participants were approximately 90% cis men, 8% cis women, and 2% non-binary/trans people – the two of us; among the teaching staff, there were 95% cis men and 5% cis women – 0% non-binary/trans people. However, all on-the-field practices were carried out by male teachers.

We have decided to build and write this case study following a queer and feminist methodology of talking dialogue (Freire, 1970; Mcphie & Clarke, 2020; Platero & Drager, 2015). We conceive an 'in movement methodology' which recognises how 'lived experiences, perception, and meaning-making are constructed through place and spatial practices of sociality and positionality' (Springgay & Truman, 2019, p. 3). Furthermore, a walking methodology produces an ontological shift that fosters us 'to think about experience differently, to experience differently, and to experience difference in experiencing' (Clough & Calderaro, 2019, p. xiii). That's why we go outdoors; we walk to 'produce different knowledge and produce knowledge differently' (Lather, 2013, p. 643), delving into an embodied way of knowing in which the mind, body, and environment connect.

Both of us currently live in the Black Forest, Germany, where we often go for hikes or runs in the mountains around the area. Quite often, the topics we talk about concern the issues of how privileged we are to live in this place and to be able to enjoy the mountains and forests, how they make us feel in the moment, and afterwards when going back to our daily jobs.

Elía: Don't forget your water bottle!

Bart: Ah, yes. Thanks! Now that you said this, it made me think about our experience during the mountain guide course. Do you remember what they used to say in this kind of situation?

Elía: How can I forget it: 'The sight of the gypsy'. I hadn't heard this expression before, and I didn't even realise it when I heard it. It is so normalised to joke about other ethnicities …

Bart: We were doing one of the field trips, and after a break, one of the students – who is already a climbing guide – said it. He meant that we should look behind in order not to forget anything, which is always very important when hiking in groups. However, the expression he used for it was very racist, and when we continued hiking, I brought up the topic. Not even the course leader acknowledged my point, and the only response I got when I pointed out that it was a racist statement was 'I'm not racist'. It felt very frustrating trying to explain to them why we, as mountain guides, should never use this kind of expression if we want to welcome everybody into our activities. They didn't even try to understand it.

Elía: Yes, and it was not an isolated case. On another field trip to Sierra Nevada (Spain), we were doing a geological and botanical hiking route to identify terrain shapes and plants that live at high altitudes. It was very interesting; do you remember the moraine? The plants were different from what we are used to seeing, their structures, their colour, their positions. … Also, the shape of the moraine, which was in the past a glacier, forges the thalweg and gives the plants a place to grow. It was very enriching to learn about all of that, and it was in the south of Spain! People from other countries or even from Spain do not expect these kinds of vegetation and terrain there. We took turns practising guiding the group, and we were supposed to use our knowledge to explain the geology and vegetation of the surroundings; what a huge challenge!

At one point, one of our colleagues, a cis straight man, older than the rest of the group, was guiding the end of the hiking route, and when there were just a few hundred meters to complete the route, he said: 'Faggot the last one!' His intention was to encourage the group to finish with energy and get back to the cars faster. I was just shocked, but nobody reacted; everybody laughed and was motivated to continue walking until the end. It made me feel out of the group, out of that amazing place, out of the activity, out of the terrain shape, and the differences between *Junipers* and *Savins*. I was thinking about all the people who would participate in a guided hike and identify within the LGTBQ+ spectrum, and their feelings if they heard that from their leader. Am I less able? Am I weaker because of my sexual orientation? Are straight people stronger or more able than me? Shouldn't I be here? Nature was supposed to be a kind environment for everybody, but am I not able enough to enjoy it?

Afterwards, at the final reflection of the day, we formed a circle to give feedback and share our thoughts about the activity. We couldn't avoid saying it, highlighting that comment and how it made us feel. I think nobody expected it, maybe because we both have a cis passing and people usually expect heteronormativity.

Bart: We then started a short debate, and one of the course leaders – a cis, straight, white man – argued that 'now nothing can be said', contending that political correctness doesn't allow him to make comments. I think that was the hardest part, that even when we tried to make a difference, it was for nothing because the course leaders didn't want to give importance to our feedback on these matters. From his cisheteronormative point of view, those kinds of comments are the order of the day in the adventure tourism industry, and they are treated as 'harmless'. However, this kind of hate speech is a clear example of how social normativities and hierarchies are constructed and embedded in our everyday lives (Nielsen, 2002). One of our colleagues – who had experience working with diverse groups in public administration – did support us, arguing that it is necessary to care about sexist expressions used in our everyday life and that we need to learn to avoid them because they can be harmful to some of the participants in our activities.

I felt really unsafe, and I told them – teachers and participants – that I would never hire a mountain guide that says that kind of expression, that I wouldn't feel safe with that person in the mountains. This topic never came out again in any final joint reflection, which always took place at the end of every field practice, but it gave us more motivation to complete the training and change the landscape of adventure tourism.

Elía: Those kinds of expressions don't make us feel safe and comfortable in the natural environment. Even though we feel freedom in nature, and we perceive it as a space where we are not being judged in the same way as in urban spaces, they are making us feel 'out of place' (Ahmed, 2004; Puwar, 2004). Nevertheless, that should be the first aim for everyone who wants to participate in an adventure activity, to feel comfortable and safe while enjoying the place and company. Therefore, we should – adventure tourism professionals – work to create a welcoming environment and be an example for our participants. Don't you think?

Bart: Exactly, and even activities or spaces that are less gendered while being outdoors became uncomfortable. Public bathrooms have been highly contested within trans geographical studies (Browne, 2004; Doan, 2010; Preciado, 2006), considered one of the scariest and most dangerous spaces for gender-variant people. I usually really enjoy the fact that when being outdoors, the 'bathrooms' are not gendered; a bush or a tree doesn't have a gendered bathroom sign! This makes it so much easier for trans and gender-non-conforming people, knowing that we won't need to confront any other bathroom user that questions our gender. However, all the course leaders and participants – cis men – were 'going to the toilet' next to the group, just two or three meters apart from us. This felt very uncomfortable for me because I had to go somewhere behind a bush or a tree, which made me feel quite unsafe from what they would think, not to mention the fact that I don't enjoy having strangers peeing close to the group. This got even worse when one of the leaders of the group, during a snow field trip, painted his name in the snow with his pee, two meters away from us, and then he showed it to us and joked about it. It was such a toxic masculine attitude that I could almost not believe it.

Elía: Really? You didn't mention it when we were there! I would have never expected such an attitude from a course leader, it doesn't matter if he is a cis, straight man. How can something like that even happen? How is our society built that makes our kids learn that cis men need to show how masculine they are? Or even worse, how can this kind of behaviour be a sign of masculinity?

Following this masculinity topic, I would also like to share with you and highlight my experiences in the course as a person with a cis woman's passing. Everything was different compared to our male and male passing colleagues; we were even evaluated differently on many occasions, and I am not talking about physical abilities. I was treated as a delicate lady who is supposed to need help on exposed/ dangerous points of the route, accentuated by the fact that I was mainly surrounded by cis, straight men. Besides that, when learning how to use a GPS, they presumed that I would need more time to understand how it works and if any of the women students had a question, the course leaders would dedicate more time to explain it to us separately from the group. On the other side, when one of our male colleagues had a doubt, they would say that it is something very simple and would explain it quickly in front of the group. What makes me even angrier is that our female colleagues were also joking and behaving as more delicate, less adventurous, or capable than our male colleagues with expressions like 'Yes, you know, we are a bit bad with technologies, could you help me?' – when talking to the leader with a forced smile.

    There is much to improve in the mountain guide training courses. These were some of our experiences from the course in Andalusia, which may or may not differ from the mountain guide courses in other places in Spain or even other countries in Europe. Further research is needed to record the experiences of LGBTQ+ people on courses that will enable people to become outdoor leaders and the experiences of participants in adventure tourism activities with a professional leader. The adventure tourism industry is still very cisheteronormative, and LGBTQ+ people, quite often, don't feel welcome to participate. The lack of representation in adventure media increases this feeling. The presence of LGBTQ + people in the outdoors becomes a defiance and challenges expectations of where our bodies belong (in the cities; Bloem Herraiz, 2019). These acts of occupying public spaces in adventure recreation contribute to the production of the space (Rodó-de-Zárate, 2016). Despite these constraints, adventure activities offer many benefits to LGBTQ+ people (Bloem Herraiz, 2019; Meyer, 2010; Meyer & Borrie, 2013).

LGBTQ+ people's experiences in outdoor recreation and adventure tourism

Following the results of the two studies about LGBTQ+ people's experiences in outdoor recreation and adventure tourism (Bloem Herraiz, 2019; Meyer & Borrie, 2013) and our own experiences in adventure tourism, we now highlight the main benefits of adventure activities and the particularities that these activities have relating to gender performativities and expectations

Freedom from social constraints

Many people try to find in nature the freedom they don't have in their towns (Li, 2018). This becomes especially significant when we talk about the LGBTQ+ community, which encounters daily social constraints related to their gender and/or sexual orientation. LGBTQ+ people are looking for a place to escape from society's judgment, and natural environments are the perfect place for it, where animals and plants don't care about our identities, how our bodies are or how we are dressed. When this judgement decreases, when there is no gender policing, we find physical and mental relaxation to enjoy what is around us. As Meyer and Borrie (2013) say, 'the absence of judgment is an important part of refuge, particularly for transgender people' (p. 314).

In both studies, participants mention how being alone in natural environments allows them to break or escape social normativities, experiencing a feeling of freedom, of being able to be themselves because there are no established identities, bodies, or gender expressions; there are just relationships between living beings and the environment.

Feelings of safety

Nature is frequently conceived as a dangerous place where wild animals could be life-threatening, and changes in geographic terrain or the weather are out of human control. In contrast, the outdoors is frequently perceived as a place of refuge for LGBTQ+ people, a space to be safe and perform as we are, making us feel protected against societal oppression. As Meyer and Borrie (2013) conclude: 'oppression alienates and excludes people from full participation in society and thus we wonder if the search for refuge, connection, and belonging might be common links among how "Othered" populations can experience wild nature' (p. 314). Participants (Bloem Herraiz, 2019; Meyer & Borrie, 2013) mentioned that cities often become more dangerous regarding gender identities, sexual orientations, or performativity, making them feel vulnerable. However, in nature, the danger is not related to our identities, but rather, the vulnerability comes from the weather, the predators, or the environment. It is not a kind of violence targeted directly at us.

Empowerment

Adventure tourism contexts are challenging, and we have to focus on confronting those challenges. One of the advantages of being outdoors that was stressed is that our problems become less significant because there are other things to be concentrated on: obstacles, orientation, weather, animals, or *just* contemplating the environment; being outdoors is about being in the moment. 'I'm here, and I have the right to be here' says a participant in one of the studies (Bloem Herraiz, 2019, p. 42). In a society where there is no place for Otherness (Ahmed, 2004; Butler, 1990; 1993; Stryker, 2006), these feelings of competence and usefulness when confronting those challenges make us love and trust our bodies. Moreover, the possibility of finding safe spaces in adventure activities also boosts feelings of empowerment, as Doan (2010) states: 'finding safe spaces in which to perform my gender empowered me and sustained my ability to continue to express my gender variance and be openly transgendered in the wider community' (p. 647).

Less gender policing

The feelings of freedom, safety, and empowerment allow our bodies to become physically and mentally relaxed. We slow down to be more conscious about what we are doing, being in the moment and in our bodies (Bloem Herraiz, 2019; Meyer & Borrie, 2013).

Although it was noted that surveillance is minimal outdoors because there are fewer people, adventure activities contest this situation, hindering the previously mentioned benefits. Some participants (Bloem Herraiz, 2019; Meyer & Borrie, 2013) noted that they preferred to go alone, thereby avoiding having a strategy in case of violence or harassment when going with a group of strangers. This is the case of organised activities in adventure tourism, in which participants have to perform cisheteronormativy to prevent questioning or judging sights and comments.

Recommendations for creating safe spaces in adventure tourism for LGBTQ+ people

We have outlined the benefits and particularities of adventure activities for LGBTQ+ people, highlighting aspects such as the possibility of escaping gendered normativities and expectations, finding 'ecological belonging' (Meyer & Borrie, 2013), and empowering ourselves in our bodies. Thus, we consider that accessing outdoor and adventure recreation is about building livable lives (Butler, 2004a; 2004b) for LGBTQ+ people.

Meyer (2010) and Meyer and Borrie (2013) point out that, because of the particularities of the outdoor setting, which offers opportunities to escape from gendered societal normativities, these spaces facilitate feelings of ecological connectedness and belonging. This, we would contend, is a transecological belonging (Bredford, 2020; Gaard, 2020; Straube, 2020; Stryker, 2020; Thorsteinson & Serenity Joo, 2020):

> Because I am there, and because I am trans, this is a transecology. It is an environment capable of accommodating naturally techno-cultural me and my kind and our tools and fuel, even as it accommodates so little else of life.
>
> (Stryker, 2020, p. xvi)

Stryker argues further that we transfolk 'are home/not home in nature' (Stryker, 2020, p. xvi). However, LGBTQ+ people do not always feel safe and at home when being outdoors, and feelings of discomfort[3] arise when being with strangers; therefore, choose to go on adventure activities alone or in small groups of friends. Hence, it is urgent that adventure tourism professionals address this matter. What do we need to change in adventure tourism for that strategy not to be necessary anymore? How can we make the industry welcoming for everybody, regardless of gender identity, sexual orientation – bodily ability, race, ethnicity, and the like?

Throughout the chapter, we have traced and unmapped (Razack, 2000) the counter-geographies (Sassen, 2003) that LGBTQ+ people are creating when participating in adventure recreation. Unmapping is the process of looking at the topographies of exclusion that maintain our bodies invisible in the production of space. By pointing out this 'inclusion through exclusion' (Mountz, 2011), counter-geographies are revealed. Sassen

(2003, p. 49 our translation), describes counter-geographies as alternative transnational maps of communication and intersections; as she explains:

> These counter-geographies are deeply imbricated with some of the main constitutive dynamics of globalization: the formation of global markets, the intensification of transnational and translocal networks and the development of communication technologies that easily elude conventional control practices.

Hence, counter-geographies make visible the time-space of Other subjectivities. LGBTQ+ people, by participating in adventure activities, are reclaiming their space outdoors, finding less judgement – from nature – and refuge from gendered normativities, thus mapping counter-geographies in these spaces.

Adventure tourism is currently mapped by cisheteronormative gazes that result in exclusion processes. Earlier in the chapter, we pointed out the importance of representation for creating imaginaries of what it is possible to do; it is urgent that we start creating other imaginaries of what an adventure person can look like – other than white, male, cis, straight, muscular, and able-bodied – by, for example, displaying a diversity of bodies and identities in the promotion of adventure activities – flyers, advertisement, social media, and so on. By doing this, we will not only be changing the social imaginary but also creating welcoming spaces for all of us. Furthermore, the creation of safe spaces needs to be actively acknowledged, pointing out that the space is free of LGBTQphobic attitudes, for example.

For this, we also need to strengthen the quality of the mountain leader training. Figure 27.1 proposes actions to enhance LGBTQ+ inclusivity.

Using LGBTQ+ inclusive language

- Avoid gendered language and phrases.
- Refrain from assigning gender labels to objects or facilities
- Introduction rounds where everybody states their name and pronouns – no matter their identity.

Diverse teaching staff

- Having a diverse teaching staff with women and LGBTQ+ representation that acknowledges issues from a feminist perspective.

Training the teaching staff

- Training the teaching staff in gender, feminist, and LGBTQ+ topics.
- The leaders of the activities need to be trained on how to create safe spaces for LGBTQ+ people.

Curriculum

- Including a subject in the curriculum that addresses gender and LGBTQ+ experiences in adventure tourism from a feminist perspective.

Figure 27.1 Actions to enhance LGBTQ+ inclusivity in mountain leader training.

These proposals would be a great start for the change that is needed in the training of future mountain leaders. However, we need to keep in mind that the adventure tourism industry needs to start taking steps toward this inclusion to have safe and welcoming spaces for the LGBTQ+ population.

Conclusion and future research

While many people go outdoors to disconnect, we have shown that for us – the LGBTQ+ population – it is about finding transecological belonging and connectedness. We argue that we also belong to the outdoors/nature, and maybe that is the reason we try to find refuge there when we are the target of violence; it is like kids going back home: they feel untouchable, protected.

We have highlighted the many benefits that adventure activities have for us: freedom, less gender surveillance, possibilities to escape gender normativities, empowerment in our bodies, self-confidence, feelings of safety, and refuge from the gender system. Condemning LGBTQ+ people to the cities creates disempowered subjectivities, which has also negative effects on our health, thus *killing us softly* (Shakhsari, 2014).

Future lines of studies would be to keep researching the experiences of LGBTQ+ practitioners in adventure tourism, which would help create safe and welcoming spaces for everyone, regardless of their gender identity and expression, race, class, body type, and the like. Furthermore, research into the training of adventure leaders and deepening into proposals of improvement toward inclusion needs to be done.

Review questions

1 What unique characteristics does adventure recreation possess regarding gender performativities and expectations?
2 How can adventure tourism support LGBTQ+ communities?
3 How can we strengthen the quality of mountain leadership training?
4 What steps should be taken in adventure recreation to create safe spaces for LGBTQ+ people?

Notes

1 As we are both part of the LGBTQ+ community, throughout the chapter, we use the term *we* because the experiences that we talk about are also our reality.
2 All authors except Bloem Herraiz were cisgender. In the same way, as heteronormativity works, by which society assumes that everyone is straight until they prove otherwise, and if one is straight, this is not relevant to point out, cisnormativity is the assumption that all individuals are cisgender and acknowledging one's identity is not seen as necessary, not even when researching about trans issues. Therefore, the author's identity was not addressed by themselves in the articles but rather assumed through the lack of proving otherwise.
3 As Ahmed (2004) describes discomfort as 'a feeling of disorientation: one's body feels out of place, awkward, unsettled' (p. 148).

Further reading

Bloem Herraiz, B. (2019). *Bodies out/in place? Unmapping Trans people's experiences in outdoor activities* University of Gothenburg. http://hdl.handle.net/2077/61669

Doan, P. L. (2010). The tyranny of gendered spaces – reflections from beyond the gender dichotomy. *Gender, Place & Culture, 17*(5), 635–654. https://doi.org/10.1080/0966369X.2010.503121

Lundin, M., & Bombaci, S. (2022). Making outdoor field experiences more inclusive for the LGBTQ+ community. *Ecological Applications*. https://doi.org/10.1002/eap.2771

Meyer, A. M., & Borrie, W. T. (2013). Engendering wilderness: Body, belonging, and refuge. *Journal of Leisure Research, 45*(3), 295–323. https://doi.org/10.18666/jlr-2013-v45-i3-3153

Stanley, P. (2020). Unlikely hikers? Activism, Instagram, and the queer mobilities of fat hikers, women hiking alone, and hikers of colour. *Mobilities, 15*(2), 241–256. https://doi.org/10.1080/17450101.2019.1696038

References

Ahmed, S. (2004). *The Cultural Politics of Emotion* (Second ed.). Edinburgh University Press.

Allen-Craig, S., Gray, T., Charles, R., Socha, T., Cosgriff, M., Mitten, D., & Loeffler, T. (2020). Together we have impact: Exploring gendered experiences in outdoor leadership. *Journal of Outdoor Recreation, Education, and Leadership, 12*(1), 121–139. https://doi.org/10.18666/jorel-2020-v12-i1-9937

Argus, S. M. (2018). LGBTQ girl scouts reflect on their outdoor experiences. In T. Gray & D. Mitten (Eds.), *The Palgrave International Handbook of Women and Journalism* (pp. 529–543). Palgrave Macmillan. https://doi.org/10.1057/9781137273246

Bloem Herraiz, B. (2019). *Bodies out/in Place? Unmapping Trans People's Experiences in Outdoor Activities*. University of Gothenburg. http://hdl.handle.net/2077/61669

Boniface, M. (2006). The meaning of adventurous activities for 'women in the outdoors'. *Journal of Adventure Education & Outdoor Learning, 6*(1), 9–24. https://doi.org/10.1080/14729670685200711

Bredford, A. (2020). Introduction. Transecology-(re)caliming the natural, belonging, intimacy, and impurity. In D. A. Vakoch (Ed.), *Transecology. Transgender Perspectives on Environment and Nature* (pp. 1–16). Routledge.

Browne, K. (2004). Genderism and the bathroom problem: (Re)materialising sexed sites, (re)creating sexed bodies. *Gender, Place & Culture, 11*(3), 331–346. https://doi.org/10.1080/0966369042000258668

Butler, J. (1990). *Gender Trouble. Feminism and the Subversion of Identity*. Routledge.

Butler, J. (1993). *Bodies that Matter. On the Discursive Limits of "Sex"*. Routledge.

Butler, J. (2004a). *Precarious Life: The Powers of Mourning and Violence*. Verso.

Butler, J. (2004b). *Undoing Gender*. Routledge.

Clough, P. T., & Calderaro, B. (2019). Foreword. In *Walking Methodologies in a More-than-Human World: WalkingLab* (pp. xiii–xiv). Routledge.

Díaz Carrión, I. A. (2012). *Género y turismo alternativo: aproximaciones al "empoderamiento"* [Thesis]. Universidad Complutense de Madrid.

Dignan, A. (2002). Outdoor education and the reinforcement of heterosexuality. *Journal of Outdoor and Environmental Education, 6*(2), 77–80. https://doi.org/10.1007/bf03400759

Doan, P. L. (2010). The tyranny of gendered spaces – reflections from beyond the gender dichotomy. *Gender, Place & Culture, 17*(5), 635–654. https://doi.org/10.1080/0966369X.2010.503121

Doran, A. (2019). *Women's Mountaineering Tourism Experiences: The Constraint Negotiation Process and Benefits of Participation* [Doctoral, Sheffield Hallam University]. http://shu.a.shu.ac.uk/24981/

Doran, A., Schofield, P., & Low, T. (2018). Women's mountaineering tourism: An empirical investigation of its theoretical constraint dimensions. *Leisure Studies, 37*(4), 396–410. https://doi.org/10.1080/02614367.2018.1452283

Freire, P. (1970). *Pedagogy of the Oppressed*. Continuum.

Frohlick, S. (1999). The "Hypermasculine" landscape of high-altitude mountaineering. *Michigan Feminist Studies, 14*, 83–106.

Frohlick, S. (2005). 'That playfulness of white masculinity': Mediating masculinities and adventure at mountain film festivals. *Tourist Studies, 5*(2), 175–193. https://doi.org/10.1177/1468797605066926

Gaard, G. (2020). Preface. In D. A. Vakoch (Ed.), *Transecology. Transgender Perspectives on Environment and Nature* (pp. xx–xxv). Routledge.

Graham, R. (1993). Gay outdoor club climbs over stereotypes. *The Boston Globe*.

Guzmán, M. (1997). "Pa "La Escuilita con Mucho Cuida'o y por la Orillita": A Journey through the Contested Terrains of the Nation and Sexual Orientation." In F. Negrón-Muntaner & R. Grosfoguel (Eds.), *Puerto Rican Jam. Rethinking Colonialism and Nationalism* (pp. 209–228). University of Minnesota Press. https://doi.org/10.5749/j.ctttsqd0

Halberstam, J. (2005). *In a Queer Time and Place: Transgender Bodies, Subcultural Lives*. New York University Press.

Haraway, D. (1988). Situated knowledges: The science question in feminism and the privilege of partial perspective. *Feminist Studies, 14*(3), 575–599. https://doi.org/10.2307/3178066

Humberstone, B. (2000). The 'outdoor industry' as social and educational phenomena: Gender and outdoor adventure/education. *Journal of Adventure Education & Outdoor Learning, 1*(1), 21–35. https://doi.org/10.1080/14729670085200041

Jordan, D. J. (2018). Ongoing challenges for women as outdoor leaders. In T. Gray & D. Mitten (Eds.), *The Palgrave international handbook of women and outdoor learning* (pp. 217–234). Palgrave Macmillan.

Kling, K. G., Margaryan, L., & Fuchs, M. (2020). (In) equality in the outdoors: Gender perspective on recreation and tourism media in the Swedish mountains. *Current Issues in Tourism, 23*(2), 233–247. https://doi.org/10.1080/13683500.2018.1495698

Lather, P. (2013). Methodology-21: What do we do in the afterward? *International Journal of Qualitative Studies in Education, 26*, 634–645.

Li, D. Q. (2018). *El poder del bosque. Shinrin-Yoku: Cómo encontrar la felicidad y la salud a través de los árboles*. Roca Editorial.

Little, D. E., & Wilson, E. (2005). Adventure and the gender gap: Acknowledging diversity of experience. *Loisir et Societe, 28*(1), 185–208. https://doi.org/10.1080/07053436.2005.10707676

Lundin, M., & Bombaci, S. (2022). Making outdoor field experiences more inclusive for the LGBTQ+ community. *Ecological Applications*. https://doi.org/10.1002/eap.2771

Lynch, J. (2020). Mobile methods in outdoor studies: Walking interviews with educators. In B. Humberstone & H. Prince (Eds.), *Research Methods in Outdoor Studies* (pp. 207–217). Routledge.

Martín Talavera, L., & Mediavilla Saldaña, L. (2020). Diferencias de género en el perfil y los hábitos de practicantes de actividades en el medio natural (Gender differences in the profile and habits of practitioners of outdoor activities). *Retos, 2041*(38), 713–718. https://doi.org/10.47197/retos.v38i38.78499

Mcphie, J., & Clarke, D. A. G. (2020). Post-qualitative inquiry in outdoor studies: A radical (non-) methodology. In B. Humberstone & H. Prince (Eds.), *Research Methods in Outdoor Studies* (pp. 186–195). Routledge.

Meyer, A. M. (2010). *Gender, Body, and Wilderness: Searching for refuge, connection, and ecological belonging* [The University of Montana]. https://scholarworks.umt.edu/etd/480

Meyer, A. M., & Borrie, W. T. (2013). Engendering wilderness: Body, belonging, and refuge. *Journal of Leisure Research, 45*(3), 295–323. https://doi.org/10.18666/jlr-2013-v45-i3-3153

Mitten, D., & D'Amore, C. (2017). The nature of body image: The relationship between women's body image and physical activity in natural environments. In D. A. Vakoch & S. Mickey (Eds.), *Women and Nature?: Beyond Dualism in Gender, Body, and Environment* (pp. 96–116). Routledge. https://doi.org/10.4324/9781315167244

Moscoso-Sánchez, D. J. (2008). The social construction of gender identity amongst mountaineers. *European Journal for Sport and Society, 5*(2), 187–194.

Mountz, A. (2011). Where asylum-seekers wait: Feminist counter-topographies of sites between states. *Gender, Place & Culture, 18*(3), 381–399. https://doi.org/10.1080/0966369X.2011.566370

Nielsen, L. B. (2002). Subtle, pervasive, harmful: Racist and sexist remarks in public as hate speech. *Journal of Social Issues, 58*(2), 265–280. https://doi.org/10.1111/1540-4560.00260

Overholt, J. R., & Ewert, A. (2015). Gender matters: Exploring the process of developing resilience through outdoor adventure. *Journal of Experiential Education, 38*(1), 41–55. https://doi.org/10.1177/1053825913513720

Pedersen, K. (2003). Discourses on nature and gender identities. In K. Pedersen & A. Viken (Eds.), *Nature and Identity. Essays on the Culture of Nature* (pp. 121–150). Hogskoleforlaget.

Platero, R. (Lucas), & Drager, E. H. (2015). Two trans* teachers in Madrid. *TSQ: Transgender Studies Quarterly, 2*(3), 447–463. https://doi.org/10.1215/23289252-2926419

Pomfret, G., & Doran, A. (2015). Gender and mountaineering tourism. *Mountaineering Tourism*, 138–155. https://doi.org/10.4324/9781315769202

Preciado, P. B. (2006). Basura y Género. *Mear/Cagar. Masculino/Femenino. Parole de Queer*, Errancia, La palabra inconclusa, Revista de Psicoanálisis, Teoría Crítica y Cultura, núm. 0, 14–17.

Puwar, N. (2004). *Space Invaders. Race, Gender and Bodies Out of Place.* Berg.

Razack, S. H. (2000). When place becomes race. In S. Razack (Ed.), *Race, Space, and the Law: Unmapping a White Settler Society* (pp. 1–20). Between the Lines.

Rodó-de-Zárate, M. (2016). ¿Quién tiene Derecho a la Ciudad? Jóvenes lesbianas en Brasil y Cataluña desde las geografías emocionales e interseccionales. *Revista Latino-Americana de Geografia e Genero*, 7(1), 3–20. https://doi.org/10.5212/Rlagg.v.7.i1.0001

Sassen, S. (2003). *Contrageografías de la globalización.* Traficantes de Sueños.

Saz, D. (2020, December 3). "Sexilio" rural: cuando las personas LGTBIQ+ abandonan su lugar de origen por su orientación sexual. *Eldiario.Es.* https://www.eldiario.es/aragon/sociedad/ sexilio-medio-rural-personas-lgtbiq-abandonan-lugar-origen-identidad-u-orientacion-sexual_1_6480592.html

Shakhsari, S. (2014). Killing me softly with your rights. Queer death and the politics of rightful killing. In J. Haritaworn, A. Kuntsman, & S. Posocco (Eds.), *Queer Necropolitics* (pp. 93–110). Routledge.

Springgay, S., & Truman, S. E. (2019). *Walking Methodologies in a More-than-Human World: WalkingLab.* Routledge.

Stanley, P. (2020). Unlikely hikers? Activism, Instagram, and the queer mobilities of fat hikers, women hiking alone, and hikers of colour. *Mobilities*, *15*(2), 241–256. https://doi.org/10.1080/ 17450101.2019.1696038

Straube, W. (2020). Posthuman ecological intimacy, waste, and the trans body in Nånting måste gå sönder (2014). In D. A. Vakoch (Ed.), *Transecology. Transgender Perspectives on Environment and Nature* (pp. 54–78). Routledge.

Stryker, S. (2006). My words to Victor Frankenstein above the village of Chamounix: Performing transgender rage. In S. Stryker & S. Whittle (Eds.), *The Transgender Studies Reader* (pp. 244–256). Routledge. https://doi.org/10.4324/9780203955055-29

Stryker, S. (2020). Foreword. In D. A. Vakoch (Ed.), *Transecology. Transgender Perspectives on Environment and Nature* (pp. xvi–xix). Routledge.

Thorsteinson, K., & Serenity Joo, H.-J. (2020). Coming out, camping out: Transparent's eco-ethical approach to gender. In D. A. Vakoch (Ed.), *Transecology. Transgender Perspectives on Environment and Nature* (pp. 31–53). Routledge.

Warren, K. (2016). Gender in outdoor studies. In *Routledge International Handbook of Outdoor Studies* (pp. 360–368). Routledge. https://doi.org/10.4324/9781315768465

Warren, K., Risinger, S., & Loeffler, T. (2018). Challenges faced by women outdoor leaders. In T. Gray & D. Mitten (Eds.), *The Palgrave International Handbook of Women and Outdoor Learning* (pp. 247–258). Palgrave Macmillan.

Weatherby, T. G., & Vidon, E. S. (2018). Delegitimizing wilderness as the man cave: The role of social media in female wilderness empowerment. *Tourist Studies*, *18*(3), 332–352. https://doi. org/10.1177/1468797618771691

Wilson, J., & Lewis, S. (2012). Transgender-specific programming: An Oasis in the Storm. In B. Martin & M. Wagstaff (Eds.), *Controversial Issues in Adventure Programming* (pp. 229–234). Human Kinetics. https://doi.org/10.1080/14729679.2013.870820

28 Disability and accessible adventure tourism

Jasmine Goodnow and Kristen Chmielewski

Chapter learning outcomes

1 Identify the need for increased accessibility in adventure travel.
2 Explain the differences between adopting the medical model of disability and adopting the social model of disability when planning and running inclusive adventure travel.
3 Address physical, informational, and attitudinal barriers disabled adventure travellers encounter through universal design, activity analysis and adaptation, and the provision of accessible information and customer service.

Introduction

My office phone rang, and the hopeful voice on the line asked, 'Is there still room on the Central American EcoAdventure trip?' When I replied that there was, the caller was momentarily silent before hesitantly questioning, 'What about for a person who sometimes uses a wheelchair?' 'Absolutely', I responded, launching into a plan in which we would rely heavily on the caller's expertise to proactively address accessibility on this trip, acknowledging some completely inaccessible activity would necessitate substitutions. As I asked the caller if they were open to providing their input throughout the entire process, they apologised for getting emotional. 'I expected a no', they said. 'I've heard no so many times'.

While not every single adventure tour is for everyone, adventure travel *is* for everyone. Adventure tourism at its core is a beneficial experience involving elements of uncertainty, skill development, novelty, risk, and natural environments leading to an array of positive physical and psychological outcomes (Brymer & Schweitzer, 2013; Houge Mackenzie & Brymer, 2020). Personal, interpersonal, and environmental factors (Card, Cole, & Humphrey, 2006; Devile & Moura, 2014) and the persistent belief that adventure is exclusively reserved for those who are white, male, wealthy, and non-disabled (Doran, 2016) often prevent disabled travellers from participating in adventure opportunities and

DOI: 10.4324/9781003393153-33"

promote exclusionary atmospheres and attitudes. Furthermore, disabled people face discrimination due to intersections with age, gender, income, race and ethnicity, and sexual identity. Disabled would-be adventure tourists must navigate a broad range of access needs..

Inclusivity and equity have recently become a significant part of the dialogue in the adventure tourism literature. Calls for *tourism for all* have multiplied, promoting the ideal that everyone has the right to travel and leisure, thus all spaces and activities, including that adventure should be accessible to all people (Beltramo, Duglio, & Cappelletti, 2022; Devile & Moura, 2014; Pigram, 1984; Var, Yeşiltaş, Yaylı, & Öztürk, 2011), yet there are few empirical studies about disability and adventure travel. What research exists typically focuses on participants' physical disabilities as barriers often without presenting solutions (Beltramo, Duglio, & Cappelletti, 2022). Therefore, the purpose of this chapter is to not only identify common barriers to participation but also discuss existing yet underutilized techniques and to persuade and, perhaps more importantly, to inspire students, practitioners, and academics to develop innovative inclusion approaches for adventure travel products, experiences, and environments. The overall goal of this chapter is to promote universal design, a system of creating products, experiences, environments, and systems to be usable by all people (regardless of age, ability, or life circumstance) without the need for adaptation (Steinfeld, Maisel, & Levine, 2012). Adventure travel itineraries, experiences, and destinations that are accessible and inclusive from the start, rather than as a special accommodation, should be the norm. The industry needs to develop and utilize adapted adventure tourism products that respond to latent disabled adventure tourism market needs (Devile & Moura, 2014). Furthermore, adventure tourism market segmentation must be reframed and expanded to include those who are disabled, without assuming homogeneity (Blichfeldt & Nicolaisen, 2011) or engaging in *handicapitalism*, exploitation due to the large untapped market share and potential for profit (Devile & Moura, 2014; Ray, 2003).

Disabled adventure travellers

Our shared excitement and nerves evident, the traveller and I determined what barriers might arise and how to best minimise them on the trip to Central America. This was a once-in-a-lifetime event for them. They were the expert on their own needs, strengths, and abilities, but they had never travelled internationally or engaged in what they described as adventure. The airline could not accommodate their wheelchair, so our tour company bought a manual wheelchair. We anticipated rough surface roads, no sidewalks, uneven hiking trails (see Figure 28.1 for an example of the landscape), and inaccessible public transportation, so we negotiated an additional guide with the destination management organisation (DMO) for the purposes of physical support during transport and activities. Aware that there would be excessive heat and inaccessible lodging and restrooms, we requested lodging on the ground floor with accessible bathrooms. Finally, the tour company, DMO, and traveller developed a short list of alternative fun activities if any glitches arose. Even with this planning in place, I found myself scrutinising our detailed itinerary repeatedly. Can we truly meet their needs? What if we miss something? What if they have a horrible experience? What if? What if?

Figure 28.1 An accessible environment? A river and tree roots in Central America. Photograph by Jasmine Goodnow.

Defining disability in adventure tourism

The World Health Organization (WHO, 2021) estimates that approximately 1 billion people worldwide experience disability, which corresponds to approximately 15% of the world's population. While most people tend to think of disability as a definite, concrete diagnosis, definitions of disability vary from organisation to organisation. WHO, instead of defining disability as a diagnosis, states that 'disability refers to the interaction between individuals with a health condition (e.g., cerebral palsy, Down syndrome, and depression) and personal and environmental factors (e.g., negative attitudes, inaccessible transportation and public buildings, and limited social supports)'. The landmark Americans with Disabilities Act (ADA), civil rights legislation signed into law in 1990 which guaranteed Americans certain protections from disability discrimination, provides a similarly amorphous definition, stating that 'disability is a physical or mental impairment that substantially limits one or more major life activities, a person who has a history or record of such an impairment, or a person who is perceived by others as having such an impairment'. The ADA does not specify which impairments qualify as disabilities under the law. The American Centers for Disease Control and Prevention (CDC, 2020) provides a definition that approaches the idea of disability as a diagnosis independent of societal interaction and perceptions writing that 'disability is any condition of the body or mind (impairment) that makes it more difficult for the person with the condition to do certain activities (activity limitation) and interact with the world around them (participation restrictions)'.

These varying definitions of disability illustrate the differences in how societies, professions, organisations, and individuals attempt to understand and categorise disability and exemplify the differences between the medical model and the social model of disability. While other more nuanced models of categorisation and understanding exist and continue to evolve, for our purposes of introducing disability studies to the field of adventure tourism, we focus on attempting to persuade practitioners to abandon the medical model in favour of the social model of disability. The medical model frames disability as an individual deficit in need of treatment or correction. This view of disability is employed throughout disciplines, shaping education and practice in medicine, rehabilitation, special education, the social sciences, and the humanities. The popularity of the medical model leads to modes of analysis which cannot explain the social and economic disadvantages experienced by disabled individuals; when educators, researchers, doctors, and practitioners view disability as an individual defect, it is the individual who needs fixing, curing, and rehabilitating. When the field of adventure tourism subscribes to the medical model of disability, either consciously or unconsciously, we are ultimately limiting our participant field and abdicating our power in creating accessible experiences. If it is solely the disabled individual who needs rehabilitation or is required to acquire specialised equipment before embarking on an adventure, we relay the message that the benefits of adventure tourism are not and never will be for everyone. By espousing this message, we also place an expiration date on every traveller's ability to continue adventuring as every person, if they live long enough, will acquire a disability.

If we as a field aspire to make the benefits of adventure tourism more widely available and believe we have an ethical imperative to work towards greater accessibility, we must shift our conceptualisation of disability to a social model. The social model of disability, along with disability activists and disability studies scholars, assert that, yes, differences exist in how individual bodies and minds function but that there is nothing inherently wrong or bad about those differences. Disability – according to this model – is not the product of bodily difference but is largely a social construction: society creates ideas of normality and then marginalises those whose bodies and minds do not fit those standards. Proponents of this social model seek to reframe our understanding of disability as a complex interaction of social, cultural, political, and economic variables and to challenge the idea that disability disadvantage is an inevitable outcome of a biological condition. This frame 'centers the study of disability on its social construction, the processes that have accorded particular meaning to disability and that have determined the treatment and positioning of people with disabilities in society' (Davis & Linton, 1985, p. 2–3). This model empowers us to be proactive in changing the conditions, environments, and attitudes that exclude disabled individuals from adventure travel. As proponents of this social model of disability and of destigmatising language about disability, we have opted to use identity-first language in this chapter and our practice. Person-first language—that is, a person with a disability or a person with autism—gained popularity amongst practitioners in different fields over the past few decades in an effort to see the disabled person as an individual independent of their disability. Disability activists and scholars have noted that many person-first advocates tend to be non-disabled individuals who unintentionally convey that disability is a negative part of the human experience that needs to be tucked away in prepositional phrases or euphemisms. Identity-first language—that is, a disabled person or an autistic person—asserts that there is nothing inherently bad or shameful about disability, that disability is a part of that person's identity.

As disability intersects with other marginalised identities, planning for greater access for disabled travellers can also benefit other groups historically excluded from adventure travel. For the purposes of illustrating these intersections, we use examples from the United States due to its diverse population and laws prohibiting disability discrimination, yet disparities persist. The CDC (2020) reports that Indigenous Americans and Black Americans are more likely to be disabled than white Americans at disability rates of two in five, one in four, and one in five, respectively. Socio-economic status and disability also intersect, as disabled Americans live in poverty at more than twice the rate of non-disabled Americans, and studies have found that disabled Americans are underemployed, with about 70% of disabled working-age adults unemployed (National Council on Disability, 2017). Other research has shown that lesbian, gay, bisexual, and transgender Americans are more likely than cisgender, heterosexual Americans to have a disability and face systemic challenges in finding employment (Fredriksen-Goldsen, Kim & Barkan, 2012). As many acquired disabilities are age-related, disability and age clearly intersect; the CDC (2020) reported that two in five adults over the age of 65 are disabled. Many intersections between marginalised statuses are likely the result—despite legislation in the United States aimed at protecting groups from discrimination—of unequal access to medical care, employment, housing, safety, and healthy, affordable food. Regardless of the cause of these intersections, though, rethinking adventure tourism in terms of greater accessibility for disabled travellers will also benefit other historically excluded groups.

Accessibility for disabled travellers—in the largely inaccessible world of adventure tourism—is a large, challenging undertaking. Disabilities are as varied as they are prevalent; they can be physical or mental, visible or invisible. A traveller can have been disabled from birth and be well aware of their own capabilities and needs, or they may have acquired their disability a year ago and be navigating a period of transition and adjustment. The needs of a blind adventure traveller will be vastly different from the needs of a paraplegic adventure traveller. And we are, whether we want to acknowledge it or not, a field deeply rooted in exclusivity and inaccessibility, in ideas of who deserves and is worthy of life-changing, peak experiences and who is not. Reframing our ideas about who adventure travel is for, who can benefit from adventure travel, and what adventure planning and travel can entail is a moral imperative for a more just, equitable tourism industry.

History of accessible adventure travel

The history of accessible adventure travel is brief and limited as adventure travel historically has been designed for nondisabled people (Beltramo, Duglio, & Cappelletti, 2022; Moura, Eusébio, & Devile, 2022; Pigram, 1984), and to date, there are no published research articles specifically on disabled adventure travel. However, this limited history is not the result of disabled individuals' lack of a desire to travel. The desire to travel and adventure has been present throughout history, even though accessibility and corresponding research has been lacking. Disabled people first participated in early travel for the purposes of pilgrimage and medical care, and these forms of medical pilgrimages continue to this day to Lourdes, France, the healing waters of Bath, England, and other destinations around the world (Van Horn & Isola, 2014).

The roots of accessible leisure tourism began after World Wars I and II when European and U.S. governments sought to provide benefits to wounded veterans, such as preferred seating on public transportation, specially designed outdoor recreation camps, and wheelchair sport competitions, which eventually led to the England's premier veterans' rehabilitation centre, the first International Wheelchair Games, and to the first Paralympic

Games in 1960 (Van Horn & Isola). Regional disabled sport competitions emerged around the world in Asia and the South Pacific. International and regional sport competition spurred accessibility within host cities, including alterations to public streets, hotels, attractions, and some mass transit.

Unfortunately, the tourism industry and most especially the adventure tourism niche have relied primarily on existing disability laws and policies of the countries in which adventure takes place instead of designing and managing adventure tourism to be inclusive and accessible. Most existing laws are lacking, providing little to no access for disabled travellers. The calls to reimagine and address accessibility in the tourism industry are plentiful. The United Nations Convention Article 30 states that 'parties recognise the rights of persons with disabilities to participate on an equal basis with others in a cultural life and must take appropriate measure to ensure that persons with disabilities have access to venues for cultural events or services' (Beltramo, Duglio, & Cappelletti, 2022, p. 2). The World Tourism Organization (WTO, n.d.) claims that inclusion and accessibility are part of the overall concept of sustainable tourism 'that takes full account of its current and future economic, social, and environmental impacts, addressing the needs of visitors, the industry, the environment and host communities' (p. 12). As early as the 1970s, there have been movements to create more inclusive recreation and leisure opportunities. In the United States in 1979, the Third Nationwide Outdoor Recreation Plan included a task force on inclusive recreation needs which led to a guide to designing accessible outdoor recreation facilities including campgrounds, trails, swimming, and fishing and outlined how transportation, lodging, activities, natural places, cities, and destinations in general need to be accessible (Pigram, 1984). Canada, China, Dubai, Egypt, India, Japan, South Africa, Thailand, and many other countries have been developing and promoting access yet research finds that access is still limited for a variety of reasons, and rarely does the tourism industry go beyond the bare minimum of local laws and guidelines to provide true accessible opportunities (Aguilar-Carrasco et al, 2022; Van Horn & Isola, 2014; Pigram, 1984).

Disabled adventure travel segmentation

Research is still absent, despite Devile and Moura (2014) calling for this nearly 10 years ago and a notable interest in disabled people engaging in adventure travel since 1995 (Rose, 1995).

Most studies regarding accessible tourism focus on mass tourism; currently, there are no peer-reviewed studies on disabled adventure tourism and only a few on alternative forms such as nature-based and ecotourism (Garrod & Fennell, 2021). The following market segmentation infers from all accessible tourism studies, not just adventure.

Disabled tourism is an umbrella term encompassing great heterogeneity, including physical disability, sensory disability, and intellectual disability (Blichfeldt & Nicolaisen, 2011; Garrod & Fennell, 2021). Assuming homogeneity amongst disabled travellers would be a mistake. As the research shows, there may be differences between travellers with disabilities acquired with age and those with congenital disabilities or disabilities acquired at a younger age (Card, Cole, Humphrey, 2006; Moura, Eusébio, & Devile, 2022; Var et al., 2011).

Currently, the billion disabled individuals worldwide have approximately half a trillion dollars in disposable income (Kelly, 2022). Disabled travellers generally spend US$95 billion a year on travel, yet in the United States alone, and 27 million self-reported disabled Americans took 81 million trips and spent US$58.7 million (Kelly, 2022). Travability (2018) states that 88% of disabled travellers take an overnight holiday each year and that if both domestic and inbound totals of disabled travellers are added together (US$10.8 billion), the total accessible tourism market for Australia would rival the entire Chinese market for the same period (US$10.4 billion). The growth and scope of the disabled traveller market is significant. Furthermore, the majority of disabled travellers (87%) travel with at least one other person. Disabled travellers and their travel companions value inclusivity, access, and great customer service and when they do not encounter these features, 75% of disabled travellers and their families transfer to another provider who does (Ayling-Smith, 2020). The most important decision-making factor when booking travel is a truly accessible destination (Morris, 2020).

The mature market—those 55 years and older—is considered the fastest growing tourism market segment (Blichfeldt & Nicolaisen, 2011; Kelly, 2022) and although not yet confirmed, in 2018, it was predicted to become 50% of the total tourism spending (Travability, 2018). Many mature travellers not only have high disposable incomes but have or will also acquire age-related disabilities (Blichfeldt & Nicolaisen, 2011; Travability, 2018). Interestingly, as the mature market ages, those who self-identify as nondisabled travellers will continue to identify as nondisabled even as they acquire age-related disabilities (Travability, 2018). They, however, demand that the travel and tourism industry provide a new level of accommodation above what was provided to mature markets of previous generations. They expect accessible, engaging, and active tourism services provided by mainstream tourism providers and that information about accessible attractions, transportation, lodging, and destinations be provided in the exact place as all other information, not segregated. This refusal to identify with the 'traditional disability sector' while demanding increased accessibility (Travelability, p. 1) highlights the negative and exclusionary attitude towards disability in tourism, often stemming from tourists themselves. Without a dramatic shift in these attitudes, the ideal of true access and inclusivity will remain an unrealized ideal.

Disabled travellers' motivations

Disabled travellers often identify themselves as risk-takers (Ray, 2003) and exhibit similar travel motivations as non-disabled travellers (Blichfeldt & Nicolaisen, 2011; Moura, Eusébio, & Devile, 2022; Var et al., 2011). Common adventure activities are rafting, climbing, sailing, canoeing, off-hill downhill racing, skiing, scuba, soaring, and nature activities (Ray, 2003), and disabled travellers seek adventure, challenge, physical activity, and contact with nature to experience an array of positive benefits (Table 28.1). Gains in

Table 28.1 Disabled travellers' motivations

Social Connections	Closeness to family
	Reduced feelings of social isolation
	Opportunities for new social contacts
	Increased trust in others
Well-being	Perceptions of increased strength, physical endurance, energy, vitality, knowledge, self-confidence
	Reduced feelings of weakness, vulnerability and lack of control
Freedom, novelty, and escape	Change in routine
	Independence from caregivers

the perception of self-efficacy were also realised by family members along with an increased level of trust, confidence, and understanding of the disabled individual's ability (Rose, 1995).

Barriers and constraints to adventure travel

> *The initial conversations with the DMO were characterised by worry about legal liability, reluctance to provide accommodations for disabled travellers for the first time, and overall nervousness. Their initial response was no. So, we threatened to use a different DMO if they could not meet all our travellers' needs. The DMO agreed to work with our tour company, but, even with our careful planning, we encountered a host of issues in Central America. The manual wheelchair was difficult to manoeuvre due to terrain that was often rocky, steep, narrow, or uneven. When the roads were paved, cars would not let us safely utilise them, and there were no sidewalks. The traveller became exhausted by the end of the trip and was not able to complete a hike through a national park to a beach, so one guide stayed back to keep them company. During other activities, the disabled traveller was not able to keep pace with the rest of the group, and sometimes the guides had to carry the traveller. This placed an undue burden on guides as well as the disabled traveller, and other travellers felt slighted as they did not receive similar levels of one-on-one attention. Some travellers individually approached tour guides to express dissatisfaction and irritation at the slower pace, claiming their trip was being negatively and unfairly impacted and demanding a partial refund.*

Most common barriers that disabled travellers face can be sorted into physical, attitudinal, and informational categories during the travel experience within the domains of transportation, lodging, food, and attractions (Card, Cole, & Humphrey, 2006; Garrod & Fennell, 2021; Pigram (1984); Yau, McKercher, & Packer, 2004).

Physical barriers

Physical barriers refer to the structural or environmental factors that prevent physical access to transport systems, buildings (e.g., hotels, attractions, restaurants), and activities (Ray, 2003). Physical barriers present in the lives of disabled individuals are heightened when they travel. Planes are inaccessible to people with a variety of disabilities including mobility, sight, hearing, and sensory impairments. Airports have long waits, signage can be difficult to read, and intercom information is confounded by the din of the airport. Once travellers board the plane, the aisles of planes are too narrow for typical wheelchairs, so special aisle chairs are necessary for transport to seats and washrooms. Aisle chairs are not mandatory on flights, thus many travellers who use wheelchairs do not drink or eat prior or during the flight as they might be confined to their seats for the duration of the flight. At the destination, many travellers must wait, sometimes for hours, tired, hungry, and dehydrated, for an available aisle chair in order to depart the plane (Mann, 2022; Ray, 2003). Off the plane, many find that their wheelchairs have been damaged or lost en route so they are without their personal wheelchairs for the duration of their trip. Trains, subways, other forms of public transportation, and buildings share many of the same barriers, including stairs, uneven surfaces, narrow aisles, a lack of braille, and sensory overstimulation. Many travellers require a companion or assistant to navigate such barriers, but the extra cost may be prohibitive (Kelly, 2022).

The monetary costs disabled adventurers incur for access is another component of navigating physical barriers (Kelly, 2022). Most services and products in travel list a base price and then any additional features or accommodations are additional costs. Services that may seem like luxuries to non-disabled clients can be necessary to the health of disabled travellers, such as access to service animals, air-conditioning, and refrigerators to keep medications cool. If a traveller needs the extra customer service and support only offered in business class, the transport expense might be quadrupled.

Attitudinal barriers

Disabled adventurers must negotiate a range of attitudinal barriers. Negative perceptions about disability are present in most aspects of the travel experience. Tourism providers (e.g., transportation, lodging, food/beverage, and attractions sectors) often view accessibility adaptations, renovations, and universal design of physical spaces and programmes as costly and unnecessary (Bi, Card, & Cole, 2007; Garrod & Fennell, 2021; Travability, 2018; Yau, McKercher, & Packer, 2004). Accessible rooms and adaptive programmes are viewed as a 'waste of productive space' and resources (Garrod & Fennell, 2021, p. 3) as providers fear non-disabled travellers will not book them (Beltramo, Duglio, & Cappelletti, 2022; Garrod & Fennell, 2021; Travability, 2018). Disabled travellers have been 'regarded as a risk management issue and the accessible facilities that have been created as cost imposition and not a viable commercial asset … those assets are excessive as they have below industry average utilisation' (Travability, 2018, p. 3). Common negative attitudinal barriers exhibited by tourism staff towards disabled travellers include avoidance, obsolete and derogatory labelling, paternalistic behaviour, and apathetic behaviour (Card, Cole, & Humphrey, 2006). Disabled adventurers experience severely negative attitudes at attractions (Bi, Card, & Cole, 2007; Card, Cole, & Humphrey, 2006) and negative attitudes from other travellers (Kelly, 2022). Finally, historical, cultural, and natural environmental preservation and conservation are prioritised over the accessibility needs of disabled travellers (Garrod & Fennell, 2021; Pretto, 2022).

Informational barriers

Disabled adventure travellers need readily available, up-to-date information regarding accessibility for each stage and component of their trip. Disabled travellers must plan their trip like a problem-solver and each stage and component of the trip must meet their functional level or they might become stranded, without a bathroom, or unable to engage in the activity (Beltramo, Duglio, & Cappelletti, 2022). Unfortunately, information regarding physical accessibility and attitudes is often incorrect, incomplete, or inaccessible (Kelly, 2022). Eighty-eight per cent of disabled travellers cited that hotels do not provide enough information about accessible room types and features on their websites (Morris, 2020). The tourism industry expects disabled travellers to contact each travel entity to inquire about accessible information or accommodations and wait for a response that rarely comes. Furthermore, disabled travellers report that they are unable to depend on the expertise of typical travel agents because they lack the knowledge and expertise regarding accessible destinations, activities, facilities, and transportation (Blichfeldt & Nicolaisen, 2011; Fennell & Garrod, 2022; Garrod & Fennell, 2021; Kelly, 2022; Ray & Ryder, 2003). Too often, travel agents have erroneously informed disabled travellers about available accommodations, so the disabled community must instead rely on word of mouth, the internet, and travel guides (Ray, 2003). Those with higher incomes might utilise luxury travel advisors and operators, but the cost of access, along with the potential need for a travel companion often is unfeasible (Kelly, 2022).

This lack of information is further compounded by the lack of marketing in adventure travel showing diversity. Few adventure tourism destinations or operators mention accessibility or accommodation, and only 1% of travel marketing is representative of disabled travellers (Kelly, 2022). This lack of representation can lead to further mistrust that needs for access, comfort, and safety can be met.

Accessible design strategies

> *Despite the issues with accessibility and attitudes, our planning did allow us to successfully negotiate barriers that might have otherwise derailed the adventure. The travellers were able to participate in all adventure activities – jungle hiking, a zip-lining and canopy tour, visits to wildlife sanctuaries, a local cooking class, canoeing, waterfall viewing, kayaking, and whitewater rafting – except for the last day. All lodging and accommodations were on the ground floor so energy was not unnecessarily expended going up and down stairs. All toilets were easily accessible with the help of their travel companion. The DMO suggested that this experience helped them realise that they can accommodate a greater range of abilities and travellers. Throughout this process, they developed better ideas about what they would do next time and were excited to expand their offerings to a wider range of travellers. The travellers told me that this was the most amazing experience of their life.*

While adventure travel providers often have little control over the accessibility of particular destinations, they have a tremendous amount of control over the accessibility of the products and experiences they create (see Case study 28.1). The four key themes to fostering assessable adventure travel are designing barrier-free travel experiences, making information clear, providing inclusive customer service, and involving disabled people in the design (Ayling-Smith, 2020).

Universal design

A major way to increase access to adventure travel and create barrier-free experiences is to adopt the principles of universal design (see Figure 28.2 for an example). Architect Ronald Mace – who used a wheelchair after the age of 9 – originated the concept: the idea that buildings, environments, and products should be useable by everyone to the greatest extent possible, regardless of their age, disability, or status in life (Connell et al., 1997). Implementing the principles of universal design while creating new adventure travel itineraries and products increases their accessibility and potential for profit, as the beauty of universal design is that it benefits all users.

Figure 28.2 Example of universal design accessible walk/rollway. Photograph by Jasmine Goodnow.

The principles of universal design are as follows, as adapted from the Center for Universal Design (1997):

- Equitable use
- Flexibility in use
- Simple and intuitive use
- Perceptible information
- Tolerance for error
- Low physical effort
- Size and space for approach and use

Products created with universal design in mind provide the same means of use for all participants and avoid segregating or stigmatising users. They should provide adaptability to the user's pace and choice in the methods of use. These products should eliminate unnecessary complications and maximise the legibility of the most important information. While the principles about tolerance for error and level of effort may seem antithetical to adventure, remember that error is not synonymous with risk and that not every adventure must be physically exhausting. The adventure guide should provide warnings of hazards and common errors to facilitate travellers' success, and certain adventures can allow travellers to maintain a neutral body position and minimise sustained physical effort. The spaces and equipment that providers use should ideally provide every traveller with adequate space for the use of assistive devices or personal assistance. Automatic doors, audiobooks, rolling suitcases, ramps with handrails into swimming pools, and closed captioning on videos are all examples of products that exemplify the principles of universal design.

Activity analysis, adaptation, and modification

Another way to increase the accessibility of travel is to assist adventure travellers in selecting experiences that provide them with the best challenge levels while also aligning with their needs and strengths. Performing activity analyses of the various programme components offered in a tour allows providers to assist disabled travellers – and any traveller – in selecting the adventure best suited to their travel goals and determining what modifications and adaptations of the activity might be most successful (Table 28.2). Activity analysis is a systemic process for identifying the knowledge, skills, and materials people need to participate in a selected activity (Creighton, 1992).

Understanding these aspects of the experience will assist travel providers with both making information clear and providing inclusive customer service. Activity analysis will also allow a travel provider or guide to understand what adaptations and modifications might reasonably make aspects of an adventure more accessible. Adaptations to activities should maintain the integrity of the experience, increasing participation in the activity and making it more enjoyable for participants. Adaptations should be individualised to the traveller needing the modification; there is no one-size-fits-all approach that will be appropriate for every traveller in a particular disability group, which is one of the many reasons why communication with the disabled traveller is vital. Common adaptations in adventure travel can include the equipment used for an activity, the time allotted to complete an activity, multiple ways of presenting instructions, and adjustments to the environment and space for an activity.

Accessible information and customer service

Understanding what aspects of an activity are best adaptable and embracing the idea that adventure travel providers are also foundational to making information about travel

Table 28.2 Activity analysis domains

Physical	Make note of body parts utilised, as well as range of movement, necessary sensory abilities, and required strength, balance, flexibility, endurance, coordination, and speed.
Cognitive	Determine if the activity requires long-term or short-term memory, concentration, abstract thinking skills, or the ability to follow complex instructions.
Social	Determine the types of interaction, communication, cooperation, or competition necessitated by the activity. Does it require any level of intimacy or physical proximity amongst participants?
Affective	Consider what types of feelings – both positive and negative – may result from the activity.

clear and providing inclusive customer service. A disabled adventurer needs comprehensive information about the trip so they can decide if the experience will meet their needs, be comfortable, and be safe (Blichfeldt & Nicolaisen, 2011; Fennell & Garrod, 2022; Garrod & Fennell, 2021; Ray & Ryder, 2003). The majority of disabled travellers utilise websites or mobile apps when booking their trips, and 69% of disabled travellers will click away from websites if they encounter barriers (Ayling-Smith, 2020). Disabled travellers are more likely to spend money on websites that are the easiest to navigate rather than the cheapest option. Only 10% of website users who encounter difficulties will contact the provider (Ayling-Smith, 2020). An accessible website with comprehensive descriptions of the experiences offered is vital and must include photos or video—with alternate text and captions—of all aspects of the trip. The travel industry needs to normalise providing such information to travellers; the measurements of doorways and paths, the type of dietary restrictions a company can accommodate, and the presence (or lack) of accessible restrooms and lodgings should be as easy to find as dates of upcoming trips and prices. Information about air-conditioning, lighting, sound level, and steps in provided lodging is crucial (Kelly, 2022), as is information about the physical demands of the trip, such as terrain, altitude above sea level, expected temperature, and the amount of time needed to complete the activity (Kelly, 2022; Travability, 2018.) Warnings about heights, small spaces, handrails, unpaved paths, and steep or hazardous terrain empower travellers to plan for a successful trip. The website for the Olympic National Park in the United States does a particularly thorough job of detailing the accessibility of its facilities and the condition of its trails. Figure 28.3 provides examples of what can be seen from the trails.

Figure 28.3 Accessible Temperate Rainforest, Olympic National Park, USA. Photograph by Jasmine Goodnow.

An accessible adventure travel website should offer multiple ways for potential travellers to contact tour providers. The website and other promotional materials also must feature disabled travellers, as 92% of consumers report it is important for travel providers to meet the accessibility needs of all travellers (Ayling-Smith, 2020). Spending on structural modification should be followed up with inclusive imagery to illustrate changes. Such imagery can increase patronage and loyalty. As adventure travel providers increase the visibility of disabled travellers, it is also essential that they continue to work to make their products and services more accessible. Providers should solicit after-trip reviews, photos, and feedback from all travellers regarding accessible and inaccessible experiences so future trips can be modified and information on websites and brochures is accurate and reliable (Kelly, 2022).

Microadventures

One immediate possibility for a dramatic increase in the accessibility of adventure lies in microadventures. Microadventures are an inclusive and accessible form of adventure travel and may be the preferred way to adventure for the disabled and nondisabled alike (Mackenzie & Goodnow, 2022). Microadventure is rewarding 'adventure close to home, cheap, simple, short, and ... effective. It still captures the essence of big adventures, the challenge, the fun, the escapism, the learning experiences and the excitement' (Humphreys, 2014, p. 14). Growing evidence suggests that microadventures facilitate similar psychological benefits to traditional adventure experiences while its inherent nature minimises the common barriers and constraints to typical adventures such as mobility, access, time, and money (MacKenzie & Goodnow, 2022; Roberts, 2018).

Adventure that is close to home eliminates barriers associated with plane travel. Mackenzie and Goodnow (2022) suggest that instead of relying on exotic or remote destinations or extended durations and hardship to increase the feeling of adventure, microadventures can create a sense of travel and freedom by increasing novelty, liminality, and other threshold experiences (e.g., border crossings, bridges, ferries, city to nature) and focusing on participant mindset (e.g., the degree to which participants' cognitively or emotionally disconnect from everyday life). Microadventures lend the opportunity for many different transportation options, some of which might become the adventure itself. Beyond their existing personal transportation (e.g., personal automobile, public transportation, and taxi), disabled adventures might engage in mode-shifting and human-scaled mobility or self-supported human-powered travel (e.g., biking, rowing/paddling, walking). Adventure arises from the challenges and uncertainty inherent in travelling, being, and recreating in new ways. Other techniques Mackenzie and Goodnow (2022, p. 5) emphasise to increase adventure is an emphasis on

> simplicity, personal skill development, immersion in nature, curiosity, and personal insight to facilitate a return to the core of what adventure is about, elements increasingly lost in modern day adventure travel. Rather than pursuing 'more, further, faster' with advanced equipment and technology, microadventures present opportunities for enhanced community connections in local places by going 'deeper' not further.

Conclusion

> *If I were to go back in time, talk over the itinerary with the traveller again, our biggest desire for the adventure travel experience would have been an attitudinal shift. The industry must normalize everyone possessing the right to travel. We must normalise the ideas of slowing down, working together to assist fellow travellers, and engaging in slow tourism to appreciate the experience, gain meaning, and create community rather than checking activities off a list. An adventure wheelchair would also have sped up the pace, minimised fatigue associated with carrying or moving the manual wheelchair and offered greater independence for the traveller. We can't always shift or change the destination's structural or environmental factors, but we can always improve our attitudes.*

The access needs of disabled community members must be prioritised even when concepts of sustainability and cultural/historical/environmental conservation and preservation are concerned. Equal access is part of the sustainable tourism ethos and access to cultural and historical sites has already been modified and normalised for nondisabled populations to protect the safety of travellers from accidents, natural environments from tourist trampling, and cultural/historical sites from tourist graffiti and other types of destruction, and tour agencies and attractions from legal liability when accidents occur. Why do the safety needs of nondisabled tourists result in handrails, even walkways, and barriers along cliffsides and rivers, deep within caves, in old medieval castles, and along the spine of precipitous mountain trails, yet we baulk at making similar types of modification to increase the access of disabled travellers? A drastic attitude shift must occur at the industry level about who deserves access to adventure. No longer should the tourism industry and destination meet only the minimum disability regulations but instead should do the morally right thing and capitalize on the untapped disabled travel market.

Case study 28.1   Adventure provider imperatives for accessibility

View each disabled traveller as autonomous and as the expert on their own abilities and needs. Involve them in the planning process at all levels and integrate their suggestions and ideas into the trip.

Never assume someone's disability status—many disabilities are invisible. If someone informs you that they need a certain accommodation, believe them. If you ask questions about the accommodation, be sure you are asking to obtain information that will enable you to provide the best experience possible, not to force the traveller to prove their need for accommodation or to prove that they are disabled.

Watch the language you use about disability, both in your promotional material and conversations. Using phrases like 'he is confined to a wheelchair' or 'she suffers from cerebral palsy' conveys your own negative beliefs about disability and makes assumptions about other people's experiences. Many wheelchair users do not feel confined to their chairs at all; their chairs allow them the freedom to move and explore. If it is necessary to talk about someone's disability status or accommodations, use phrases like 'he uses a wheelchair' or 'she has cerebral palsy'.

Provide current, reliable, detailed, and comprehensive information regarding the activity components and facilities so disabled travellers can gauge whether the destination, facility, and activity meet their needs.

Work to change the attitudes of yourself, your adventure tourism agency, other adventure travellers, and destinations to believe that disabled travellers are part of the adventure travel norm and are an adventure travel segment that must be wooed, designed for, and accommodated.

People of all ages, races, and disabilities should be featured on marketing material as a typical adventurer.

Integrate the easy-to-change and low-cost strategies and techniques discussed in this chapter. Make a realistic plan for your next season, year, and 5 years. Start saving, investing, and applying for grants to acquire more costly adaptive equipment and implement infrastructure changes.

Collect traveller information for everyone regarding their needs, abilities, and diet and normalise conversations about accommodations and adaptations for all people without singling out or stigmatising disabled travellers.

Research instead of relying on the labour of disabled travellers to teach or inform. There is a wealth of research and 'how-to' guides readily available depending on your participants' needs.

Utilize new technology and equipment including freewheel wheelchair extensions, off-road handcycles, road handcycles, all-abilities sailboats, adaptive fishing equipment, paddle boards designed for wheelchairs, adaptive canoes and canoe launchers, in addition to traditional hearing loops, tactile markers, and so on.

Review questions

1 Create an activity analysis of your favourite adventure activity.
2 Find five resources on how to adapt your favourite activity for a disabled adventure traveller with a mobility disability, hearing impairment, or developmental disability.

Further reading

Adams, R., Reiss, B., & Serlin, D. (Eds.). (2015). *Keywords for disability studies* (Vol. 7). NYU Press.

Longmore, P. (2003). *Why I burned my book* (Vol. 12). Temple University Press.

Wong, A. (Ed.). (2020). *Disability visibility: First-person stories from the twenty-first century.* Vintage.

References

Aguilar-Carrasco, M. J., Gielen, E., Vallés-Planells, M., Galiana, F., Almenar-Muñoz, M., & Konijnendijk, C. (2022). Promoting inclusive outdoor recreation in national park governance: A comparative perspective from Canada and Spain. *International Journal of Environmental Research and Public Health, 19,* 2566. https://doi.org/10.3390/ijerph19052566

Ayling-Smith, V. (2020). Breaking down barriers to travel: Championing disability inclusive and accessible travel. Expedia Group and Leonard Cheshire. Retrieved January 9, 2022, from https://s27.q4cdn.com/708721433/files/doc_downloads/Breaking-Barriers-to-Travel-Report.pdf

Beltramo, R., Duglio, S., & Cappelletti, G. M. (2022). Should I stay or can I go? Accessible Tourism and Mountain Huts in Gran Paradiso National Park. *Sustainability, 14,* 2936. https://doi.org/10.3390/su14052936

Bi, Y., Card, J. A., & Cole, S. T. (2007). Accessibility and attitudinal barriers encountered by Chinese travellers with physical disabilities. *International Journal of Tourism Research, 9,* 205–216. https://doi.org/10.1002/jtr.603

Blichfeldt, B. S., & Nicolaisen, J. (2011) Disabled travel: Not easy, but doable. *Current Issues in Tourism, 14*(1), 79–102. https://doi.org/10.1080/13683500903370159

Brymer, E., & Schweitzer, R. (2013). Extreme sports are good for your health: A phenomenological understanding of fear and anxiety in extreme sport. *Journal of Health Psychology, 18*(4), 477–487. https://doi.org/10.1177/1359105312446770

Card, J. A., Cole, S. T., & Humphrey, A. H. (2006). A comparison of the Accessibility and Attitudinal Barriers Model: Travel providers and travelers with physical disabilities. *Asia Pacific Journal of Tourism Research, 11*(2), 161–175. https://doi.org/10.1080/10941660600727566

Center for Universal Design. (1997). *The Principles of Universal Design, Version 2.0.* Raleigh, NC: North Carolina State University. Retrieved October 1, 2022, from https://design.ncsu.edu/research/center-for-universal-design/

Centers for Disease Control and Prevention. (2020, September 16). *Disability and health overview.* Centers for Disease Control and Prevention. Retrieved October 1, 2022, from https://www.cdc.gov/ncbddd/disabilityandhealth/disability.html

Connell, B. R., Jones, M., Mace, R., et al. (1997, April 1). *Universal Design Principles.* The Center for Universal Design. Retrieved October 1, 2022, from https://projects.ncsu.edu/ncsu/design/cud/about_ud/udprinciplestext.htm

Creighton, C. (1992). The origin and evolution of activity analysis. *The American Journal of Occupational Therapy, 46*(1), 45–48.

Davis, L., & Linton, S. (1985). Introduction to disability studies. *Radical Teacher, 47,* 2–3.

Devile, E. L., & Moura, A. (2014). Adventure tourism for people with disabilities in Portugal: Opportunities and challenges. *Sport Tourism Conference.* Proceedings E-Book, 9–21.

Doran, A. (2016). Empowerment and women in adventure tourism: A negotiated journey. *Journal of Sport and Tourism, 20*(1), 57–80. https://doi.org/10.1080/14775085.2016.1176594

Fennell, D., & Garrod, B. (2022). Seeking a deeper level of responsibility for inclusive (eco)tourism duty and the pinnacle of practice. *Journal of Sustainable Tourism, 30*(6), 1403–1422.

Fredriksen-Goldsen, K. I., Kim, H. J., & Barkan, S. E. (2012). Disability among lesbian, gay, and bisexual adults: Disparities in prevalence and risk. *American Journal of Public Health, 102*(1), e16–e21.

Garrod, B., & Fennell, D. A. (2021): Strategic approaches to accessible ecotourism: Small steps, the domino effect and not paving paradise. *Journal of Sustainable Tourism.* DOI: 10.1080/09669582.2021.2016778

Horn, L. V., & Isola, J. (2014). Toward a global history of inclusive travel. *Review of Disability Studies, 2*(2), https://rdsjournal.org/index.php/journal/article/view/346/1064

Houge Mackenzie, S., & Brymer, E. (2020). Conceptualizing adventurous nature sport: A positive psychology perspective. *Annals of Leisure Research, 23*(1), 79–91. https://doi.org/10.1080/11745398.2018.1483733

Humphreys, A. (2014). *Microadventures: Local discoveries for great escapes.* Williams Collins.

Kelly, H. (2022, August 9). Adventure for all: Adapting to accessibility in adventure travel. Adventure Travel News. Retrieved September 19, 2022, https://www.adventuretravelnews.com/accessibility-in-adventure-travel

MacKenzie, S. H., & Goodnow, J. M. (2022). Adventure in the age of COVID-19: Embracing microadventures and locavism in a post-Pandemic world. *Leisure Sciences.* https://doi.org/10.1080/01490400.2020.1773984

Mann, J. (2022, July 24). A disabled woman as stuck on a plane for an hour while the cleaners cleaned around her. *Insider.* Retrieved January 18, 2022, from https://www.businessinsider.com/disabled-woman-left-waiting-on-plane-for-an-hour-2022-7

Morris, J. (2020). Lessons for Destinations and Travel Providers from the 2020 Accessible Travel Study. WheelchairTravel.org Retrieved September 18, 2022, from https://wheelchairtravel.org/accessible-travel-study-2020-lessons-for-destinations-travel-providers/

Moura, A., Eusébio, C., & Devile, E. (2022). The 'why' and 'what for' of participation in tourism activities: Travel motivations of people with disabilities. *Current Issues in Tourism*. https://doi.org/10.1080/13683500.2022.2044292

National Council on Disability. (2017, October 26). National disability policy: A progress report. https://ncd.gov/progressreport/2017/national-disability-policy-progress-report-october-2017

Pigram, J. J. (1984). Tourism for the handicapped. *Tourism Recreation Research*, *9*(1), 27–30.

Pretto, A. (2022) A study on accessibility in an Old Italian City: When the past is worth more than the present, *Disability & Society*, *37*(3), 496–521. https://doi.org/10.1080/09687599.2020.1829552

Ray, N. M., & Ryder, M. E. (2003). "Ebilities" tourism: An exploratory discussion of the travel needs and motivations of the mobility-disabled. *Tourism Management*, *24*, 67–72.

Roberts, J. W. (2018). Re-placing outdoor education: Diversity, inclusion, and the microadventures of the everyday. *Journal of Outdoor Recreation, Education, and Leadership*, *1*(10), 20–32.

Rose, S. (1995). Adventure for all: Disability is no handicap. *Journal of Adventure Education and Outdoor Leadership*, *12*(3), 16–17.

Steinfeld, E., Maisel, J. L., & Levine, D. (2012). Defining universal design. In *Universal design: Creating inclusive environments* (pp. 27–44). Wiley.

Travability. (2018, December 7). A blueprint for the development of a successful accessible tourism strategy. Retrieved September 18, 2022, from https://travability.travel/2018/12/07/a-blueprint-for-the-development-of-a-successful-accessible-tourism-strategy/

Var, T., Yeşiltaş, M., Yaylı, A., & Öztürk, Y. (2011) A study on the travel patterns of physically disabled people. *Asia Pacific Journal of Tourism Research*, *16*(6), 599–618. https://doi.org/10.1080/10941665.2011.610143

World Health Organization. (2021, November 24). *Disability and health*. World Health Organization. Retrieved October 1, 2022, from https://www.who.int/news-room/fact-sheets/detail/disability-and-health

Yau, M. K., McKercher, B., & Packer, T. L. (2004). Traveling with a disability: More than an access issue. *Annals of Tourism Research*, *31*(4), 946–960.

Index

Pages in *italics* refer to figures and pages in **bold** refer to tables.

For Product Safety Concerns and Information please contact our EU
representative GPSR@taylorandfrancis.com
Taylor & Francis Verlag GmbH, Kaufingerstraße 24, 80331 München, Germany

www.ingramcontent.com/pod-product-compliance
Ingram Content Group UK Ltd.
Pitfield, Milton Keynes, MK11 3LW, UK
UKHW050438020726
472812UK00007B/223